DISCARDED

D1116438

TECHNOLOGY AND SOCIETY

TECHNOLOGY AND SOCIETY

THE

New-England Farmer;

OR

GEORGICAL DICTIONARY.

BY SAMUEL DEANE

ARNO PRESS
A NEW YORK TIMES COMPANY
New York • 1972

Reprint Edition 1972 by Arno Press Inc.

Reprinted from a copy in The Massachusetts State Library

Technology and Society
ISBN for complete set: 0-405-04680-4
See last pages of this volume for titles.

Manufactured in the United States of America

Library of Congress Cataloging in Publication Data

Deane, Samuel, 1733-1814.
The New-England farmer.

(Technology and society)
Reprint of the 3d ed., published in 1822.
1. Agriculture--Dictionaries. I. Title. II. Series.
S411.D282 1972 630'.3 72-5043
ISBN 0-405-04695-2

THE

New-England Farmer;

OR

GEORGICAL DICTIONARY.

CONTAINING

A COMPENDIOUS ACCOUNT

OF THE

WAYS AND METHODS

IN WHICH THE

IMPORTANT ART OF HUSBANDRY,

IN ALL ITS VARIOUS BRANCHES,

IS, OR MAY BE,

PRACTISED, TO THE GREATEST ADVANTAGE,

IN THIS COUNTRY.

BY SAMUEL DEANE, D.D.

VICE-PRESIDENT OF BOWDOIN COLLEGE, AND FELLOW OF THE AMERICAN ACADEMY OF ARTS AND SCIENCES.

Third Edition,

CORRECTED, IMPROVED, GREATLY ENLARGED AND ADAPTED TO THE PRESENT STATE OF THE SCIENCE OF AGRICULTURE.

" Frigoribus parto agricolæ plerumque fruuntur,
Mutuaque inter se læti convivia curant :
Invitat genialis hyems, curasque resolvit."—*Virgil.*

BOSTON :

WELLS AND LILLY—COURT-STREET.

....

1822.

Advertisement

TO THE PRESENT EDITION.

No work upon the subject of Agriculture has, perhaps, ever been published in the United States, which has sustained so great, and so well deserved reputation as Dr. Deane's Georgical Dictionary. And its being adapted to our own soil and climate, must give it a decided advantage, in point of practical utility, to the American Farmer, over foreign publications, which, otherwise, might be of equal value.

The Compiler of the present edition has made such alterations, additions, and obliterations only, as were suggested by important improvements in agriculture, since the publication (in 1797) of the last edition of the original work; and hopes, as it now stands, it will be thought worthy of a place in the library of the practical as well as the scientific agriculturist.

September, 1822.

Introduction

TO THE SECOND EDITION, PUBLISHED IN 1797.

It is much to be regretted, that the most complicated of all the arts, in which the brightest genius may find sufficient room to exert and display itself, should be slighted and neglected, by a people not generally wanting in ambition. And it is equally strange and unaccountable, that the most useful and necessary of all employments should have been considered, even by the enlightened people of New-England, as below the attention of any persons, excepting those who are in the lowest walks of life; or, that persons of a liberal or polite education should think it intolerably degrading to them, to attend to practical agriculture for their support.

Perhaps, one occasion of the low esteem in which husbandry has been held, in this country, may have been the poor success which has most commonly attended the labours of those who have embraced the profession. Not only have most of them failed of rapidly increasing their estates by it, but too many have had the mortification of making but an indifferent figure in life, even when they have used the strictest economy, and worn out their constitutions by hard and incessant labour. The misfortune has been, that a great proportion of their toil has been lost by its misapplication. To prevent this evil in future is a leading design of the present publication. And since many among us begin to be convinced of the urgent necessity of having the attention of the publick turned to agriculture, it is hoped that the following attempt to promote the knowledge of its mysteries, and a spirited attention to the operations of it, will meet with the greater approbation and success. And as a very respectable Society in the Commonwealth of Massachusetts have undertaken to propagate the knowledge of husbandry, the day may be at hand, when the employment of the farmer shall no more be treated with contempt; when the rich, the polite, and the ambitious, shall glory in paying a close attention to their farms; when respectable persons shall confess it is one of the noblest employments to assist nature in her bountiful productions; when it shall be our ambition to follow the example of the first man in the nation, who does not think an attention to husbandry degrading; and when, instead of being ashamed of their employment, our laborious farmers shall, as a great writer says, "toss about their dung with an air of majesty."

Amidst the laudable efforts that are now making to promote so excellent a design as the revival of agriculture, the writer of the following sheets is humbly attempting to throw in his mite. He has been more prompted to engage in so arduous an undertaking, by an opinion he has long entertained of the need of a work of this kind, adapted to the state and circumstances of this country, than by any idea of his being thoroughly qualified to undertake it.

European books on agriculture are sufficiently plenty in the world, some of which are extremely well written; and this country is not wholly unfurnished with them. But they are not perfectly adapted to a region so differently circumstanced. Though the productions of English writers may be perused by the judicious to great advantage, it would be unadvisable, and perhaps ruinous, for our farmers to adopt the methods of culture in gross, which they recommend to their countrymen. Local circumstances so widely differ in the two countries, that, in many cases, the right management in the one must needs be wrong in the other. Britain, being generally liable to too much wetness, the English methods of culture must in many respects be different from those of a region that is mostly annoyed, as ours is, with the opposite extremity of drought. Difference of heat and cold must require a correspondent variation in the suitable crops and management. Difference of seasons and climates vary the fit times for sowing the same kinds of seed; and the manures that prove to be most profitable in one country, cannot always be rationally expected to prove so in another, although they were equally obtainable. And though Americans speak the English language, yet the diction peculiar to different farmers on the east and west of the Atlantick, and the manner of their communicating their ideas on husbandry are so little alike, as to render it highly expedient that we should be instructed in it by our own countrymen, rather than by strangers, if any among us can be found capable of doing it in a tolerable degree.

The writer confesses he has never had sufficient leisure to attend very closely to the study of agriculture. But, having always had a high relish for natural philosophy. and particularly for this most profitable and important branch of it, he has paid all possible attention to it for a number of years, employed many of his vacant hours in perusing what has been published by the best writers, and in making useful experiments in husbandry. He flatters himself, therefore, that he shall not have the unhappiness of grossly misleading any of the most ignorant of his readers. Many things are written from his own experience, and from that of others in this country, on whose veracity in their communications he can rely. Things which are not certainly known are mentioned only as opinion or conjecture. Extracts are made from some of the best authors, and marked as such. He has not wilfully asserted any thing which he does not know to be fact. And though he has adopted the ideas of others, he has not passed any thing on the publick as his own, which has been published by others, unless it be through inattention or mistake. Whether the reasonings be just, every intelligent reader must judge and determine; and to the candour of such the whole is submitted.

Long and particular accounts of experiments, such as abound in many European publications, are generally omitted, lest they should take up too much room, in a book that is meant to be comprehensive, and cheap to the purchaser, at the same time that it is designed to contain a whole system of husbandry. Neither would the intention of comprehending much in a little room permit the pages to be filled with lengthy bills of the cost of culture, and computations of profit, which many writers have too much run into; and in which any writer in this country, where the price of labour is variable, would be in danger of deceiving both himself and his readers. Our farmers have a sufficient knowledge of arithmetick to do these things for themselves; and it would not be amiss for them to amuse themselves in this way, in some of their moments of leisure.

That the writer has been excited to treat on the present subject by a tender concern for the welfare of his country, more than by any selfish and sinister view, those who are best acquainted with him are sufficiently convinced. At the same time, he will not pretend to deny his feeling of an ambition to be one of the first of his nation, who has thus endeavoured to lighten the labours, and promote the happiness of his countrymen. Yet he most sincerely wishes, that other writers on the subject may soon carry the system nearer to perfection, as they undoubtedly will. But the disadvantages he is under by being so early, and having an unbeaten way to explore, will doubtless apologize for him with all who are candid and considerate, and partly atone for his errors and imperfections, from which it would be strange if he were wholly free.

Though agriculture, strictly considered, has nothing to do with the breeding and management of tame animals, yet it is so closely connected with those employments, in practice, that the farmer cannot be complete without a considerable knowledge thereof. It is by the assistance of labouring beasts, such as horses and oxen, that he must carry on his tillage, and send the produce of his lands to market. By the help of milch kine his grass, hay, and other fodder, are to be converted into butter and cheese. Bullocks, poultry and swine must be fed and fattened with the produce of his farm, that he and his family may be fed with their flesh, and the markets supplied with meat. And the sheep must assist him in the transmutation of the fruits of his ground into materials for clothing and food. Therefore the rearing, tending, and whole management, of all these sorts of animals, are attended to in the following work; including the methods of preventing and curing the most common distempers to which, in this climate, they are liable.

Noxious animals, such as beasts of prey, ravenous birds, and devouring insects, have too much connexion with agriculture, as the farmer knows by his sorrowful experience. He ought therefore to be instructed in the most effectual methods of defending his property against them. This arduous task, to which no one perhaps can pretend to be fully equal, the reader will find attempted, and it is hoped, in some good degree performed, in the following pages.

As fruit trees are of essential importance to the farmer, the rearing of them from seeds and otherwise, as also the grafting, transplanting and pruning them, are attended to in this work.

And as agriculture cannot be carried on to the best advantage, without a variety of suitable tools and machines; the most important and useful of farming implements are treated of. Much of the ease and comfort of the labourer, as well as the profit of the farmer, depends upon their being well constructed. Their construction, therefore, is minutely attended to, although the art of the mechanic is the branch to which it most properly belongs.

The author attempted to arrange the parts of his subject analytically. But the variety of the materials he had collected was so great, and their heterogeneousness so obvious, that he found it not easy to do it to his own satisfaction; which is one of the reasons why the book makes its appearance in the lexicographical form. And when he considers that what he is doing is not principally for the instruction of critical scholars, but for the direction of the common people, it appears that the want of a systematical arrangement is a matter of no great consequence. On the present plan, he has saved himself the trouble of writing a long index, which must have added several pages to the volume, and increased its price to the purchasers, which he wishes may be as low as possible, for their encouragement. Perhaps it need not be added, that the fashionableness of an alphabetical method is a further apology for the form in which this book appears; nor the advantage the most illiterate reader will have of readily turning to any particular part of the general subject.

It is hoped that an acquaintance with this volume, if it should be perused by the generality of our farmers, will enable them to communicate their ideas to each other, and to learners in husbandry, with the greater perspicuity and propriety, and lead them to use nearly the same language in doing it, in the various parts of the country. For the writer has endeavoured that his diction should not only be concise, but plain and intelligible to ordinary readers; such as is most suitable to the subject, and not adapted to lead any into the use of absurd and ungrammatical language. How far these designs are accomplished the learned and judicious reader will be able to determine.

As a number of vulgar errors and prejudices are detected, and new methods of management proposed, it is expected that what is written will be censured by many, who have confirmed themselves in wrong practices by inveterate habits. But if persons will only be so fair as to allow, that there is a possibility of some want of perfection in their present established practice; which is at least highly probable, as this is a country where husbandry as an art has not been taught, nor much attended to; they will then see it is reasonable to give a candid hearing to any new scheme of improvement suggested, and to plausible arguments offered in support of its utility; and allow themselves to be influenced by them. If those who are in low circumstances should fear they may suffer loss, by trying any new practice in husbandry, it is hoped the richer sort will be inclined to do it by love of their country. For others will undoubtedly inquire concerning their success; and when they are convinced by experiments made by their neighbours of the advantage of any new practice, one would think they can need no other motive to induce them to adopt it.

On the other hand, let not the book be reprobated for containing so many things as it does, which are already well known to farmers. The farmer may find reasons for his good practice which he has not before thought of, and be induced to persevere in it. And besides, all useful knowledge ought to be recorded, that it may be retained, and be in no danger of being lost, as a great deal has been in the world. It should also be remembered that things which are well known by some may be quite new to others; especially to young persons, and to all those who have newly turned their attention to husbandry.

The writer has had more zeal and courage in attempting to promote improvements in agriculture, since the happy termination of the late struggle for independence than before. Our holding the rank of a free and independent nation allows us to consider the country as indisputably our own, and ourselves as monarchs over our farms. Nor does it appear probable, that we shall soon meet with any thing that will give us a material interruption, in pursuing the arts, or enjoying the blessings of peace. If great improvements were now to be made, we might have reason to hope we should enjoy the benefits of them through life, and that posterity would not be deprived of them.

But the most forcible reason for our cultivating this art, is the indispensable necessity of it, to enable us to live as becomes an independent people. The alarming effect of the present low state of husbandry is, that we are necessitated to import much of our food, and clothing, while we are incapable of making proportionable remittances in the produce of the soil, or in any thing else. As a good system of national government is now established, I see no reason to doubt but that a spirited attention to husbandry and manufactures accompanied with a more general practice of frugality and economy, would put us on a respectable footing; so that such a foundation would be laid for our increasing wealth, that we should be able, in a short time, to cancel our publick debts; and might reasonably hope ere long to become an opulent, respectable and very powerful nation.

As to the present edition, its appearing so soon after the first is occasioned by the rapid sale of the book, arising from the general acceptance it has obtained; and the increasing demand could not otherwise be supplied.

The author has taken the opportunity to correct a great number of small errors. Some few things are suppressed in this publication. The diction in many parts is much improved. Many articles are more largely, and more accurately treated of than they were before; and a number of new and important ones are added, with a view to render the work a more complete directory for husbandmen. And that the vegetables that are treated on may be known to persons in other countries, as well as in remote parts of our own, where they are probably called by different names from those English ones he had given them, he has now added the botanical names, which are extensively known by persons of erudition. On the whole, he thinks the book is far more increased in value than in size. If, in its present improved state, it shall be found to contribute towards reviving and continuing the spirit of husbandry, and towards the increasing advantage of those who are employed in it, he will consider it as the most happy reward he can have for his labour.

WORCESTER, MARCH, 1797.

NEW-ENGLAND FARMER;

OR

GEORGICAL DICTIONARY.

AGRICULTURE is the most ancient, the most honourable and the most useful of the arts. Its origin was prior to the invention of letters, and in attempting to trace it, we are lost in the fables which obscure the annals of antiquity. The most wise and most powerful nations have ever been the most assiduous in their attention to this art, and the degree of estimation in which it has been held, has ever, we believe, presented an accurate criterion of the morals, prosperity and civilization of any age or country. The Chaldeans, the Egyptians, the Phoenicians, the Carthaginians, the Greeks and the Romans, were as greatly, and much more honourably distinguished by their attention to agriculture than by their military achievements. Mago, a Carthaginian, wrote books on agriculture, which were not more highly prized by his own countrymen than by their conquerors.

The ancient Romans venerated the plough, and their greatest generals and most illustrious senators guided it with their own hands. Cato, Varro, Virgil, Pliny and Columella wrote upon agriculture, and Cicero has bestowed the highest eulogies on this art. But husbandry declined with the decline of the empire, and, as was well observed by an able writer, "In the Campania of Rome, where, in the time of Pliny were counted twenty-three cities the traveller is astonished and depressed by the desolation that surrounds him."

In some parts of continental Europe the science of agriculture is taught in universities, and a regular education is thought to be as necessary to qualify a man to become a complete farmer as it is to fit him for the practice of law or physic. In Great Britain a very large proportion of the wealth, the talents and the influence of men who occupy the first ranks in society, is directed to agricultural improvements, and the munificence of the government encourages and rewards the patriotic exertions of individuals.

In the United States a spirit has of late years been exerted, which has produced a new era in the annals of American agriculture; and the most fortunate results do not rest altogether in anticipation, but have been already in some measure realized.

AGRICULTURAL SOCIETIES are now becoming common in the United States, and the prejudices which existed against these institutions are subsiding, as the benefits resulting from them are too obvious not to be generally acknowledged and highly appreciated. Probably the wisdom of man could not devise means more effectual than those presented by these societies, for stimulating and rewarding that skill and industry, which are the corner stones of national as well as individual prosperity. The following sketches, which we have extracted from "*An Address delivered before the Massachusetts Agricultural Society*," may serve to exhibit some of the advantages which have accrued, and may be anticipated from these institutions.

"The first public society of this description, we believe, was founded in Great Britain; where agriculture, even at the time of its foundation, was in a more improved state than that of any other European nation, except, perhaps, Flanders. The Bath and West of England Society, had the honour of leading the van in this generous attempt to improve and elevate the most important and interesting, as well as most innocent human art. That art, which free from vice or a tendency to produce it, is productive of none of the injurious effects, from which not many other extensive human arts are wholly exempt. The British nation soon became awakened to the importance and value of such institutions, and by a parliamentary provision, with noble munificence, founded a National Society. The state of Massachusetts had the glory of early appreciating the value of such institutions, and this society can trace its origin to a period, when those of Europe were still in their infancy."

In speaking of the advantages which result from these societies, the writer observed;

"Men and their feelings and motives are the same in whatever situation they may be placed. They may be taught to encourage the ferocious passions, or to cultivate the kindly affections. It depends on rulers and on the laws to give them a virtuous and useful, or vicious and pernicious direction. The plain and too much forgotten and neglected husbandman, contributes *directly* or *indirectly* more than seven-eighths of all that is effected towards the support and prosperity of a state. Indeed, their number, in this happy country is nearly in that ratio. It cannot be questioned, unless the laws of nature are reversed as to this great mass of society, our chief stay and support, that attention to their profession, the holding it in deserved respect, the rewarding examples of successful and peculiar industry in individuals must have an influence on the advancement and progress of the art. The spirit of emulation is as capable of being excited in this art, as in the calamitous, and what we all acknowledge to be the unprofitable one of War. A farmer's breast is as open to generous feelings, is as alive to applause and approbation, and as much chilled by neglect, as that of the proudest spirit that ever wore a sword. Our defence then of these exhibitions and these rewards, rests on the immutable laws which regulate the human heart." *See Mass. Agr. Rep. vol. V.* p. 85. 92. 215. 237. 374.

AIR. Since the time in which Mr. Deane wrote, great discoveries, relative to the constituent parts, and essential qualities of atmospheric air have been made by Scheele, Priestley, Lavoisier, Davy and others. These celebrated men have proved that its principal elements are two gases, oxygen and azote, or nitrogen ; and that it likewise contains small quantities of aqueous vapour and of carbonic acid gas ; and Lavoisier proved that this last body is itself a compound elastic fluid, consisting of charcoal, dissolved in oxygen.

Oxygen gas is necessary to some functions of vegetables, as well as indispensable to animal life. If a seed germinates in confined air, the oxygen, or a part of it, is absorbed. The azote, or nitrogen remains unaltered. But as vegetation proceeds in the open air, oxygen gas is given off, and the carbon, or charcoal of carbonic acid gas, is absorbed, and becomes a part of the organized matter in the vegetable.

The effects of nitrogen in vegetation are not exactly known. It is ascertained, however, that it prevents the oxygen from acting with too much violence, and is, perhaps absorbed by some vegetables.

Carbonic acid gas combines with many different bodies, and furnishes vegetables and animals with carbon or charcoal. When a growing plant, the roots of which are supplied with proper nourishment, is exposed in the sunshine to atmospheric air, containing its due proportion of carbonic acid, the carbonic acid, after a time, is destroyed, and a certain quantity of oxygen is found in its place. "Carbonic acid gas," says Sir Humphrey Davy, "is formed in a variety of processes of fermentation and combustion, and in the respiration of animals. And as yet no other process is known in nature, by which it can be consumed, except by vegetation. Animals produce a substance, which appears to be a necessary food of vegetables : vegetables evolve a principle necessary to the existence of animals ; and these different classes of beings seem to be thus connected together, in the exercise of their living functions, and to a certain extent made to depend on each other for their existence."

Water, likewise always exists in the atmosphere ; and the quantity is greater or less in proportion to the heat of the weather. When the thermometer stands at 50° of

Fahrenheit, according to Sir Humphrey Davy, air contains about one fiftieth of its volume of vapour or rarefied water; and as the specific gravity of vapour is to that of air, nearly as 10 to 15, this is about one-seventy-fifth part of its weight. At 100° if there is a free communication with water, air contains about one-fourteenth part in volume, or a one-and-twentieth part of its weight. It is the condensation of vapour by the diminution of the temperature of the atmosphere, which is probably the principal cause of the formation of clouds, and of the deposition of dew, mist, snow or hail.

The leaves of plants and many other substances have a power of absorbing moisture from the atmosphere. In some vegetables, such as the house-leek, and different kinds of the aloe, this power is so great that they will increase in weight, when suspended in air, and unconnected with the soil. Water is likewise absorbed from the atmosphere by the soil, and that soil is said to be most fertile, which has the greatest power to absorb water. See *Davy's Elements of Agricultural Chemistry*.

Air and earth mutually transmit fertility, like the transmission of heat and cold in all bodies in contact; or at least there is a certain point of saturation of fertility, beyond which exposure to the air carries off the overplus. In the case of fresh dung, exposure soon lowers the fertility of its surface down to the point of saturation; and the process of the transmission of some volatile substance is disagreeably sensible to the nostril. On the other hand, air is a cause of fertility, and in general it is said that the greatest quantity of the beneficial influence of air will be imbibed by the poorest mould, and that which has been least exposed to the action of the atmosphere. Hence the utility of deep ploughing, which increases the quantity of soil fit for cultivation, by exposing the lower and more barren strata to imbibe fertilizing principles from the atmosphere.

ANTICOR, "a swelling in the gullet and throat of a horse, and is the same which in man is called angina. It proceeds from the same causes that bring on many other diseases on horses, from hard riding, exposing a horse to the cold, giving him cold water to drink when he is hot, full feeding, and whatever else may cause a stagnation of the blood.

"The signs of this disorder are all those that accompany a fever; for an *Anticor*, while it is internal, is never without fever: but when it shews itself externally, the fever begins to abate, unless it continue to be both external and internal.

"So long as the inflammation continues in the gullet, the horse forsakes his food: and though he has frequent inclinations to drink, the first gulp deters him from meddling with it again, until he has forgotten the pain and agony it put him into. And the pain in the gullet is yet more manifest from this, that whenever a drench is given him, he staggers and seems as if he would fall down, and makes short interrupted groans, and sometimes will have a cold clammy sweat about his ears.

"The cure must be begun by bleeding, and that needs not be very sparing: for this disease seldom happens to horses that are poor and low. And here we also approve of sticking one or other of the veins in the hind parts, to make revulsion.

"After bleeding, the following clyster may be given:

"Take two handfuls of barley, two ounces of sal polychrest, reduced to fine powder: boil them in two quarts of water for a quarter of an hour: add to the decoction a pint of wine, four ounces of fresh butter, and two ounces of oil of rue. Let this be given bloodwarm, and repeated twice a day, or oftener.

"If he takes to food, nothing must be given him but moistened hay, and scalded bran; or whatever else must be chiefly such things as are proper to keep down the heat and inflammation, and abate the feverish symptoms; for which purpose we recommend, after blooding, those things that are proper to promote sweat. Therefore, let the following drench be prepared for him:

"Take treacle water and carduus water, of each one pint; dissolve in these two ounces of Venice treacle: and after this has been given, clothe him well, and give him a little warm water to drink. Instead of the treacle and carduus water, a pint of stale beer, mixed with small beer, may be used. Nothing is so effectual to remove inflammation, especially after bleeding, as sweating: therefore, if you find it difficult to promote sweat, you may give the following ball:

"Take old Venice treacle two ounces, volatile salt of hartshorn fifteen grains, Mathews' pill one drachm, camphor in powder six grains, powder of liquorice, or sassafras in powder, as much as is sufficient to make it into a paste. Let this be given after the operation of the clyster is over.

"If the symptoms begin to abate, you may venture to give your horse a gentle purge.

"If the swelling appears outwards, and if the other symptoms abate, you may leave off purging: for what is intended by that evacuation, is chiefly to disperse the inward disorder. Next. you are only to apply ripening cataplasms, allowing him sal prunellæ, nitre, or the sal polychrestum, dissolved in his drink.

"Cow's dung alone, applied warm to the part, with lard or ointment of marsh mallows, may be sufficient to bring the swelling to maturity.

"When the matter seems ready for a discharge, it may be opened in the dependent lowermost part, by the application of a hot iron; afterwards keeping a dossel in the mouth of the wound till the running abates; and applying compresses and convenient bandage to keep the elevated skin close to the flesh, that it may be the sooner united. But if the cavity of the imposthumation be large, it will not be amiss to lay it open an inch or more.

"The cure may be finished with applying only the unguentum basilicum; or a digestive made with turpentine, the yolks of eggs, or honey, with a moderate mixture of brandy, or spirit of wine. And if any foulness appears, or if it heal too fast, or if spungy soft flesh arise, pledgets dipped in copperas water, or a solution of blue vitriol, may be applied, which will keep it smooth and even.

"But if the swelling increase fast, with no tendency to digestion, and if it rise up towards the neck, affecting all the muscles of the part, the horse will be in danger of suffocation, unless a course different from the former be taken.

"Besides repeated bleeding, if he is not too much worn out, take a hot searing iron, and apply it to five or six places on the lower part of the swelling, cauterizing those parts, that they may be speedily brought to matter, which may be dressed with flax dipped in tar and turpentine, mixed before the fire, and applied warm. For by giving pain in those dependent and inferior parts, you cause the humours to flown downwards from the swelling; and by making vents you prevent excessive violence of pain. Nor need you be afraid of the swelling that may happen in the fore-legs, &c. by cauterizing; for that cannot be of so ill consequence as when it is upon the neck and throat; nor will it be of any consequence, if care be taken of the vents.

"*Solleysell* recommends the making of small incisions with a fleam or lancet, in eight or ten places, on the swelling; and to thrust into the holes, between the skin and the flesh, pieces of the root of black hellebore: and if the tumour be very large, he recommends the use of white hellebore; at the same time chaffing the part with the ointments of agrippa and marsh mallows. The root, by their hot quality, draw down and increase the swelling; and the ointments are to ripen the inclosed matter, and fit it for a discharge.

"The same author also recommends the use of Ruptories, for drawing an immediate flux of moisture from the diseased part. These are ointments of the same nature as those made to draw blisters on the human body, and composed of the same materials. The way to apply them is, to spread them by little at a time on the part affected, holding a bar of hot iron to make them sink in." *Gibson's Farriery.*

ANTS, an insect, which sometimes annoys fields. "They will destroy barley, rye, hemp seed, flax seed, and rape seed. Other grain is either too large, double skinned, or too bitter and ill tasted for them. When you find them in quantities near home, pour hot water upon them. The farmer, when he dungs his land, if he uses ashes, lime, or salt sand, he may be certain no ant will ever stay upon the ground where any of the three is spread." *Scot's Farmer.*

"The ant is very destructive to fruit, especially the peach when ripe; you will frequently see these insect travelling all over the trees, and sometimes the fruit will be filled with them. The best method that I have found to destroy them is, to get a sharp pointed wooden stake, or an iron crow, if the ground be hard, and with it bore a hole not far from the stem of the tree, and as deep as the ground will permit. By stirring the earth, you will set the ants in motion: then work your stake or crow round the sides of the hole, making them as smooth as you can; the ants will come to the mouth of the hole and tumble in, and by the shape of the hole and smoothness of its sides will be prevented from climbing up again. When you see a great many in the bottom of the hole, pour in some water from a watering-pot; and thus you may drown thousands of them.

"You may likewise destroy many of them by mixing quick lime with soot, and laying it along their roads, where you see them thickest; but where you can come at their nests, the best way is to put a piece of quick lime into it, and pour as much water over the lime as will slake it, the heat of which will destroy them: when you have poured in the water, cover the lime with a turf or a little earth, which will render it more effectual by confining the heat. You may slake the lime with a mixture of urine and soap suds, which will render it still more effectual." *Forsyth's Treatise on Fruit Trees.*

The small garden ants, it is said may be destroyed by placing among them a number of large ants, commonly found in woods, which fall upon and destroy, or drive from their neighbourhood, the smaller kinds. If the walls of an apartment are washed with a painter's brush, dipped in a solution made of four ounces of sublimate, in two gallons of water, both the ant and the red spider will be dislodged. See article *Ant* in the Domestic Encyclopædia, 2d American edition.

APPLES. Mr. Forsyth's Treatise on the Culture and Management of Fruit Trees gives a list of 44 different kinds of apples which had come to the author's knowledge, with a description of the qualities of each, and the author has added several lists besides from the catalogues of nursery men in England and Scotland. See p. 49, *American edition.* The Massachusetts Agricultural Repository, vol. iii. p. 95, gives the following. The *Rhode Island Greening*, a good *fall* and early *winter apple.* The *Nonsuch*, a red apple, very late keeping apple. The *Nonpareil*, a Russet apple early in winter. The *Newtown pippin*, a good, hard, late keeping fruit. The *Spitzenberg*, a fine fruit, which keeps sound till May or June. The Roxbury *Russeting*. This is one of the best known, and most valuable of fruits. It is not fit to eat till February, and is very easily preserved till June. The *Baldwin apple*, a very valuable fruit, beautiful, fine flavoured, and will keep to the last of March.

Apples keep best in a low temperature, and may be well preserved in an *ice house*.

An English Journal recommends the use of dry pit sand, for preserving pears and apples. Glazed earthen jars are to be provided, and the sand to be thoroughly dried. A layer of sand an inch thick, is then placed in the bottom of the jar; above this a layer of fruit, to be covered with a layer of sand an inch thick; then lay a second stratum of fruit, covering again with an inch of sand. An inch and an half of sand may be placed over the uppermost row of fruit. The jar is now to be closed and placed in a dry situation, as cool as possible, but entirely free from frost. Some assert that apples may be kept in casks through the winter in a chamber or garret, by being merely covered with linen cloths. Apples which are intended to be preserved for winter's use should be suffered to hang on the tree as long as they are safe from frosts.

A letter from Ebenezer Preble, Esq. published in the Massachusetts Agricultural Repository, Vol. iv. No. I. p. 84, contains the following instructions, relative to gathering and preserving fruit.

"The general method of gathering apples for cider, is, shaking the tree, and thrashing the branches with poles. The former will answer when the fruit is at maturity; they will then drop without injury to the buds. Poles should never be used, but with a hook at the end covered with cloth or matts to prevent wounding the bark; they then serve to shake the small limbs. Particular attention is required in gathering winter fruit. They should be taken in the hand, the fingers placed at the foot stalk, and by bending it upwards the fruit is gathered with ease, and without injury; they should be removed from the gathering baskets with great care," &c. The same writer says, "The injudicious method practised in gathering fruit is more destructive in its consequences, than is generally understood; the blossom buds of the succeeding year are placed at the side of the foot stalk of the fruit, and if the spurs are broken, no fruit on that part will be produced the next season." See *Orchard.*

APPLE TREE, *pyrus malus*, a well known fruit tree of great importance to mankind. The way to propagate them is, by sowing the pomace from cydermills, digging, or hoeing it into the earth in autumn. The young plants will be up in the following spring, and the next autumn, they should be transplanted from the seed bed into the nursery, in rows from two to three feet apart, and one foot in the rows where the ground has been fitted to receive them. The ground for a nursery should not be very rich, but mellow, and well pulverized, and cleared of the roots and seeds of weeds. It is a good rule, That the young trees, at their final transplanting into orchards, should not be put into poorer, but rather into richer ground, than that to which they have been accustomed. For by not finding their usual supply of nourishment, they will be stinted in their growth, and never become good trees.

It is said, that when an appletree has become barren, its fruitfulness may be renewed by stripping off all the bark from its body, and from some part of the largest limbs; and that this operation must be performed at the time of the summer solstice.

The following has been found by experience to be an excellent mode of setting out apple trees, and other fruit trees on a light soil.

Dig a hole sufficiently large to prevent the root of the tree, when it is to be transplanted from being doubled or placed in an unnatural position, and to give room for the young shoots to extend themselves. Place about the roots of each tree, together with the mould, about half a bushel of small stones, the size of an ordinary apple, or less, which will give stability to the soil, and prevent the roots from being loosened by the wind. See *Fruit Trees*, *Orchards*, and *Transplanting*.

Rotten leaves of trees are recommended by Mr. Forsyth, as the best manure for fruit trees, which, he says, is "much better than dung, which I by no means approve of for trees, unless it be perfectly rotten, and mixed up with mould." It is better, however, not to make use of such leaves as manure for fruit trees, till they are rotted to a fine black vegetable mould. See *Forsyth's Treatise on Fruit Trees*, p. 62, 63. American Edition.

"It is a general complaint, that the finest apple trees of this country have degenerated, and that many of the best sorts have entirely disappeared from our gardens and orchards. It would not be difficult to show that every successive grafting deteriorates the part engrafted; or to point out an effectual method of retaining good apples in this country without the trouble of grafting, as in every perfectly ripe apple there will be found one and sometimes two round seeds; the others will have one or more flatted sides. The round ones will produce the improved fruit from which they are taken, and those with the flatted sides will produce the fruit of the crab upon which the graft was inserted. It requires not a long time to ascertain the difference; for if a circle is drawn in rich ground, and the flat-sided seeds planted therein, and the round seeds in the centre, the variation of quality will be discovered in two or three years. The first will throw out the leaves of a crab, and the latter the leaves of an improved tree, distinguished in shape and fibre, and with a woolly appearance; and in due time the fruit of each will put every thing beyond doubt. It is observed that the seeds of crabs (being originals) are mostly if not altogether *round*." *European Magazine*.

APPLE TREE. The laws, which govern the alternate or annual productiveness of apple trees, and other fruits, are dependent upon and connected inseparably with the varieties—some kinds are full bearers only every alternate year. Others bear every year, nearly alike, with occasional differences, dependent on the seasons.

ASHES, a dust, consisting of the earthy and saline parts of wood, and other combustibles, which remains after burning.

Ashes are commonly accounted a manure most suitable for low and moist lands. A cold and sour spot certainly needs them more than any other. But I have found them to be good in all sorts of soil.

They are not only a valuable manure, but an excellent antidote to the rapaciousness of worms and other insects. Therefore they are a more proper manure for all those plants, which are liable to suffer by worms and insects; such as cabbages, turnips, cucumbers, melons, peas, and other pulse. They should be spread evenly, and not in too great quantity.

Wood ashes is an excellent nourishment for the roots of trees.

Ashes of all kinds are a good ingredient in composts which are kept under cover. But when they are laid upon land unmixt, they should be spread as evenly as possible. They are thought to do better on the top of the surface than when buried in the soil ; for there is nothing in them that will evaporate. Their tendency is only downwards ; and their salts will soon sink too low, if they be put under the surface. If they be spread upon ground, which has tender plants, it should be done just before a rain, which will dissolve and soften their acrimony : For tender plants, when the weather is dry, will be apt to be injured by them ; at least, if they are in contact with the stems or leaves.

Ashes in their full strength are certainly best for manure ; and they will not be in full strength, unless they be kept dry ; nor will it be easy to spread them properly. And they should not be laid on lands long before there are roots to be nourished by them, lest the rains rob them of their salts, by washing them into the hollows, or by sinking them to too great a depth in the soil. A few bushels on an acre are a good dressing for grass lands that are low, and inclining to be mossy. But ashes from which *lie* has been drawn have no small degree of virtue in them. The earthy particles are but little diminished; and some of the saline particles remain in them, especially in *soaper's waste* which has lime mixed with it.

A handful of ashes, laid about the roots of a hill of Indian corn, is good to quicken its vegetation. But it should not much if any of it be in contact with the stalks. The best time for giving corn this dressing, is thought to be just before the second or third hoeing : But some do it before the first, and even before the plants are up. Like other top dressings, it is of most service when applied at the time when plants need the greatest quantity of nourishment. This happens, in Indian corn, at the time when the plants are just going to send out ears and spindles.

ASPARAGUS, a valuable plant, the young shoots of which are a pleasant and wholesome food ; of more account for the table than any other greens which the spring produces. They come up early, and are consequently of the greater importance. The fruit is a spherical, red berry, which ripens in autumn, containing two black seeds.

The root of this plant is esteemed in medicine, as an opener and diuretic.

To cultivate asparagus in the best manner, open a trench three feet wide, and twelve inches deep. If it be close to the south side of a garden wall, it will be up the earlier in the spring. Fill the trench half full of good dung ; make it level, and sprinkle a little rich earth over it, and lay on the roots, in their natural position, eight or nine inches apart. Or, if you cannot get roots, place the seeds at half the distance from each other. Cover them by filling up the trench with the blackest of the earth which was taken out. If you plant roots, the shoots may be cut the second year after ; if seeds, they will not be fit to cut till the third year. All the shoots which come up before the middle of June, may be cut off without injuring the roots : After which time, the late shoots should be left to run up, and seed ; otherwise the roots will be weakened. The seeds may be well preserved on the branches through the winter, hung up in a dry situation.

This plant grows well in ground that is shaded. The sprouts will be very large and tender ; but they will not be so early. It is not amiss to have one bed in a shady place, to supply the table, after the season is over for cutting the first. In autumn, after the tops are turned white by the frost, they should be cleared off, and a layer of dung, or rich soil, an inch thick, laid over the bed. This should be done yearly, and the bed kept clear of weeds. If the bed should get too high by this management, the surface may be taken off with a spade early in the spring, to the depth of two inches, before the young shoots are in the way. But when this is done, a thin dressing of rotten dung or compost should be laid on.

The soil for asparagus should be the best which the garden affords, not wet nor too strong and stubborn, but such as is moderately light and pliable, and will readily fall to pieces in digging or raking, &c. Mr. M'Mahon recommends the dividing of a plantation of asparagus into beds four feet and an half wide, with alleys two feet wide between each bed. And at each corner of every bed, let a firm stake be driven into the ground, to serve as a mark for the alleys.

Four rows of asparagus, according to this writer, are to be planted in each bed, and ten or twelve inches distance to be allowed between plant and plant in the row ; and the outside rows of each bed should be eight inches from the edge. The plants should not be *more* than two years old, and most good gardeners prefer those of one, and Mr. M'Mahon recommends the latter in preference to the former, from experience.

The following is Mr. M'Mahon's method of planting asparagus.

" Strain your line along the bed eight inches from the edge ; then with a spade cut out a small trench or drill close to the line, about six inches deep, making that side next the line nearly upright ; and when one trench is opened, plant that before you open another, placing the plants upright ten or twelve inches distance in the row.

" In planting these plants, observe that they must not be placed flat in the bottom of the trench, but nearly upright against the bank of the trench or drill, and so that the crown of the plants may also stand upright, and two or three inches below the surface of the ground ; let them be all placed an equal depth, spreading their roots somewhat regular against the back of the trench, and at the same time drawing a little earth up against them with the hand as you place them, just to fix the plants in their due position, till the row is planted; when one row is thus finished, immediately with a rake draw the earth into a drill over the plants, and then proceed to open another drill, or trench, as before directed; plant it in the same manner, and cover in the plants as above, and so on till the whole is planted; then let the surface of the bed be raked smooth and cleared from stones."

The plants should be kept free from weeds, and no early crops sowed among them, as has been sometimes injudiciously practised. A plantation of asparagus, with proper management, will continue to produce good buds ten or twelve years, or longer.

Mr. M'Mahon says asparagus' seeds may be sown about the middle or towards the latter end of March, calculating probably for the latitude of Philadelphia. In the more northern States, perhaps a somewhat later period would be proper for that purpose.

Asparagus beds should be completely loosened to a moderate depth every spring, as soon as the frost is out of the ground, with a proper fork, having three short tines, six to eight or nine inches long. But care must be taken not to go too deep, so as to wound the crowns of the roots. The beds, being loosened in every part to a moderate depth, should be raked even, before the buds begin to advance.

In the 13th vol. of the "*Repertory of Arts*," &c. a new method of rendering asparagus more productive, is communicated by Mr. Richard Weston, who observes that the male plants yield a greater number of shoots than the female ones; though the former are of an inferior size. He consequently advises that males only be selected for the formation of beds; and to prevent mistakes, they should not be planted from the seed bed, till they have flowered. After having grown twelve months, Mr. W. directs them to be removed into beds, at the distance of six inches from each other, where they ought to remain another year, in which they generally flower; a small stick must then be driven into the ground, contiguous to each of the male plants, in order to separate them from the females, the latter of which are then to be pulled.

This plant is found growing naturally on the borders of salt marshes, and even upon such marshes. This is considered to be its natural situation; and this fact has led to the employment of *salt* as a manure to it with very good effect. To a bed 50 feet by 6, a bushel of salt may be safely applied before the plants start in the spring.

AXE, a necessary tool for farmers. A narrow axe is meant; for a broad axe is a carpenter's tool. A narrow axe should have a thick poll, as in that part it commonly fails soonest. It should be made of the best of iron and steel, be quite free from cracks and flaws, and nicely tempered; not so soft as to bend, nor so hard as to break.

Take care that you do not grind your axes thin at first, till you learn by using them what their temper is, and whether they will bear it. A rounding edge is best for chopping large logs, a straighter one for smaller wood.

Let the helve of an axe be made of the toughest of wood, either walnut or white oak. Let it be set in the centre of the eye, and at right angles with the outer side of the axe ; let it be small near the eye, that the hands may not be too much jarred by the strokes in chopping, and gradually larger towards the other end. Three feet is the greatest length that almost ever will be needful : Shorter for chopping sticks not uncommonly large. It should never be less than 32 inches.

A good deal of rubbing with a whetstone, (after an axe is ground on a coarse grindstone,) is best ; not only to bring it to a good edge that will not crumble, but chiefly to make the blade very smooth, that it may enter the wood easily, and not stick too fast when entered.

B.

BARLEY, *Hordeum*, a well known grain of which malt is made. In some countries, it is also much used for bread. If it be kept long before grinding, it will be the better for this use, as a certain bitter taste, which it has when new, is abated by age. Barley is accounted cooling and detersive ; a broth of it is therefore given to persons in fevers : But it must be hulled before it is fit for this use.

It is a sort of corn very suitable for cultivation in this region, as it seems liable to no distemper, in our northerly part of Massachusetts especially ; bears the drought well, and never fails of yielding a crop. I have commonly gained 40 bushels per acre, without any extraordinary tillage, and without much manuring. It will grow in any soil : Even a soil so clayey that it is fit for scarcely any other grain, will answer well for this, as I have found by long experience. But it does better on some other soils.

It should be sowed as early as the season and soil will admit.— About the beginning of May is a suitable time. The quantity of seed for an acre is two bushels, if the grain be small ; if larger, more in proportion. A correspondent of the Bath Agricultural Society writes : " The last spring (1783) being remarkably dry, I soaked my seed barley in the black water, taken from a reservoir, which constantly receives the draining of my dung heap and stables. As the light corn floated on the top, I skimmed it off, and let the rest stand 24 hours. On taking it from the water, I mixed the grain with a sufficient quantity of sifted wood ashes, to make it spread regularly, and sowed three fields with it.— The produce was 60 bushels per acre. I sowed some other fields with the same seed dry ; but the crop, like those of my neighbours, was very poor, not more than 20 bushels per acre, and much mixed with green corn and weeds, when harvested. I also sowed some of my seed dry on one ridge in each of my former fields, but the produce was very poor in comparison of the other parts of the field." The ground should have two ploughings at least. It should be well harrowed after sowing; and then a roller passed over it to close the soil about the corns, that they may not fail of vegetating. And rolling prepares the surface for mowing the crop, and raking it up clean, which is a matter of great importance. For it is impossible to rake it up clean, when the ground has been laid rough at sowing.

In Scotland, after the grain is up, the farmers, near the sea coast, give it a top dressing of sea weeds, which has an excellent effect. This practice I would recommend to those of my countrymen who farm near the sea.

I should have observed, that barley must be sowed soon after ploughing, lest the moisture of the soil be too much evaporated. It being a dry husky grain, a considerable degree of moisture is requisite to make it vegetate. If the ground should be very dry at sowing time, and the season late, steeping the seed in lie would not be amiss. Steeping it in the wash of a barn yard has an excellent effect.

Some have got an opinion, that barley should be harvested before it is quite ripe. Though the flour may be a little whiter, the grain shrinks so much, that the crop seems to be greatly diminished and wasted by early cutting. No grain, I think, requires more ripening than this; and it is not apt to shatter out when it is very ripe. It should be threshed soon after harvesting: And much beating, after it is cleared from the straw, is needful to get off the beards. Let it lie a night or two in the dew after it is cut, and the beards will come off the more easily.

I had gained the idea of the necessity of barley's being well ripened before cutting, from my own experience. I have been more confirmed in the opinion, by the following passage in an English writer, who appears to have been well acquainted with the culture of this corn. "This grain," says he, "may be greatly damaged, or spoiled, by being mown too soon; which may afterwards be discovered by its shrivelled and lean body, that never will make good malt."

The same writer says, "This grain I annually sow in my fields on different soils, whereby I have brought to my knowledge, several differences arising therefrom. On our red clays, this grain generally comes off reddish at both ends, and sometimes all over, with a thick skin and tough nature, somewhat like the soil it grows in; and, therefore, is not so valuable as that of contrary qualities. Nor are the black, bluish, marly clays, of the vale much better: But loams and gravels are better. On these two last soils the barley acquires a whitish body, a thin skin, a short plump kernel, and a sweet flour."

Barley is a corn that is very apt to degenerate, unless prevented by a frequent changing of the seed.

If ever so few oats are sown among barley, the crop, in a few years, will come to be mostly oats; because oats increase more than barley. Swimming the barley before it is sown, will in great measure prevent this inconvenience. Almost every oat, and a few of the worst of the barley corns, will be on the surface of the water, and may be taken off.

But the speedy degeneration of barley is a good reason for changing the seed very frequently. In some parts of the country, the barley, for want of changing, has come to produce little or nothing.

Not only changing seed, but sorts of barley, should be attended to. Some sorts are at least more productive than others, if not of a better quality. The two rowed barley has seldom more than 32 corns on an ear : The six rowed has sometimes 72, that is 12 in a row. Of the latter sort one pint produced me three pecks in a single drill row. It was at the rate of about three pecks of seed, and forty bushels crop to the acre, on a poor gravelly soil. This sort is called Bear, Bere, or Barley big. It is a winter grain in England and Ireland. But I must mention one inconvenience attending the six rowed barley, which is, that the seeds are apt to break off and fall, if the corn stands till it is fully ripe. I now cultivate a four rowed barley, which has not this inconvenience attending it : And it yields as plentifully as any other.

I would recommend the drill and horse hoeing method of raising barley, when it is designed for hulling, as the corns will be the more full and plump, and have a less quantity of hull in proportion to the flour.

Sir John Sinclair, in speaking of the agricultural practice of Mr. Coke, of Holkham, says, "He uses the Rev. Mr. Coke's drill, which sows six rows at a time, and an acre an hour, drawn by a single horse. His wheat he sows at nine inches asunder, his barley at six inches three quarters. He considers any attempt to save seed by drilling founded on erroneous principles ; and the quantity of seed he sows per acre is four bushels of wheat, three of barley, and six of oats. On his farm it is a practice, not generally known, but which ought to be attended to, on rich soils, to draw the drills from north to south (if the nature of the ground will admit of it), and on poor weak soils from east to west."

The author of the 'Synopsis of Husbandry, an English publication, observes that it is improper to sow clover among barley on rich land, because the natural fertility of the soil hastens on the vegetation of the grass, which will, before harvest have advanced to a considerable height among the corn, and occasion a longer time to be necessary for drying the swath. But on lands where there is not the danger of so luxuriant an increase, clover and other grass seeds, he thinks, may often be sown among barley ; and if a favourable time can be procured for harvesting it, the straw may be greatly improved by the mixture of clover or other grasses, and become then a valuable fodder in the winter ; but barley straw simply is, he says, the most ordinary cattle food of any.

The following has been recommended as a proper steep for seed barley : Dissolve three pounds of copperas in a pail of boiling water. Add to this as much dung puddle water as will cover three or four bushels of barley. Stir it, and let it steep twenty-four hours ; when the seed is drained and spread, sift on fine lime, which fits it for sowing.

BARN, a sort of house used for storing unthreshed grain, hay and straw, and all kinds of fodder. But the other uses of barns in this country are, to lodge and feed beasts in, to thresh grain, dress flax, &c. A barn should be large enough to serve the farmer for all these purposes : For there is always more lost by stacking of hay and grain, than enough to balance the expense of barn room.

Regard must be had to the situation of a barn. It should be at a convenient distance from the dwelling house, and other buildings ; but as near as may be without danger of fire, if the shape of the ground permits. Too low a spot will be miry in spring and fall. Too high an eminence will be bad for drawing in loads, and on account of saving and making manures. If other circumstances permit, it may be best to place a barn in such a manner as to defend the dwelling house from the force of the coldest winds.

The most considerable parts of a barn are, the floor, the bay, the cow house, the scaffolds, the stable. See *Cow House*, and *Stable*. The threshing floor should be laid on strong and steady sleepers, well supported beneath ; otherwise carting in loads upon it will soon loosen it, and render it unfit for the operation of threshing. It should be made of planks, well seasoned, and nicely jointed ; and care should be taken to keep it very tight. If it should be so open as to let grain, or any seeds, pass through, the grain will be worse than lost, as it will serve to feed and increase vermin. A floor of boards should therefore be laid under the planks.

The sills of a barn should be made of the most durable kind of timber, as they are more liable to rot than those of other buildings, on account of the dung lying about them. White oak is very fit for this use. The sills must be laid rather low, not only for the convenient entrance of cattle and carts, but because the ground will be lowered round barns, by the yearly taking away of some of the surface with the dung. They should be well underpinned with stones laid a little below the surface of the ground ; and well pointed with lime, to prevent loss of manure. And dung should not lie fermenting against the sides of a barn ; but be speedily removed when warm weather comes on.

In order to prevent hay or grain from heating in a large mow, the farmers in Pennsylvania set up four poles or pieces of timber in the middle, so as to form within them a square space of about two feet. The poles are braced by cross-pieces at certain distances. Through the aperture, thus made, the extra moisture in the hay or grain will evaporate so as to prevent its being mow burnt.

BARN YARD, a small piece of inclosed ground, contiguous to a barn, in which cattle are usually kept. It should have a high, close, and strong fence, both to shelter the beasts from the force of driving storms, and to keep the most unruly ones from breaking out. By the help of this yard, a farmer may prodigiously increase his quantity of manure, if he will be careful to take the right methods.

The ground of a yard for this purpose should be of such a shape as to retain all the manure, or prevent its being washed away by rains. It should be lowest in the middle; or at least so high on all the sides, that even the greatest rains shall not carry away any of the manure. This is a matter of so much importance, that it may be well worth while to form the ground to the right shape, where nature has not done it.

A yard should be larger or smaller in proportion to the stock that is kept in it. A small one is bad, as the cattle will be more apt to push and hurt one another. A large one is more favourable to the design of making abundance of manure. Not only should the yard be contiguous to the barn, but as many of the other out houses as conveniently may be should be placed on the sides of the yard, especially those of them which afford manure or rubbish, as the hog-sty, &c.

Many, who have good farm yards, are not so careful as they should be to make the greatest advantage by them, by confining the cattle continually in them, during the foddering season. The practice of driving cattle to water, at a distance, is attended with great loss of manure. Instead of continuing in this absurd practice, the well that serves the house, or one dug for the purpose, should be so near the yard, that a watering trough may reach from it into the yard. Some have a well in the yard; but this is not so advisable, as the water may become impregnated with the excrements of the cattle, and rendered less palatable. He that has a large stock, may save enough in manure in this way, in one year, to pay him for making a well of a moderate depth: Besides securing the advantage of having his cattle under his eye; and of preventing their straggling away, as they sometimes do. Innumerable are the accidents to which a stock are exposed, by going to watering places, in winter, without a driver, as they commonly do. And oftentimes, by means of snow and ice, the difficulty is so great, as to discourage them from going to the water; the consequence is, that they suffer for want of drink, and the owner is ignorant of it. All these things plead strongly in favour of the mode of watering I have here recommended. They should not be let out, even when the ground is bare: For what they get will make them to winter the worse; and they will damage the fields.

There should be more yards than one to a barn, where divers

sorts of cattle are kept. The sheep should have a yard by themselves, at least; and the young stock another, that they may be wholly confined to such fodder as the farmer can afford them. But the principal yard may be for the cows, oxen, calves and horses. And the water from the well may be led into each of these yards by wooden gutters.

If the soil of the yard be clay, or a pan of very hard earth, it will be the more fit for the purpose of making manure, as the excrements of the cattle will not be so apt to soak deep into it. Otherwise a layer of clay may be laid on to retain the stale, and the wash of the dung, which otherwise would be almost entirely lost.

Some farmers seem well pleased to have a wash run away from their barns upon the contiguous sloping lands. But they are not aware how much they lose by it. A small quantity of land, by means of it, may be made too rich. But the quantity of manure that is expended in doing it, if otherwise employed, might be vastly more advantageous; especially if it were so confined as to be incorporated with a variety of absorbent and dissolvable substances; and afterwards laid on those parts of the farm where it is most wanted.

It is best, in this climate, that a barn yard should be on the south side of a barn. It being less shaded, the manure will make the faster, as it will be free from frost a greater part of the year, and consequently have a longer time to ferment in. The feet of the cattle will also mix the materials the more, which are thrown into the yard, and wear them to pieces, so that they will become short and fine.

After the yard is cleaned in the spring, the farmer should embrace the first leisure he has, to store it with a variety of materials for making manure. For this purpose, he may cart into it swamp-mud, clay, brick dust, straw, thatch, fern, weeds, leaves of trees, turfs, marsh mud, eel grass, flats, or even sand and loam. If he cannot get all these kinds of rubbish, he may take such of them as are the most easily obtained. Any of these substances, being mixed with the dung and stale of cattle, will become good manure. But some regard may be had to the nature of the soil on which the manure is to be laid. If it be clay, the less clay and the more brick dust and sand will be proper: If a sandy soil, clay, pond mud, and flats, will be better ingredients.

All the materials above mentioned, and many more that might be named, will in one year become good manure, by being mixed with the excrements of the cattle, and prevent the waste of them. And this is thought, by the best writers on husbandry, to be the cheapest method a farmer can take to manure his lands, considering the small cost of the materials made into manure.

If water should stand long in any part of the yard, the manure must be raked out of the water, and heaped round the borders of the puddle, that it may be dry. For there will be no fermentation where there is too much wetness: The materials will not dissolve, but turn sour. As these heaps grow dry, the water should be scooped up, and thrown upon them from time to time. This will increase the fermentation in the heaps, and they will grow mellow the faster. It will be of service to shovel the whole of the manure into heaps, a few days before it is carted out, as it will bring on a brisk fermentation, and make it fitter to be laid upon the land. Or if shovelling be thought too laborious, turning it up with a plough will be advantageous. Or if there be not a deep layer, tearing it with a harrow may be sufficient.

Some modern agriculturists of eminence, with Sir Humphrey Davy, are of opinion, that unfermented manure is most profitable. If so, manure in a yard need not be stirred unless it be necessary to *prevent* fermentation, or what is called the dry rot. The best part of the manure it is said is dissipated, and goes off in gas, during fermentation. See *Dung*, *Manure*, *and Stercorary*.

BEANS. The following mode of planting beans has been recommended by an English writer. The rows are marked out one foot asunder, and the seed planted in holes two inches apart: the lines are stretched across the lands, which are formed about six feet over, so that when one row is planted, the sticks to which the line is fastened, are moved by a regular measurement to the distance required, and the same method pursued till the field is completed.—The usual price for this work is 9*d.* sterling per week, and the allowance two bushels per acre.

Sir John Sinclair, in "The Code of Agriculture" recommends cutting the tops of beans in order to accelerate their podding. This eminent writer states that "it was begun about the year 1804, and has already been tried on more than 200 acres. The operation is performed by means of a sharp edged instrument or knife, 12 or 14 inches long exclusive of the handle; but it may be done by a sickle or reaping hook. The expense has never exceeded 3*s.* per acre, and it is done by contract.—At a certain stage of its growth the head of the bean stalk does not seem essential to the purpose of vegetation, but by its luxuriance to exhaust the strength of the plant. The proper time to cut them off is when the first blossoms begin to drop: if done sooner, a fresh shoot will put forth. As soon as the tops are cut off, the pods rapidly increase in size, and the period of ripening is accelerated. The timely removal of these parts, where the insects chiefly lodge, materially contributes to the health and vigour of the plant, and probably increases the weight of the crop. The harvest is by this means advanced *at least* a fortnight. In the ordinary mode of managing a bean crop, their tops are green when reaped, consequently they absorb and retain moisture, and require a considerable exposure in the field to prepare them for the stack; whereas without their tops, the crop is sooner in a condition to be carried, and less risk is incurred from the effects of frost and wet seasons. The tops are left to rot on the ground."

The following is taken from Mr. M'Mahon's Treatise on Gardening, &c. "The early Mazagan, long-podded Windsor, and all the varieties of that species of bean, should be topped, when arrived at full bloom, and the lower pods beginning to set; this will greatly promote the swelling of the pods, as well as their early maturity; for having no advancing tops to nourish, their whole efforts must go to the support of the fruit.

This should be performed on the beans in general when in full blossom; observing to let the stems be first advanced to such a due height, as to have a *sufficient quantity* of pods; the early Mazagan bean may be topped when about two feet high, and the larger sorts when from about two feet and a half or yard, to three and a half high, according to the growth of the different varieties, and may be done with the finger and thumb.

But with respect to the small early beans, if you would have them come in as soon as possible, you should top them when the blossoms at the bottom of the stalks begin to open." *American Gardener's Calendar.*

A gentleman who is a practical, as well as a scientific farmer, is of opinion, after repeated experiments, that the *Field White Bean* is the only kind, which can be cultivated in this country on a great agricultural scale.

BEER. Much has been published for the direction of those who undertake large breweries. It is much to be wished that many such were carried on in this country, where barley for making malt can be so easily raised. The use of ardent spirits, which are most costly, and less wholesome than beer, might thus be lessened. They who are disposed to undertake brewing, may supply themselves with volumes on the subject. I shall only undertake to direct farmers, who may be disposed to brew beer for their own consumption.

Almost any householder may brew, without putting himself to much if any charge for an apparatus. Instead of a large copper, which is necessary in a brew house, a large kettle or two may answer the purposes of heating the water, and boiling the wort. Hogshead and barrel tubs, and other vessels, may serve for mashing tubs, backs, coolers, and tuns.

The water used for making beer or ale, should be soft, and such as is fit for washing. For this will better penetrate the malt, and cause it to discharge its spirituous virtue. Some recommend throwing a spoonful of salt into a kettlefull, which will cause any foulness contained in the water to rise to the surface when it boils, which may be skimmed off. When the water is very good this will be needless. But let the water be ever so pure, a little bran, or malt, should be thrown upon the top, while it is heating ; to be taken off when the water begins to boil. If malt be used, throw it into the mash tub. The design of thus covering the water is, to prevent the best, most subtil and volatile particles of the water from evaporating, or going off in steam. The water, for the same reason, should but just boil ; after which it should not be left to cool gradually, as the evaporation would be too great : But as much cold water should be thrown in, and mixed with it in the mash tub, as will bring it to the right temper, perhaps about three gallons to half a barrel. For the malt should not be scalded, but steeped in water, as warm as it can be without scalding ; because the scalding of the malt would rather close up its pores, and prevent its impregnating the water with its virtue, so much as it will in a tepid menstruum. It will also render it glutinous and adhesive, so that the water will not have a free passage through it. The cold water should be put first into the mash, and the hot after it.

The mash-tub should have a cock, or a tap and faucet, fixed into its bottom, and the hole covered within with a little flat shaped inverted basket, fastened with nails, that it may not get out of place by the mashing, and a close straining cloth may be put over it, and fastened in the same manner.

The water being in the mash tub, one person should put in the malt by little and little, and another should stir it about with a stick or paddle, that it may not remain in lumps, or fail of being thoroughly wetted. This is all the stirring that is needful. For too much stirring would cause the malt to thicken, so as not to give a free passage to the water that is to pass through it.

Some of the last of the malt, instead of being stirred into the water, should be strewed loosely over the surface, to serve as a coat for the rest, and prevent the copious passing away of the spirit in steams. Besides, the tub should be closely covered with sacks, or other cloths, that none of the steam may escape. In this situation it should stand for two or three hours. Then with a small stream draw off the wort, upon a handful or two of hops, into the back, which is placed under the mash tub. Fill with water again, and mash ; in half an hour run it off ; in the mean while be pouring hot water into the mash as it is running. It should be poured in on that side of the tub which is most distant from the cock, or so that all the malt may be washed with it as equally as possible. This water may be almost or quite boiling hot, as mixing it with that in the tub will so cool it as to prevent scalding. Continue thus to pour in water and run it off, till you have the quantity in the back which you design for your strong ale or beer. Then stop the cock, and fill the grains with a sufficient quantity of cold water, for small beer, or it may be hot if the weather is cold, so that there be no danger of souring. Let it stand, covered as before, and boil your first run. When it has boiled smartly for half an hour, put in your hops, and boil it another half hour, or till it breaks or curdles, as it will when it is sufficiently boiled. Or you may put your hops into a thin coarse linen bag, leaving room for them to swell, and boil them the first half hour in the wort, which I take to be a better method.

When your wort is boiled enough, strain it into your coolers, in which the thinner it lies the better, as it will cool the faster.

The next thing is to put the wort into the tun, an open vessel, to ferment. If very fine and clear drink is desired, the sediments in the coolers should be left behind, and strained through a flannel bag: For the less of the grounds go into the tun, the purer the beer may be expected to be in the cask, and the more easily fined.

That which is intended for long keeping, should be almost or quite cold before it is put into the tun, because a slow fermentation will be most proper for it. But ale, or small beer, for speedy use, may be put up a little warm.

Then stir in your barm, or yest, a pint of which is enough for a barrel. If the fermentation be too slow, beat in the yest once or twice, but not oftener, lest the drink should be injured by it.

In two or three days the beer will purify itself by throwing up the lighter parts to the top in a white curled foam, and precipitating the heavier and fouler parts to the bottom. It should then be tapped just above the lees, and, having taken off the yest, the beer must be drawn off into the casks in which it is to be kept: Which should stand with the bungs open, till the fermentation ceases, and be kept constantly full, not by pouring in that which runs over with the yest at the bung hole; but with some of the same beer kept in a vessel by itself. Thus it will throw off the yest, and deposit a dreggy part sufficient for the beer to feed upon in the cask. Reserving the yest for use, bung the casks close as soon as the working ceases. If the brewing be done in October, the bungs should not be taken out till spring. Then open the vent holes: For the coming of warm weather will cause a new fermentation. This being over, keep the casks well stopped till September following: Then fine it with isinglass, first racking it off, if it be not pretty fine.

But for ales and small beers, it may answer well enough, to omit the tunning, and remove the wort from the coolers directly into the casks; observing to keep them full, that they may purge themselves of the yeast.

Butt beer of the strongest kind, takes eight bushels of malt for a barrel. But a smaller quantity will make a pleasanter and wholesomer drink. The same quantity will make a barrel and a half of good strong ale; or six barrels of small beer.

RECEIPT *for brewing for a private family.*

Take four bushels of malt, and

from ten ounces to a pound of hops, as you wish your beer to be more or less bitter. Brew according to the above method. You will have one barrel of good ale, and another of small beer. For the small beer half a pound of hops will be enough. Some use the hops that have been boiled before: But fresh hops will be far better and wholesomer.

Spruce Beer.

Take a sufficient quantity of spruceboughs; boil them in water about half an hour, or till the outward skin, or rind, peels off: Strain the liquor, and stir in at the rate of two quarts of molasses to half a barrel. Work it with beer grounds, or emptyings; or rather with yest.

Instead of spruce, some use juniper, and prefer it. It is the low species, commonly called *savin*. A little wheat bran should be boiled in this beer to give it a briskness.

Molasses Beer;

according to a method said to be practised in Philadelphia.

"Take five pounds of molasses, half a pint of yest, and a spoonfull of powdered race ginger: Put these ingredients into your vessel, and pour on them two gallons of scalding hot, soft and clear water: Shake them well till it ferments; and add thirteen gallons of the same water cold, to fill up the cask: Let the liquor ferment about twelve hours, then bottle it off, with a raisin or two in each bottle."

A good Household Beer.

Take a heaped half peck of wheat bran, and three or four ounces of hops: Boil them a quarter of an hour in fifteen gallons of clear water: Strain it through a close sive, and sweeten it with two quarts of molasses: Cool it quick till it is no warmer than new milk, and fill your half barrel. Warm water may be used to fill up the cask if needful. Leave the bung out for 24 hours, that the drink may work, and throw off the yest, and it will be fit for use.—About the fourth or fifth day, bottle off what remains in the vessel, especially if the weather be hot, that it may not turn sour or stale. If the cask be new, or not before used for beer, apply yest or beer grounds to ferment it: Otherwise it will not be necessary.

The practice, which is common in this country, of fermenting our small drinks, with the sediments, or dregs of the same, ought to be laid aside. For this is undoubtedly the foulest, and most unwholesome excrement of liquor. Practice is apt to reconcile the minds of people to the most absurd and unwholesome things. Would not a man be considered as insane, who should take the emptyings of cyder, and put it into his new cider to ferment it? But how much better a practice is it, to ferment our small beers in this manner, with the sediments of small beer? It is true, that yest is also an excrementitious part; but that which is white, is evidently far lighter, and freer from filth, and contains much of the volatile and spirituous parts. As I had rather receive the breath or perspiration of cattle into my body, than their dung, or stale, so I prefer the white scum in my drink to the ponderous dregs of liquors. These observations will as well supply to the fermenting of dough.

To mend disorders in beer, and improve it, the London and country brewer gives the following directions.

To cure a butt of ropy beer.—Mix two handfuls of bean flour with one handful of salt, and stir it in.

To feed a butt of beer.—Bake a rye loaf well nutmeged, put it in pieces into a narrow bag of hops with some wheat, and put the bag into the cask at the bung hole.

To cure musty drink.—Run it through some hops that have been boiling in strong wort, and afterwards work it with two parts of new beer, to one of the musty old. This is called vamping, and is a cure for musty, or drinking beer.

To feed and give a fine flavour to a barrel of beer.—Put six sea biscuits into a bag of hops, and put all into the cask.

To fine or clarify beer in twenty four hours.—Put in a piece of soft chalk burnt, about the bigness of two hen's eggs, which will disturb the liquor, and cause it afterwards to be fine, and draw off brisk to the last, though it were flat before. This will do for a kilderkin, or half barrel.

To fine and feed butt beer.—Cut isinglass into small pieces, and soak it in some stale beer; then boil sugar in small beer or ale to a thin syrup, and mix it with some of the isinglass beer, which put into a butt of beer, stirring it briskly together. It will fine and preserve the drink well.

To recover a kilderkin of stale small beer.—Put two ounces of good hops, and one pound of mellow fat chalk, broke into a dozen pieces, in at the bung hole, and stop it up close. It will prove sound and pleasant to the last.

To fine a kilderkin of ale or beer, and preserve the same sound and pleasant for a long time.—Take a large handful of hops, boiled in a first wort only half an hour, and dried; half a pound of loaf sugar dissolved in some of the ale or beer; one pound of chalk broke in six pieces; the white part of oystershells, calcined in a clear charcoal fire to a whiteness, and the stems of tobacco pipes, that have been used and are burnt again, of each in powder four ounces. Put in your hops first, with the pieces of chalk; and then mix your two powders and loaf sugar in some of the ale or beer, and pour all in immediately after the hops and chalk, stirring them well about with a staff, and bung down.

Some put these into ale quickly after it has done working; others will rack off their October or March beer into another cask, and then put in these ingredients, and stir it well with a staff: Or give the vessel a roll or two, that the bottom may be turned up. You may tap it at a week's end: You will have a clear wholesome ale or beer.

It is said that beer tasting of the cask may be freed from it, by putting a handful of wheat in a bag, and hanging it in the vessel. *Rees' Cyclopædia.*

The pods of green peas, after being dried, are likewise recommended as ingredients in beer to which they afford spirit, and an agreeable flavour.

BEES, an industrious and profitable species of insects. Rural economy is incomplete where bees are wanting. The cost of keeping them is nothing, after the house and boxes are made; and the care that is required about them is but trifling, affording an agreeable amusement.

There are three sorts of bees in a hive: 1. The queen bee, which is larger, and of a brighter red, than the rest. Her business is to conduct the new swarm, and lay eggs in the cells for a new brood: And her fertility is so great that she brings forth many thousands of young ones in a year. 2. The drones, which have no stings, are of a darker colour than the rest, and are supposed to be the males. 3. The honey bees, or working bees, which are by far more numerous than the other two kinds.

A bee-house should be situated at a good distance from places where cattle are kept, especially from hogsties, hen and dove houses, and remote from filth and dunghills. It should be defended from high winds on all sides, so far as may be, consistently with admiting the heat of the sun. The house should be open to the south, or southwest, and the backside should be very tight; with a tight roof projecting, that driving rains may not injure the bees. If snow lodges upon or about the hives, it should be brushed off without delay. The bench on which the hives stand, should be a little canting outwards, that if wet should fall on it, it may run off without entering the hives. Mr. Bromwich proposes, "that a bee-house be boarded in front: And that the backside should consist of three doors, which, opened, give a full view of the hives, and give opportunity to assist or lift them. All seams are to be stopped, which would admit insects, from which the house is often to be brushed.

"If the house should be in danger of being too hot, when thus inclosed, it may be occasionally shaded with boughs of trees. As winter approaches, all the seams of the house are plaistered with clay. In very cold climates, the house should be filled with straw, to keep the bees warm, watching against mice, and removing the straw in the spring.

"Cut a hole through the front, of the same size as the mouth of the lower hive, and directly against it. Under this passage, on a level with the floor, is a lighting board, at the mouth of each hive, of about five inches long, and three wide. It is a little shelf for the bees to land upon after their excursions. These being separate, not in one piece of the length of the house, is to prevent intercourse between colony and colony; but is more essential to prevent mice, snails, and other intruders. These alighting boards are sometimes painted of different colours, to direct each bee to his home more readily.—A long shelving board should be placed over the alighting boards, to shelter the bees in a rainy time. It should be twelve inches wide, and placed nine inches above the mouths of the hives."

Broom, clover, and mustard, are said to afford bees an excellent pasture; and they appear very fond of the flowers of poppies. Gardens, and any places where flowers abound, and especially where there is a succession of flowers through the greater part of the year, are most favourable to them: For they undoubtedly draw the principal part of their honey from the nectaria of flowers. Fields of buck-wheat are good, as they continue in bloom for a long time. In Germany they move their bee hives in boats to the neighbouring fields of buck wheat.

Bees are wont to send out new swarms in May and June. Much has been written concerning the management of them on these occasions. But the new mode of managing them renders all this less necessary. It is this: Let the bee house be made so tall as to admit three tier of hives, or boxes, one above another. The hives should not be tall, but rather broad and short, that they may take up less room. A hive of such dimensions as to be equal to a cube of 13 inches, will be sufficiently capacious. Mr. Thorley directs that they should be 10 inches deep, and from 12 to 14 inches broad in the inside. If hives be made larger, the swarms will not multiply so fast. An under hive is made with a round hole through the top of three inches diameter, covered with a sliding shutter. Each hive or box should have a passage at the bottom for the bees to pass in and out, four or five inches long, and about one third of an inch deep. One of these hives should be placed directly under an inhabited hive, before they are disposed to send out a new swarm. This will prevent the going out of a swarm, and save trouble and watching: For instead of swarming, when the upper hive is full, they will build and deposit their honey in the one that is below: and when that is full, let them find another beneath it; they will take possession of the lowermost. It is their manner always to begin at the top, and build downwards.

When the top hive is well filled with honey, it may be discovered by lifting it, or more accurately by weighing it gently with a steelyard, in a cool morning, when the bees are stiff, and not apt to come out.

When a hive is taken up, there is no need of murdering the poor insects with fire and brimstone, as has been the usual practice. Only drive in the shutter, and run a thin long knife round, to part it from that which is below it; slip the hive off upon a smooth piece of board, or slide the board under, and carry the hive into your dwelling house, which you may do in a cool morning without any danger from their stings. Lay the hive upon its side, and have a window of the room open. As the sun gets up, and the air grows warmer, they will quit the hive, and go into the hive next to the place whence they were taken. When you take out the honey, which should be done speedily, the bees that are found among the honey, stiff and unable to fly, should be thrown into a tub of water. They will soon recover their activity, and go after their companions.

Some practice feeding bees. But, says one, " There is but little use in it, because those which have not a good stock of honey to serve them through the winter, are not fit to keep."—He adds, " There are some stocks of bees in the spring time, that may seem worthy of our care to preserve; such as have but little honey, and a good number of bees, by means of a cold and dry spring, yet in all probability may prove an excellent stock, and may be worth consideration."

" The best method of supplying bees with food, is by small canes, or troughs conveyed into their hives and beginning in March, when they begin to breed and sit on their young, it must be daily continued, till the season affords them ease and provision abroad.

" Honey is not only the best, but the most natural of all food, and will go much farther mixed well with a moderate quantity of good sweet wort. Some prescribe toasts of bread sopped in strong ale, and put into the hive, whereof they will not leave one crumb remaining."

Mr. Thorley advises when stocks of bees are weak, to double them, which he thinks the most effectual way of preserving them in common hives. He does it by the help of a fume, or opiate, which will so stupify them for a time that they may be handled at pleasure. Having done this, the queen must be searched for and killed. And examine whether the stock to which you intend to join the bees of another, have honey enough to maintain the bees of both: It should weigh 20 pounds.

" The narcotic, or stupifying fume, is made with the large mushroom, commonly known by the name bunt, puckfist, or frog cheese. It is of a brown colour, turns to powder, and is exceeding light.—Put one of these pucks into a large paper; press it therein to two thirds, or half its former bulk, and tie it up very close: Then put it into an oven, after the bread has been drawn, and let it remain there all night: When it is dry enough to hold fire, it is fit for use. The manner of using it is thus:

" Cut off a piece of the puck, as large as a hen's egg, and fix it in the end of a small stick slit for that purpose, and sharpened at the other end, which place so that the puck may hang near the middle of an empty hive. This hive must be set with the mouth upwards, near the stock you intend to take. This being done, set fire to the puck, and immediately place the stock of bees over it, tying a cloth around the hives, that no smoke may come forth. In a minute's time, you will hear the bees fall like drops of hail, into the empty hive. You may then beat the top of the hive gently with your hand, to get as many of them as you can: After this, loosing the cloth, lift the hive off to a table, knock it several times against the table, several more bees will tumble out, and perhaps the queen among them. She often is one of the last that falls. If she is not there, search for her among the main body in the empty hive, spreading them for this purpose on a table.

"You must proceed in the same manner with the other hive, with the bees of which these are to be united. One of the queens being secured, you must put the bees of both hives together, mingle them thoroughly, and drop them among the combs of the hive which they are intended to inhabit.—When they are all in, cover it with a packing or coarse cloth, which will admit air, and let them remain shut up all that night. and the next day. You will soon be sensible they are awaked from their sleep.

"The second night after their union, in the dusk of the evening, gently remove the cloth from off the mouth of the hive, and the bees will immediately sally forth with a great noise: But being too late they will soon return. Then keep them confined for three or four days; after which the door may be left open."

It is convenient to have a pane of glass in each hive, in order to watch the motions of the bees, and to know by inspection when is the right time to take up a hive. The Rev. Mr. White says, "In the back part you must cut a hole with a rabbit in it, in which you are to fix a pane of the clearest and best crown glass, about five inches in length, and three in breadth, and fasten it with putty. Let the top of the glass be placed as high as the roof within side, that you may see the upper part of the combs, where the bees with their riches are mostly placed. You will, by this means, be better able to judge of their state and strength, than if your glass was fixed in the middle.

The glass must be covered with a thin piece of board, by way of shutter, which may be made to hang by a string, or turn upon a nail, or slide sideways between two mouldings. Such as are desirous of seeing more of the bees' works, may make the glass as large as the box will admit, without weakening it too much. Or they may add a pane of glass on the top, which must likewise be covered with a shutter, fastened down with pegs to prevent accidents.

"Be careful to fasten the shutter so close to the glass, that no light may enter; for the bees seem to look upon such light as a hole or breach in their house, and on that account may not so well like their habitation."

"It is no unusual circumstance for one colony of bees to attack and plunder the hive of another.—This happens chiefly in the spring and autumn. The most effectual way to guard against their incursions, is to lessen the entrance into the hive, so as to leave room for only two or three bees to pass abreast,or to stop up the hives that are attached, till the rovers disappear." *Rees' Cyclopœdia.*

The Massachusetts Agricultural Repository for May, 1812, Vol. III. No. 2, contains a valuable "*Memoir on the Management of Bees; by Rev. Thomas Noyes,*" from which the following is extracted.

"I have the satisfaction to state from actual experiment, a simple and safe process, which I have adopted, of making my bees pay an annual tribute, without waging an exterminating war upon them. In order to effect this, I find it necessary to construct my hives very differently from what has been the general plan of their formation. I have substituted boxes, and give each swarm two, three, or four boxes, as I find their situation requires. Their introduction from one box to another is easy, and the removal of either of the boxes is safe and practicable, at any season of the year, or in the middle of the day, when they are most active. With very little injury to them, I can take their hive into as many parts as it consists of boxes, put it together in the same, or different order ; take which box I please ; avail myself of its contents, without injuring the bees ; give them another box, or return the same empty ; place it at the top, bottom, or centre, at pleasure. And, in this way make them more industrious, by affording them constant employment,and completely remedy their necessity for idleness, which is sometimes the case, when they have filled their hives, and have no where to bestow their goods.

"The boxes, which I use, are constructed in the following manner. I take a clean inch board, six inches wide, and saw it into pieces fifteen inches long. Four of these pieces, when put together at right angles, will form a box, whose dimensions within will be fourteen inches square and six inches deep. At the centre of the bottom of that side which I design for the front, I make, for a door or passage for the bees, an opening, one and an half inch long, and one third of an inch deep. Then cover the top and bottom with pieces of boards, whose thickness does not exceed one sixth or one eighth of an inch. These covers ought to project in the front about one inch to accommodate the bees with a place or stage, on which they may alight and rest. They serve as partitions between the boxes, and the lower cover ought to be confined with small screws, that it may be easily removed, when the honey is to be taken from the box. The door or passage for the bees being already made, I proceed to open a large hole for the purpose of introducing the bees into the box, and as a communication from one box to another. For this purpose I cut two holes, between three and four inches square, through the centre of the two thin covers.—Having all my boxes made in the same form and size; with holes corresponding when placed one above another; it is immaterial which is used for the top, bottom, or centre one, as in every position there will be a correspondence and uniformity.

"The top box is always to be covered with a board not less in its dimensions, than the top surface of the hive; with a weight upon it to keep it in its place, and prevent its warping. By having these partitions thin, the bees in each box come nearer in contact; and by having the boxes so completely closed, united only by a communication in the centre, they can be separated by a very little breakage of the comb and honey, as all the cells are parallel with the surface of the boxes; and in separating them they are cut horizontally and not transversely.

"When my bees swarm, and having alighted on some branch of a tree, and become quiet, I am generally ready to receive them; but am careful to introduce them into a clean box, which is prepared by faithfully rubbing the inside with the leaves of sweet balm, hazel nut, or balm of gilead; moistened in a strong brine, made by the solution of clean salt. Having prepared two boxes in this way for their reception, I place a clean table in the shade; and with a knife or saw cut off the branch on which the bees have alighted, and place them on the table, and having united two boxes, place the hole in the lower box directly over the centre of the bees; with care that it does not press so hard as to injure them. They generally indicate, by their movements, that they know what is designed, and take possession of their new habitation. If they do not immediately retire to their provided tenement, I take a small branch of a tree that is clothed with leaves, and gently thrust it in among them, and they will leave the branch of which they first took possession, and retreat into the box for protection without offering to resent any want of civility towards them. I gradually lower the box, as they retire into it, till it comes in contact with the table, and leaves them no communication to pass out of the box, but by the door.

"If my bees alight on the trunk of a tree, or on a branch that is too unwieldly to move, or too valuable to be cut, I set a table under them. If they are high, I suspend a table between ladders, put my boxes on it, elevated a little so as to give them a passage between the lower box and the table; and with a clean wing or something that is soft, brush the bees in the direction of the table, and they will not fail to take possession of the tenement provided for their reception. Proceeding in this way, I meet with no difficulty in introducing my bees into my new constructed hives. The bees ought to be removed as soon as they are quiet, to the place where they are to remain, before they go abroad, or at evening when they are all in their habitation; or else many of them will be lost and never again join the swarm. In hiving and moving bees, particular care ought to be taken not to irritate, or injure them by wounding or killing them.

"Bees require particular attention during the season in which they usually swarm; and at every season, they ought to be secured from wet, and kept remote from swine, geese, and fowls, that no disagreeable stench arise from any thing among them. Their house ought to be so constructed and situated that the rising sun may animate them to early industry, and at the same time shield them from his meridian beams during the warmest of the season, for too much heat is unfriendly to their comfort and activity. When the season is nearly past for collecting honey, the doors of the hive must be contracted, to enable them better to defend themselves against invaders. When the resources of nature fail they sometimes invade other hives with a view to make conquest; but in doing this they not unfrequently attack a late swarm, that has not a sufficient store of provision to preserve them through the winter. In this case their conquest seems to be humane, for they receive the vanquished into their own dominions, and incorporate them with all the privileges of their own subjects.

"By contracting the aperture of the hive in the autumn, the bees can not only defend their territories against their enemies, but their hive will be rendered more comfortable for winter. The extremes, cold and heat, are unfavourable to the prosperity of bees. Before cold weather commences enclose the hives on every side, except the front, with straw to defend them against the piercing cold. In the month of February, when the weather is moderate, the bees will leave their hive, and many of them light on the snow, and unless straw be spread in front of the hive on the snow, many of them will never rise again. They will soon chill and die, unless they find something, by which they may crawl from the snow, and so rise and return to the hive. It will be necessary to spread straw every new snow, or else their numbers will be greatly diminished by chilling and dying on the snow in front of the hive.

"The process of diminishing or increasing the number of boxes, is both simple, easy and safe. In collecting their tribute or enlarging their habitation, two things I carefully observe; one is not to kill or wound them; the other is to put it out of their power to injure me. For this purpose, if it be in the morning before they have left their habitation, I close the door upon them, and confine them at home. If it be in the middle of the day, when the bees are most active, I approach them with the smoke, that arises from burning leather, which has the best effect. This will make them retreat into their hive at any time, or leave the branch or trunk of a tree when hiving them, should they attempt to regain the place they at first occupied. Besides, by this act of fumigation another important advantage results. It will disarm them of all resentment, and render them harmless. Whenever they assume a hostile attitude in hiving them,

I first fumigate them, and they are immediately transformed into peaceful subjects, and my reception is friendly. But to return to the process of collecting their tribute.

"I have already observed that the bees are confined in the hive. Supposing the hive to consist of three boxes, and I wish to avail myself of the honey in the middle box. I approach them with four sheets of tin, on sliders made of a board one eighth of an inch thick, twelve inches wide, and eighteen inches long—made sharp at one end, and the other secured by a narrow piece of board to prevent its warping and as a handle to the slide. I raise the upper box in front a little, and insert one of the slides, which cuts off all communication between the two upper boxes; then insert another slide directly under the first. This being done, take the top box off, lifting it by the upper slide, which will prevent the bees escaping out of the top box, and the other slide remaining on the top of the second will confine them in the other boxes. I then insert two more slides between the two remaining boxes, in the same manner, and the hive is prepared to be separated into three parts. I then raise the middle box by lifting it by the slide immediately under it, carry it a little distance from the others, place it on a little carriage made with four low wheels, or trundles, simply connecting the two axles by the two side pieces, of such dimensions as to receive the box and confine it, when the carriage is in motion. Then put the two remaining boxes together, insert an empty one in the middle or on the top; remove the slides, let the boxes come in contact; open the doors and give them their liberty. Remove the slides from the box on the carriage, stand at a distance and draw it by a cord; and the gentle motion will not break the comb, but will cause the bees to come out and return to the hive, for they will not long pursue the box that is continually receding from them. In this way, I have taken a box of honey in the middle of the day without injury, when the bees are the most active, and most susceptible of resentment. The process of removing the top or bottom box, requires but two slides, and is much more simple, and does not render it necessary to cut off the communication between the two remaining boxes.

"In putting in the slides, the boxes must not be raised so high as to have the bees escape; and in selecting a box, if the one that contains the monarch of the kingdom be taken, the bees will not so willingly relinquish their habitation, unless she take her departure from it. They will appear to be unwilling to leave her without some attendants. Experience has convinced me of this. I removed a box last July, from the hive which was filled with the choicest honey; but on removing the honey from the box, I found at one corner of it about forty or fifty bees collected together, and they appeared unwilling to separate, yet as harmless as though they possessed no weapon of defence. I suspected that the monarch was a prisoner. With a spoon I took out the bees and soon found evidence to confirm my supposition. The monarchs of these well regulated kingdoms, I had, in several instances before, been favored with opportunities to inspect. I found that she was much longer than the other bees. Her wings were of a lighter hue, and the rest of her much darker colour. Her motions were nimble, and she displayed an activity not common to her subjects.—After having inspected her to my own satisfaction, and exhibited her to the view of my family, I returned her safe to her own dominions. In doing this I placed her several inches from the door of the hive to ascertain whether any particular attention would be paid their sovereign on her return. To my astonishment I beheld several bees crawl to her, and with evident marks of joy and congratulation attended her to the hive.*

"I shall now attempt a numerical statement of the advantages, which result from the method proposed of managing these industrious insects, superior to that which is generally practised.

"First. It contributes to their preservation. Bees are profitable insects. They richly reward us for all our attention to them, and whatever they collect that is useful and pleasant, adds so much to the real wealth of the community; for the sweets they extract would evaporate in the air, or decline with the fading blossoms. Surely then, we have motives, arising from interest to spare and protect the industrious bee, whose honey is not only grateful to the taste, pleasant and wholesome for food, but highly valuable in a medicinal point of view; and whose wax, when extracted from the comb is valuable for light, is almost indispensable in some arts, and is highly beneficial in medicine.*

"Secondly. Another advantage arising from my mode of managing bees, is that you can draw from the hive what is more than necessary to supply them during the winter. Instead of leaving them honey to remain in their cells year after year in a candied state, you may take *annually* what you deem superfluous. Besides, if you find there is a deficiency in some of your hives, and you are apprehensive their stock of provisions will not supply them through the season, in which they cannot find resources abroad, you may give them a box of honey

* It is remarked by the editors of the Massachusetts Agricultural Repository, that "Mr. Noyes, following Virgil, seems to consider the monarch of the bees *a male*. There is no point more clearly settled than that the monarch is a female, and the mother of the whole new swarm or progeny." We have therefore altered Mr. Noyes's text to conform to received opinion.

* The generality of mankind appear not to be sensible how much they are indebted to this insect for many of their comforts and luxuries, in the ample supply of the various kinds of fruit to regale them. I believe it is well ascertained, that unless the bee or other insects visited the flowers, the whole class of vines, fruit trees, and many culinary plants would be unproductive. There evidently appears to be two kind of blossoms, which may be distinguished by the terms male and female. Insects, by visiting both the male and female flowers, convey the farina, or subtle particles to the stigma, which impregnates the flower, and thus it is rendered fruitful.

from a hive that can spare it; and in this way make up the deficiency of one hive from the redundancy of another.

"Thirdly. A small swarm of bees put into a larger hive than they can fill, appear to be discouraged, and will not be so industrious, as they would be, were they put into a single box, and then another added, when their situation should require it.

"Bees put into a larger hive than they can fill in the season will not keep warm in the winter; and should they fall down when the weather is cold, which is often the case, they would be likely to chill and become inactive, and so be unable to regain the place they left. The comb not extending to the bottom of the hive would not afford them a conductor to re-ascend; they would be under the necessity of climbing by the surface of the hive. It is a fact that a hive of bees thus situated, generally lose double the number of bees during the winter, that a swarm does, whose comb affords them a conductor to re-ascend when they meet with this misfortune. Besides the bees have further to travel after entering the hive to reach the repository of their treasures, for they carefully avoid that confusion, which flying would create in the hive. Neither can the bees work to so good advantage in a large hive, as they can in one that is smaller. I find from several years inspecting them through glass windows, and the hole at the top of the upper box, that they always keep in the hive bees enough to cover the surface of their work, in order to keep their wax warm enough to mould it into their sexangular cells; besides a number of bees are generally employed as a chain for the others to ascend and descend. In a large hive, the surface of their work is more extensive than it could be in a small one. The consequence is, more bees must be kept at home, and a less number spared to collect the sweet vegetable juice, or other necessary materials to enrich and adorn their hive. Neither can the bees so effectually defend themselves against the attacks of lawless invaders, when their residence is so remote from the entrance of their camp.

"Fourthly. Honey deposited in boxes is not so liable to melt, waste, and destroy the lives of no inconsiderable proportion of the bees, as that which is laid up in large hives. The comb in large hives is suspended from the top, and forms a contact by the sides, and does not rest on the bottom, even when the bees have filled the hive as full as they ever do. They leave room to pass under the comb. Long and wide pieces of comb filled with honey, supporting many bees, suspended when the weather is warm, become tender, separate and fall. The position of the comb being changed, the honey attenuated by heat, the sexangular cells discharge the sweet glutinous liquor, and the bees chained in the rich melliferous flood are borne away in death by the materials they collected to preserve life. Whereas, honey deposited in boxes seems not to be subject to this misfortune. The comb is constructed in flakes, supported at the top, sides and bottom generally, except the lower box, or the first they occupy, by short legs or braces, in order that the bees may encircle it on every side; and the pieces being small there is not that pressure or stress of weight; it retains its position, and consequently the honey does not burst from the comb when the weather is warm.

"Fifthly. Another advantage resulting from having honey deposited in boxes, is that it may be taken out of them, without breaking the honey to any considerable degree. I took from a box in the month of July last, thirty pounds of honey-comb, and several pieces, from one to four pounds each, without uncapping one cell. The honey was transparent, and completely closed, nor was there a single cell in the whole box, but what was filled with the delicious fluid. It was superior to any that I ever saw taken from large hives; and if kept cool, to all probability it might be preserved months or years unbroken in the comb. Besides the method, which I have proposed, affords an opportunity of taking honey before the bees collect any from the flowers of buck wheat, which honey is far inferior in flavour and delicacy, to that which is collected earlier in the season.*

Sixthly. The usual time to take up bees is in the autumn, but I do not confine myself to that season. My method of managing bees will admit of taking honey fromthem any time when they have it to spare; but in selecting a box, care must be taken to avoid that which contains the young bees. The object of the new colony is to add to their strength by increasing their numbers; consequently their early attention is directed to provide cells to deposit their eggs. The lower box, which they first enter, becomes their place of residence, till they have filled it with comb, and young bees, if it be an early swarm. The honey is to be sought in the second box, of which they gradually take possession. In filling the second box they begin at the communication between the two boxes, and raise their comb in the form of a frustrum or segment of a globe, and proceed in this way till they have raised it to form a contact with the cover.

"Seventhly. The method of keeping bees in boxes that may be separated at pleasure, affords an opportunity of changing the comb in the hive before it becomes dark coloured, and apparently rusty, and an unsuitable receptacle for that sweet, delicate and nutritious fluid, which was the emblem of plenty in ancient Canaan. The purest honey, deposited in such combs loses its transparency and delicate flavour, and partakes in no incon-

* Honey, collected from buck-wheat flowers will do to preserve the bees, but it is several shades darker in its colour, less sweet in its nature, possessing something peculiar in its taste that is unpleasant, and emitting an effluvium which excites unpleasant sensations; and thus affording an evidence by three senses, what proportion of honey in the hive was collected after the bees had access to the flowers of buck-wheat.

siderable degree, of the colour of the comb.

"Eighthly. This method of constructing bee-hives affords an opportunity of inspecting them, watching their various movements, and witnessing the progress they make in filling their hives. For this purpose the bees must first occupy the bottom box; and then by lifting up, or removing the top cover, you may through the hole on the top, see them without disconcerting them, until they have nearly filled the upper box, unless you breathe upon them; in this case they will soon let you know that you are an unwelcome visitor. I have inspected them in this way, and through glass windows inserted in their hives, for several years past; and I have beheld with astonishment their industry, economy, singular instinct and ingenuity.

"Ninthly. By accommodating my bees with several apartments, I have it in my power to regulate their swarming. If bees are amply accommodated they will not swarm; but it being an object to make them swarm, I therefore in the autumn reduce the number of their boxes, and leave them, generally, but two, which, if well filled will amply supply them with sustenance through the winter.

"I do not at first give an early swarm more than two boxes; for they will often send forth a young colony; and when they have done this, add another box, or in a short time take from them the one they have filled, and give them one that is empty. Thus, this method of managing bees will contribute to increase the number of swarms, and eventually make them more profitable; for experience evinces that multiplying the number of swarms makes them more industrious and productive."

For a description of several sorts of newly invented hives, and the manner of using them so as to take the honey and wax, without injuring the bees, see Farmer's Cyclopædia, *Art. Apiary* and *Bees*.

The following method of taking honey without destroying the bees is recommended in the "American Farmer."

In the evening, when the bees have retired, take the hive gently from the stand; spread a table cloth on the ground; set the hive on it, placing something under to raise it three or four inches; then draw up the corners of the cloth, and fasten them tight around the middle of the hive, leaving room sufficient between them and the hive. Then raise the lid of the hive a little, and blow in the smoke of a cigar: a few puffs of which, as it is very disagreeable to them, will drive them down. Continue raising the lid gradually, and blowing in the smoke all around, and in a few minutes it will be found that they have all gone out of the hive. You may then take off the lid, and cut away as much honey as you think proper. If the operation be performed the beginning of July, you may take nearly all, as there will be time enough to provide a sufficiency for their support during the winter. As soon as you have taken the honey, put on the lid, loosen the cloth, and spread it out, and in an hour or two the bees will have returned into the hive. It may then be replaced on the stand, and on the following day they will be found at work as usual."

When a person has been stung by a bee, the sting should immediately be extracted with a steady hand, for if any part of it breaks in, remedies will, in a great measure prove ineffectual. A strong solution of salt or potash, or liquid laudanum will then speedily effect a cure.

We shall add to this article a few facts relative to the natural history of these wonderful insects, and the best mode of managing them, their diseases, &c.

Every hive must have its queen bee, who is the mother of the whole. If she be lost or destroyed the whole commonwealth is broken up. Besides these, are the drones and the working bees. Drones are larger than the working bees.—They die in the latter part of summer, and are dragged out of the hive by the working bees.

When more than one swarm leaves the hive in a season, those, following, consist of those which were abroad when the first swarming took place, together with young ones, which have been hatched since the departure of the first swarm.

The kind of swallow called martins, will destroy bees. "It is advisable to have large swarms; small ones never thrive so well; and for this purpose two or more small ones should be put together. The swarm should weigh from four to six pounds. To ascertain their weight, that of the empty hive should be first known, and marked on it; and then the addition to that will be the weight of the bees.

"For joining two or more swarms, take a full hive, at night, and set it bottom upwards; then set an empty one of the same size, with its bottom exactly on the other, and let there be cross pieces in the empty one for the bees to light on. Then strike gently on the two sides of the full hive to which the edges of the comb are fastened, and the bees will leave it and ascend into the upper. Then repeat the operation with another full hive, the one with the bees being set uppermost, as before, and you have two swarms together. Repeat it again, as before, and you have three; and so on, if more swarms are to be added. Then set the hive with the bees in it where one of the full ones stood, and they will go to work together. The queens however, must be first searched for and all destroyed but one.

"Another method recommended is to take a full hive, set it on a cloth with the bottom downwards, and then give it a smart stroke, which will cause all the bees to fall; search for the queen and destroy her; have another full hive ready and put it over the bees, and they will soon crawl up into it, and become incorporated with those of that hive. This may be repeated for the purpose of adding a third swarm to the other two." *Farmer's Assistant.*

Bees have enemies and diseases, of which we shall take some notice.

Butterflies are said to conceal themselves in the hives, and annoy the bees: these intruders may easily be exterminated, by placing lighted candles in deep tin pots between the hives: as the flame will attract them to their destruction.

Hornets may be destroyed by exposing shallow vessels near the hive with a little water, in which these insects will drown themselves.

But the most pernicious enemy to bees is a kind of *tinea*, or moth, which is thus described by *M. Reaumur*.

" These creatures are of the caterpillar kind, and have sixteen legs. They feed on wax, and for food enter the beehives; where they boldly engage the bees, and are not to be prevented by them from feeding, though at the expence of their habitations; so that it is no uncommon thing for a swarm of bees to be forced to change their places leaving this contemptible victor in possession of the hive.

" All the authors who have written on bees have complained of this destructive animal. It never eats the honey, but feeds only on the wax; attacking principally those waxy cells where the female bee deposits her eggs for the future progeny.

" The bees would readily destroy these creatures were it not for the armour they are covered with. They form themselves a coat of armour of a double matter. The first next to the body is a kind of silk of their own spinning; and the outer covering is of bees wax, laid on considerably thick. The creature just thrusting his head out to feed, goes on devouring the cells; while the bees are buzzing about him, attempting, in vain to pierce him with their stings. He never forsakes his covering, but lengthens and enlarges it as he goes; and gnawing down the sides of the cells in his march, without staying to cut them one by one, the destruction he occasions is scarcely to be concieved.

" When the time of change approaches, it contracts its body within its double covering, and there changes into the nymph state; whence, after a proper time, it comes forth in the form of a moth, with granulated horns, and a crooked probocis.

" The bees know their enemy in this new form, and destroy all the moths they can meet with. They are seldom so fortunate, however, as to kill the whole race as soon as produced; and if only one escape, it is able to lay a foundation of revenge for the death of its brethren.

" All the flies of the moth kind lay a vast number of eggs; and the young ones produced by one female are sufficient to destroy many hives of honey combs. The moth produced by this catterpillar flies but little; but is very nimble in avoiding danger by running, which it does with great swiftness."

It appears that this insect began its career of destruction in some place to the southward of New York, and Mr. Van Schaick, a writer, who is quoted in *The*

Farmer's Assistant, asserts that the most effectual mode of destroying the insect, is "to raise the hive about an inch above the floor, and keep it there, when the moment the bees discover their unmasked enemies, whether in the shape of eggs, or of catterpillars, in different stages of formation, they attack them with fury, and toil incessantly until they have destroyed or removed every vestige of them off the board."

He further very judiciously recommends the construction of the floor or plank on which the hive stands, and also the rim of the hive, to be "of such materials and dimensions as would afford no place of concealment for the fly or its eggs." Dr. Low, another writer, quoted on the same work, advises to suspend the hive, by a cord fastened in the top, and have the plank forming the floor moveable up and down. During cold weather, the plank is brought up close to the rim, to keep the bees sufficiently warm; but on the approach of spring, or when the weather was become suitably moderated, the plank was let down about four inches, and kept in that situation during the warm or growing season. At particular cold spells, during the first of the spring or latter end of the fall, the floor, no doubt ought to be raised up, until the return of warmer weather, but it should only be kept in this position while the comfort of the bees requires it.

Dr. J. Anderson in one of his practical papers "On *the Management of the Dairy*," communicated to the Bath and West of England Society, observes in a note, that bees, in that variable climate are a very precarious stock, though extremely profitable where they thrive. During the frequent mild days of winter, and the warm mornings of spring which are suddenly succeeded by a nipping frost, or sleety rain, these creatures are roused from their torpid state; and being unable to obtain food abroad, they are obliged to consume and exhaust their stores, and to perish from want. And as the warmth of the weather invites them to search for flowers in vain affording them nourishment, they are often chilled by cold before they are able to return to the hive.

To prevent such fatal accidents Dr. Anderson is of opinion, that no method would be so effectual as that of placing the hives in an ice house, at the approach of winter. Here they may be kept till the spring has so far advanced, that no danger is to be apprehended from bad weather. During the whole winter, they will remain in a state of torpor, and require no food. As soon as the mild weather invites them to appear, they will commence their labours with vigour. The intense degree of cold which bees sustain without the least injury, in Poland and Russia, where even quicksilver is sometimes frozen, removes every doubt or anxiety, concerning the safety of bees in an ice-house."

BEET, *Beta*, a well known esculent root.

There is a sea beet which grows

in salt marshes; and a white beet cultivated in gardens for the sake of its leaves, which are sometimes used in soups. The root is small, and commonly hard and tough.

But the sort which is most valuable, is the red beet, with a large, pyramidal, fleshy root; the leaves of which are large, thick and juicy. The larger these roots grow, the more tender they are: And the deeper their colour, the better. The best of red beets have reddish leaves. In some of the varieties the leaves are all over red.

Beets require a mellow and warm soil, moderately rich, and well pulverized to a good depth. For as they naturally run deep, in shallow ground they will be short, stringy, and irregularly shaped.

Beets should be sown early. A good method is, to set the seeds in squares of about eight or nine inches in poor ground; in rich ground they should be at least a foot asunder. If a fourth part of the seeds should fail, the crop will not be lessened.

As the capsule, commonly called a beet-seed, contains several cells, *each* of which has a seed, they are apt to produce four or five plants in a bunch.—These should be carefully reduced to one, leaving the healthiest plant. If this precaution is omitted, the plants will be inevitably small and frequently the roots will be intertwisted.—Those which are taken out may be transplanted; but they are not so apt to make good roots. Though they may be thick, they will be apt to be wanting in length.

The ground should be hoed two or three times, after which the leaves will so cover the ground, as to stop the further growth of weeds.

The under leaves may be broken off towards fall, and thrown to the swine, which are very fond of them. This will not injure the roots at all; for if they are left on, they will soon decay. Taking away part of the leaves will let in the sun and air, which will be of advantage to the roots.

The roots should be taken up before any severe frost comes; none of the fibrous roots should be taken away; nor the heads cut very close. In this state, also, they should be boiled, that none of their rich juice may escape.

They may be used in autumn, and kept good all winter. But if any frost touches them, though they will not presently rot, they will become tough, and unfit for the table. And, in the spring, their early sprouting depreciates them.

"*The Mangel Wurtzel Beet*, or *Root of Scarcity* is raised from seed sown annually in the spring, the same as the other sorts, in any open situation, but should generally be sown thinner, either in drills one or two feet asunder, or broad-cast on the general surface, and raked in; and when the plants are come up one, two, or three inches in growth, they should be thinned to a proportionable distance, to give room for the full expansion of their large leaves. Some however, advise transplanting, when the young plants are of two or three inches growth, setting them in rows of one or two feet asunder: This seems, however, unnecessary: especially as they have long, downright, tap roots, which generally are the most successful when they remain where sown; the method may however be practised occasionally by way of experiment.

"The leaves, which if the plants have large scope of room, grow twelve or fifteen inches broad or more, and of proportionable length, are exceedingly good, when young to use as the common white and green beet, and the young, thick, fleshy stalks, stripped of the leafy part, peeled or scraped, then boiled and served up with butter, are tender and agreeably tasted; also the leaves to boil occasionally as spinach and other small greens; and of which the root is remarkably productive in quick growth, so as to afford frequent successional gatherings all summer and autumn. Only the larger outward ones, should be broken off, and care taken not to injure the small ones in the centre."

Beets and other roots may be preserved in winter by the following method, "Take off the tops and expose the roots for a few hours till sufficiently dry. On the surface of a very dry piece of ground in a well sheltered situation, lay a stratum of sand two inches thick, and on this a layer of roots, covering them with another layer of sand (the drier the better) and so continue layer above layer of sand and roots till all are laid in, giving the whole on every side a roof like slope; then cover this heap or ridge all over with about two inches of sand over which lay a good coat of drawn straw up and down as if thatching a house in order to carry off wet and prevent its entering to the roots; then dig a wide trench round the heap, and cover the straw with earth so dug up, to a depth sufficient to preserve the roots effectually from frost. An opening may be made the south side of this heap, and completely covered with bundles of straw, so as to have access to the roots at all times, when wanted either for sale or use.

"Some people lay straw, or hay, between the layers of roots and immediately on the top of them; this I do not approve of, as the straw or hay will become damp and mouldy, and very often occasion the roots to rot, while the sand would preserve them sweet and sound.

"All these roots may be preserved in like manner in a cellar; but in such a place they are subject to vegetate and become stringy earlier in spring. The only advantage of this method is, that in the cellar they may be had when wanted more conveniently during winter, than out of the field or garden heaps."

NOTE. All the above roots will be preserved better in sand than in common earth, but when the former cannot be had the most sandy you can procure must be used.—

American Gardener's Calendar.

The following remarks respecting the Mangel Wurtzel or *Root of scarcity*, are from a paper communicated to the Trustees of the Massachusetts Agricultural Society, by J. Lowell, corresponding

Sec'ry. and published in Vol. III. No. 3 of the Massachusetts Agricultural Journal. They are derived, principally, from a French publication by the *Abbe Rosier*.

This root is very little affected by changes of weather. It is attacked by no insect; drought affects but little its vegetation. It prepares the land extremely well for other crops. It may be sown and treated precisely like the common beet, except that it ought to stand eighteen inches asunder.

In good land they often weigh nine or ten pounds, and are stripped eight or nine times. In a light sandy, but well manured soil they sometimes weigh fourteen and even sixteen pounds each!

The first crop of leaves in France is taken off in the latter end of June, or the beginning of July. In this country, probably, the latter period would be preferable. The lower leaves, those which incline towards the ground are those which are taken away, and care must be taken to preserve the top leaves, or the crown of the plants. The leaves may be taken off every fifteen days after the first gathering. Oxen, cows, and sheep devour them greedily, and fatten readily upon them. All domestic poultry eat them readily, when chopped fine and mixed with grain. Horses will feed upon them very well, mixed with chopped straw. Hogs also fatten upon them.

Cows, fed upon this root solely, give a greater quantity of milk and cream, and of better quality for the first fifteen days, after which they grow too fat, and the milk lessens. The food of cows must therefore be varied. Oxen and sheep fatten very well upon them. Cows should have grass in proportion of one third to the beet leaves, or every third day they should be turned to grass. In this mode their milk will be excellent. The trouble of gathering the leaves is less than that of gathering any other green fodder. It may be done by children, while men are required to cut other green food for cattle. It is the surest crop, since the plant will stand the largest droughts. The roots are gathered and treated like those of the common beet. The skin is very tender and care should be taken to handle them so as they may not be wounded, as they will, in that case, not keep so well. In order to preserve the seed in purity, care must be taken to change the ground in which the seed-beets are planted. The seed can be preserved after it is gathered three or four years without injury. In giving these roots to cattle for food, they are first washed and then cut up into pieces about the size of a nut. It is always best to accompany them, when given to horned cattle with clover, or other hay or straw, and if the hay or straw has been previously cut fine, it will be preferable. If horses are fed with this root, with a proportion of hay or cut straw, (half of each,) they will be fat, vigorous and healthy. If they are worked severely, a little oats or corn may be added. It is thus they are treated in Germany, where this root stands in the stead of meadows or grass lands, and whose excellent horses are well known.

Hogs, fed upon them raw, after they have been cut up fine and mixed with milk or other drink, fatten as well upon them as upon boiled potatoes, by which the fuel and trouble of boiling is saved.

As to the quantity given to animals, much will depend on the proportion of other fodder, which you allow them. Cows fed twice a day in winter upon eighteen pounds of these roots at each time, together with four pounds of hay or chopped straw, will give as much and as good milk as in summer, and they will be kept in the best possible state.

Oxen fed with forty weight of these roots per day, with ten pounds of hay for one month, and after that with fifty weight per day of the roots alone, will be fat enough for sale in two months more.

Any person disposed may, from the facts above stated, calculate how many cattle will be supported by a single acre of land on which this plant is cultivated.

Men can eat this vegetable throughout the year, it is agreeable and healthy. No insect attacks it, and it suffers but little from the variety of the seasons. The leaves of this plant form alone an excellent food for every species of domestic quadruped during four months in the year. Turnips and other vegetables are besides liable to be destroyed by insects, whereas this beet is not. The roots can be preserved eight months in a sound state, while turnips are of little value after March. In some soils turnips will not grow, particularly in those which are very stiff or strong. The root of scarcity grows every where. The milk of cows fed on turnips has a bad taste. That of those fed on this plant is excellent, as is also the butter made from it. This forage on green fodder comes also at the hot seasons, when almost all other green food is scarce, and sometimes not to be procured. Cattle never get tired of it. In many parts of Germany where it is raised with success, they prefer it to every thing else to fatten those large herds of cattle which they annually export to France. In feeding cattle with beets the same dry food must be given which is usually given with turnips.

Sugar may be made from this root, and for this purpose it is said the white beet excels. For details respecting the mode of manufacturing sugar from the beet, see *Rees' Cyclopœdia*, Art. *Beta*.

We shall add to this article a brief account of the mode of cultivating this plant in Bedford, in England.

In the middle or latter end of April, the furrows two feet apart, are double ploughed, that is, the plough returns on the furrow to the point whence it set out, forming a ridge between the furrows. In these furrows, the manure, which should be well rotted is deposited, at the rate of six cubic yards to an acre. The ridges are then split by the plough going and returning the same way as before mentioned, leaving the manure directly under the middle of the new ridges. A light roller is then passed along the ridges in the middle of which the seed is dibbled, about an inch deep, covered with a garden rake, and the light roller is again passed along the ridges. The plants are hoed when about the size of radishes ; where missing are filled up by transplanting, care being taken to set the points of the roots straight, and left about twelve inches apart in the rows.

In the year 1820, John Prince, Esq. of Roxbury, received a premium from Mass. Agr. Soc. for the best crop of Mangel Wurtzel, amounting to six hundred and seventy bushels and an half on one acre.

BIRD GRASS, *or Fowl Meadow Grass.* See Grasses.

BLASTING OF STONES, the operation of tearing asunder large stones or rocks. The mode of performing this operation by gunpowder, is too well known to require a particular description. We shall merely mention some improvements, which we believe are not generally practised. An English publication states that a small portion of quick lime, in fine powder is found to increase the force, and consequently to diminish the expense, of blasting stones. It is likewise said that saw dust, particularly of soft wood, mixed with gun-powder, in equal parts, has thrice the strength of powder alone in blasting rocks.

BLOOD. The blood and offals of animals is one of the most powerful manures, and no farmer who slaughters animals should suffer it to be wasted on his barn floor.—They should be bled on his manure heap, or in his barn-yard, where substances, calculated for manure, are collected.

BOG, a piece of land with a wet miry soil, or a swamp. Some bogs, when they have a sward of grass roots, will shake and tremble under the foot. Such land is unprofitable, or even a nuisance, until it be drained. But after draining, it becomes the best of soil, producing the greatest of crops, without any manure. The way to drain a swamp effectually, is to pass a ditch through the middle of it ; and another ditch round the border, to cut off the springs which come from the upland.

In order to judge whether a bog will pay the expense of draining, the depth of the drain which will be necessary at the outlet, and its length, must be considered, and also the depth of the soil in the bog. If the soil be very thin, it will not be of so much value when drained. It will be thinner after drying than before ; but it should have depth enough for the deepest ploughing, after it is dried and settled. Otherwise the operation of draining may as well be omitted. See *Eliot on Field Husbandry.*

If, however, on inspection the soil be found to consist of *peat*, or earth closely interwoven with the roots of vegetables, all hopes of making it productive as a soil must be abandoned. In places where fuel is scarce it may be of great value, when dug up in the summer and left to dry. But, of itself, and without admixture, it will not repay the cost of cultivation.—Peaty soil may, however, be very usefully transplanted in considerable quantities into the dung heap, where it will prove a most admirable aid in the composition ; it being known that a proportion of peat facilitates and encourages the growth of all plants ; and for some plants it is of inestimable value.

BOTTS, are a peculiar kind of grubs, found in the stomach of the horse, of a cylindrical figure, pointed at one extremity and obtuse at the other, and beset every where with numerous spines. There are several kinds, all belonging to the same family. Some live in the grub state under the skin of the backs of oxen, the deer and reindeer ; and others take their abode in the frontal cavities of the sheep, and the deer, and cells at the base of the horns.

Volumes have been written on the natural history of these insects, but we have only room for stating some remedies, which experience has proved to be useful against that species of this insect which so often proves fatal to the horse.

Scrape off the eggs of the horse bee, when laid on the horse, every eight or ten days with a sharp knife. This practice must be continued during the season of them. The eggs should not be scraped off where the horse can feed, as in that case the young botts may be taken in. It is difficult to remove those eggs which are laid under the throat, with a knife, but they may be destroyed with a hot iron made for the purpose.

Blood letting, and the copious use of mild oils will always palliate, and sometimes cure the disorder.

The following remedies have been recommended. Take a table spoonful of unslacked lime,and let it be given with the water or feed of a horse at night and morning for three or four days. Or, take of copperas two table spoonfuls : water, milk warm, one pint : dissolve the copperas, and give it as a drench. If the horse is not relieved in fifteen minutes, repeat the dose. Molasses and milk, of each a pint, are likewise recommended, to be given as a drench. An active purge should follow either of the last mentioned remedies. But the remedy which experience has tested as most efficacious is the following :

Apply spirits of turpentine to the outside of the breast and stomach of a horse, and the botts will immediately let go their hold, and will soon be expelled. My informant assures me that he has seen horses, which seemed to be in the last agonies with this complaint, relieved instantly, and apparently restored to full health in five minutes by this remedy. Spirits of turpentine are sometimes poured down the throats of horses, suffering by botts, but the external application of this pungent substance is more safe, easy, and at least equally efficacious.

BROWSE, young sprouts from wood, twigs of trees, and bushes. In a new country, browsing is a considerable part of the food of cattle. They will eat browse all parts of the year, unless when the snow is so deep that they cannot wander in pursuit of it. Late in autumn, and early in spring, much hay may be saved by turning out cattle to browse. In the former part of summer, when the young shoots are in the most tender state, some cattle will even grow fat upon browse. Salt hay is found to give cattle an extraordinary appetite for this kind of food.

BUCK WHEAT, *Polygonum*, a dark coloured grain, shaped like the seed of onions, but much larger, and of a dark brown colour. It yields plentifully, and is said to be better than barley for fattening of hogs and poultry. It should not be sown in this climate, till after the middle of May. One bushel is enough to seed an acre, if sown broad cast; less than half that quantity, if drilled.

In the State of New-York, farmers sow it with their winter wheat about August. It affords them a ripe crop in the fall, and is no damage to the crop of wheat which grows with it, and succeeds it.—When the plants are green, they are large, sappy and soft. European writers, therefore, greatly recommend sowing it for a green dressing, and ploughing it into the ground, in its most green and juicy state.

In light lands buck-wheat may be raised to great advantage as a lucrative crop. When green it is a fine feed for milch kine. It fattens pigs with great economy, and, passed through a mill is, with carrots, a good feed for work horses. The seed is excellent food for poultry.

It is said in Dr. Willich's Domestic Encyclopædia, that "hogs feeding upon buck-wheat, are very liable to scabby eruptions."

BULL, the male of the ox kind. The marks of a good one for propagation, according to Mortimer, are these. He should have a quick countenance, his forehead large and curled, his eyes black and large, his horns large, straight and black, his neck fleshy, his belly long and large, his hair smooth like velvet, his breast big, his back straight and flat, his buttocks square, his thighs round, his legs straight, and his joints short.

One good bull will answer for a large number of cows. But to mend our breed of cattle, more attention should be paid to the properties of bulls. Those calves which are not large, or not well shaped, should be castrated while they are young, that a mean race of cattle may not be propagated. Neither should the practice of suffering bulls that are too young, to go to the cows, be continued. For either the cows, through the insufficiency of the bull, will go farrow, which is a great loss to the farmer, and a breach upon the dairy; or at best, the calves will be small, and scarcely worth rearing; as some of our best farmers are now fully convinced.

Sir John Sinclair observes, "A bull may be first used at fourteen or eighteen months. He then shows most vigour, and more energy may be expected in his produce. At two or three years old, they frequently become ungovernable, and are killed. Many contend, that the offspring of a bull, if well bred, becomes generally better till he reaches seven or eight years, and indeed till his constitution is impaired by age. This doctrine, however, does not agree with the practice of Mr. Vandergoes in Holland; nor can the question be finally decided without a regular course of experiments. *Code of Agriculture* p. 61: Am. Edition.

It was formerly believed that a frequent change of animals, from one country, or one section of a country, to another, was indispensable to the preservation of a pure and excellent breed of cattle, but this idea is almost exploded. Bakewell, Princeps, and other noted graziers and raisers of stock in England have proved, that it is by careful selection of individuals of good farms and properties, that the race of cattle can be best improved. The system of "breeding in and in," which confines the improved races or animals to their own families, is now much more in fashion, and the success of the British graziers, and the best experience of our own country seem to justify the opinion. Importations of the best animals of Great Britain have been made, it is true, and they have essentially and manifestly improved our stock. But this does not disprove the utility of "breeding in and in," because these excellent imported animals were produced, and bought to their present perfection by the principle of selection from the *same* excellent flock. And when imported here, the principle of "breeding in and in," or preserving the stock, has been scrupulously adhered to with highly beneficial effects.

Before we quit this article, it may be adduced as a proof of the soundness of the modern doctrine, as opposed to the old practice of *crossing* the breeds constantly, that, as to horses, the interest in which is much greater in some countries on account of the rage for the pleasures of the chace and of the turf, it is an established law, that the race should be kept pure and the pedigree of a race-horse is as accurately preserved as that of a line of kings.

BURNING THE SOIL. In this process the turfs of swarded land are cut up with a kind of hoe, called a beating axe, which, after drying, are piled and burnt. The ashes and burnt soil are spread over the surface, from whence the turfs were taken, by way of manure; then ploughed in, and mixed with the soil; first with a shoal furrow, and deeper at the second ploughing.

The Marquis of Tourbilly says, "The paring mattock, or beating axe, should have an edge like an adze, of well tempered steel, and about nine inches wide, that the iron part should be six inches in length, growing narrower towards the handle; that the hole to receive the handle should be two inches in diameter; that the handle should be of wood, about three feet long; that the instrument without the handle should weigh from ten to twelve pounds; that the turfs raised will be about 18 inches long, a foot broad, and four inches thick ; that they must be set up to dry, leaning against each other ; that when the season is not very wet, they will be dry enough to burn in about three weeks ; that when dry, they must be piled up in the form of ovens, the mouths to the most windward side ; that a hole should be left in the top for the smoke to go out ; that as soon as they are piled, they must be set on fire with some straw or heath ; that if they burn too fast, earth must be thrown on to deaden the flames ; and that they will continue burning some days. When the burning is ended, he advises, that the ashes be piled up in round heaps ; that when it is time to sow winter grain, the ashes should be spread, and the corn sown on them, and then the ground ploughed with a shoal furrow, and harrowed."

He says, "half the usual quantity of seed will be sufficient ; and that it ought to be sowed two weeks later than other ground." The reason is, because the grain will grow rapidly, and be uncommonly large.

I conceive this must be a good method of culture for our cold lands, inclining to moss, which can no other way be made to produce well the first year after breaking up. But this method will not readily be adopted in a country where labour is dear. The work, however, might be greatly diminished, by paring the surface with a very sharp ironed plough ; though in order to do this, the ground must have an extremely even surface, and be free from stones. I have said so much of this culture, in hopes of exciting some, who are curious, to make trial of it. See *Paring* and *Burning*.

BURNET, *Pimpinella*, a perennial plant, which was brought into notice and highly praised by a Mr. Roque, of England, about 40 years since, as a productive and valuable article of fodder. It was also pretty strongly recommended in the first and second editions of this work, but it has not sustained its reputation, and it is believed it is no where extensively cultivated as a grass.

BURNT CLAY. See *Burning the Soil*.

BUSHES, shrubs. These are apt to spring up and increase in pasture lands, which have never been tilled, if timely care be not taken to destroy them. Eradicating them requires so much labour, that farmers are most commonly content with cutting them once in a few years. But the more cuttings they survive, the longer lived they are apt to be ; and the harder to kill, as the roots continually gain strength.

Keeping cattle short in pastures will cause them to browse the more ; and this will have a tendency to subdue many kinds of bushes. Those which grow on high ground are oftener subdued this way than those which grow in swampy low lands, the latter being less palatable to the cattle.

It is undoubtedly true, that cutting bushes in the summer will do more towards destroying them, than doing it in any other season particularly in August ; Other circumstances being equal, the wettest weather is best for destroying shrubs by cutting; because the sap vessels of the stumps will continue open the longer; there will be the greater discharge of sap through them, and the roots will be the more weakened.

Bushes which grow in clusters, as alder, and some other sorts, may be expeditiously pulled up by oxen; and this is an effectual way to subdue them. The expense of it I suppose will not be more than that of cutting them twice would amount to.

Elder is a kind of bush which spreads fast in some soils, and has been accounted harder to subdue than almost any other. Mr. Eliot says, "He knows by experience, that mowing them five times in a year will not kill them." This has been proved by the experience of other farmers. The roots of the shrub oak will not be killed, but by digging them out, or by pasturing goats on them.

The bushes in swamps are in general more hard to conquer, than those which grow upon upland.—Flooding a swamp, where it is practicable, or can be done without too much cost, is perhaps the most approved method which can be taken. Flooding for two or three summers will totally destroy them, root and branch.

But if a swamp cannot conveniently be flooded, the next thing is, to consider whether it cannot be drained to advantage. Draining will so alter the nature of the soil, that the shrubs which it naturally produced before, will not be any longer nourished by it. Therefore

they will mostly die without cutting, or it may be expected that one cutting will be sufficient. But if draining were not serviceable on any other account, perhaps it would not answer to go to the expense of it merely for the sake of clearing a swamp of the bushes.

After all, extirpation, by digging them out, and by fire, is cheapest and most effectual.

BUTTER, a fat unctuous substance, prepared from milk by churning. If the dairy consist of three or four cows, they should be milked in the summer thrice a day; in the morning, at noon, and in the evening. In winter, however, the cows are to be milked only twice a day. The dairy house should be kept neat, should never front the south, south east or south-west. It should be situated near a good spring or current of water. The proper receptacles for milk are earthern or tin pans. In warm weather milk should remain in the pail till nearly cool before it is strained, but in frosty weather it should be strained immediately, and a small quantity of boiling water may be mixed with it, which will cause it to produce cream in great abundance, and the more so if the pans or vats have a large surface.

In hot weather the cream should be skimmed from the milk at or before sun-rise, before the dairy gets warm, nor should the milk, in hot weather, stand in its receptacles longer than twenty-four hours. In winter, milk may remain unskimmed thirty-six or forty-eight hours. The cream should be deposited in a deep pan, kept during summer in a cool place, where a free air is admitted. Unless churning is performed every other day, the cream should be shifted daily into clean pans, but churning should be performed at least twice a week in hot weather, and this should be done in the morning before sunrise, taking care to fix the churn where there is a good draught of air. If a pump churn is used, it may be plunged a foot deep in cold water, and remain in that situation during the whole time of churning, which will much harden the butter. A strong rancid flavour will be given to butter if we churn so near the fire as to heat the wood in the winter season. In churning for butter, always have an orifice sufficient for the air to have access to the cream. Butter is produced by the union of oxygen with the cream, and more butter will be made and of a finer flavour, if the churn is sufficiently open.

After the butter is churned, it should immediately be washed in many different waters, till it is perfectly cleansed from the milk; and it should be worked by two pieces of wood, for a warm hand will soften and make it appear greasy. It will require more working in winter than in summer.

Those who use a pump churn must keep a regular stroke: nor should they permit any person to assist them, who does not keep nearly the same stroke; for if they churn more slowly, the butter will in the winter go back, as it is called; and if the stroke be more quick, it will cause a fermentation, by which means the butter will acquire a very disagreeable flavour.

Cows should never be suffered to drink improper water; stagnated pools, water wherein frogs spawn, common sewers, and ponds that receive the drainings of stables, are improper.

The operation of churning may be very much shortened by mixing a little distilled vinegar with the cream in the churn. The butter being afterwards well washed in two or three changes of water, the whole of the acid will be carried off; or if any remain it will not be perceived by the taste. A table-spoonful or two of the vinegar to a gallon of cream. See *Churn*. Likewise *Remarks on the Management of the Dairy*, &c. Mass. Agr. Rep. Vol. III. p. 192.

To take the rancid taste from Butter.

When fresh butter has not been salted in proper time, or when salt butter has become rancid or musty, after melting and simmering it, dip in it a crust of bread well toasted on both sides; and in a few minutes the butter will lose its disagreeable taste.

Butter made from scalded Cream.

As soon as the milk is taken from the cow, let it be placed on a steady wood fire, free as possible from smoke, and scalded for thirty minutes; particular care must be taken not to allow it to boil. It must then be placed in a cool situation, and on the following day a thick rich cream will appear on the surface of the milk (which is excellent for desert purposes) this may be taken off and made into butter in the common way. This method is practised in England, and it is said that a greater quantity of butter, and of a better quality, can be made by this than by the common mode.

Directions for putting down Butter.

Wooden vessels are the most proper for containing salted butter, and oak the best kind of wood.—Iron hoops should not be used, as the rust of them will in time sink through the wood and injure the colour of the butter. To season a new vessel for the reception of salted butter requires great care; it should be filled *frequently* with scalding water, allowing it to remain till it slowly cools. Let the vessel be rendered as clean and sweet as possible, and be rubbed all over on the inside with common salt; and let a little melted butter be run into the cavity between the bottom and the sides at their joining, so as to fill it, and make it every where flush with the bottom and sides; it is then fit to receive the butter.

The butter may then be put down with the following composition:

Take two parts of the best common salt, one part of sugar, and one part of salt petre, beat them up together so as to blend the whole completely: take one ounce of this composition for every sixteen ounces of butter. Mix it thoroughly with the butter, as soon as it has been freed from the milk (which should be done effectually,) and put it without loss of time into the vessels prepared to receive it, pressing it so close as to have no air holes, or any kind of cavities within it; smooth the surface, and if you expect it will be more than two days before you add more, cover it close up with a piece of clean linen, and over that a piece of fine linen that has been dipped in melted butter, that is exactly fitted to the edges of the vessel all round, so as to exclude the air as much as possible, without the assistance of any watery brine.—When more butter is to be added, remove the covering, and let the butter be applied close above the former, pressing it down, and smoothing it as before, and so on till the vessel is full. When full, let the two covers be spread over it with the greatest care, and let a little melted butter be poured round the edges, so as to fill up every cranny, and effectually *exclude the air*. A little salt may then be strewed over the whole, and the cover firmly fixed down, to remain closely shut till opened for use.

Butter cured in this manner, does not taste well till it has stood at least a fortnight after being salted: after that period has elapsed, it eats with a rich marrowy taste that no other butter ever acquires. Butter thus cured will go well to the East or West-Indies.

"*To make salt Butter fresh.* Put four pounds of salt butter into a churn, with four quarts of new milk, and a small portion of arnatto; churn them together, and in about an hour, take out the butter, and treat it exactly as fresh butter, by washing it in water, and adding the customary quantity of salt." *Willich's Domestic Encyclopædia.*

CABBAGE, *Brassica*, an esculent plant in high estimation, which, when well boiled, is a very wholesome food. Many sorts of cabbages are cultivated. The common white and red cabbages, the savoy, the cauliflower, and the low dutch cabbages are common in this country. The savoy, for keeping in the winter, seems to be equal to any. Besides these, other sorts are cultivated in Europe, as the borecole, the broccoli, the battersea, &c.

Cabbages require a rich soil, rather moist than dry. A clay soil well mixed with other matters, is very proper for them. They are said to grow well in drained swamps without manure. Hog dung well rotted, door dung and ashes, are suitable manures for them.—Each plant should have at least four feet of ground: In other words, the plants should be two feet asunder. In gardens and small yards this is a good distance. But in fields, where they are to be cultivated by the plough, a greater distance is necessary. The rows may be three feet apart, and the plants two feet in the rows; or perhaps a foot and a half may answer, unless it be for the largest sort.

Some think cabbages will not answer more than one year on the same spot. But this is an erroneous opinion. I have raised them for eighteen years in the same part of my garden, being an unfavourable soil, dry and gravelly: And the crops are better than they were at first, though the ground has been but little manured.

Some drop the seeds where the cabbages are to grow. By this they escape being stinted by transplanting. For winter cabbages, the latter part of May is early enough to put the seed into the ground, whether the plants are to be removed or not. I have tried both ways, and on the whole, I prefer transplanting. They are otherwise apt to be too tall, and to have crooked stems. Covering of plants with leaves is not a good practice. They will be much heated through some sorts of leaves, the free circulation of air about them will be prevented, and their perspiration partly obstructed. If a hot sun cause them to droop, a shingle stuck into the ground will be sufficient shelter, if it be on the south side of the plants. I commonly allow each plant two shingles, one on the southeast side, and one on the southwest, meeting at the south corner.

The principal things which prevent the growth of cabbages, are, the fumble-foot, so called, grubs, maggots and lice. Manuring with ashes and lime tends to prevent the first, as the roots become mis-shapen by means of being wounded by insects, to which the hot qualities of ashes and lime are an antidote.

The grub, or black worm, travels in the night from plant to plant, eats off the stalks just above the ground, and buries itself in the soil when the sun is up. To guard against this worm, a little circle of lime, or rockweed round the plant is of service. But the best preservative is to seek for the insect itself which will be readily found at the root of the plant last destroyed.

To destroy lice on cabbages, they should be washed with strong brine, or sea water, or smoke should be made among them with straw, sulphur, tobacco, &c. But the hard frosts in autumn do not fail to subdue them. A moderate frost will very much thin them.

In washing plants with brine, care should be taken not to make the solution too strong, lest you kill the vegetable as well as the insect. See *Insect*.

If cabbages grow near to a barn yard, or other yard where cattle are lodged, the under leaves, when they begin to decay, may be taken off, and thrown to them. The plants will not be at all injured, and they are an excellent food for cattle, and will increase the milk of cows. But the least decayed of them should go to the cows, lest they give the milk an ill taste. Much account is made of cabbages in England for feeding cattle in the winter. But the difficulty of preserving them alters the case with regard to us. They can gather them there as they have occasion to use them, through the winter, and in the spring.

In procuring seeds for raising young cabbage plants, great care should be taken, that it be obtained from the most perfect plants of the different kinds, and such as have seeded without any other variety of the same tribe blowing near them, as it is, perhaps, only in this method, that they are capable of being kept of a true kind. It is therefore best to have the plants, intended for seed, planted out by themselves, at a distance from others. New seed is to be preferred, as it not only vegetates much quicker, but is more to be depended on. Care should be taken that young cabbage plants are properly thinned out, whenever they come up too thick. Mr. Young advises to sow three ounces of seed to each square perch, well raked in. For the manner of transplanting cabbages, see Mr. Cobbett's method of setting Ruta Baga, under the article *Turnip*.

In regard to the distance of planting, it must depend in a great measure upon the strength and goodness of the soil, and the natural size of the variety of the cabbage that is employed. An English writer says, it is the practice in some districts, where this culture is well performed, to set them out regularly at the distance of three feet each way. Mr. M'Mahon directs to set all the early heading kinds at the distance of two feet and an half every way, and all the late sorts at that of three feet.

It is asserted in Dr. Rees' Cyclopædia that " cabbages possess the property of fattening cattle, not only more expeditiously, but in less proportion than turnips; an acre of the latter having been found to fatten one in four more than the same extent of the latter crop."

Mr. M'Mahon recommends the following method for preserving

cabbages for winter and spring use. "Immediately previous to the setting in of hard frost, take up your cabbages and savoys, observing to do it in a dry day; turn their tops downward, and let them remain so for a few hours, to drain off any water that may be lodged between their leaves; then make choice of a ridge of dry earth in a well sheltered warm exposure, and plant them down to their heads therein, close to one another, having previously taken off some of their loose hanging leaves. Immediately erect over them a low temporary shed, of any kind that will keep them perfectly free from wet, which is to be open at both ends, to admit a current of air in mild dry weather. These ends are to be closed with straw when the weather is very severe. In this situation your cabbage will keep in a high state of preservation till spring, for being kept perfectly free from wet, as well as from the action of the sun, the frost will have little or no effect upon them. In such a place the heads may be cut off as wanted, and if frozen, soak them in spring, well, or pump water, for a few hours previous to their being cooked, which will dissolve the frost and extract any disagreeable taste occasioned thereby."

Mr. M'Mahon prefers this mode of preserving cabbages to placing them in the ground with the roots upwards, and says that the application of straw immediately round the heads is a bad practice, as the straw will soon become damp and mouldy, and will of course communicate the disorder to the cabbages.

Mr. Derby of Salem, Mass. states his mode of preserving cabbages as follows. "I have selected one of the most airy situations on the farm, spread a few leaves on the ground, to keep them clean, and placed them upside down, close to each other, and shook in among them leaves sufficient to cover them, leaving part of the root projecting out, then threw on just enough sea weed to prevent the leaves blowing away." Mass. Agr. Rep. vol. vii. p. 57.

"*To produce early Cabbages.*—In the spring, as soon as the sprouts on the cabbage-stalks have grown to the length of a plant fit for setting, cut them out with a small slice of the stalk, about two inches long; and if the season permit, plant them in a garden, and the usual care will produce good cabbages. *Domestic Encyclopædia.*

Mr. Francis Winship of Brighton, Mass. received a premium from the Massachusetts Agricultural Society in 1820, for the best crop of cabbages, being thirty two tons and two hundred weight, raised on one acre of land.

CALF, the young of a cow, whether male or female. The method of managing calves to advantage is of no small importance to a farmer; for on the raising of young stock, his living and wealth in great measure depend. when calves are designed for veal, they should be taken from the cow the next day after they are calved. Let them suck only two teats during the first week; three during the second; and let them have the whole of the milk during the third and fourth weeks; and then kill them. If they have all the milk at first, they will grow so fast that they will soon need more than all: The natural consequence is, that they will grow lean, and not be fit for veal. Many kill them at three weeks old; but the veal is not commonly so good, and the skins of calves so young, are of but little value.

When calves are to be reared, some permit them to run with, or at least to suck the cow, during the first season, and it is a prevalent opinion that they are improved by it. It is, however, certain that the best raisers of valuable stock in England, and even here, do not concur in this sentiment. Experience has shown that the finest possible animals have been raised in great numbers, without taking any milk from the cow after three days. They may go with the cows the first three or four days. They should have milk, more or less, for about twelve weeks. They may be fed with skimmed milk, or water porridge, after the first fortnight; or hay tea may be mixed with their milk; or their milk may be mixed with meal and water. After a calf has sucked, or drunk milk, for the space of a month, take some of the freshest and sweetest hay, and put little wisps of it into some cleft sticks, stuck up in such a manner that he can easily come at them, and he will soon learn to eat.

"If skim milk is given to calves, it should be boiled, and suffered to stand till it cools to the temperature of that first given by the cow. It is better boiled than when warmed only. If the milk be given too cold it will cause the calf to purge. If this is the case, put two or three spoonfuls of runnet in the milk and it will stop the looseness. If the calf is bound, pork broth is said to be a good and safe thing to put into the milk.

It is not true that calves are best weaned upon grass; and the reason assigned is both unfounded and absurd viz. that when raised on hay they become big-bellied. They do not in fact become so. They are much more docile when raised in the barn, thrive faster, and are as lively; and even if their bellies did grow larger, we are yet to learn that such a circumstance would be a blemish in a milk cow. It is thought by many to be an excellent mark.

Mr. Budd, of Massachusetts directs to take the calves from the cows, when three days old, and feed them with gruel, composed of one third barley and two thirds oats, each ground fine, and the mixture sifted. A quart of this gruel is to be given to each calf, morning and evening. The gruel is made by taking one quart of the flour, and twelve of water, and boiling them together for half an hour, and is to be given while milk warm.

Mr. John Gordon says that "calves should not be suffered to eat any grass the first year, and from experience I find it much the cheapest to keep them shut up and feed them, as the land sufficient to pasture one will well produce hay sufficient to keep two calves through the year, and pay the expence of cultivation, and one year's growth will certainly be added to the cattle. Mass. Agr. Rep. vol. v. p. 78.

Mortimer says, "The best calves for bringing up, are those calved in April, May, and June: Because it is seldom that those which come later acquire sufficient vigour to support them during the inclemency of the following winter; and the cold causes them to droop, and many of them to die." Much oftener may this be expected to be the case in this country, where the cold in winter is so much more intense.

Those which come earlier are preferred in this country, being more hardy, and better able to endure the rigour of the first winter. But the cost of rearing them is greater. All things considered, April may be as suitable a time as any.

"When calves are weaned, they should not be suffered to be with their dams any more till fall: Neither should they be pastured within sight or hearing of them. It will cause them to neglect their feeding; and they will not forget their sucking.

"At the setting in of cold nights in autumn, calves must be nightly housed: And not be out early in the morning, nor late in the evening. And as the pinching cold of winter will be extremely detrimental to them, they should be kept very warm in their house, well supplied with water, and let out only in the warmest days. A great deal of care is necessary to bring them through the first winter, which is the most dangerous period of their lives. They will acquire so much strength during the following summer, that they will have nothing to fear from the cold of a second winter." *Buffon; Histoire Naturelle.*

CANKER. Mr. Forsyth is of opinion that canker is caused by injudicious pruning, from the fruit being left on the trees, from injuries sustained by applying ladders in gathering the fruit, from leaving dead shoots on a tree throughout the summer, &c. but not from any thing peculiar in the soil; and says,

"When by accident, or improper treatment, trees receive large wounds, and the cure is left to nature, they are frequently overrun with gum and canker, which, if not checked, will in a short time totally ruin them.

"In this case you must carefully pare off, with a draw-knife, or any other convenient instrument, all the diseased part of the bark. The inner white bark is frequently affected; this must also be cut away till no appearance of infection remains. The infection in the inner bark appears like dots made with a pen, all of which must be cut clean out; for if any part of the canker be left, it will infect the new wood and bark. Wherever you see the gum oozing out, you may rest assured that the canker is not quite eradicated; which if suffered to remain, will spread till the whole tree becomes a mass of gum and canker, and will be killed in a short time. When the trunk is become hollow, cut the loose rotten part clean out, till you come to the sound wood, taking care to round the edges of the hollow part; then apply the composition in a liquid state, laying it on with a painter's brush, wherever the cankered bark has been pared off, or the dead wood cut out, till these places are entirely covered with it: When that is done shake some of the powder of wood ashes and burnt bones over the composition, and pat it gently down with your hands." See *Composition*.

He contends "that if the foregoing directions be carefully followed, the canker will be completely eradicated, and the hollow trunk in time be filled up with sound wood. When the stem is much decayed, he thinks it will be absolutely necessary to open the ground, examine the roots, and cut off all the rotten parts. When you have cut out all the rotten and decayed parts below ground, and scraped the hollow clean, make up a mass of the composition mixed with some clay, like what is used for grafting; then fill the hollow part with it to within two inches of the surface of the ground, treading it with your foot, or pressing it in with the hand as close as you possibly can, to prevent the wet from penetrating to the roots, and leave the surface of the composition sloping from the tree towards the outside of the border to throw the wet off, which will prevent the fresh part of the root from rotting; then cover the root over with mould level with the rest of the border;" and when you have examined all the old wounds, where large limbs have been cut off, you should next examine the old bark; and if you should find the outside of it wrinkled and cracked, pare it off, as it is always, when in that state, very much hurt by the canker. This should be done with the draw-knife, or other sharp instrument; then apply the composition as before directed, which will bring on a fine smooth bark under it. In the succeeding winter or spring you will see all the plaster with the old part of the bark that was left in the hollow parts of the tree, or where old branches had been amputated, peeling off and shewing the smooth bark underneath. You should then scrape off, with a wooden or bone knife, what old bark remains in the hollows where the draw-knife could not reach without cutting too much away. When that is done, mix up some fresh cow dung with soap-suds and urine, making it very thin, and give the tree a coat of this mixture all over where the bark has been scraped off; the cow-dung will adhere to it, and heal the parts where you were obliged to scrape to the inner bark. This wash will remain till the fresh bark comes on; when it will be discharged of itself, during the summer or the next spring, leaving a new fresh bark where the old and cankery was taken off.—Next spring if any of the old bark remains, you may repeat the same operation, which will cause all the remaining old bark to slough off like a scab from a wound of a human body.

"If any of the trees become bark-bound, you must scarify them, by taking a sharp knife, and running the point of it straight down the middle of the stem from top to bottom, taking care to run your knife through the outer bark only; then with a brush, or your finger, rub in some of the composition to prevent the incision from bringing on the canker."

Sir Humphrey Davy in his "Elements of Agricultural Chemistry," attributes canker to an "excess of alkaline and earthy matter in the descending sap;" and says, "Perhaps the application of a weak acid to the canker might be of use; where the tree is great it may be watered occasionally with a very diluted acid."

The Massachusetts Agricultural Repository, Vol. III. p. 280, gives the following

"*Remedy for the Canker and other Wounds in Trees.* The damaged parts of the tree must be cut or peeled off in the spring, and the places rubbed in a fine sunny day with turpentine, which becomes a sort of varnish, so that the wounds will be hermetically closed, and the tree will speedily recover." It is said also that bees-wax and tar will answer as good a purpose as Mr. Forsyth's composition. See *Fruit Trees*.

CANKER WORM, an insect, so called, I suppose, from its having much the same effect upon apple trees as canker. This worm is produced from the eggs of an earth-coloured bug, which having continued under ground during winter, passes up on the bodies of apple trees early in the spring.—They are hatched as early as the end of May, and are so voracious, that in a few weeks they destroy all the leaves of a tree, prevent its bearing for that year, and the next, and give it the appearance of its having been burnt. As the perspiration of trees is stopped by the loss of their leaves, they sicken and die in a few years.

The worms let themselves down by threads in quest of prey, like spiders; by means of which, the wind blows them from tree to tree; so that in a close orchard, not one tree will escape them. But trees which stand singly are seldomer infested with these insects. As they are the most pernicious kind of insects with which New-England is now infested, if any person could invent some easy, cheap, and effectual method of subduing them, he would merit the thanks of the public, and more especially of every owner of an orchard.

Several methods have been tried, with some degree of success: 1. Tarring. A strip of canvas, or linen, is put round the body of a tree, before the ground is open in the spring, and well smeared with tar. The females, in attempting to pass over it, stick fast and perish. But unless the tarring be renewed every day, it will become hard, and permit the insects to pass safely over it. And renewing the tar in season is too apt to be neglected, through hurry of business and forgetfulness. If birdlime were to be had, it might answer the purpose better, as its tenacity will continue for some time. 2. Some tie straw round the bodies of the trees. This serves to entangle and retard the insects, and prevents the ascent of many of them. But they are so amazingly prolific, that if ever so few of them get up, a tree is greatly damaged, at least for an ensuing season or two.

The pasturing of swine in an orchard, where it can conveniently be done, I suppose to be an excellent method. With their snouts and their feet, they will destroy many of the insects, before they come out of the ground, or while they are coming out. And I have never known any orchard, constantly used as a hog pasture, wholly destroyed, or even made wholly unfruitful by these worms. But this method cannot always be taken ; and if it could, I do not suppose it would be quite effectual.—When the trees are young, the swine will be apt to injure them by tearing the bark.

But as tarring the trees is the best antidote that we yet know of, and as many persons of experience believe it is possible that the insects may be thus quite prevented passing up the trees, I shall here give directions how to perform it in the most effectual manner.

In the first place, it is necessary to begin the operation very early in the year. Not observing this caution, has occasioned the want of success which many have complained of: For it is certain that the bugs will begin to pass up as soon as the ground is so much thawed, that they can extricate themselves from the soil ; which is, in some years, as early as February. Therefore, to make sure work, it is best to begin as soon as the ground is bare of snow in that month, that the first thawing of the ground may not happen before the trees are prepared ; for, beginning after ever so few of the insects are gone up, the labour will all be lost.

Another thing to be observed is, to fill the crevices of the bark with clay, before the strip of linen or canvas is put on, that the insects may not find any passage for them under it.

Having put on the strip, which should be at least three inches wide, drawn it close, and strongly fastened the ends together, a thumb rope of tow should be tied round the tree, close to the lower edge of the strip. The design of doing this is, that the tar may not drip, nor run down on the bark of the tree, which would injure it.

When all the trees of an orchard are thus prepared, let the strips be plentifully smeared with cold tar, put on with a brush. Perhaps tar mixed with a small proportion of fish oil would be still better. It would not harden so soon as tar alone. And oil is known to repel most kinds of insects. The smearing should be renewed once a day without fail. The best time is soon after sun-set ; because the insects are wont to pass up in the evening, and the tar will not harden so much in the night as in the day, because of the dampness of the air. The daily task must be renewed, and performed with the greatest care, till the latter end of May, or till the time when the hatching of the worms is commonly over, which will be earlier or later, according to the difference of climate.

Another mode of tarring, and which bids fair to be preferred to the foregoing, is as follows :—Take two pretty wide pieces of board, plane them, make semi-circular notches in each, fitting them to the stem, or body of the tree, and fasten them securely together at the ends, so that the most violent winds and storms may not displace nor stir them. The crevices betwixt the boards and the tree may be easily stopped with rags, or tow. Then smear the under sides of the boards with tar. The tar, being defended from the direct rays of the sun, will hold its tenacity the longer ; and, therefore, will not need to be so frequently renewed. And the trees may be more secured in this way from the dripping of the tar, as a margin of two or three inches, next to the tree, may be left unsmeared.

If the trees are small, the stems may be encircled with cartridge paper, in the shape of an inverted funnel. The outside of the paper should be well smeared with fish oil. The insects will proceed to the brim of the paper, but will not be able to pass it ; as the oil will hang on that part.

Another expedient, much recommended is, to put a strip of raw sheep or lamb skin round the body of each tree, the wool outwards. It is asserted, that, though the insects can pass over hair and straw, they cannot pass over the wool. But, to render this the more effectual, it will be proper to open the fibres of the wool now and then, with a coarse comb.

When it so happens that the worms are permitted to prevail in an orchard for two or three years, the limbs will be so-corrupted, that the trees are not apt to recover their fruitfulness, although the ascent of the worms should be afterwards prevented. In such a case, it is advisable to cut off all the limbs from the trees, near to the stock where they are produced, that so the tops may be wholly renewed by fresh shoots, as they will be in a few years.

It is not less than about fifty years, since this insect began its depredations in New-England, in the parts which had been longest cultivated. But perhaps there is some reason to hope that Providence is about to extirpate them : For a kind of little bird has lately made its appearance in some parts of the country, which feeds upon the canker worms. Should these birds have a rapid increase, the insects will be thinned, so as to be less formidable, if not wholly destroyed.

The Massachusetts Agricultural Journal Vol. III. No. 4, contains some remarks on the Canker worm by J. Lowell, Esq. from which we extract the following.

"I had the turf dug in around sixty apple trees and the earth laid smooth. I then took three hogsheads of *effete* or air slacked lime and strewed it an inch thick round my trees to the extent of about two or three feet from the roots, so that the whole diameter of the opening was from four to six feet.

"I tarred these trees as well as the others, and although I had worms or grubs on most that were not limed, I did not catch a single grub, where the trees were limed.

"I do not mean to speak with confidence, I am however strongly encouraged to believe the remedy perfect. It was assertained by Professor Peck, that the insect seldom descended into the ground at a greater distance than three or four feet from the trunk, and to the depth of four inches, or that the greater part come within that distance. The lime is known to be destructive of all animal substances, and I have little doubt that it actually decomposes and destroys the insect in the chrysalis state, at least I hope that is the case.

"There are many reasons which should encourage the repetition of this experiment. The digging round the trees is highly useful to them, while tarring is very injurious. The expense is not great. A man can dig round fifty large trees in one day. The lime is a most salutary manure to the trees. After the spot has been once opened and limed, the labour of keeping it open will not be great. Three hogsheads of air slacked lime, or sweepings of a lime store, will suffice for fifty trees, and will not cost three dollars. As it is done but once a year, I think it cannot be half so expensive as tarring."

Mr. Forsyth directs that the body of the tree be covered with a composition of *old* urine, kept some time for that purpose, soap suds, and fresh cow dung, which he says, if laid on plentifully, will keep off all insects. A strip of oiled paper put round the tree, with the lower end projecting so as to form a rim is recommended. Digging the earth from the roots of trees, early in the spring and returning it, mixed with a small quantity of sulphur, and sprinkling a little of it, on the surface is said to be effectual for several years.

It has been ascertained that a few canker worms often rise in autumn as well as in the spring. Professor Peck therefore advises turning up the ground in October as far as the trees extend to the depth of five inches, and breaking the clods and smoothing the surface by passing a heavy roller over it, so as to make it very hard and without cracks, which will prevent the canker moths from rising from the earth. See *Mass. Agricultural Repository*, vol. iv. page 91.

John Kenrick Esq. of Newton, Mass. proposes the following "New and cheap method" of destroying this insect.

From any time in June, after the worms have entirely disappeared, until the twentieth of October, let the whole of the soil surrounding the trees, to the extent of at least four feet from the trunks, and to a suitable depth be dug up and carted away; and placed at a distance from any trees the canker-worms are in the habit of feeding upon; and let these be returned in the same cart, an equal quantity of compost, or rich earth intermixed with manure. Or the soil returned may be taken from ponds, roads, or the ridges which often accumulate on the margins of fields; or if there happen a ploughed field near the orchard, in which there are no apple trees, the soil may be exchanged load for load without any injury to the field: but in either case let a good quantity of manure be spread at the bottom of the soil returned to the trees. *Massachusetts Agricultural Repository*, vol. iv. p. 168.

CARROT, *Daucus*, a well known and useful root for food.—The seeds are carminative and diuretic, and the root is useful to abate the malignity of cancers. A sandy soil is very proper for carrots; but they do very well in gravelly, and even in loamy soils, when made rich and loosened to a sufficient depth. The largest I have ever raised has been in gravel. The ground should be ploughed, or dug, more than twelve inches deep, and well pulverized.

I have found by long experience that carrots should be sowed early. The last week in April is late enough, when intended for feeding o cattle; and they may be sowed earlier, if the ground be in good order, and so dry as to be made light and loose. The earliest sown will be the largest, and, in the northern parts of this country, nearly as tender and good tasted as if sown later.

A small quantity of dressing will be sufficient for them. But whatever manure is used, it should be well rotted, and made fine, or putrefy very soon in the ground; otherwise the little obstacles in it, will cause the roots to divide, and become forked. I have known carrots, manured with old hog dung, grow to a surprising bigness. But if a large quantity of this strong manure be used, they will grow so fast as to burst open. It is a crop that bears drought well, as it draws its principal nourishment from a considerable depth. Nor is the ground apt to be exhausted by continued crops.

In the garden I sow them in drills, or little furrows, made an inch deep with the head of a rake, from 9 to 12 inches apart, across beds four feet wide. This prevents treading the ground hard too near to the roots; greatly facilitates clearing them of weeds with a hoe, and keeping the earth loose to a sufficient depth. I do not thin them much, till I begin to pull them for use, about the beginning of July; from which time I pull them, not only for the table, but to feed swine, as that sort of animals are so fond of them, that they will greedily devour both roots and tops. The spaces between the beds may be kept clear of weeds, by turning over the soil with a narrow spade, once or twice in May and June. It will not only subdue the weeds, but increase the pasture of the nearest plants.

But the field culture of this root begins to prevail in the country: As carrots are found to be valuable, for feeding not only swine, but horses and cattle, and for fattening them. But to fatten swine on them, they should be boiled, or parboiled. They are so easily cultivated, and so hardy, that they may be raised in fields to great advantage. They will grow well in a soil that is but moderately rich, if it be ploughed deep, and made mellow. And there is no difficulty in keeping them through the winter, in good order for feeding cattle. The ground should be ploughed in the fall preceding, and ploughed very deep. If the plough do not go deep enough at once, it should be trench ploughed; that is, the plough should pass twice in the furrow. And if some of the earth, which was never before stirred, should be thrown up to the surface, it will be no damage, provided it be such earth as crumbles easily, and does not remain in lumps, after the winter frosts have acted upon it.

If the lands incline to much wetness, it should be water furrowed, after the autumnal ploughing, that so it may be dry, and fit to be ploughed again very early in the spring. It must be well harrowed before sowing, first with a heavy harrow, and afterwards with a lighter one, with shorter teeth placed near together. After the seed is sown, the ground should be raked. When sown in the broad cast method, they should stand so far apart after thinning,as to have each half a foot of soil. There will be no danger of thinning them early, as they are a plant which is seldom diminished by insects.

It is with pleasure that I find the attention of some of my countrymen turned to the field culture of this excellent root. They who have but little land may probably enable themselves to keep considerable stocks by means of it.

This root has greatly the advantage of turnips, not only in its being a richer and more nourishing food, and in yielding a larger produce, but also in its being never,or very seldom, annoyed or hurt by insects. This crop, rightly managed, I have never known to fail, as it is well known the other often does.

The drill method, sowing on narrow ridges, raised by the cultivator, is preferred by some, and is that which I use. But the labour will perhaps be increased a little. The seeds will do best sown by hand, as their shape will not well admit of their being drilled. To prepare them for sowing at all, they should be well rubbed, and passed through a sieve. The first hoeing of carrots in rows must be also by hand; at which hoeing they should be thinned to one or two inches asunder, if large ones are desired.—The after hoeings may be expeditiously done by the horse hoe and cultivator alternately. It is not amiss, if they grow large and rank, when they are chiefly designed as food for cattle, though small sized ones are preferred for the table. For this use they need not be thinned to more than half an inch asunder in the rows: And perhaps not so much in good ground. The way to keep carrots good for eating through the winter is to bury them in a dry sand of the yellow kind from pits. Or, if they are put into casks, covering them with fresh turfs may be sufficient.

It has been maintained by some agriculturists that the tops of carrots might be cut off, as soon as the lower leaves begin to wither, without injury to the root. But the Massachusetts Agricultural Journal, Vol. III. No. 3, p. 181—2, gives the details of certain experiments made by the Hon. Josiah Quincy, in which a certain number of beds of carrots were *cut*, and the same number of similar beds were left *uncut*, and the advantage was in favour of the latter, about as 8 to 5. From the result of these experiments Mr. Quincy concludes "that the carrot forms no exception to the usual analogy of nature in the growth of vegetables," which depends nearly as much upon the leaves as the root.

The best way to keep carrots through the winter, for family use, is to bury them in moist sand. Or they may be kept in the same manner with beets. See *Beets*.

Mr. Cooper, Editor of the second American edition of Dr. Willich's Domestic Encyclopædia, says that horses often prefer carrots to oats, and that "they are decidedly the best food for broken-winded horses, and as I believe for asthmatic men."

Mr. Quincy gives the following statement of the mode of cultivating carrots, made use of by Mr. Samuel Wyllys Pomeroy, Esq. and which he prefers to all others.

"Plough as deep in the fall or spring as the state of the land will permit. Cross plough in the spring and harrow level. Put on fifteen, twenty or twenty-five buck loads of the most rotten compost to the acre, as the heart of the land may be. Spread and harrow it fine.—Then with a horse plough strike it into two-bout ridges, as near together as four back furrows will make them, and if the two first back furrows are narrow the other two being deep, the ridge will be nearly to a point, and should be eighteen or twenty inches from the bottom of the furrow if it be well cleared out. To do which make another bout in the furrow, if necessary. Then with the head of a rake strike off the crown of the ridge, till it is three or four inches wide, and with it, or a hoe, open a drill in the usual manner. Sow the seed pretty thick, cover and press down a little with a hoe or shovel. When the weeds appear, run a small plough through the furrows. Hand weed the crop, and hoe the weeds from the sides of the ridge. The orange carrot is best."

"In harvesting, a plough with one yoke of oxen should be run near the side of the range of carrots, and as deep as possible. This loosens the dirt and clears one side of the carrots almost entirely from the earth. The labourers then with great facility take them by their tops out of the beds, and throw them into carts, with only an occasional use of the hoe to plants which the plough has not loosened.

"I have no question that conducted in this mode a carrot crop may be made more productive, and much less expensive than the potatoe crop usually is. In sowing I use a small hand drill, which lays the seed with great regularity, a circumstance very important both to facilitate weeding and harvesting; since if the carrots stand straggling and not in a line, the plough, when harvesting, leaves the more to be loosened by the hoe or the fork. Massachusetts Agricultural Repository, Vol. IV. p. 24.

A mode of cultivating carrots differing slightly from the above, is described by Mr. Quincy, in the same work, Vol. IV. p. 212.

For other modes of cultivating this root, see Mass. Agr. Rep. Vol. V. p. 20, 255, 347.

Mr. Ebenezer Thrasher, of Salem, Mass. in 1820, received a premium from the Massachusetts Agricultural Society, for the best crop of carrots, weighing exclusive of the tops, twenty-one tons, four hundred and an half hundred, equal to eight hundred and forty-nine bushels of fifty-six pounds each, raised on one acre of land.

The same year Ezekiel Hersy Derby, Esq. of Salem, Mass. raised on three quarters of an acre of ground seven hundred and twenty bushels of carrots, exclusive of their tops, which were estimated by competent judges, to weigh five tons.

CART, a wheel carriage, of essential importance to the farmer, to carry his manures, remove stuff for fences, get in his crops, &c. Horse carts are sometimes used; but ox carts generally. Of the latter some are short, some long. The short cart is eight feet long, four feet wide, and two feet high. The long cart is used for carting hay, straw, and other bulky matters; therefore it is made from ten to twelve feet, or more, in length, four feet in breadth; and instead of sides, it has only long, sharp pointed stakes. In some parts of the country they lengthen out a short cart, with what are called ladders, when they cart hay. But this is not a good practice. The load lies higher than in a long cart, and is more liable to be overset.

The greatest excellencies of a cart, are lightness, strength, and durableness. Therefore, it is very proper to construct carts of ash timber. But as white oak is not so apt to decay, the principal parts are commonly made of that. A cart should be kept under shelter when it is not in use. It will last the longer.

The axle and wheels, should be of the toughest of oak. Wheels to be used on a farm only, need not be shod with iron. A wooden rim, well made, will last several years, and is easily renewed, and it will do less injury to the grass grounds in passing over them. The softer the soil is, the wider the rim of a wheel should be. Some have the rims a foot wide, to cart upon marshes.

A great improvement has recently been introduced in hay carts, which consists of a floor, longer than the old cart bodies, with perpendicular posts, or slats connected together at the top by a light frame, and so high as to contain as much hay as the cattle will draw. This saves much labour in packing the hay—requires no ladders—and no ropes ; and a load is pitched in and prepared for removal in much less time. The frame is laid upon the axle-trees, and secured to them by the same pins or fastening which confine the ordinary cart or waggon body. It is easily changed when occasion may require. It is also well calculated for the removal of brush, or any other light article.

CATERPILLAR, a worm that feeds on leaves and fruits. These insects differ in colour and size according to their situation, and according to the different matters on which they feed. The principal inconvenience the farmer meets with from caterpillars, is the damage they do to his orchard. A hairy kind of caterpillars build their nests on apple trees in May, and are gone entirely in June, about the summer solstice. But they feed so industriously on the leaves, as to destroy a great part of them, if they be not timely prevented.—As they are far less mischievous than the canker worm, so they are more easily subdued. Some destroy them by firing gun-powder at their nests. The same may be effected with a match of brimstone on the end of a pole.

Nests which have been neglected till the insects have forsaken them should be destroyed. A nest will be found to contain several of the insects in a chrysalis state.

The Hon. Timothy Pickering, in a letter to the Corresponding Secretary of the Massachusetts Agricultural Society, has recommended an instrument, which he has found simple and more convenient than any he had used for the destruction of caterpillars. It is made by inserting some hog's bristles between a twisted wire, in such manner as to form a cylindrical brush, which will present bristles on every side. This is attached to a pole of such length as the trees may require, and the caterpillars are brought down by it, and then crushed.

It is affirmed that caterpillars, and other insects, which infest our fruit trees may be destroyed by casting over the tree a few handfuls of ashes, in the morning before the dew is dissipated from the foilage, or after a shower of rain.—The former the preferable time.

A strong white-wash of fresh stone lime, applied by the means of a mop, or sponge, fixed to the end of a pole, has by some been recommended. Spirits of turpentine, or a little oil of any kind, particularly blubber oil, are likewise fatal to these insects. But, perhaps, the most effectual remedy is the *hand*, by which the nests may easily be removed at an early stage, but if this be neglected, it is thought that the next best remedy is that mentioned above, communicated by Col. Pickering, an excellent practical as well as scientific farmer.—In applying either of these remedies care must be taken to choose that part of the day, when the caterpillars are in the nests. They rarely quit it till 9 o'clock, and generally return to it again about 12.

It is said that caterpillars will

take shelter under woollen rags. placed near their haunts, from which they may be taken and destroyed.

CATTLE, a name applied to all quadrupeds, which are used for tilling of ground, and for the food of man. But under this article we shall confine ourselves to animals of the ox kind, to which the word is most usually applied.

As soon as a calf is weaned, it should either be permitted to run in the best pasture, or confined in the barn, and fed by hand, and be carefully tended, kept warm, and live upon the best of fodder, through the first winter. Afterwards it will become so hardy as to require less care. But cattle should be frequently looked to and examined ; that so, if they be overtaken with any sickness, hurt, or lameness, suitable remedies may be seasonably applied. And in order to do this, they should be accustomed to come home, and be shut up in the yard every night. By this method, a farmer will save a much larger quantity of dung. And, in case of an uncommonly cold storm, the cattle may be housed with very little trouble, as the yard is contiguous to their house.

Cattle, from one year old to three, will usually get a living in summer, and even thrive, upon the commons, or in the meanest, and most bushy pastures. And in winter the poorest fodder will keep them alive. And, as our farmers know these things, they are very apt to treat their young cattle in this manner. Those which are so treated, may oftentimes become as hardy cattle as any ; but they will be small, and therefore not so profitable. Farmers in general are too ambitious to keep a large stock of cattle : A necessary consequence of which is, that they are pinched in their food, and never come to their full growth. Another ill consequence is, their growing unruly and mischievous through hunger, learning to leap over fences, or break through them.

It would certainly be more conducive to the interest of farmers, to keep smaller stocks of cattle than they do : For then they would be able to feed them to the full.—Their oxen would be much larger and stronger than they are, and their cows would give plenty of milk, and bring larger calves : Not to say how much they would save in taxes, by reducing their number of rateable cattle.

Farmers should allow their young stock the best of pasture. This would keep them out of mischief, prevent their learning bad tricks, and prevent many ill accidents which befal them. And it would be no small advantage always to know where to find them. But, in the common method of treating them, it is too common a case for them to straggle so far from home, that the owner entirely loses them ; or else spends as much time as they are worth in seeking after them.

If a young stock were well fed at all seasons, the heifers would commonly have calves at two years old, which is no small advantage, and steers would be fit for labour earlier in proportion. And when they come to be killed off, the quantity of beef would make amends for their being so fed as to be well grown. If the farmer's view in increasing his stock, be to make as much dung as possible, he should be reminded of what he ought to know already, that the dung of a small stock will be equal to that of a large one, if it consume the same quantity of fodder. If a farmer make this objection to pasturing his young flock, that his farm is not large enough to admit of it; he may find an answer, by turning to the article, *Mowing Grounds*, where diminishing their number of acres,and increasing that of pasture ground, is recommended, and sufficient reasons assigned.

In the winter, cattle should be housed, to defend them from the inclemencies of the weather. For though nature furnishes them with a thicker covering of hair in winter than in summer, the difference is not near so great as that of the weather in this climate. Working oxen, and milch cows, will suffer more than the rest by lying abroad. If the farmer cannot conveniently house all his cattle, those should be left out which are between the age of one and three years. And those that lie out should have a shed, open only to the south and west, to shelter themselves under in stormy weather.

The injuries which cattle receive from one another, when they are lodged together in a yard, is an additional reason for tying them up in the barn. To which it may be added, that a great part of the fodder given them is wasted, even when it is given them in racks; much more, when it is thrown upon the ground. They trample it into the dung with their feet,which is no inconsiderable loss.

Cattle will bear to be cold much better than to be wet. If they be left out in cold storms of rain, it pinches them exceedingly; so that they will not look so well again for several days after it. The sides of the house where they are lodged, need not be very tight. It might be apt to make them too tender. It will certainly abate the freshness of the air they breathe in,and hurt the agreeable flavour of their fodder. But the covering of their house should be perfectly tight.—No window should be open,through which snow or rain may drive in upon them. The floor they lie on should have a gentle descent backward, that they may be wetted as little as possible by their stale; and they should always have straw or litter under them, not only to soften their lodging, but to lay them the more warm and dry, and absorb the wetness. The better they are littered, the more manure will the owner make for his farm. This is an object of high importance.

It would be a good method for cattle that are tied up, to fodder them in racks. They would not be so apt to rob one another; nor to get their fodder under their feet; nor to render it unpalatable by their breathing upon it.

Where salt hay can be had, cattle should now and then be treated with a little of it. It will so increase their appetite, that they will eat poor meadow hay, and straw with it, or after it. But farmers, who are far from the sea, and not furnished with salt hay, should now and then sprinkle some of their meanest fodder with salt dissolved in water, which will answer the same valuable purpose. And at no season of the year should cattle be kept, for any long time, without salt. They are greedy after it, and it conduces to keep them in health.

As to summer feeding, it is not fit that a whole stock go promiscuously in the same pasture. Some would be overmuch fed, and some not enough. A farmer's pasture grounds should be made into a number of separate inclosures ; the greater the number the better.—Milch kine and cattle fattening for slaughter should have the first of the feed in each inclosure ; then working oxen ; afterwards, young stock, horses and sheep. When each kind have had their turn, for two or three days, or perhaps a week, the apartment may be shut up, till it be sufficiently grown for the milch cows. By such a rotation much may be saved ; but little of the grass will be wasted by trampling ; and what one sort leaves another will eat ; so that none of the grass will be lost.

Oxen should not live to be more than eight years old, nor cows more than ten or eleven. When they are kept longer, they do not fatten so easily ; and the beef is not so good. Cattle to be fattened should have the best of pasture during the whole grass season, or they will not be fat so early as December ; and they should lose a little blood, when they are first turned to grass. In autumn, when grass grows short, or is corrupted by frosts, their fattening should be promoted by feeding them morning and evening with the stalks of Indian corn, pumpions, potatoes, or carrots ; and especially with ears of corn, if the owner can afford it. Indian meal is supposed to be still better to complete their fattening. Oil cakes from the linseed mills are much recommended in English books, as conducing to the speedy fattening of cattle. See *Stock*.

The Philadelphia edition of the Domestic Encyclopædia observes, that feeding cattle with turnips does not succeed so well in the United States as in Great Britain, and mentions an unsuccessful experiment, which resulted in that conclusion. It is possible that the want of a proper mixture of dry food, such as hay or straw, might be the cause of the failure.

Cattle are apt to be *hoven* or *swollen* in consequence of having eaten too much green succulent food. The common remedy for this disorder has been to stab the infected animal with a pen-knife or other sharp instrument, under the short ribs, and put into the orifice a tube of ivory, elder, a quill, or something of the kind, to give vent to the confined air. The wound is then dressed with some sort of adhesive plaster, and thus, in general the cure is easily effected.—This, however, is a rough and dangerous remedy, and we therefore give place to others more safe and gentle.

The 33d volume of the Annals of Agriculture announces the following recipe for hoven cattle, which it assures us will effect a cure for hoven cattle, in the most desperate cases in half an hour. Take three quarters of a pint of olive oil; one pint of melted butter, or hog's lard; give this mixture by means of a horn or bottle, and if it does not produce a favourable change in a quarter of an hour, repeat the same quantity, and walk the animal gently about. For sheep, attacked with this malady the dose is from a wine-glass and an half to two glasses.

Besides these remedies, flexible tubes, and canes, with knobs at their ends, have been used to force a passage from the mouth to the stomach, to let the confined air escape upwards from the trunk of the animal affected. Descriptions of these instruments may be seen in the second American edition of the Domestic Encyclopædia, vol. 1. p. 409, 410. The following simple remedy we have been assured is effectual. Make about a pint of lie either with hot embers thrown into a sufficient quantity of water, or by dissolving therein about an ounce of pot or pearl ash, and turn it down the throat of the ox or cow affected. A proportionably less quantity will answer for a sheep. This is said to give immediate relief by neutralizing the carbonic acid gas in the stomach of the creature, which causes the swelling, and other symptoms of the complaint to subside.

"When oxen are long and hardly driven, in muddy roads, particularly where the soil is calcareous, they are liable to a soreness between the claws. This will make the beast lame, and, when discovered, the part should be cleansed, and healed with some proper ointment. Sometimes from inattention to this, the part becomes horny; in this case the hard parts must be cut away, and the wounded flesh cured.

"A general indication of health in neat cattle is a moist or wet nose, and when this is found dry it is a certain symptom of disease of some kind or other." *Farmer's Assistant.*

Hoof Ail, or Hoof Distemper. A writer for the Massachusetts Agricultural Repository gives some account of this disorder, vol. 4, p. 339, from which we extract the following :—

"*Symptoms.* When an animal is at all lame, its foot should be carefully felt. The first indication is usually an uncommon degree of warmth, and a soft and puffed feel of the parts immediately connected with the slit between the hoof, either before or behind the foot and generally just above it. If in the hind foot, and not easily handled, a fullness may generally be perceived, by standing behind the animal and carefully comparing the appearance of the two feet, between the dew claws and the hoofs (for it very rarely commences its attack in more than *one* foot.) In the fore foot it generally swells forward; and in taking up the foot, the slit between the hoofs will have an appearance of dryness, easily distinguishable to a person used to cattle; and the animal frequently licks the front part of the foot. Instances often occur of sudden and extreme lameness, without any appearance of heat or swelling in the foot; and these are often the worst cases; but one symptom rarely fails to accompany the disease, which is extreme restlessness and appearance of anguish, attended with loss of appetite and flesh; but without in the least affecting the brightness of the eye, and, perhaps, sometimes, unnaturally increasing it; but the eye has a peculiar cast. As a general rule it is safest to attribute all lameness of the foot, which cannot be traced to a sufficient cause, to the hoof ail. Lameness of the foot can generally be distinguished from that of the leg, hip or shoulder, by making the animal step over a stick or rail, and carefully watching its motions.

"*Remedies*. The foot should be carefully washed and cleansed, and thoroughly examined, to be sure that the lameness does not arise from a nail casually run into the foot, or a prick in shoeing, or from a wound from a stump or other substance between the hoofs (a case frequently occurring.) If no appearance occurs of any break in the skin, while the foot is still wet, apply, as nearly as may be, to the centre of the slit, between the hoofs from one to three grains of corrosive sublimate (reduced to a fine powder) the dose to be proportioned to the size of the animal and the violence of the attack.—Care must be used that the powder is put completely in this slit, for it is a very strong poison, and the animal, as soon as at liberty, will begin to lick the foot, if a sore one. The moisture left by the washing, makes the powder adhere; and the effect is produced in a very short time. Some prefer mixing the powder with hog's lard, which answers; but is thought less powerful: it has one advantage, however, as being less dangerous to keep in a house (for no one takes salve inwardly.) Where corrosive sublimate cannot be obtained, any other violent stimulant may be applied. Common salt is often effectual in very slight attacks; but it is of the greatest importance to lose no time. The application is to be repeated every twenty-four hours, till a cure is effected, or till the foot shews unequivocal signs of a gathering which will break."

We shall add to this article a few brief sketches, descriptive of some of the most famous *Breeds*, known in Great Britain, where great pains have been taken to improve the different races of this valuable animal.

1. The *Short-horned*, sometimes called the Dutch breed. The different families of this race are distinguished by the names of the *Holderness*, the *Teeswater*, the *Yorkshire*, *Durham*, *Northumberland*, and other breeds. The Teeswater breed from the head of the vale of York, and the banks of the Tees is in the highest estimation. The bone, head, and neck of these cattle are fine, the hide very thin, the chine full, the loin broad, and the carcase throughout large, and well fashioned; and the flush and fatting quality equal, or perhaps superior to those of any other large breed. The cows give a greater

quantity of milk than any other cattle, each yielding 24 quarts a day, and three firkins of butter in the grass season. Their colours are varied, generally red and white mixed. A bull of this species was imported in 1817, by Stephen Williams, Esq. of Northborough, Mass. received from his brother, Charles Williams, Esq. of London, and is said to have cost five hundred dollars, besides expenses of transportation.

2. The *Long-horned, or Lancashire* Breed. Distinguished for the length of their horns, the thickness, and firm texture of their hides, the length and closeness of their hair, the large size of their hoofs, and coarse leathery thick necks. Deeper in their fore-quarters, and lighter in their hind quarters than most other breeds. They give less milk, but of richer quality than some others. Their colours are various; but they have in general a white streak along their backs, and mostly a white spot on the inside of the hoof. The improved breed of Leicestershire was commenced by Mr. Webster, and afterwards farther improved by Mr. Bakewell, of Dishley.

3. The *Middle-horned* breeds comprehend several local varieties, of which the most noted are the *Devons*, the *Sussexes*, and the *Herefords*. These cattle are the most esteemed of all the English breeds. The Devons, or Devonshire cattle are of a high red colour (if any white spots, they reckon the breed impure, particularly if those spots run into one another,) with a light-dun ring round the eye, and the muzzle of the same colour; fine in the bone, clean in the neck, horns of a medium length bent upwards, thin faced and fine in the chops, wide in the hips, a tolerable barrel, but rather flat on the sides; tail small and set on very high; they are thin skinned, and silky in handling, feed at an early age, or arrive at maturity sooner than most other breeds." (*Culley on Live Stock*, p. 51.) Oxen of this breed are most proper for the yoke. It is said that the common cattle of this country most nearly resemble this breed.

4. The *Sussex*, and *Herefordshire* cattle are of a deep red colour, with fine hair, and very thin hides; neck and head clean; horns neither long nor short; rather turning up at the points; well made in the hind quarters, wide across the hips, rump and sirloin, but narrow in the chine, tolerably straight along the back; ribs too flat; thin in the thigh; and bone not large. They are next in size to the Yorkshire short horns, and are very useful as dairy stock.

5. The *Polled* or *hornless* breeds. The most numerous and esteemed variety is the Galloway breed, so called from the province of that name. The true Galloway bullock "is straight and broad on the back, and nearly level from the head to the rump; broad at the loins, not, however, with hooked bones, or projecting knobs, so that when viewed from above, the whole body appears beautifully rounded, like the the longitudinal section of a roller. He is long in

the quarters, but not broad in the twist. He is deep in the chest, short in the leg, and moderately fine in the bone, clean in the chop and in the neck. His head is of a moderate size, with large rough ears, and full, but not prominent eyes or heavy eye-brows. His well proportioned form is clothed with a loose and mellow skin adorned with soft glossy hair. The prevailing colour is black or dark-brindled, and though they are occasionally found of every colour, the dark colours are uniformly preferred, from a belief that they are connected with superior hardiness of constitution. The Galloways are rather under-sized, not very different from the size of the Devons, but as much less than the long-horns, as the long-horns are less than the short-horns.

6. *Argyleshire* breed, or *Kyloes*. A bull of the Kyloe breed should be of a middle size, black, dark, or reddish brown, without any white or yellow spots. His head should be rather small, his muzzle fine, his horns equable, not very thick, of a clear green and waxy tinge ; his general appearance should combine agility, vivacity and strength ; and his hair should be glossy, thick and vigorous, indicating a sound constitution, and perfect health." Bulls of this description have been valued at upwards of 200 guineas.

For a further description of the most noted English breeds of cattle, see *The Complete Grazier ;* Supplement to the third edition of The Encyclopædia Britannica. Art. Agriculture, and the Farmer's Cyclopedia.

CHAFF for FEEDING HORSES. Mr. Thomas Williamson in a communication to the Bath Society on the use of chaff, or cut hay for feeding horses, remarks, that " one hundred weight of hay was found to yield 20 bushels of chaff pressed into the measure and piled as high as it could safely be carried ; consequently each bushel weighed about 5½ lbs. It was found that five horses would eat twelve bushels of chaff in 24 hours, and that somewhat more than half their usual quantity of food was saved by having it cut."

Mr. Benjamin Hale, proprietor of a line of stages running between Newburyport and Boston, has given a statement, published in the Massachusetts Agricultural Repository, vol. 10, p. 400, by which it appears that the total saving in using Hotchkiss's Straw Cutter nine months, viz. at Newburyport four months, and at Salem five, in the years 1816—17 was $780,97.

CHANGE of CROPS, a method of cultivating different sorts of vegetables in succession, on the same piece of ground, with a view to make tillage lands more profitable in the long run ; and, at the same time, to prevent exhausting them of their strength.

Those who believe that the food of different plants is different, cannot but look on the changing of crops as a matter of essential importance. For, on their hypothesis, land which is worn out with one sort of vegetables, may be in good order to produce a large crop of another sort. But there are other reasons for the changing of crops, which are more substantial, being founded in undoubted fact, and proved by experience.

Some plants are known to impoverish land much faster than others: Such as Indian corn, flax, hemp, &c. And it would not be a prudent method to scourge a piece of land, with such crops, year after year, till its strength were all exhausted. For it has been justly observed, that it is easier and cheaper to keep land in heart, than to restore it after it is worn out.—It is advisable, therefore, in general, to take but one crop of flax from a piece of land; and not more than two of Indian corn, in succession; nor indeed more than one, unless abundance of manure be applied.

Again, some plants take the principal part of their nourishment near the surface of the soil, and others draw it from a greater depth: And a regard must be had to this in choosing a rotation of crops. For it will be found that after land has been much worn by plants, the roots of which chiefly consist of either long or short lateral fibres, it will be in good order to produce plants which are tap rooted. Clover, for instance, will grow rank and good, on a spot which will not answer for wheat, barley, or oats. The clover will draw its principal nourishment from those parts of the soil, to which the roots of preceding crops have not reached. And if grounds have been dunged for several years, the nutritive particles of the dung may have penetrated deeper into the soil than any roots have reached. In a loose sandy soil, used for grain. this will often be the case: So that part of the manure laid on it will be entirely lost, unless a crop of tap rooted plants should overtake it in its descent.

Preventing the prevalence of weeds is another good reason for the changing of crops. Weeds will so increase, especially in old farms, as almost to spoil a crop, unless a hoed crop intervene to check them once in two or three years. And a green hoed crop helps to prepare land for producing other crops, by enriching it. The weeds, which are several times cut to pieces, and hoed into the soil, during a summer, answer much the same end as green dressings: And by keeping the soil loose, the enriching particles floating in the atmosphere, are plentifully imbibed. See *Rotation of Crops*.

Also, a change of crops, judiciously managed, supersedes the necessity of fallowing, which is no small advantage. Instead of an expensive resting of the soil, during a year of fallow, land may yield an unintermitting profit to the owner. Wheat land, for instance, may be recruited, and cleared of its weeds, by a crop of beans, or potatoes, as effectually as by fallowing. If such a crop only pay the cost of culture, it may be considered as gain.

What particular routine of crops is best, is not easy to determine. Green and white crops, alternately, are in general recommended. I suppose one course may be best in one county, and another in another. In the county of Bristol,

Massachusetts, it is called good husbandry, to plant Indian corn the first year after ground is broken up; to sow rye, wheat, oats, or barley, the second year; and lay it down with clover. After two or three crops of clover are taken off, the land is broken up again, and managed as before.

But in the counties of Cumberland and Lincoln, in Maine, this course would not answer so well. Indian corn is not found to be the most beneficial crop for the first year, in this climate. It will be backward, and in danger of not ripening well, unless it be on a sandy spot, with a southern exposure. And when land is broken up, it will not be subdued enough to lay down for grass so soon as the third year, on account of the toughness of the sward.—But when laid down, it may lie six or seven years before it will need breaking up again; for the lands are so natural to grass, that the crops will continue to be good.

An eligible course of crops in these northern counties may be, pease, oats, or potatoes, the first year; Indian corn, much dunged, the second; barley, or rye, the third; and the fourth, herdsgrass and clover mixed, and so on to the tenth. As the clover diminishes, the herdsgrass will increase, which is a more valuable grass for fodder. But every judicious farmer must judge for himself in these matters. Soils differ so greatly, even in fields which lie contiguous, that the course of crops which is suitable for one, would be unsuitable for another. Change of crops is no new doctrine among farmers.

"Repeated observations convinced the Romans, that besides the alternate resting of the land, wheat may, as Pliny observes, be sown after lupines, vetches, beans, or any other plant which has the quality of fertilizing and enriching the soil. A judicious change of crops is of great importance in the common tillage husbandry, as it enables the farmers to save the expence and loss of a crop, in the fallow year; and to conquer his great enemies, the weeds, by attacking them at different seasons of the year, and in different periods of their growth; especially when the intermediate crops are hoed, as those of pease and beans; for the repeated hoeings, not only destroy the weeds, but also very much enrich the land. The benefit of changing crops appears to arise from these circumstances, rather than from any different food that the several crops are supposed to extract from the soil." *Complete Farmer*.

CHANGE OF SEEDS. One of the most important, and still unsettled questions is, whether frequent and regular changes of seed, from different soils, climates, and fields, are necessary, or at least important for the attainment of good crops. While some contend that such changes are necessary, and proceed to assign certain philosophical reasons, which are much less satisfactory than a very few examples would be; others, among whom might be mentioned the deservedly celebrated Mr. Cooper, of New-Jersey, maintain that no such changes are either necessary or expedient; that seed may not only be sown in the same land indefinitely as to time, and without any deterioration of the quantity or quality of the crops, but that they will improve, provided a careful selection is made of the plants reserved for seed, and provided the earliest ripe, and fairest, and in all particulars the best, are uniformly and carefully selected. It is probable that both these parties are partially right. It is possible that some plants may require a change of soil oftener than others. It seems to be the nearly universal opinion, founded on such a variety of facts, and remarked during so long a period, that it may be considered settled, that different plants require different qualities of food; that the soil may become exhausted of the particular species of food requisite for one, while it may abound in matter suited to the healthy and vigorous growth of another description of plants. The yearly course of nature, open to the eyes of every intelligent man, proves this.—Plant after plant succeeds spontaneously, and when lands, by overcropping, and neglect, refuse to yield any vegetable, useful to man or animals, they will still enable noxious or useless weeds to sustain themselves, and when all others are gone, the mosses, and mushroom tribes will vigorously flourish. It is probable, therefore, that a change of seeds from one part of a farm to another, may be useful; but it is not so easy to see why transplantation of seeds from one climate to another should be useful, unless the climate from which the seeds were obtained, was not so congenial to them as the one to which they are transferred. Our own experience, after twenty years careful examination and enquiry, leads us to believe that there is no advantage in a change of climate, but on the contrary that plants resist such changes, and suffer by them. The same remark applies to different soils. It is certainly true, no truth is more settled. than that almost every plant flourishes better in one particular species of soil than in any other. But happily they will submit to great changes, and acquire new habits; and with the aid of more, and judicious observation and treatment they may be made to flourish very well in a soil not naturally adapted to them.—But it is not easy to see any advantage in seeking the seeds of a vegetable raised in an uncongenial soil, to introduce it into a congenial one. It would seem to be wiser in this case to prefer the seeds grown upon soils and in climates which were congenial to them.—On the other hand, if we were about to propagate a plant in a soil and climate not congenial or natural to it, we should prefer the seed or the root, which had been long since transplanted into the uncongenial soil and climate, and had by slow degrees been naturalized there. For example, if we were to attempt to grow cotton in New-England, we should take the seed from the utmost northern verge of its present growth rather than from Louisiana. Sir Joseph Banks has some admirable remarks on the subject of acclimating plants, and shews that plants, which one century since were treated as tender exotics in England, now openly brave their winter frosts.

On the whole, as a practical rule, we should say that Mr. Cooper's is the best general one ; and that is to select with great care, the best part of your own seeds and roots for future cultivation. This rule is within the power of every farmer to apply—but in an extended scale it is not practicable for farmers generally to change their seeds either annually, or even every few years. This must be left to speculative and more opulent cultivators, and when they have succeeded after several years trial, the effects of their efforts will be felt and enjoyed throughout the whole country. If they fail, the loss is their own, and they are able to sustain it.—We would, however, observe,from the result of 20 year's experience, that it is highly inexpedient to transplant seeds or plants from a high northern to a southern climate, and the reverse. The potatoes of Great Britain and Ireland, and even of Nova-Scotia, do not succeed generally with us. Some exceptions may be made to the rule. but they are rare. The potatoe,taken from the south, appears to do better transplanted to the north. The River Plate, or long potatoe, has done admirably well, though the fact that it originally came from South America is by no means settled on good evidence. This fact, if it be settled, which we believe it to be, that the potatoe is not improved by transplantation from a colder and more moist climate to a more southern and drier one, and on the other hand is improved by transplantation from a southern one to a more northern climate, may be accounted for from the fact, that its natural indigenous location, the one in which it was first found, and has recently been discovered in a natural state, is the elevated lands of South America, at the foot of their highest mountains, and but little below the region of perpetual snows. It is certain that the potatoe loves moisture, and is rapidly checked by extreme heat and drought.

On the other hand the Indian corn will not bear transplantation from a southern to a northern climate. It is familiar to us all, that the flat corn of Carolina, though it will grow to a great size with us, will never ripen its seed, but with great precaution, and then but imperfectly. The flint corn of Cuba will not even form its ears in our climate. The same remark applies to wheat. It has been proved by experiments so numerous as to put the question at rest, that wheat from southern climates, though it will grow vigorously at first, will not come to perfection. The only exception we have yet heard of is that of the wheat of Leghorn used by the Italians in the manufacture of straw bonnets. This has succeeded in one or two instances, but these are not sufficient to enable us to pronounce it to be an exception from a general rule.

It is confidently affirmed that flaxseed constitutes an exception, and that it uniformly is improved by change. It may be so ; it is important that this should be tested by frequent trials—but we believe that the best general rule is to select our best seeds and roots, or to buy them of our successful neighbours, rather than to rely on foreign productions.

CHARCOAL, wood charred, or burnt with a slow, smothered fire. The making of charcoal is a business mostly performed by farmers. And in clearing new lands, making their wood into coals is better than burning it to waste, unless the distance of a market for coals be too great. One cord of wood will make forty bushels of coals : And those farmers who are not distant from populous towns, or who are near iron works, may turn their coals to considerable profit.

I have long observed, that where coal kilns have been burnt, the ground has discovered a remarkable fertility for many years after ; and more especially when it has been naturally a cold and wet soil. The dust of the coals and that of the burnt turf, have conspired to produce this effect. Hence I have concluded that small coals, or the dust from coal kilns, spread over sour meadow lands, would answer the end of a good manure. Being extremely porous, the pieces of coal imbibe much of the superfluous water, as well as increase the heat on the surface, as all black substances do. And when the weather becomes dry, they discharge the moisture, partly into the soil when it grows dry enough to attract it, and partly into the air, by the action of the sun upon it.—Autumn is, on several accounts, the best season for laying on coal dust ; and I would recommend it to all who have bottoms of kilns, to make this use of the dust.

I have been confirmed in my opinion, by reading in the *Complete Farmer* as follows : " The dust of charcoal has been found, by repeated experience, to be of great benefit to land, especially to such as is stiff and sour. It is to be used in the same manner as soot and wood ashes."

All sorts of glass vessels, and other utensils, may be purified from long retained smells of every kind, in the easiest and most perfect manner, by rinsing them out well with charcoal powder, after the grosser impurities have been scoured off, with sand and pot-ash. Putrid water is immediately deprived of its offensive smell by charcoal.

Meat which has been kept too long in summer, may be deprived of its bad smell, by throwing into the pot in which it is cooked just before it begins to boil, a quantity of live coals destitute of smoke ; after a few minutes the water should be changed, and if necessary the operation may be repeated. See the article *Meat.*

For a further account of the use of coal in agriculture, see Massachusetts Agricultural Repository, vol. III. pages 18, 19.

CHARLOCK, *Sinapis*, a well known and troublesome weed. It is known also by the names chadlock, catlock, carlock, and white rape. It is similar to radish. The young plants so nearly resemble turnips, that they are scarcely distinguishable, unless it be by the taste. Mortimer mentions a field of barley, mowed when the charlock was in blossom, which took off only the tops of the blades of barley ; and which gave the barley an opportunity to get above the weeds, and so it produced a good crop. He supposes cow dung increases it more than any other manure : And recommends feeding fallows with sheep when they are infested with this weed. It is well known that sheep will eat this weed rather than turnips. The seeds will live in the earth many years, and afterwards vegetate by means of tillage.

Grain should be sown thick, where there is danger of its being injured by charlock, so that the crop may overtop the weeds.—Barley sown thick will certainly prosper in such a situation.

CHEESE, a sort of food made of milk, purged of the serum, or whey, and dried for use.

Some cheeses are wholly made of unskimmed milk, which are called new milk cheeses, although part of the milk be kept over night, or longer. These cheeses, as any one would expect, are the fattest, and most valuable.

Another sort are called two meal cheeses, being made of the morning's milk unskimmed, together with the evening's milk skimmed. These, when well made, without the mixture of any sour milk, are not much inferior in quality to new milk cheeses.

The third sort of cheeses are wholly made of milk deprived of its cream. This kind is tough, and hard to digest ; and containing only the glutinous part of the milk, it affords little nourishment, and is scarcely worth making.

The method of making cheese, in Yorkshire in England, is as follows.—" If your milk be not just come from the cows, make it blood warm, turn it into a clean vessel for the purpose, and put in the rennet ; be sure to give it no more than what will make it come lightly. After it comes, stir it with your hand, till it be gathered, and parted from the whey. Then take the curd up in a strainer, and work it with your hands, till you get as much of the whey from it as possible : Then lay it in a clean linen cloth, and put it into the hoop : After it is lightly covered with the cloth, put it into the press : Let it stand in the press two hours ; take it out, and the cloth from it, and rub it over with fine salt : Put it in a dry cloth, and press it eight hours : Then put it in another cloth, and let it remain in the press till your next cheese be ready.—When you take it out, rub it well with salt, and wrap the round ring of the cheese with a garter made of linen yarn, and pin it at the end, which keeps the cheese in a good shape : Then let it lie in brine twenty-four hours. Add a little salt to your brine every time you put in a new cheese. When you find the brine turning unsavoury, make new brine ; and turn the cheese in the brine vat twice in twenty-four hours, always rubbing a little salt on the top of it when it is turned. When you take it out of the brine, dry it with a cloth; and turn it every day on the shelf for two months. The shelf should be a little wider than the cheeses, and the garters should continue round them five or six days."

A dairy woman in my neighbourhood, whose cheese is most excellent, is nearly in the Yorkshire practice. But lest the salt should not penetrate the whole of the cheese equally in every part, she sprinkles a little fine salt on the curd, when she breaks it, perhaps as much as an ounce to a cheese of fifteen pounds weight; and her cheeses never appear to be too much salted.

This may be partly owing to another improvement in her method. To the brine, in which she lays a cheese after it is pressed, she allows as much nitre as would lie on a shilling. She has found, by long experience, that the nitre not only gives a reddish cast to the rind of the cheese, but makes it more tender after it is thoroughly dried. It also prevents the cracking of the rind, which is a matter of much consequence. At the same time it prevents the distention of the cheese by wind, makes it mellow and soft throughout, and improves the taste.

But it is in vain to attempt the making of good cheese, unless the rennet be uncorrupted, and perfectly sweet. See the *Article*, *Rennet*.

In this country, where the summers are hot, and flies abound, cheeses are often destroyed, or greatly damaged, by maggots. To prevent this, every precaution ought to be taken to prevent fly blows getting into a cheese while it is making. For it is certain that cheeses will sometimes prove maggotty, which could not have been fly blown after they were made.—To prevent this evil, the milking pails, the cheese tub, &c. should be kept in dark places till the moment they are used, after being dried before a hot fire. And if the milk stand any time, or more than during one night, the room it stands in should be dark: Because flies are not apt to lay their eggs in dark places.

After cheeses are made, they will sometimes have little flaws in them, or cracks in drying, which the flies will be fond of depositing their eggs in. To prevent this, the cheeses should be smeared with a little tar mixed with salt butter: Or the cracks may be filled with a soft paste of flour as often as they appear.

But all this precaution will be apt to prove insufficient, unless the cheeses be dried in a dark room. As flies do not frequent dark places, cheeses dried in a dark room may be full of cracks, and yet escape maggots.

Some persons choose to medicate their cheeses with the juice of some wholesome plant, as sage, baum, mint, tansy, pennyroyal, &c. which they put into the curd. But I think this is no real improvement. To give cheese the hue of that which is made in Glocestershire in England, a little of the annotto may be put into the milk.

The following method of cheese

making is recommended in the Massachusetts Agricultural Repository, vol. III. p. 52. The milk is universally set for cheese as soon as it comes from the cow.

The management of the curd depends on the kind of cheese: thin cheese requires the least labour and attention.

Breaking the curd is done with the hand and dish. The finer the curd is broken the better, particularly in thick cheeses. The best colour of this kind of cheese is that of bees-wax, which is produced by annotto, rubbed into the milk after it is warmed. The dairy woman is to judge of the quantity by the colour of the milk, as it differs much in strength. The rennet is prepared, by taking some whey and salting it till it will bear an egg; it is then suffered to stand over night, and in the morning it is skimmed and racked off clear; to this is added an equal quantity of water brine, strong as the whey, and to this mixture, some sweet briar, thyme, or some other sweet herbs, also a little black pepper and salt petre; the herbs are kept in the brine three or four days, after which it is decanted clear from them. Into six quarts of this liquor four large calves' bags, or more properly called calves' stomachs, are put. No part of the preparation is heated, and frequently the calves' bags are only steeped in cold salt and water. Turning the milk differs in different dairies, no two dairy women conduct exactly alike.

Setting the milk too hot inclines the cheese to heave, and cooling it with cold water produces a similar effect. The degree of heat is varied according to the weather. The curd, when formed is broken with what is called a triple cheese knife. The use of this is to keep the fat in the cheese; it is drawn the depth of the curd two or three times across the tub, to give the whey an opportunity of running off clear; after a few minutes the knife is more freely used, and the curd is cut into small pieces like chequers, and is broken fine in the whey with the hand and a wooden dish. The curd being allowed about half an hour to settle, the whey is laded off with the dish, after it is pretty well separated from the curd.

It is almost an invariable practice to scald the curd. The mass is first broken very fine, and then the scalding whey is added to it, and stirred a few minutes; some make use of hot water in preference to whey, and it is in both cases treated according to the nature of the curd; if it is soft the whey or water is used nearly boiling; but if hard, it is used only a little hotter than the hand. After the curd is thoroughly mixed with the hot stuff, it is suffered to stand a few minutes to settle, and is then separated as at the first operation. After the scalding liquor is separated, a vat, or what is often called a cheese hoop, is laid across the cheese ladder over the tub, and the curd is crumbled into it with the hands and pressed into the vat, to squeeze out the whey. The vat being filled as full and firmly as the hand alone can fill it, and rounded up in the middle, a cheese cloth is spread over it and the curd is turned out of the hoop into the cloth; the vat is then washed. and the inverted mass of curd, with the cloth under it, is returned into the vat and put into the press; after standing two or three hours in the press, the vat is taken out and the cloth is taken off, washed, and put round the cheese, and it is replaced in the vat and in the press. In about seven or eight hours it is taken out of the press and salted, the cheese is placed on a board, and an handful of salt is rubbed all over it, and the edges are pared off if necessary; another handful of salt is strewed on the upper side, and as much left as will stick to it; afterwards it is turned into the bare vat without a cloth, and an equal quantity of salt is added to it, and the cheese is returned into the press; where it continues one night, and the next morning it is turned in the vat, and continues till the succeeding morning, and is taken out and placed on the dairy shelf; where they are turned every day or every other day, as the weather may be. If it is hot and dry, the windows and door are kept shut, but if wet or moist, the door and windows are kept open night and day.

Cleaning the Cheese. The cheeses having remained about ten days after leaving the press, are to be washed and scraped in the following manner; a large tub of cold sweet whey placed on the floor, the cheeses are immerged in it, where they continue one hour or longer if necessary, to soften the rind. They are then taken out and scraped with a common case-knife, with great care so as not to injure the tender rind, till every part of the cheese is smooth; they are after the last operation rinsed in the whey and wiped clean with a coarse cloth, and placed in an airy situation to dry, after which they are placed in the cheese room. The floor of the cheese room is generally prepared by rubbing it with bean or potatoe tops, or any succulent herb, till it appears of a black wet colour; on this floor the cheeses are placed, and are turned twice a week, their edges are wiped hard with a cloth once a week, and the floor is cleansed and rubbed with fresh herbs once a fortnight. They must not lie too long or they will stick to the floor.—This preparation of the floor gives the cheese a blue coat, which is always considered as of great consequence.

Stilton Cheese, how made. Take the night's cream and put it to the morning's new milk with the rennet; when the curd is come, it is not to be broken, as is done in other cheeses, but is to be taken out of the whey as whole as possible with a large dish, and placed in a sieve to drain gradually, and as it drains keep pressing it gently till it becomes dry and firm; then place it in a wooden hoop, afterwards to be kept dry on boards, and turned frequently with cloth binders round it, which are to be tightened as occasion requires; these cloths are kept on, but often shifted for clean ones, till the cheese acquires firmness enough to support itself;

after which it must be rubbed twice every day with a brush for two or three months.

Cheese, Skippers in. Wrap the cheese in thin brown paper,so that moisture may strike through soon; dig a hole in good sweet earth about two feet deep, in which the cheese must be buried 36 hours, and the skippers will be found on the outside of the cheese. brush them off immediately, and you will find your cheese sound and good. To prevent skippers, take a pod of red pepper, and put it into a piece of fine linen, moisten it with a little butter, and rub your cheese frequently. Mass. Agr. Rep. vol. —, p. 314.

To prevent Cheese from having a rancid nauseous flavour.

Put about one table-spoonful of salt to each gallon of milk when taken from the cows in the evening, for the cheese to be made the next day ; put the salt at the bottom of the vessel that is to receive the milk ; it will increase the curd and prevent the milk from growing sour or putrid the hottest nights in the summer. For further directions relative to making cheese, see Mass. Agr. Rep. vol. V. p. 201, 203.

CHICK WEED, the same as *Alsine*, a tender creeping weed,often troublesome in old gardens, and which grows luxuriantly in shady places. Swine will feed upon it when they are hungry.

CHURN, a wooden vessel, in which butter is produced by churning. It is broad at the bottom,and narrow at the top, to prevent the contents from coming out at the top during the agitation. But the shape does not perfectly answer this design.

Churns are commonly made of pine. But when they are new, they give the butter a taste of the wood ; so that oak is generally preferred. The hoops are of ash,and should be made very smooth and regular, that the vessel may be easily cleaned and kept sweet.

But on great farms, and where the dairies are large, the barrel-churn is to be preferred. Its name gives the idea of its shape ; and when it equals a barrel in size it can be easily managed. On each head of it is an iron spindle, and on each spindle a winch, by which the vessel is turned on a horse made for the purpose. A much greater quantity of cream, or milk, may be churned in this than in the common churn ; and the labour is easier.—There are two boards within this churn on each side of the centre, like shelves, which serve to agitate the cream.

The aperture in the barrel churn ought to be five or six inches square, to which a stopper must be exactly fitted, which must be kept in its place by a bar of iron across it, held fast by staples.

A great variety of churns have been contrived for the purpose of facilitating the process, (often a very tedious one) of making butter. Our limits will not permit descriptions of these machines accompanied with such drawings as would be necessary to make them intelligible. We shall therefore inform our readers where such descriptions may be found ; and would advise any person, who proposes introduce any machine of the sort, to consult the books to which we refer, to ascertain whether his invention contains any meritorious novelty.

Mr. William Bowler invented an improved churn, for which the Society for the Encouragement of Arts, Manufactures and Commerce, in England, gave him thirty guineas. This is described with drawings, in the Domestic Encyclopædia, 3d American edition, vol. I. p. 462.

Wright's Churn, described in the same volume and the same page, and made by Wright & Co. of Philadelphia.

The Rev. Mr. Willmarth, of Wilton, N. H. has invented a Patent Churning Machine, which has been thus described. The body of the churn is square at the bottom—two opposite sides are perpendicular. The other two opposite sides converge or incline towards the top. The dasher consists of two arms, hung on pivots or pins, on each perpendicular side of the churn, which projecting down, almost to the bottom, have inserted between them a number of flats, or thin pieces of wood. On the top of these arms are framed two levers at the centre, so as to project horizontally each way, and connected at each end with rounds. The dasher moves or vibrates in a manner similar to that of the patent washing machine. By this means, a powerful force is mechanically applied to the cream, uniformly straining and agitating the whole, at every vibration, with great ease to the operator. Children can use it. The principal specific improvement in this machine, is the *Pendent Swing Dasher*.

A churn invented by Mr. Andrew Halliburton, of Portsmouth, N.H. is described and recommended in the *Massachusetts Agricultural Repository*, p. 164.

An improvement in the dasher of the common churn has been made by a Mr. Fisher, in England, who obtained a premium for his discovery. The dasher turns on the handle by being fixed to it by a pivot. The dasher is two cross pieces, about three inches square, put together by being let into each other, which then form four wings. These are cut bevelling on each side, at an angle of forty-five degrees, so that they stand diagonally; the whole being very similar to the little wind-mills, (so called) which are set up on poles to be turned by the wind.

As this dasher goes down in the milk or cream, it turns one way, and as it comes up, the other. It is so efficacious, that the churning must be performed moderately, or the butter will come too soon, and be *swelted*, as it is technically called by some.

Oak is said to be better for churns than pine, as the latter gives something of its taste to the butter.

CIDER. In making cider see that the mill, the press, and all the materials be sweet and clean, and the straw free from must. The fruit should be ripe, but not rotten, and when the apples are ground, if the juice is left in the pomace from

12 to 24 hours according to the heat of the weather, the cider will be the richer, softer, and higher coloured. The fruit should be all of one kind, as the fermentation will be more regular. The juice of the fruit as it comes from the press should be placed in open headed casks or vats ; in this situation, it is likely to undergo a proper fermentation, and the person attending may with great correctness ascertain when the first fermentation ceases ; this is of great importance, and must be particularly attended to. The fermentation is attended with a hissing noise, bubbles rising to the surface, and its forming a soft spungy crust over the surface of the liquor. When this crust begins to crack, and a white froth appears in the cracks level with the surface of the head, the fermentation is about stopping. At this time the liquor is in the fine genuine clear state, and must be drawn off immediately into clean casks ; and this is the time to fumigate it with sulphur. To do this, take a strip of canvas, or rag, about two inches broad, and twelve inches long—dip this into melted sulphur, and when a few pails of worked cider are put into the cask, set this match on fire, and hold it in the cask till it is consumed,then bung the cask, and shake it that the liquor may incorporate with, and retain the fumes; after this fill the cask and bung it up.

When the apples are gathered, they should be put for some time in piles, and sorted before they are pressed. The first runnings of the press should be kept separate, being of a superior quality of cider. When cider is moved it must be racked off into other casks, that the lees may not mix with the cider. The later apples hang on the trees, the more spirit the cider will contain.

When casks have a sour smell, put in about a pint of unslacked lime for a barrel of common size, and pour in three or four gallons of hot water; shake it well, giving it vent ; let it stand till cool, and then rinse with cold water. Repeat the operation till the cask smells perfectly sweet. Wood ashes might perhaps answer as a substitute for lime.

The juice of fruit will undergo three fermentations, the vinous,the acid, and the putrid. The first fermentation is the only one which cider should undergo. To preserve it in this state, as soon as the first fermentation ceases, which may be known by the liquor ceasing to throw up little bubbles to the top, draw off and fumigate the liquor with sulphur as before directed.—After this you may further improve it by adding a gallon of French brandy to each barrel.

If the cider should require further clarifying, it may be performed with eggs, skim milk, molasses, or isinglass.

Mr. Joseph Cooper, of New-Jersey, published an article in the True American, relative to fining cider, by which it appears that the jelly made from bullock's feet will answer the purpose. It should be warmed, and mixed with a little of the cider intended to be fined, then strained, and when cold put into the vessel which contains the rest, and stirred till thoroughly incorporated.

"To refine cider and give it a fine amber colour, the following method is much approved of. Take the whites of six eggs, with a handful of fine beach sand, washed clean; stir them well together; then boil a quart of molasses down to a candy, and cool it by pouring in cider, and put this together with the eggs and sand, into a barrel of cider, and mix the whole together. When thus managed, it will keep for many years. A piece of fresh bloody meat put into the cask, will also refine the cider and serve for it to feed on.

"To cure oily cider, take one ounce of salt of tartar, and two and an half of sweet spirit of nitre, in a gallon of milk for a hogshead. To cure ropy cider take six pounds of powdered allum, and stir it into a hogshead; then rack it and clarify it.

In bottling cider it is recommended to raise the proof by putting about two tea-spoonfuls of French brandy into each bottle.

A dozen of sweet apples, sliced into a cask of cider, have been found advantageous. Three quarts of good w heat boiled and hulled put into each hogshead, is is said improves the liquor, and prevents its becoming tart.

For further directions on this subject, see Massachusetts Agricultural Repository, vol. IV. p. 170 to p. 189.

CLAY, a ponderous kind of earth, consisting of fine particles, firmly cohering when dry; and when wet, viscid and tenacious. It is of various colours in different countries. But in this country it is mostly either a dull blue, or of the colour of ashes. In Nova Scotia, the clay is of the colour of a well burnt brick.

CLAY-SOIL, land which consists almost wholly of clay, with perhaps a thin surface of dark mould over it, made by substances which have consumed upon it. This kind of soil abounds in the north-eastern territories of Massachusetts.

Clayey lands are apt to be very barren in their natural state, unless when a summer is so divided betwixt rain and sunshine, that they are kept on a medium continually betwixt drought and wetness, which seldom or never happens. In a wet season, plants growing on such a soil are drowned, as the closeness of the clay will not suffer the water to soak into the ground: And in a dry season, the ground becomes so solid that the roots of plants cannot penetrate it, some few strong feeders excepted.

This kind of earth, however, is thought to contain more of the food of plants than almost any other. But something needs to be done to bring it into action. The European farmers think their clay soils the richest and most valuable of their land. But many of our farmers despise them, for want of knowing what methods to take to render them profitable; or through fear of the labour, or expense, of doing it.

Some of these soils, without

much alteration, will bear good crops of grass, if care only be taken not to feed them close in autumn, nor to let cattle in upon them in the spring. But the farmer, who wishes to keep them in tillage, must alter them by the admixture of such substances as may serve to open the soil, and break the cohesion of its particles. When this is once accomplished, the land will become highly valuable; holding the manure to admiration, and never returning to its pristine state.

Dung is helpful towards opening a clayey soil, by the fermentation it raises, as well as by the mixture of its earthy, saline and oily particles. But dung of itself will not be sufficient, unless it were laid on more plentifully than farmers can well afford. A mixture of dung and sand is found to be a much better dressing for this sort of land, than dung alone. And if sand be not too far distant, it would be advisable to put on a layer of it two or three inches thick. Beach-sand is preferable to any other, as the saltness of it will help to make the ground fruitful. But pit-sand will do very well; or rather, that which has been washed down to low places in the roads.

In places where sand is not to be had, the ground may be loosened with other substances. Gravel, or light loam, from neighbouring spots, may be carted upon it; dust from saw-pits, chips and rubbish from the back yards of houses, straw and stubble, swamp mud, the bark of trees and rotten wood, or burnt clay. I have known a clayey spot made very fruitful, merely by the remains of a rotten log fence, when mixed with the soil.

When a clay soil is sanded, or any other thing laid on to open it, it will take several ploughings and harrowings to mix it, so as to bring the land to a good consistence. As the expense of mixing it at once would be too great, it is better to use it for two or three years after, for the growing of such tillage crops as are most suitable to clayey soil, such as barley, flax, &c. The soil will grow better year after year, till the sand, &c. is thoroughly mixed with the soil; after which it will be fruitful forever without large dressings. Hoed crops will mix it sooner than any other method, and without any expense.

A small quantity of dung, each year that it bears a hoed, or a green crop, will be proper: And the most suitable dungs are those of horses and sheep, pigeons and other fowls, which by their heat will correct the natural coldness of the soil.—Folding with sheep has an excellent effect on this kind of land.

Such a stiff soil is also mended by frequent ploughings. The Europeans allow three ploughings previous to feeding, to be enough for a free soil; but to a clayey soil they give four or five. The oftener it is stirred with the plough, the more the cohesion of the particles is broken, and the more easily the roots of plants can penetrate it in search for their food. But it never should be ploughed when it is so wet as to potch with the feet of the cattle, or to run like mortar. In this condition, the more it is worked the stiffer it will become. On the other hand, when it is very dry, it cannot well be ploughed, by reason of its hardness. Suitable seasons should be embraced, for ploughing it, when it is neither too wet nor too dry. At the first ploughing it comes up in large clods; but the oftener it is ploughed in fit times, the smaller the clods will be, and the more fine mould will be among them.

Exposing the clods to the sun and air has some tendency to mellow the soil: But a winter furrow is of very great advantage. The frost does much towards breaking the cohesion, as I have found by experience.

Clay soil, after all the melioration that can be given them, will be more suitable for some plants than for others. Those plants in general which require a great degree of heat, or a long summer, are not so well adapted to be cultivated in a clayey soil, such as Indian corn, tobacco, &c. But it may be made to produce good crops of wheat, grass, barley, oats, flax, cabbage, &c. No good eating potatoes or carrots are ever produced in such a soil.

Fruit trees, in general, and I think all sorts, excepting pear trees, answer but poorly in a clayey soil, how much soever the surface may have been mixt with other substances. The roots of trees will need to draw some of their nourishment from a part of the soil below that which has been meliorated by mixing; but the compactness of it will scarcely suffer them to penetrate it.

Fallowing and green dressing may help to pulverize a clay soil; and sowing it frequently with pease is recommended. Any crop that forms a close cover for the surface causes the soil to putrefy, breaks the cohesion of its particles, and prevents the ground from hardening by the influence of the sun.

If a clay soil lie so flat that water stands on it some part of the year, it cannot be brought to a good consistence without ploughing in ridges, and water furrowing. The ridges may be wider or narrower, according to the degree of wetness to which it is subject. Sometimes deep drains will be necessary to give it the needful degree of dryness. See *Soil.*

CLEARING OF LAND, an operation often necessary to be performed in this new country, especially in the most inland parts. Lands which were before in a state of nature, are said to be cleared, when they are so freed from their natural growth, as to become fit for tillage, mowing, or pasture.

In those parts of the country where wood is of little or no value, the method of clearing upland is as follows:—The trees are felled in one of the summer months; the earlier in summer the better, as they will have a longer time to dry, and as the stumps will be less apt to sprout. The trees lie till the following spring; when the limbs which do not lie very near to the ground should be chopped off, that they may burn the better. Fire must be put to them in the driest part of the month of May; or if the whole of May prove wet, it may be done to advantage in the beginning of June. Only the bodies of the trees will remain after burning, and some of them will be burnt to pieces. Then they are to be cut into pieces nearly of one length, drawn together by oxen, piled in close heaps, and burnt; only reserving suitable trees, which will be needed for the fencing.—The heating of the soil so destroys the green roots; and the ashes, made by burning, are so beneficial a manure to the land, that it will produce a good crop of Indian corn, or wheat, the same year, without ploughing, hoeing, or manuring. Indian corn is most commonly the first crop; and it will bear a good crop of winter rye the second year, if the seed be only hacked in with hoes in September, before the Indian corn is harvested. After which, if grass seeds be thrown in with the rye, the land will be fit for pasturage, if not for mowing. The few sprouts which spring up from the stumps in the first summer, should be pulled off. and the ground is quite subdued. But if wheat or rye be the first crop, the ground must be well harrowed; once before sowing, and once after. And it often happens that the first and second crops pay the farmer well for all the labour of clearing and fencing. It is certain that sometimes the first crop will do it.

The invention of this kind of culture has been of essential advantage to the poorer sort of people: And it has been conducive to bringing forward rapid settlements in our new towns and plantations. A farm may be thus begun in the wilderness with little or no stock.

But those persons who are able to do it, had better plough and harrow their ground after burning, before they seed it. The ashes will thus be well mixed with the soil; and the land has always been found to retain its fertility the longer, when so managed.

If new land lie in such a situation, that the natural growth may turn to better account, whether for timber or fire wood, fencing or charcoal, it will be an unpardonable waste to burn the wood on the ground. But if the trees be taken off, the land must be ploughed after clearing, or it will not produce a crop of any kind. And some warm kind of manure will be needful, if Indian corn is planted on it. This is the case at least in the most northerly parts of New-England. But rye will answer extremely well without manure.

When new lands are destitute of trees, and covered with oak shrubs, the clearing is more expensive, and the first crops not so profitable. But such lands should not remain unsubdued, as, in their present state, they are quite unprofitable, and a nuisance. The bushes should be cut with stub scythes or axes, piled in heaps, and burnt. After which the roots may be subdued by goats; or ploughed up with a strong team, the plough being proportionably strong. Some of those spots will require a team of 50 oxen to plough them. After ploughing, the roots must be taken out with a strong narrow hoe, with a good sharp edge.

Other methods must be taken for subduing low swampy lands.—See the articles *Bog*, *Bushes*, *Draining*.

CLIMATE. Climate depends on altitude as well as latitude. Sir John Sinclair observes, "It may be remarked, that land in the same parallel of latitude, other circumstances being nearly similar, is always more valuable in proportion to the comparative lowness of its situation. In the higher districts, the quality even of the herbage is less succulent and nourishing, and the reproduction slower, when in grass; while the grain is less plump, runs more to straw, is less perfectly ripened, and the harvest is also later.

"It has been estimated, that sixty yards of elevation in the land are equal to a degree of latitude; or in other words that sixty yards perpendicular higher, are in respect of climate equal to a degree more north." *Code of Agriculture*, p. 25—6, *Hartford edition.*

The above are curious facts, but perhaps would not apply to all parts of the United States. In a very warm climate, where vegetation is liable to be scorched by excessive heat, elevation of site would doubtless prove advantageous. In New-England, however, we are inclined to believe that the observations of Sir John, as it respects sixty yards of elevation being equal to a degree of north latitude, might be nearly correct. It is well known that the spring is more backward and the harvest is much later on the heights of land between Merrimack, Connecticut, and North River, than in the vallies which are pervaded by those streams.

A valuable paper respecting the difference between the climates of Sweden, England and Massachusetts, as ascertained by the flowering and leafing of plants may be seen in the *Massachusetts Agricultural Repository*, vol. IV. p. 135.

With regard to the effect of climate as relates to a change of seeds, see *Change of Seeds*.

CLOG, a wooden instrument, fastened to the neck or leg of a beast, to prevent his leaping over, or breaking fences. The best clog for the fetlock of a horse, is made of one piece of tough wood bent over at one end. Into a notch, near this end, a leathern strap, nailed to the long part of the clog, is slipped over the end. It may be put on, or taken off, in an instant.

CLOVER, *Trifolium pratense*, a species of trefoil, esteemed as an excellent grass for the feeding of cattle, both green and dried. The hay made of clover is more suitable for horses than any other that this country affords. Horses kept on it will fatten, even in the most unfavourable season of the year.—It answers well when used as soiling, or eaten out of racks without drying.

Green clover is known to be good feeding for swine. Keeping them in pastures, where there is plenty of this grass, will make them grow fast, and fatten so as to almost become fit for the knife. But when they go in pastures they should have rings in their noses. Otherwise they will root out the clover.

Red and white clover are the only sorts known and esteemed in this country ; as to the wild sort,or variety,with a rough leaf,it is of no consequence. It is the red clover that is of most importance for mowing. It is a biennial grass, and if it did not sow its own seeds, would entirely run out the third year. The white is generally too small and short, unless when it is drawn to a greater height by being mixed with other grasses.

Many farmers, instead of sowing clean seed of clover, content themselves with sowing chaff and dust from the floors of their barns.—This is a slovenly and uncertain method, oftentimes attended with great loss. Fowls are usually admitted into barns ; and when this is the case, none can tell how much, or how little of the hay seed remains among the dust : So that the farmer who sows it, may either over-seed his land ; or, which is a more common case, not seed it half enough. A consequence of which last will be, that he will have no good crop of hay from his sowing. He must either plough up his land again, for the mere purpose of seeding it with grass, or let it lie useless till the grass gradually gets in ; either of which expedients will be attended with inconvenience and loss. I am aware that farmers, especially those in the northern parts of New-England, will object, that if grass do not get in the first year, it will the second. But they should consider that the loss of the first year's crop is considerable. Not only so, but it is losing the best crop that is to be expected from a clover lay ; and the land will become bound and weedy, before it is filled with grass roots ; so that no large crop will be had from it afterwards, nor any clean or unmixed crop, from which it will answer to take clean seed.

The quantity of seed to be sown on an acre, according to the opinions of the best European cultivators, is not six or ten as with us, but from fifteen to twenty pounds. By this mode of sowing it grows less rank, lodges less, and is more profitable for soiling and for hay.

It is no small recommendation of this grass, that it is adapted to a soil, that is suitable for scarcely any other grasses, which are cultivated in this country ; to a soil that is dry, light and sandy. It does well also on gravel and loam.

European farmers recommend sowing it in the spring, after the grain is up, and harrowing it in ; and they tell us the harrowing will not damage the corn, but rather be a service to it, when it is either spring or winter grain. That it should be sowed in the spring is granted, unless it be in countries where there are no severe winter frosts. The young plants, which come up in the fall, cannot bear the frost so well as those which have had a whole summer to bring them on towards maturity. Their reason for not sowing it at the same time as the grain it grows with, is an apprehended danger of its growing so fast as to obstruct the growth of the grain. But I have never found this to be the case in fact.

Mr. Eliot recommends a different method, which is ploughing in the seed ; which, he thinks, and not without reason, will cause it to bear drought the better, and become the more strongly rooted. I believe I may add, that it will be more likely to escape in the frosts of winter, which are so intense in this country, as often to kill almost whole fields of clover.

Peat ashes are said to be a very proper dressing for clover grounds. But this grass answers so well in this country without manure, that the farmers choose to set apart the whole of the manure that they can get for other purposes.

Some think clover is so far from needing any manure, that it will recruit lands which are worn out. That it will do it more than other grasses I cannot yet see any reason to believe. It will bear no crop worth mowing, on lands which are quite exhausted. But it is probable it may produce good crops, on lands which are much impoverished near the surface, by bearing plants with short, fibrous, or horizontal roots ; because clover sends its main roots to a great depth.—And while a field lies several years in clover, the soil near the surface may be considerably recruited.—But that land on the whole will be in better heart, after several heavy crops of clover are taken from it, and no manure laid on, seems incredible.

Writers on agriculture seem, however, to be agreed, that a clover lay is proper for the culture of wheat. The rotting of its large roots and stalks may answer as a good manure, no ways liable to distemper the wheat, as some other manures are thought to be.

Some skilful farmers insist much on the propriety of sowing clover-seed with barley. I suppose it will answer well with almost any grain that we call English. But with a crop of pease, or with any other crop that forms a close shade to the soil, it will not answer. The young plants must have some advantage of the sun and air, or they will not live. And if it be sown with flax, at least in some loose soils. the pulling of the flax will be apt to eradicate much of the clover. Crops which lodge are also destructive to the young clover, by forming so close a cover as to stifle it. Therefore, when clover seed is sown, either with barley or flax, the ground should be rather under than over seeded, to prevent lodging.

Clover being an early grass, it is commonly fit to cut in June.—When half the heads are turned from red to brown, and on the decay, it is the right time to mow it. But if the seed is to be saved for use, it must stand till it is all dead ripe, both heads and stalks. It requires more care to make clover into hay than most other grasses. That which is mown in a morning, should be spread, turned, and raked up before night. The next day, if the weather be fair, it must be opened, stirred once or twice, and cocked up again. Then, after sweating a day or two, it may be put into the barn. Rank clover requires much more drying than that which is of a moderate growth. And the hay is not so good.

In the most southerly parts of New-England, land in good heart will bear two crops of clover in a year. Mr. Eliot, therefore, recommends saving the second crop for seed, the first crop having been mowed early. But two crops are not to be obtained in the northern parts of this country. And, if they were, so frequent a cutting would be apt to make the roots shorter lived. It is advisable to pasture it in May, and then let it grow for seed. It is best to cut clover for seed on land that is soon to be broken up: Because a crop of seed weakens the roots much more than a crop of hay; and it is doubtful whether it will bear any considerable crop afterwards. Indeed, no crop of clover is of any great importance for hay, after the second year. For it is a biennial plant.

The white clover, vulgarly called honeysuckle, is an excellent grass, and seems very natural to this country: But when sown by itself, it does not grow tall enough for mowing. It is good for feeding in pastures, during the fore part of summer, at which time it often appears in great plenty.

The hop clover is new in this country, but seems to appear not very promising. It is said to flourish on the most barren sands, and continue long in any soil. It is of two kinds, large and small, and the heads are yellow. I once sowed a small bed of it. It did not prosper, being almost wholly destroyed by the following winter.

The European farmers are cautious of turning neat cattle in to feed in a field of luxuriant green clover, for fear of their being *hoven* with it, as it is called, or so swelled by eating it greedily, as to be killed by it. But this is an inconvenience, which I have never known to take place in this country. The way to save the life of hoven beasts is, to stab them between the hip and the short ribs, where the swelling rises highest. It is performed with a narrow, sharp pointed knife, which makes an orifice in the maw, and lets out the air that oppresses. The wound soon heals of itself.

The author of a valuable work, entitled, "A Treatise on Agriculture," lately published in Albany, directs ten or twelve pounds of clover seed to be sown on an acre, if the soil be rich, and double the quantity if it be poor. He condemns the practice of mixing the seeds of timothy and rye grass,&c. with that of clover, "because these grasses neither rise nor ripen at the same time. Another practice equally bad, is that of sowing clover seed on winter grain, before the earth has acquired a temperature favourable to vegetation, and when there can be no doubt but that two-thirds of the seeds will perish." This writer is of opinion that clover should not be pastured the first year, and observes, that "If the *crowns* of young clover roots be nibbled, or otherwise wounded, the roots die. Sheep and horses (both of which bite closely) should therefore be particularly excluded from clover, unless intended for pasturage only."

According to the same author,

"the short period between the flowering and the seeding of clover, is that in which its use would be most advantageous, whether regarded as *forage*, or as an ameliorating crop." But "when seed is the principal object of culture, we cannot do better than to adopt the practice in Holland—where the first crop is cut *before* it flowers,and the second is reserved for seed."

Green clover is good food for swine in summer ; and when cut green, and salted, after being a little withered,with about half a peck of salt to a load, it makes good food for swine in winter. But for this latter purpose it should be steamed, or boiled.

Two sorts of machines are described in the Transactions of the New-York Agricultural Society, for gathering clover seed. They were invented in Brookhaven, Suffolk County, New-York,by Mr. L'Hommedieu. One of these machines consists of an open box, about four feet square at the bottom, and about three feet in height on three sides ; to the fore part, which is open, fingers are fixed, similar to those of a cradle, about three feet in length, and so near as to break off the heads from the clover-stocks between them, which are thrown back into the box as the horse advances. The box is fixed on an axle-tree, supported by two small wheels, two feet in diameter ; two handles are fixed to the hinder part, by means of which the driver, while he manages the horse, raises or lowers the fingers of the machine, so as to take off all the heads of the grass ; and, as often as the box is filled with them, they are thrown out,and the horse goes on as before.

The other machine is called a *cradle*, and is made of an oak board about 18 inches in length, and 10 in breadth. The fore part of it to the length of 9 inches, is sawed into fingers ; a handle is inserted behind, inclined towards them, and a cloth put round the back part of the board, which is cut somewhat circular, and raised on the handle ; this collects the heads or tops of the grass, and prevents them from scattering, as they are struck off by the cradle, which may be made of different sizes ; being smaller in proportion for women and children, who by means of it may likewise collect large quantities.

The part containing the seed must be well dried in the sun, before the seed can be threshed out. About three bushels of seed may be commonly collected from an acre.

COLE-SEED, *Brassica Rapa*. "This plant, which is generally known by the title of rape, or cole-seed, is much cultivated in the isle of Ely, and some other parts of England, for its seed, from which the rape oil is drawn ; and it hath been also cultivated of late years, in other places, for feeding of cattle, to great advantage.

"The cole-seed, when cultivated for feeding of cattle, should be sown about the middle of June.—The ground should be prepared for it in the same manner as for turnips. The quantity of seeds for an acre of land is from six to eight pounds, and as the price of seed is not great, it is better to allow eight pounds ; for if the plants are too close in any part, they may be easily thinned, when the ground is hoed ; which must be performed in the same manner as is practised for turnips, with this difference only, of leaving these much nearer together ; for as they have fibrous roots and slender stalks, so they do not require near so much room.— These plants should have a second hoeing, about five or six weeks after the first, which, if well performed in dry weather, will entirely destroy the weeds, so that they will require no farther culture.

"Where there is not an immediate want of food, these plants had better be kept as a reserve for hard weather, or spring feed, when there may be a scarcity of other green food. If the heads are cut off and the stalks left in the ground, they will shoot again early in the spring, and produce a good second crop in April, which may be either fed off, or permitted to run to seeds, as is the practice where this is cultivated for the seeds : But if the first is fed down, there should be care taken that the cattle do not destroy the stems, or pull them out of the ground. As this plant is so hardy as not to be destroyed by frost, so it is of great service in hard winters for feeding of ewes ; for when the ground is so hard frozen as that turnips cannot be taken up, these plants may be cut off for a constant supply.— This will afford late food after the turnips are run to seed ; and if it is afterwards permitted to stand for seed, one acre will produce as much as, at a moderate computation will sell for five pounds clear of charges." *Gardener's Dictionary.*

The Rev. Mr. Eliot, who made some trial of this plant, is doubtful whether it will answer for winter feeding in this country, because of the severity of our frosts. But the above author adds—" The curled colewort, or Siberian borecole, is now more generally esteemed than the former, being extremely hardy, so it is never injured by cold, but is always sweeter in severe winters, than in mild seasons."— A gentleman informs me, that, in Boston, he has made trial of this plant, and found that the winter did not injure it. It is fit for the table from December to April.

I myself made trial of three kinds of borecole the last year, in the latitude of 44. It grew very well till winter ; but not one plant in fifty had any life in it in the following spring. The sorts were the green, the white, and the red. But it is probable that in some parts of New-England, and in warm situations, this plant may be cultivated with advantage ; though not in fields, I think it may in gardens.

COMPOSITION FOR TREES, a substance discovered, prepared, and applied by Mr. Forsyth for the purpose of removing diseases, defects and injuries in fruit and forest trees. It is directed to be composed in the following manner, in his " Treatise on the Management of Trees."

"Take one bushel of fresh cowdung, half a bushel of lime rubbish of old buildings (that from the ceilings of rooms is preferable,) half a bushel of wood ashes, and a sixteenth part of a bushel of pit or river sand. The three last articles are to be sifted fine before they are mixed; then work them well together with a spade, and afterwards with a wooden beater, until the stuff is very smooth like fine plaster used for the cielings of rooms.

"The composition being thus made, care must be taken to prepare the tree properly for its application, by cutting away all the dead, decayed and injured parts, till you come to the fresh sound wood, leaving the surface of the wood very smooth, and rounding off the edges of the bark with a drawing knife, or other instruments, perfectly smooth, which must be particularly attended to; then lay in the plaster about one eighth of an inch thick, all over the part where the wood or bark has been so cut away, finishing off the edges as thin as possible: Then take a quantity of dry powder of wood ashes mixed with a sixth part of the same quantity of the ashes of burnt bones; put it into a tin box, with holes in the top, and shake the powder on the surface of the plaster, till the whole is covered over with it, letting it remain for half an hour, to absorb the moisture; then apply more powder, rubbing it on gently with the hand, and repeating the application of the powder till the whole plaster becomes a dry smooth surface.

All trees cut down near the ground should have the surface made quite smoth, rounding it off in a small degree, as before mentioned, and the dry powder directed to be used afterwards should have an equal quantity of powder of alabaster mixed with it, in order to resist the dropping of trees and heavy rains.

"If any of the composition be left for a future occasion, it should be kept in a tub, or other vessel, and urine of any kind poured on it, so as to cover the surface; otherwise the atmosphere will greatly hurt the efficacy of the application.

"When lime rubbish of old buildings cannot easily be got, take pounded chalk, or common lime, after having been slacked a month at least.

"As the growth of the tree will gradually affect the plaster, by raising up its edges next the bark, care should be taken, where that happens, to rub it over with the finger when occasion may require (which is best done when moistened by rain,) that the plaster may be kept whole, to prevent the air and wet penetrating into the wound."

"As the best way of using the composition is found by experience to be in a liquid state, ' Mr. Forsyth advises that it should be reduced to the consistence of pretty thick paint by mixing it up with a sufficient quantity of urine and soap suds, and be laid on with a painter's brush. The powder of wood ashes and burnt bones is to be applied as before directed, patting it down with the hand."

It is also further advised that "when trees are become hollow, to scoop out all the rotten, loose, and dead parts of the trunk, to the solid wood, leaving the surface smooth: then to cover the hollow, and every part where the canker has been cut out, or branches topped off, with the composition; and as the edges grow, to take care not to let the new wood come in contact with the dead, part of which it may sometimes be necessary to leave: but to cut out the old dead wood as the new advances, keeping a hollow between them to allow the new wood room to extend itself, and thereby fill up the cavity, which it will do in time, so as to make it as it were a new tree."

And if the cavity be large, to cut away as much at one operation as will be sufficient for three years. But in this to "be guided by the size of the wound, and other circumstances. When the new wood, advancing from both sides of the wound, has almost met, to cut off the bark from both the edges, that the solid wood may join, which, if properly managed, it will do, leaving only a slight seam on the bark. If the tree be very much decayed, do not cut away all the dead wood at once, which would weaken the tree too much, if a standard, and endanger its being blown down by the wind. It will therefore be necessary to leave part of the dead wood at first, to strengthen the tree, and to cut it out by degrees as the new wood is formed. If there be any canker or gum oozing, the infected parts must be pared off, or cut out with a proper instrument. When the stem is very much decayed, and hollow, it will be necessary to open the ground and examine the roots."

COMPOSTS. "The utility of composts has been proved by the experience of numbers, in various districts. The subject may be considered under the following heads: 1. The materials used; 2. The soils or crops to which it should be applied; and, 3. The effects produced.

"1. *Materials.* Unslacked lime, and earth of different sorts, are the substances commonly used. Quick lime is the proper stimulus for bringing the powers of a compost into action; operating upon a heap of earth in some degree as yeast does upon a quantity of flour or meal. Enough ought to be applied to excite a species of fermentation in the heap, and to neutralize any pernicious mineral substances which may exist in it.

"The formation and conveyance of compost being attended with much expense, where circumstances will admit of it, horses, instead of manual labour, ought to be employed in the process of preparation, and the compost should be prepared in the field to which it is afterwards to be applied.

"It has been ascertained by a number of experiments, that two bushels of unslacked lime are sufficient for each cubic yard of earth of a medium quality, and as 80 cubic yards of earth are sufficient to manure an acre, 160 bushels of unslacked lime is the quantity required. To obtain this quantity of each, it is the practice of some farmers to plough the head-ridges at both ends of the field, ten inches deep, and to collect one half of this, which can be often spared, without any loss, as the head ridges are generally too high, in consequence of the earth accumulated in the course of years, from the plough being cleared every time it turns. It has been calculated, that where a head ridge is 18 feet broad, 72 feet of it in length, ploughed ten inches deep, will produce 40 cubic yards of earth, and consequently the two head-ridges will produce 80 cubic yards of compost for the field to be manured.

"Composts are frequently made of various materials, as several sorts of earth, lime, old mortar and plaster, green vegetables, before they run to seed, soft chalk, tanners' bark, saw dust, soap-ashes, dung, &c. It is recommended, that instead of being laid in regular layers, they should be mixed as much as possible, in forming the heap. A fermentation is soon excited, and the oftener the heap is turned, so much the more will fermentation be promoted.

"A mode of making compost was suggested by the late Lord Meadowbank, of which peat is the basis. It was not unusual in various parts of Scotland, to bed cattle, and even sheep, with peat, for the purpose of increasing the quantity of manure; but Lord Meadowbank was the first individual in this country who investigated the properties of that species of manure, and explained them upon scientific principles. The result is that one ton of dung, will ferment three tons of peat, or moss earth. This is a most valuable discovery and must, if duly attended to, greatly enrich many hitherto neglected districts. The great advantage of this compost is, that it has nothing but inert vegetable matter to act upon, whereas, when lime is mixed with rich earth, it may have the effect of dissipating the gaseous matters it contains, and of rendering it proportionably less valuable.

2. Composts are peculiarly well calculated for grass lands, and ought to supersede the offensive, and wasteful practice of laying putrescent matter on the surface of the soil, by which a very large proportion of its most valuable component parts is lost in the atmosphere. They are likewise of great use to moorish lands, augmenting their staple, and adding to them a number of valuable and enriching substances. In regard to sandy or clayey soils, composts, principally consisting of articles different from their general nature respectively, will improve their texture and convert them into loams.

3. "The effects of composts are highly satisfactory. In regard to grass lands, experience has shown, that they at once improve their quality, and checks for years, the progress of moss or even unprofitable grasses. In thin moorish soils, composts properly and repeatedly applied, alter the nature of the soil; it becomes more

fertile, retains its moisture better, and does not suffer by the summer's drought, which would otherwise burn up the crops. The effects of the Meadowbank composts are still more extraordinary; a farmer in Roxburghshire, having raised as good turnips, and as productive crops of wheat, after fallow, on good soils, manured with this compost, as from dung.

"It is a circumstance not to be omitted, that lime will operate in compost, upon lands that had been exhausted by the over frequent, or too abundant application of lime or marl, even when it had not succeeded when used by itself. This is a strong recommendation of such mixed manure, as land may be thus cultivated to advantage, that would otherwise remain unproductive." *The Code of Agriculture.*

CORN-SHELLER. A machine for shelling Indian Corn has been invented in Massachusetts, and a patent obtained by the inventor. The expence of the machine is said to be about twenty dollars. Two hands with this machine, can shell, it is said, at the rate of about half a bushel a minute.

COTTON, (*Gossypium*) is a genus of plants, comprising twelve species, all of which are natives of warm climates, though four only are cultivated. In Georgia and South Carolina two kinds of cotton are planted; one of which grows upon the upland, is of a short staple, and has green seed. Another kind of a long staple and silky fineness, having black seed, is cultivated upon the islands on the coast of Georgia and South Carolina.

Pierce Butler, Esq. who successfully cultivates cotton on the island of St. Simons, State of Georgia, gives the following directions for raising it.

"If the land has been recently cleared, or has long remained fallow, turn it up deep in winter; and in the first week in March bed it up in the following manner. Form 25 beds, in 105 square feet of land, (being the space allotted to each able labourer for a days work;) this leaves about four feet, two and one half inches from the centre of one bed, to the centre of the next. The beds should be three feet wide and flat in the middle. About the 15th of March, in the latitude from 29 to 30°, the cultivator should commence sowing, or as it generally termed, planting. The seed should be well scattered in open trenches, made in the centre of the beds, and covered: the proportion of seed is one bushel to one acre; this allows for accidents occasioned by worms or night chills. The cotton should be well weeded by hoes, once every twelve days, till blown, and even longer if there is grass, observing to hoe up, that is *to* the cotton till it pods, and hoe down when the cotton is blown, in order to check the growth of the plant. From the proportion of seed mentioned, the cotton plants will come up plentifully, too much so to suffer all to remain. They should be thinned moderately at each hoeing. When the plants have got strength and growth, which may be about the third hoeing, to disregard worms and bear drought, they should be thinned according to the fertility of the soil, from six inches to near two feet between the stocks or plants. In rich river grounds the beds should be from 5 to 6 feet apart, measuring from centre to centre; and the cotton plants, when out of the way of the worms, from two to three feet apart. It is advisable to top cotton once or twice in low grounds, and also to remove the suckers. The latter end of July is generally considered a proper time for topping.—Gypsum may be used with success on cotton lands *not near the sea*. In river grounds draining is proper; yet these lands should not be kept too dry. In tide lands, it is beneficial to let the water flow over the land without retaining it. In river lands a change of crops is necessary. From actual experiment it has been proved that river tide lands having the preceding year had rice sown on them, yielded much more cotton the succeeding year than they would have afforded by a continuation of cotton.

"The mere growing of cotton is but a part of the care of the planter; very much depends on classing and cleansing it for market, after it has been housed; sorting it before it goes to the Jennies, moating and removing any yellow particles, are essential to assure a preference at a common market of competition. *Domestic Encyclopædia*.

COW, "the female of the ox kind. The marks of a good cow, according to some, are these: The forehead broad, the eyes black, the horns large and clean, the neck long and straight, the belly large and deep, the thighs thick, the legs round with short joints, and the feet broad and thick." *Dic. of Arts*.

I may add here, if the cleaning of a cow after calving be delayed, it may be promoted by giving her a pail of warm water with some ashes in it.

Cows should be milked regularly, morning and evening, and always at the same hours, as nearly as may be. At six in the morning, and six at night, is a good general rule, as the times of milking will be equidistant from each other.—But if they are milked three times a day, as a modern writer on husbandry recommends, it may be done at five, one and eight. He believes that if they are full fed, they will give half as much again milk, if milked thrice, as if only twice. At the same time, it would prevent a too great distention of their bags, to which our best cows are liable. If the milking be once omitted, they will give much more at the next milking; but it will cause them to give less milk on the whole, and tend to dry them up.

No animals that we keep are more profitable than cows. Notwithstanding this is generally known, and admitted by farmers, they are too apt to act as if they believed the reverse to be true.—They often keep two bad cows, which will not give on an average, through the year, more milk or butter than one good one, while they will consume double; and they are easily tempted, by a small extra price, to sell the best cow in their herd, not reflecting, that this excellent cow must be as profitable to them as the purchaser ; and that by this course, often repeated, they gradually deteriorate, or make worse the quality of their stock, by the loss of all the calves which such a valuable cow might have brought ; and by being thus compelled to raise stock from inferior animals.

If cows are lean when calving, no management afterwards will ever bring them to yield, for that season, any thing like the quantity of milk they would have furnished, had they been kept all winter in high condition.

" Pure water is an essential article for cows. Dr. Anderson says, he knew a man who acquired great wealth, by attending to things of this nature, and one of his principal discoveries was the importance of having a continued supply of the purest water that could be obtained for his cows, and he would on no account permit a single animal to set a foot into it, nor allow it to be tainted even by the breath of animals.

" Inflamed teats should be washed with two drachms of sugar of lead in a quart of water. Should tumours appear apply a common warm marsh with bran with a little lard.

" To prevent cows from sucking their own milk, we are informed that rubbing the teats frequently with the most fœtid cheese that can be procured has proved an effectual remedy." *Domestic Encyclopœdia.*

COW HOUSE, that apartment in a barn, in which cows and other neat cattle, are tied up and fed, during the winter, and part of autumn and spring.

Farmers may think they need but little teaching concerning these apartments, as they have been so long acquainted with them. But I shall take the liberty to give them the following directions,which they may receive or reject, as they think fit.

In the first place, it is of no small importance that the floor under a cow house be very tight, so that none of the stale may be lost,which is of great value as a manure,when mixed with other substances. A farmer would be no more blameworthy for throwing away the dung than the urine of beasts, which contains abundance of fertilizing salts and oils. But if it be suffered to run through the floor, it is entirely lost.

The stale should be conveyed through the floor into the cellar ; or if there be no cellar under the barn, let it pass, in constructed channels through the sides to the dung-heap, or stercorary ; but by no means into earth or sods, laid on the floor, and against the sides, (as directed in a former edition of this work) as the frequent repairs, thus rendered necessary, would be nearly equivalent to the value of the urine so saved.

When a farmer thinks himself not well able to be at the expense of a floor of good planks, let him get a quantity of good clay, make mortar, and lay a bed of it a foot thick or more, for a floor ; giving it a proper descent backwards,that the cattle may lie dry, and raising it at the hinder border, to prevent the escape of the urine. The floor will grow thinner by being gradually shoveled up with the dung; but it is easy to repair it; and the faster it wears away, the more the quantity of manure is increased.

Also, the floors over a cow house should be more tight than they commonly are. It would prevent the descent of dust and chaff. It would cause the hay above it to be less injured by the air; as well as less contaminated by the breath of the cattle, and the steam of their excrements.

There are different ways of tying up cattle. Some prefer one way, and some another. I prefer stanchions to bows: Not only because the cattle take up less room in this way, but are less apt to waste their fodder. They are also more secure in this way; so that they do not so often break loose, and worry and wound each other.

A cow house should be in the southerly part of a barn, when it can well be so ordered. The cattle will be less pinched with the cold northerly winds. Another advantage is, that the heaps of dung thrown out on that side, being in a sunny place, will be thawed earlier in the spring, so as to be fit to be carted out in season. On the north side, ice will sometimes remain in the heaps, or under them, till the last of May, or beginning of June.

In this climate, cattle must be housed near half the year; from the middle of November to the last of April, and occasionally at other times. Though they must have fodder for two or three weeks before and after these periods, I think it advisable to let them lie in the barn yard, and eat out of a rack, unless the weather be stormy or the air uncommonly cold. For if they be kept too warm in the fall, they will become tender, and not winter so well; or if lodged too warm in the spring, they are more apt to be lousy.

CRADLE, a frame joined to a scythe, useful in harvesting, by the help of which, three times the quantity of grain may be cut down in a given time that can be with a sickle, and laid tolerably even and regular, for binding in bundles. It is oftener used for cutting oats and rye than for wheat. There is danger, however, of too much waste in cradling, when the corn is thick and heavy, or does not stand upright; the labour also would be too painful and tiresome.

CREAM, the fat part of milk which rises to the surface.

To produce the greatest quantity of cream, the pans in which the milk is set, should be flat shaped, so that the milk may not be more than three inches deep. Those in common use are not much amiss. They should be well scalded with hot water, as often as milk is set in them, and be thoroughly cooled. If the place where milk is set be too warm, it will be apt to turn sour, before the cream has had sufficient time to ascend; and none will rise after the milk begins to coagulate. If the place be too dry, the cream will become tough and hard before it is taken off. If the place be so cold as to freeze the milk, every one knows that but little cream will be gotten from it. The time of skimming must be regulated by the weather and other circumstances: But nearly the whole will rise to the top in twenty-four hours. In large dairies it may be troublesome to let it stand longer.

Milk may be advantageously set in pans placed on a floor, covered to a small depth with spring water, which is practised in Pennsylvania.

Some set their milk in trays lined with lead. This should never be done. For the least acidity in the milk will dissolve the lead, and poison the milk. Tin pans are good, being light and handy, and wooden trays answer very well, if kept sweet.

CUCUMBERS, *Cucumis*, a cold fruit, which is pleasant to the taste of most people, and much used by those who find themselves able to digest them. They are rendered wholesomer by pickling.

The method of growing them is simple and easy. They should not be planted till after Indian corn: For the least degree of frost entirely destroys them. The dung of swine should be put under them, which makes them grow more rapidly than any other manure which I have ever tried.

Some steep the seeds, and cause them to sprout, before they are planted: But I have never found any advantage in it. It is not amiss, however, to wet them a little, and coat them with powdered soot.

Mr. Miller thinks the seeds should not be sown till they are three or four years old. Four plants are enough to stand in a hole together; therefore, when they get into rough leaf, they should be thinned to this number. The vines should be so conducted as to interfere as little as possible with each other. They who wish to raise them at all seasons of the year, may consult the *Gardener's Dictionary*.

The following directions for stopping or topping cucumbers, melons, &c. are taken in substance, from The American Gardener's Calendar.

The young plants should be stopped or topped at the first joint, by pruning off the top of the first runner bud. This operation should be performed when the plants have two rough leaves, and when the second is about an inch broad, having the first runner bud rising at its base; the sooner this is detached, the sooner the plants acquire strength, and put out fruitful runners.

You will see arising in the centre of the plant, at the bottom of the second rough leaf, the end of the first runner, like a small bud; which bud or runner, being the advancing top of the plant, is now to be taken off close, and may be done either with the point of a pen-knife or small scissors, or pinched off carefully with the finger and thumb, being careful not to go so close as to wound the joint from whence it proceeds. In ten or twelve days, each plant will begin to send forth two or three runners; which runners will probably show fruit at its first, second, or third joints; for if the main or first runner was not to be stopped as above, it would perhaps run half a yard, or two feet in length, without putting out more runners, probably without shewing a single fruit. When the said lateral shoots have three joints, if any of them do not show fruit at either of these, it will be proper to pinch off the tops of such shoots at the third joint; which will promote their putting forth a supply of two or three new shoots, some or all of which will most likely be fruitful; and after this according as they advance in growth, train the runners along in regular order, cut out casual very weakly vines, and thin others where very irregularly crowded.

To preserve cucumbers, squashes, &c. from bugs and flies, sprinkle the plants with a strong infusion of elder leaves; that of hops is likewise recommended.—Or, suspend a diamond-formed piece of white paper, shingle, or other piece of wood by a thread tied to the end of a stick stuck in the ground a small distance from the hill, so that the paper will hang directly over the hill and near the plants. The air, by constantly vibrating the paper or shingle, will have a tendency to prevent insects from alighting on the plants. It is said sprinkling the plants with the fine dust of slacked lime, while they are wet with dew or rain will preserve them from insects.

Cucumbers are rendered more wholesome by slicing them into a basin of cool spring water.

CURRANT, *Ribes*, a species of fruit tree. There are three kinds of currants produced in this country, red, white and black. The red and the white are a wholesome cooling fruit, and flourish well in this part of the country. They are easily propagated by cuttings, setting the young twigs in the ground, which will furnish themselves with roots; and will bear fruit the second year. Some plant them singly, others in clumps. The latter method is disapproved by the best gardeners. If they be set single round the borders of a garden, close to the fence, and fastened to the fence, to prevent their being bowed to the ground when loaded with fruit, they will take up little or no room, and make an agreeable appearance. And it will be easy to keep them free from weeds. When they are planted on the south, or southeast side of a wall, the fruit will be ripe in June; but on the north side, they may be kept till October on the bushes, in a sound state.

A wine that is not unpleasant is made of the red kind; but that which is made of the white is preferable; and this ought to be more attended to. This wine meliorates exceedingly by age, becoming equal to the best of Malaga wine, after being bottled a year or two.

The way to make currant wine is as follows: Take ripe currants. wash them, clear them from the stems, add a gallon of water to a gallon of currants, and bruise them well in the water. Strain it through a cloth. Then to one gallon of the mixed juice and water, put two pounds and three quarters of good brown sugar. Stir it well. When the sugar is dissolved, put the wine into a cask not stopped. When the first fermentation is over, bung it up tightly, and in six months it will be fit for bottling.

CUSTOM, an habitual manner of doing any thing. Methods of agriculture, as well as methods of doing other things, are not seldom founded merely on custom. Farmers do many things, for which they can assign no other reason than custom. They usually give themselves little or no trouble in thinking, or in examining their methods of culture, which have been handed down from father to son, from time immemorial.

In some countries, this practice answers tolerably well. It does best in old countries, where methods, which have not been found to answer well, have been gradually laid aside in a long course of years. But this customary culture has a very pernicious effect, when ignorant farmers remove to a different climate. They naturally continue in the ways to which they have been accustomed. Their crops often prove to be unsuitable to the region they inhabit. They plant, sow and harvest, at the wrong seasons. They sow seeds in unsuitable soils. The consequences are, that their labour is misapplied, their time is lost, they grow poor and disheartened. Perhaps they remove to other places, hoping to mend their circumstances; and when they come thither, their habitual methods will answer still worse, rather than better, unless they go back to their first situation, or towards it.

CUTTINGS OR SLIPS, "in gardening, the branches or sprigs of trees, or plants, cut or slipped off, to set again, which is done in any moist fine earth. The best time for this operation, is from the middle of August to the middle of April; but when it is done, the sap ought not to be too much in the top; neither must it be very dry or scanty, for the sap in the branches assists it to strike roots. If done in the spring, let them not fail of having water in the summer. In providing them, such branches as have burs, knobs or joints, are to be cut off, &c. and the leaves are to be stripped off so far as they are placed in the earth, leaving no side branch. Small top sprigs, of two or three years growth, are the best for this operation." *Dict. of Arts.*

Cuttings of the grape vine, gooseberry, willow and currants, are easily made to strike root; those taken from the quince will commonly, and the apple tree will sometimes do so, if the earth be kept very moist. It is best to set them a good depth in the earth, not less than twelve or fifteen inches, or the greater part of their whole length. In this country, the best time that I have found to set them is in April. It should be done as soon as the frost is quite out of the ground.

CION, OR SCION, a young sprig or sprout of a tree. Scions, for grafting, should always be taken from the most thrifty trees, not from those especially which are old and decaying. The time to cut them is in February or March, just before the buds begin to swell, and appear fresh, which will be earlier or later, according to the season and climate. They should be taken from the ends of limbs of the former year's growth, not from young suckers of an over quick growth; and kept moist in a cellar, with the lower ends inserted in moist clay, or mud, till the proper time for grafting. In such a position they will keep well for two months or more, before grafting.

D.

DAIRY. Butter is oftener well managed than cheese. But there are few who salt early-made butter so that it will keep good and sweet. An ounce and a half, or more, of the strongest and best salt, very finely powdered, should be worked into a pound of butter, and so thoroughly mixed that every part may be equally salt. For if ever so small a part misses of being salted, it will turn rank, and communicate its ill taste to the remainder. It should then be put into tubs that are quite sweet, and so closely packed and crowded, that no air can be in contact with the butter; which should be carefully covered with a piece of fine cloth, after dipping it in melted sweet butter. When more is to be put into the tub, take up the cloth; and after that is well crowded in, and levelled, put on the cloth again so nicely as to shut out, if possible, every particle of air. The same should be done as often as any is taken out for use. The tubs, during summer and fall, should stand on the bottom of the coldest part of the cellar. When there is occasion to carry butter to any distance for sale, in hot weather, let not the tubs, or boxes, be exposed to the heat of a traveling horse, by lying against his sides. For by this practice it is known that a great deal of butter is greatly damaged.

"The Rev. Mr. Packard of Marlborough, made an answer to a question, which we wish was pasted up in every dairy-room in the state:—The last year, said he (1799,) three cows in this town produced 278 pounds of butter. If their calves had been taken from them at a week old, they would have made 451 pounds of butter. These three cows were a more productive dairy than *six usually are*, with ordinary feed. Farmers egregiously mistake when they overstock their farms. Were dairies estimated by the *pails of milk* they produce, instead of the *number of cows*, many farmers wives, instead of asking their husbands to buy *another* cow, would urge them to *sell two* to enrich their dairies." Mass. Agr. Rep. vol. v. p. 74. See *Butter and Cheese.*

DAIRY, or DAIRY-ROOM, a house or apartment where milk is kept, &c.

Our farmers and their wives seem to think it necessary, or highly convenient, to have a dairy-room annexed to their dwelling house, partly above and partly below ground, that they may dry their cheeses in the upper part, and set milk and cream in the lower. This, in wooden houses, is certainly not the best practice, and occasions much loss. For such an apartment will be too hot in summer, and too cold in winter, to keep milk in it; neither will it be possible to keep it so sweet as it ought to be kept.

An apartment in a very sweet, and perfectly ventilated cellar is better on every account to keep milk in. As to drying of cheeses, they should never be kept to dry in the same room where milk is set; for they will undoubtedly communicate an acidity to the surrounding air, which will tend to turn all the milk sour that stands within the same enclosure. And a drier room would be better for the cheeses; only let it be kept dark, that the flies may not come at them. So that, instead of a place called a dairy, there should be a milk room, and a cheese room, in a farm house.

A room in a cellar may be kept so nearly of an equal coolness, by means of burning a few coals in it; when the weather is cold, that the milk will neither grow sour in summer, nor freeze in winter : So that nothing will obstruct the rising of all the cream. It is supposed that the warmth of the air in a milk room ought to be from 50 to 55 degrees on Farenheit's thermometer. But a few degrees over or under will produce no very disagreeable effects. The cellar should have such windows as will afford a sufficient quantity of light, and be on the most northern side ; and they should be opened now and then to let in fresh air; particularly in the coolest of the mornings in summer. The room should be ceiled with plaister, to prevent the descent of dirt ; and the top and sides white washed, to increase the light, and fill up chinks that harbor insects. Every part should be kept extremely clean and sweet, and nothing should enter into it, which can corrupt the air. The floor should be made of stones, bricks or tiles, and be frequently washed in summer with the coldest water, to cool and sweeten the air in the room ; and milk should not be suffered to stand in it till it becomes sour, lest the sourness be communicated to that which is sweet. For the same reason, cream which is put by for churning, ought not to be kept in that apartment which contains the milk. Because acidity in cream is expected, and necessary before butter will come.

Those who have large dairies, in hot climates, having a spring or brook near the dwelling-house, might find it worth while to build a milk room over it, with a stone floor, and a channel in the floor to pass the water all round, near the insides of the walls. The pans may be set in the channels, and water let in at pleasure, to cool the milk in the hottest season. An arch of brick should be turned over the building. The windows, to let in light and air, should be on the northerly side, or end. To shelter the arch from the weather, a story of wood may be erected over it, for a cheese room. The arch will be the cooler in summer and warmer in winter, as well as more durable, as it will be defended from rain, &c.

If rats and mice cannot enter the milk room, there will be no need of having shelves in it. The floor is the best place to set the vessels of milk on, it being coolest in summer, and perhaps warmest when the weather is frosty. See *Butter and Cheese*.

DARNEL, *Lolium*, a troublesome weed, which sometimes appears among grain, and is often so fruitful as to spoil a crop. The seeds of it resemble corns of blasted rye, but are more light and chaffy. These weeds should be pulled up before they go to seed. But grain for sowing may be mostly cleared of the seeds by swimming it in water.

DIBBLE. A tool, used by gardeners and farmers in Europe, chiefly in transplanting. It consists of a stick about a foot or eighteen inches long, slightly sharpened at the end, with which a hole is made to receive the plant or seed. Gardeners prefer an old spade or shovel handle that has been broken, retaining the upper part of the handle, and sharpening the lower end. This is very handy, and with care will last many years.

DITCH, a narrow channel, of great use in agriculture. Ditches serve two purposes, to enclose grounds and to carry off superfluous water. When used for fences, a hedge should be planted on their sides, otherwise they will not keep out cattle or sheep. The earth taken out of ditches should not be suffered to lie in heaps on the borders to the great loss of time in mowing, and the crop. It is too precious a material for compost heaps, to be suffered to remain in a state worse than useless.

A ditch should be three times wider at the top than at the bottom, to prevent the falling in of the sides.

DITCHING, the making of ditches. This work is most commonly performed in summer, or early in autumn. When this work is to be done in very low and wet land, a hot and dry season is best ; that the water may not prove troublesome, neither by its quantity, nor by its coldness. When it is to be performed in a salt marsh, not only a dry and warm time should be chosen for the business, but it should be done also at a time when the tides are lowest. On high lands, ditches may be made at almost any season, when the ground is not frozen. But in the spring the digging will be the easiest, the ground being softened by the preceding frosts. But as the ground is driest in autumn, then is the best time for ditching in most of our low lands. At least, the month of September is a good season. But farmers must be governed, as to the time, in some measure, by their own conveniency. It must be done when no other business of greater importance demands the whole of their attention and exertion.

When bushy ground, full of strong roots, is to be ditched, the Rev. Mr. Eliot wisely recommends beginning the ditch in the winter, when the ground is frozen two or three inches deep. The surface may be chopped into pieces by a broad axe with a long helve, and the sods pulled out with an instrument made like a dung croom.—The farmer may probably hit upon a good time for this work in December, when there happens to be no snow, and when it will not interfere with other farming business. The lower part of the ditch may be done in the following summer, or autumn. In a free and firm soil, a ditch may be begun with a plough, drawn by an orderly team that will keep to the line. This saves labour.

To make a ditch straight, and equal in all its parts, it is recommended that the work be regulated by a frame of slit deal, nailed together, to the exact size of the intended ditch. It may be a rod or more in length, and as wide as the intended ditch.

DIVISIONS, *of a Farm*, lots enclosed for the convenience of tillage, pasturing, mowing, &c.

The judicious dividing of a farm into lots, may save much labour, especially much travelling from one part to another. The more square lots are made, the more is saved in fencing. Crooked fences should if possible be avoided, not only to save expense, but to add beauty to a field, or plantation.—All tillage lots, and especially small ones, should be nearly of equal dimensions on all sides; for if a lot be out of square, the labour of ploughing will be increased, as there must be a number of short furrows. If a lot be long and narrow, cross ploughing will be either prevented, or the labour of it much increased.

When it can conveniently be so ordered, the lots designed chiefly for tillage should be nearest to the house and barn, to save labour in carting manure, and to prevent loss in getting in the crops. The nearer grain is, the less it will shatter out in carting. The mowing lots should be next to the tillage, if the soil permits; as these must be dunged, and their crops carted: The lots for pasturage should be contrived to be next, and the wood lots farthest of all the lots from the house, that so the view of the other lots may not be obstructed too much by trees.

DOOR DUNG, a manure taken from the back yards and doors of dwelling houses.

Though it may seem to be made up of chips, saw dust, and several other matters that appear unpromising, yet there are various substances intermixed with them, and soaked into them, which contain food for plants in abundance. A large proportion of the dung of swine and of fowls, which are excellent manures, are contained in the compost. It has, besides, the sweepings of the house, blood, small bones, shells, and other animal substances; also suds, ashes, soot, urine, together with salt particles, which are some of the best of manures.

It is, therefore, no wonder if this filth is found to be very conducive to the growth of plants, as it really is.

Some think it best to let it lie year after year in the yard, that it may grow fine and mellow. But it is wasted by this practice, the sun, air, and rain, depriving it of its volatile, fine, and most fructifying particles. I choose to be rid of its putrid steams, and place it where it may do good and not hurt. I, therefore, have it scraped up clean every spring, clearing it of the largest and brightest chips; and after it has lain in a heap for a few days to ferment, apply it to the soil in the field, though it be not fine enough for the garden; or else add it to the compost dunghill. I find it to be a very proper manure for land that is stiff and clayey; and it will do great service in any soil.

Those farmers are certainly guilty of bad husbandry, who take no care to avail themselves of this excellent manure, of which they all have more or less; and that is commonly best, where the greatest number of swine are permitted to run, which, however, is a slovenly and wasteful practice.

DRAIN, a channel made in the soil to carry off superfluous water, or divert its course.

Drains are of the highest importance in agriculture: For, by means of them, lands that are so wet and fenny as to be entirely useless, may oftentimes become by far the most valuable part of a farm. It would be happy for this country, if the husbandmen were fully convinced of the vast utility of them. The real value of some estates might be doubled, by a small expense in draining.

Drains used in farming are of two kinds, open, and hollow, or covered. The open drains are mostly used, because more easily made. But if the first cost be less, the expense, in the long run, may not be less, but greater than that of covered drains. For they will be continually filling up; and, therefore, will often need to be mended.

Open drains are to be shaped like other ditches, wider at the surface than at the bottom. And, for a general rule, they should be carried through the lowest and wettest parts of the soil, though it should cause them to be crooked and unsightly. The water will be carried off more effectually; and some labour in digging will be saved; for if they pass through the higher parts, the ditch must be deeper, at least in some places.—But where a plain is incommoded with too much water descending from an adjacent height, the water must be cut off by an open drain drawn along at the foot of the high ground, and the earth which is taken out should be laid on the side towards the plain.

Open drains serve well enough in swamps, if the soil be not too loose, so as to fill them up soon. In this case the covered drains are certainly best, especially where materials for making them are easily obtained.

The earth that is thrown out of open drains in swamps should not lie in banks by the side of them. This will tend to prevent the water from passing freely into them, and conduce to their filling up the sooner. It should be spread over the surface of the drained land, which will make it drier, and sometimes answer as a good manure. And, in this operation, there will be often a mixing of soils, attended with considerable advantage.

To judge rightly, whether it will be worth while to attempt the draining of a swamp, it is first to be considered what will be the cost of digging at the outlet, where it will, in some cases, be necessary to go very deep. If large rocks should be found in the way, they may be blown to pieces with gun-powder. But doing this, is somewhat expensive. Also, the depth of the black soil in the swamp must be examined, and the stratum next under it. If the under stratum be clay, the swamp may be well worth draining, though no more than six inches of till or mud be above it; for the mud and clay mixed, will make an excellent soil. But if the under stratum be gravel, or white sand, it will not be best to undertake draining, unless the depth of black mud be as much as from fifteen to eighteen inches. For it is to be remembered that the soil will settle after draining, and be not so deep as it was before. If, after draining and hardening, there should be a sufficient depth for tillage, the soil will be most excellent; and will pay well for an expensive draining.

The manner of draining a swamp is as follows: Beginning at the outlet, pass a large ditch through it, so as mostly to cut the lowest parts. Then make another ditch quite round it, near to the border, to cut off the springs which come from the upland, and to receive the water that runs down from the hills upon the surface, in great rains. These ditches are to be larger or smaller in some proportion to the bigness of the swamp, having a regular descent for the water, that not much of it may stand in them. If the swamp be large, it may be necessary that some smaller cross drains should be cut in several of the lowest parts. The bottom of the main ditches, when the soil is not of an extraordinary depth, must be lower than the bottom of the loose soil; otherwise the soil will never become sufficiently dry and firm. When the swamp comes to be sufficiently dry for tillage, such of the drains may be converted into hollow ones, as cannot profitably be kept open for fences. Thus the quantity of improveable land will be increased.

If a bridge over any of the drains should be wanted, the best way to make one will be by filling up a short piece of the drain with stones, or wood, that is, by making it hollow in that part. This will be less expensive than a common bridge, and answer the purpose better.

Those who are willing to be convinced of the amazing fruitfulness of drained swamps, should read Mr. Eliot on the subject. He represents them as producing turnips, clover, oats, &c. to great advantage; English hay, four tons per acre, and Indian corn at the rate of more than ninety bushels per acre, without manuring.

Such lands are highly advantageous, as they require no dung; and cannot be easily, if at all, worn out by cropping: Also, as they bear drought remarkably well. As this country very often has its crops greatly diminished by dry seasons, it would be well if every farmer had continually some of this kind of soil in tillage, or mowing, or in both.

Covered or hollow drains are more used for the drying of springy, wet and spungy uplands. They may be used with advantage on gentle declivities, where the soil appears spewy and cold, by means of springs. They will cause the soil above and below them, to be more dry and fruitful. But if the descent be very steep, or if the wetness of declivities be owing only to water running down on the surface, the open drains are to be preferred : For if they were covered, the water would pass over them, and the drain would be of little advantage.

To make a hollow drain, dig a channel between thirty and thirty-six inches wide at the top, and six inches, or the breadth of a spade, at the bottom, and three feet deep, giving it just descent enough to make the water run briskly. Fill it half full, or more, with small stones, thrown in at random, and cover them with a layer of straw, leaves, or the small branches of trees with the leaves on them ; then fill it up to a level with the surface, with the earth that was thrown out. Such a drain, as it will not choke or fill up, will never need repairing. If the descent should be but just so much as to make the water run slowly, there may be some danger of its choking up, and ceasing to run at all. But this danger will be greater or less according to the difference of soils. There will be no danger of it, in a soil that does not easily dissolve in water.

If stones be scarce, long faggots, or fascines, laid in the trench, will answer as well, so long as they last ; which being secluded from the air, will not rot soon. Some say they have known them to answer well for forty years, but this must only be in places where they are always kept wet. In situations exposed to wet and dryness, they perish in five or six years.

If a plain piece of ground be too wet to be made fit for tillage by ridge ploughing, it should be made drier by hollow drains. If no lower place be adjoining, where the drains may have an outlet, holes should be dug in some of the lowest parts of the plain, to examine what strata are under the soil. It is likely that a stratum of clay, or of some other earth not easily penetrated by water, is the real cause of the wetness of the soil. If you find it so, then dig through the stratum, and below it, till you come to loose gravel, sand, or something that will easily imbibe water : Fill up the hole with stones, and direct your hollow drains to it. It will serve for a perpetual outlet ; and conduce much to the drying of the soil.

The peculiar advantages of hollow drains are, that they will not need repairing, as they do not fill up ; that no soil is wasted, or rendered useless by them ; that a plough may pass over them to as great a depth as is necessary in any kind of tillage ; and carts and other carriages are not obstructed or incommoded by them. So that these drains may pass across roads without detriment, when the descent requires it. It is often necessary to hollow drain roads to lay them dry, and found to be of great advantage.

The draining of a marish, or shaking meadow, which seems to be a soil floating on the water, is sometimes practicable. I shall give the reader the method of doing it in the words of the ingenious Mr. Dickson.—" To drain a marish," says he, " it is necessary, in the first place, to convey away all the stagnating water: And this water can be conveyed away in no other manner, than by a large open drain, with a sufficient fall. This fall must be such, as to carry off the water from the bottom of the marish ; otherwise little advantage is to be expected from it. By conveying away all the stagnating water, some land on each side will be gained : For the water being removed, the earth by degrees will subside, and become firm and solid. By this, likewise, the bottom will become firm ; which will allow the drain, by degrees, to be carried forward through the middle of the marish. If the springs, by which the marish is supplied, arise near the middle, this principal drain, with some branches cut from each side, where the springs are largest, or most numerous, will be sufficient. But if there are springs in all places, as is frequently the case, it will be necessary to make drains at the sides, as nearly parallel to the principal drain, as the situation of the marish will allow, to intercept the water that comes from the heights, and supplies the springs. It will be necessary, likewise, to make communications, by cross drains, betwixt the parallel drains at the sides, and the principal drain in the middle." It is no wonder if the cost of draining a shaking meadow should be considerable, as it seems like a soil floating upon water.—But there is no reason to doubt its becoming some of the best soil, when so drained as to give firmness to it.

DRAY, or *car*, a slight kind of carriage drawn by one horse. It consists of a pair of thills, connected by two or three cross bars. The hinder ends of the thills slide along on the ground. It draws heavily on bare roads, but on grass land much more easily. The horse must carry much of the load on his back. In case of necessity, it is better than no carriage.

DRESSING, the application of dung, or other manures, to soils, to increase their fruitfulness. Dressing differs from manuring in general, only as it is chiefly intended for the increasing of one single crop. Not only are dressings necessary for poor and weak soils ; but they are profitably applied to those which are rich and strong ; especially when seeds are sown which need much nourishment, or will make good return for it.

There are four things chiefly to be regarded in dressing ; the suitableness of the dressing to the soil, and to the crop ; and the manner and the season of applying it.

To light, warm, or sandy soils,

the coldest manure should be applied; such as the dung of hogs, cows, oxen, &c. Dung that is much mixed with straw does best in such a soil, as the straw soon rots and becomes food for plants. Cold and stiff soils should be dressed with the hottest and driest manures, as the dung of horses, sheep and fowls. Wet soils should have manures that have the greatest power of absorbing moisture.—Lime, where it is cheap and plenty, may be used with great advantage; ashes, coals, and saw dust, are also very proper.

Some kinds of dressing should be well mixed with the soil, by the plough and harrow; especially such as are apt to lose their strength, by being exposed to the air. Of this sort are dungs in general, and some other manures.—Dung is to be ploughed in with a light furrow. Composts, which consist of dung, earth, and other substances,need only to be harrowed. If dressings are laid too deep, as under deep furrows, they will be in a manner lost; the roots of most kinds of annual plants will scarcely reach them; and, before the next ploughing, the strength of them will be sunk still deeper into the earth.

There are other manures which should be used only as top dressings. Their exposure to the air takes away little or none of their virtue, being of an alkaline nature, such as ashes, lime, and the like. They are speedily settled into the soil by rains, and melting snows; and afford a more kindly nourishment to the roots of grass and grain, than if they were buried in the soil. Being laid lower than the surface, their strength would be more apt to be carried lower than the roots of plants commonly reach.

Some dressings are thought to be more successfully applied some time before sowing. Such a one lime is said to be, as being apt to burn, or too much heat the seed. But this, I think, can be only when it is laid on unslacked, and in large quantities.

Other dressings answer best at the time of sowing. This is the case as to most kinds of dung that are used, and of several other manures.

But those manures which exert all their strength suddenly, are allowed to be best used only as top dressings, after the plants are up, such as soot, ashes, certain warm composts, and malt dust. If they are laid on winter grain in autumn, there will be danger of their causing too rapid a growth: In consequence of which, the grain will be afterwards stinted, and languish, unless another and larger dressing be given it in the following spring, or summer. It is probably best to apply these dressings just before the time when the plants will need the greatest supply of vegetable nourishment, which is when their growth is most rapid, or near the time when the ears are shooting out.

The adapting of dressings to the nature of plants will be found, in those parts of this work, where the most useful plants are treated of.

DRILL HUSBANDRY. Sir John Sinclair has the following remarks on this subject. "In comparing the drill and broad-cast systems, the following maxims may be laid down : 1. That it is of little or no advantage to drill autumn,or winter-sown wheat, as the crop is rarely injured by annual weeds, and they can be easily and completely extirpated by hand-hoeing and weeding ; and when the soil is much infested by root-weeds, as the crop, in ordinary cases, will completely meet in the rows, before the root-weeds make much appearance, it is impossible, without injuring the crop, to render much benefit to the land by the hoe. 2. On all lands where annual weeds are abundant, and where a number of hands can be procured for using the hoe, white crops, sown in the spring, may be drilled, for the advantage of clearing the land more effectually. By weeding broad-cast crops carefully, an addition of four and an half bushels of wheat, fifteen bushels of barley, and ten of oats has been obtained. The adoption of that practice. therefore, would probably equal the average advantage from drilling ; so that the difference between broad-cast crops, properly weeded, and the drilled, may not be material. 3. That though the hoeing of the crops, and the stirring of the soil, has no injurious effects in light, poor, or even strong soils, like those of Norfolk and Suffolk, where drilling has succeeded ; yet where the soil is fertile by nature, or enriched by art, drilling forces up the straw, and consequently diminishes the quantity and quality of the grain in so great a degree,as to render the crop unproductive. Hence, in rich soils, drilling has been so frequently given up, and hence it is that Tull probably considered dung as prejudicial.

"There doubtless must be some radical defect in the system of drilling, when after a full trial of its merits, by a number of intelligent farmers, in several districts, but more especially in the counties of Hertford,and Lincoln,it has been given up,with hardly an exception, in so far as respects crops of grain. These facts confirm the almost general result throughout the kingdom, that drilling will be found to answer to a certain extent, and with a certain degree of skill and attention ; but when the scale is extended, the attention is apt to flag, and the conclusions drawn from one or two fields are not found applicable, unless where the operation is very skillfully conducted, to a whole farm." See *New Husbandry*. Likewise Massachusetts Agricultural Repository, vol. IV. p. 217, to p. 254, in which opinions in some respects different from the above are maintained.

For a description of the different kinds of drills, see *Willich's Domestic Encyclopædia*, 2d American edition, vol. II.

DROUGHT, such a continuance of dry weather, that plants cannot draw a sufficiency of nourishment from the earth, to give them their full growth and perfection.

Some countries are much more liable to this inconvenience than others. New-England,for instance, is oftener troubled with it thanGreat

Britain; one occasion of which is, the greater heat of our summers, by which lands grow dry faster here than there. Another cause may be our having a greater quantity of fair weather. And our being more liable to drought, makes it necessary that our methods of culture should be different from those practised in that country. Heating manures are generally more needful there than here; and ridge ploughing is a more proper kind of tillage for the English than for us—though it might be of great service in many of our fields. I have found considerable advantage from it in land that is flat and wet.

To plough our driest lands, in ridges, would undoubtedly be lost labour, unless for certain particular crops, as it would cause a drought to be more hurtful to the crops, and there is no danger of too much wetness. And yet it may be, that when an over dryness of soil on the side of a hill, is owing to the rain's running off before it has time to soak into the soil, ploughing the land into ridges, and making the gutters nearly parallel with the horizon, may cause the soil to retain moisture the better.

It is in the power of the farmer in good measure to guard against the ill effects of drought. It is a matter that certainly ought to be attended to in this country, in which almost half of our summers are complained of by many, as being very dry. The best method is, to have more of our lowest lands under the best improvement in tillage. If this were the case, we should not so often hear of a scarcity caused by drought. If it were become customary to plant and sow on drained lands, and in those which are so low and wet as to need laying in ridges, possibly our dry summers would be as fruitful on the whole as our wet ones. But, as we manage our lands at present, the case is far otherwise. A great number of people are always reduced to a distressed condition by a dry summer. And they are too ready to consider the shortness of their crops in a dry year as a divine judgment, though they might have prevented it by a more prudent management.

Another way to guard against having our crops pinched by drought is, to have a variety of different crops on a farm each year, some that are least injured by a drought, and some that require the most rain. Thus, let a season happen as it will, we may hope to gain in one crop, what we lose in another; or at least that some of our crops will be very good, if others should fail.

Sometimes land is so shaped by nature, that the water of a rivulet, or of a plentiful spring, may be led by gutters, or narrow channels, to moisten places which would otherwise suffer by drought. When it can be performed without too much expense, it will be found to be an excellent piece of husbandry. In some cases it may be a double advantage, making a wet place drier, by diverting the water to places that need it. Deep tillage is also of very great importance to prevent the ill effects of a dry season. For the dryness of three or four inches in depth would scarcely alter the condition of the plants. But if the plough has gone only to this depth, a severe drought will be fatal to the crop.

It would greatly advantage the farmer, if he could foretel whether a season will be dry or wet. But as he knows this is impossible, he should so conduct his crops, and other matters, that he may be prepared for either extreme.

The earlier a drought begins, the more distressing it generally proves in this country. For, after the grass crops and English grain have nearly got their growth, a drought is less detrimental than before, because the Indian corn, by means of the tillage given while the plants are growing, bears it so well as to be seldom cut short by it: And Indian corn is the principal of our late crops.

Pasture lands are often so dried up, that both the meat and drink of the cattle are cut off at once.—This shews the propriety and necessity of having some low lands in pasture, when it is practicable.—And a few trees, growing at proper distances in a pasture, will partially shade the soil, and prevent its drying so rapidly. More grass will be produced; and the cattle will be refreshed by the shade; besides the advantage the farmer will gain in fuel and timber. In those parts of the country where trees have become scarce, the omission of planting quick-growing trees in our pasture grounds is unpardonable. When a pasture is destitute of water, Mr. Eliot advises to dig a well on the side of some hill in the pasture, and having come to water, to dig a trench below, level with the bottom of the well, and bring the water through a hollow drain out to the surface, where it may be kept in a little basin, made in the soil, for a watering place. This practice is now common in some parts of our country, and even the conveyance of water in aqueducts.

DUNG, the excrements of animals, used to increase the fertility of land. Dung may be said to be almost of the same importance to the farmer, as stock in trade is to the merchant. There are but few lots or pieces of lots, in this country, which can be tilled to any great profit, in the common way of culture, without manure; and dung is of all manures the most useful. The very best of soils, when dunged, will more than pay for it, by the increase of their crops, and the poorest will produce next to nothing without manure. Some think it more profitable to apply dung to their best soils than to their poorest, as they think the increase from it to be greater in the former case than in the latter. This opinion is probably founded in truth.

The sorts of dung which are, or may be used, are that of black cattle, sheep, horses, swine, goats, hens, pigeons, ducks, geese and rabbits, besides human ordure.

The dung of animals consists of oils, fixed and volatile salts, together with nitrous and earthy particles. But in different sorts of dung these principles are differently compounded; so that the dung of one animal is a proper manure for one kind of soil, and that of another for another. And yet there is no kind of soil that may not be enriched, in some degree, by any kind of dung.

The dung of oxen and cows is a cool, mild and oily substance; and is, therefore, most suitable for warm, sandy, and gravelly soils. It tends to prevent the soil's becoming too dry, and keeps the plants on it from being pinched for want of moisture.

The dung of sheep is more hot and fiery than that of black cattle; it ferments quicker; it is fitter, therefore, for cold, heavy lands.—Perhaps the best way of applying the dung of sheep to land is by folding, in countries especially which are not greatly infested by wolves. For in this method their urine is all saved, as well as their dung. But it ought to be turned in with the plough as soon as possible, that the sun and air may not deprive the land of it.

In Flanders, it is the practice to house their sheep at night, under slight sheds, the ground being spread with dry sand, about four or five inches thick, laying on a little more fresh every night.—This is cleared out once a week, and carried to a dung-hill, or applied to the soil. This mixture of sand and hot dung, makes a very excellent dressing for cold and stiff land. For there is scarcely a richer manure than the dung and urine of sheep. M. Quintinie thinks it the greatest promoter of fruitfulness, in all sorts of ground. This method of folding sheep in a covered fold, and of mixing their dung with stiff earth or sand, according to the nature of the soil it is intended for, is, also, with much reason, recommended by Mr. Mortimer; who also says, "that he has known vast crops of rye upon barren lands, that have been old warrens, well dunged by rabbits. and large oak and ash trees upon the same, though the soil was very shallow."

Some have recommended the reducing of sheep-dung to powder, by pounding it with mallets, and using it as a top dressing for grain, perhaps half a dozen bushels on an acre. But this is a tedious piece of work, and of no lasting advantage: Whereas too much can hardly be said in praise of the Flanders method of using it. A prodigious quantity of good manure may thus be obtained from a flock of sheep, by housing them regularly every night.

If a light soil is intended to be manured with this compost; instead of sand, clay, pond mud, or the mud of flats may be used, these substances having been first mellowed by the frosts of winter. The dung of goats is supposed to be nearly of the nature of sheep's dung.

Horse dung is a still hotter manure, as appears by its quick fermentation in heaps, even in cool weather. It is consequently fittest for hot beds, when it is new, and for nourishing those plants which require the greatest degrees of heat. The dung of horses that are fed on grain, is a richer manure than that of those fed only on grass and hay.

Great care should be taken that

horse dung be not spoiled, by being overheated, or burnt in the heaps, before it is used. For, in this country, it is very commonly the case. When it has been so heated as to give it a white and mouldy appearance, the virtue of it is gone. It is difficult to give it age, without mixing it with other substances. A mixture of horse and cow dung is very proper for land that is neither too light nor too stiff. Horse dung is a much stronger manure than it is supposed to be by those whose constant practice is to suffer it to be spoilt by overheating in the heaps. This manure, when used as an ingredient in composts, has an excellent effect, as by its quick and strong fermentation, it speedily dissolves other substances that are mixed with it.

Mr. Miller says he has frequently seen new horse dung buried as it came from the stable in very cold, moist land; and always observed that the crops have succeeded better than where the ground was dressed with very rotten dung.

The dung of swine is a very rich and fat manure, and so cool as to ferment very slowly. It is so rich and oily, as to be double in value to neats' dung. It will render the moist dry and hungry soils exceedingly fruitful in a wettish season, as I have found by experience. It resists the ill effects of drought, and does most service in a hot country. By its steady and gradual supply of a rich nourishment, it is peculiarly adapted for the growing of hops, pumpions, running beans, and every plant which has long vines. Nothing can equal it for the growing of potatoes. It has produced me more than a peck in a hill on the poorest hungry sands. Or rather I might say, straw only a little impregnated with the dung of hogs has done it. This is so strong a manure, that it answers well, when mixed with a large proportion of earth, weeds, straw, or other such substances. It is almost incredible how great a quantity of good manure may be obtained, by supplying a hogsty with rubbish to mix with the dung. I have heard of 40 loads of manure being made in a year by means of one hogsty. And I have no doubt of its being practicable.

The dung of ducks and geese, is deemed too hot and burning. But if the farmer would gather it in a heap, and mix it with the dung of cattle, he would bring it to a temperate heat, and draw from it such advantage as would indemnify him for the pains he should take. The virtue of this method is known by experience. A farmer having abandoned a piece of ground to his geese for twelve years, afterwards turned them out to let the grass grow, and it rose so thick and strong that the scythe would scarcely pass through it. Hen dung is recommended to be scattered in small quantities upon land intended to be sown, and on account of its heat it is never used, unless when rain is foreseen. It is an excellent manure for meadows. Pigeon's dung is much the same with that of poultry, the only difference being its superiour heat." *Scots Farmer.*

I should think it better to mix the dung of poultry and pigeons with other substances, to allay their heat, before they are applied to the soil. And thus qualified, they would be an excellent top dressing for corn, especially in cold and wet lands. On old mowing grounds, I have found the grass abundantly increased, by a sprinkling of earth taken out of an apartment used as a hen house, though there was little or none of their dung visible amongst it.

"Human ordure is a very fat and hot manure, full of fertilizing salts and oils; and, therefore, extremely proper for all cold, sour soils; especially if it be mixed with other dung, straw, or earth, to give it a fermentation, and render it convenient for carriage. Some do not like the use of it, on account of its bad smell; and others imagine, that it gives a fetid taste to plants. But in this they seem to carry their delicacy too far. Mr. Bradley says, it is kept in pits made on purpose, in foreign countries, till it be one, two, three or four years old: That of four years old is accounted the best, that of three years tolerable. Perhaps it may owe great part of its richness to the urine with which it is mixed; for though the human urine be destructive to vegetables, whilst it is new, by reason of its burning sal ammoniacal spirit, as Glauber terms it, yet time will digest the urine, and render it an extraordinary fertilizer of every kind of soil." *Complete Farmer.*

This kind of manure should be compounded with a large quantity of earth, and lie one or two summers at least, that it may be thoroughly mixed. The contents of an old vault would thus make a surprising quantity of excellent manure.

As dung in general is so important a manure, every possible method should be taken to prevent its being wasted, as indeed a great proportion of it is, by the common management of our farmers. In no way is it more wasted, than by its being too much exposed to the sun, air, and rains. Mixing of dry earth, or other absorbent substances, with heaps of dung, will do much towards preventing this loss. Or slight sheds may be made over them to prevent their strength being too much wasted by heavy rains; and at the same time, to prevent a too great exhalation from them. Some cover them with turfs, when they choose to keep dung till it be old. This is not a bad practice; for the turfs in that situation will become good manure. I would hope farmers need not be told, that the grassy side should be laid on the dung. Otherwise, instead of consuming, it will produce a crop of grass.

If the heaps lie at the sides of barns or under the eaves, the least that should be done to prevent the wasting of the dung, is to put up gutters, that the heaps may not be washed with the streams from the eaves. And besides, some loose boards should be set against the sides of the barn, in such a manner as to prevent the greatest part of the rain from falling on the heaps of dung.

"The quality of the dung of every animal, will in a great measure be proportioned to the richness or poverty of its food. Hence the dung from the stables and cow-houses is preferred to that of young cattle kept in the farm yard, on food of a less nourishing quality. Next to animal excrement, straw is the chief material of a dung hill; and too much attention cannot be paid to collecting it. For that purpose, the greatest care is taken by the diligent husbandman to cut his crops low. When reaping is carelessly executed, it has been calculated, that one fourth part of the straw is left upon the ground, where its strength and substance is wasted by the rains and storms of autumn and winter; whereas by attention to the reaping process, one ton, and even more of additional manure *per acre*, can be obtained. The value of straw is great, not only in consequence of its own substance, but from the quantity of liquid matter it absorbs. By an experiment, carefully made, it appeared that dry wheat straw had increased in quantity from 300 to 719 stone, making an increase of not less than 419 stone during a period of seven months.

"As in heavy rains, notwithstanding every precaution, some water will run from the yard, there ought to be a covered reservoir to receive it, so situated that the liquid can either be pumped upon the dunghill if it requires it, or upon heaps of earth, collected for that purpose. The stables and cow-houses ought also to be regularly washed, as is done in Flanders; and much useful matter might in this way be collected, and conveyed to the reservoir. Where land lies conveniently beneath the farm yard, the contents of this reservoir, or the overflowings of the dung-yard, may be conveyed to it for the purposes of irrigation." *Code of Agriculture.*

There are different opinions relative to the preparation which dung should undergo before it is applied to the soil. It is remarked by Sir Humphrey Davy, that straw fermented is a more manageable substance, *and furnishes more manure for a single* crop, than unfermented. In the latter state it decomposes more slowly, and consequently, though its influence will be more lasting, yet it produces at first less effect.

It is observed by Sir John Sinclair, that "the old practice of *frequently* stirring, turning, and mixing the dung, without in general even covering the heap with earth, when it was turned, is now very properly laid aside by every judicious farmer, as the finer and more volatile parts of the soluble matter, escaped in a gaseous form, into the atmosphere, and left behind only the earthy and coarser parts of the mass."

"In the Netherlands," says the same writer, "the more opulent pave and line with brick the receptacles for their dung, which is thus kept constantly plunged in a mass of liquid matter. The fibrous parts of the vegetables, are thus completely decomposed, and four tons of this manure, go as far as five, collected and kept with less precaution."

"In Switzerland the farmers consider liquid manure as the most efficacious, and after the dung is fermented, they dilute it in water, and the liquid alone is carried to the field, and scattered over it. The straw that remains, after the dung is thus washed, is applied as manure for potatoes.

The Hon. Richard Peters, of Penn. is opposed to the use of unfermented manures. In a letter to Josiah Quincy, Esq. published in the Agricultural Repository, vol. iv. page 328, Mr. Peters observes, "I insist that *all* the hot muck ferments over violently in the earth. My opponents say, (and so does Sir H. I believe,) that the earth checks fermentation. And so it may to a certain degree. But strawy muck cannot be regularly spread. The animal matter is not mixed with the straw, but lies in masses *per se*; this over ferments, and throws up your crops, in bunches or spots, over luxuriantly; and it lays, smuts, or mildews. The earth prevents fermentation in the straw; and this dry rots and becomes a *caput mortuum*. I have experienced this effect over and over again. I think reasonably fermented dung goes further. All the straw and additional matter is impregnated, and, being decomposed, spreads with the animal ejections more equally, and to infinitely better advantage, assisting your crops in more points, and operating equally and more efficiently on the whole."

DUNGHILLS, heaps of manure laid up to ferment, consisting of dung and earth, together with lime, or marle, and any animal or vegetable substances, which easily putrify and consume.

It would be well if every farmer had some of them preparing, to be carted out in autumn, or to lie two summers, when it is found convenient. He would avail himself of much manure that might be collected between spring and fall; for, in the summer, the crops on the ground must prevent carting it; so that it should be preserved in the best manner to prevent waste. And this can in no way be so well prevented, as by mixing it with other substances.

Farmers should have such dunghills, some at their barns, or cowyards, one at a hogsty, when swine are shut up, and another not too far from the back door of a house. They may be tended, and augmented at odd times, when no other business stands in the way. That at the back door, especially, may be very easily made up, of a variety of rich and fertilizing ingredients, besides dung; such as the scrapings of the yard after rain; soot and ashes; shells, lime and bones; the sweepings of the kitchen; oil dregs, and any fat things; woollen rags; bloody water, in which meat or fish has been washed; greasy water; suds; ashes, although the lie has been drawn from them; old useless brine; urine; and, in short, any animal or even vegetable substance, that has not too much acid. Or, even acids, if they be overbalanced by plenty of alkaline substances.

To prevent the heaps being too

much torn and spread about by swine, or by the scratching of dunghill fowls, the heaps may be included in pens made with wide boards; or some rocks may be laid round them. Turfs may be laid over them, to prevent their evaporating; as well as under them, to prevent their soaking into the earth.

The heaps should have such a degree of moisture as best promotes fermentation and corruption. A cavity may be made close to the lower side of the heap, to receive the superfluous moisture as it runs from it after rain; and this liquid, highly impregnated with the strength of the manure, should be thrown, from time to time, on the top of the heaps, with a scooping shovel. In a wet season, the heaps will need some slight sheds over them. Indeed it would be best to cover them in all seasons, and to apply water to them when they need it.

Heaps about the barn or cow-yard, may be augmented with some of the nearest earth, swamp mud, straw, weeds, &c. those at the hogsty with the same, together with the dung of fowls, or other hot manures, as the dung of swine is naturally cold. But the farmer should acquaint himself with the nature of different manures; and always let that ingredient in his heaps be predominant, which is best adapted to correct and meliorate the soil on which it is to be laid. If it be destined for a sandy soil, clay will be an excellent ingredient in the composition of the heaps. If it be designed to lay on a clayey soil, sand is proper.

DUTCH HOE, sometimes called a *Scuffle;* an iron instrument, with a sharp steeled edge, nearly in the shape of the letter D. with a shank from the rounding part, five or six inches long, which passes into a handle of about six feet in length. It is of use to clean walks and avenues in gardens. No gardener should be without one of these instruments.

DYKE, OR DIKE, a sort of dam, constructed of earth, timber, fascines, &c. to oppose the entrance of water from rivers and from the sea.

Dykes made to exclude the sea from marshes, are built with sods cut out of the marsh, so as to make a ditch near the dyke, or else a ditch on each side. The sods are laid as a wall sloping on both sides; they should be laid very close,that the water may not enter; and some slender bushes should be laid between them, that the work may hold together the better. Some of the bushes should have roots to them,that they may grow,and more strongly bind the sods together.—Shrubs without roots will not live placed in the dykes at mid-summer, the time when dykes should be built. But they may be inserted afterwards, at a proper season.

A dyke, seven or eight feet wide at bottom, and three a-top, and made a little higher than the highest spring tides rise, will be sufficient on high marsh. When a dyke passes through a low place,or through a creek, it must be wider at bottom in proportion to the depth of the hollow, or creek, so that the sides of the dyke may be perfect inclined planes. Though this will make it very thick at bottom, it is necessary, that it may resist the greatest pressure of water against that part.

When we build on an oozy, soft spot, it is best to fill the mud with piles, driven as deep as they will easily go, and then cut off even with the surface or a little above it. This will give stability to the foundation,and prevent the water's undermining the dyke. On a sideling place, stakes should be driven through the dyke into the marsh, to hold the sods in their places.— There should be many of them, and they should be strong.

In the creek, or creeks, there must be sluices, larger or smaller in proportion to the quantity of fresh water that will need to pass out. See *Sluice*.

E.

EARTH, the soil, or land, in which the roots of plants find nourishment. There are several simple kinds of earth, considered only with respect to husbandry ; as clay, marle, loam, gravel, sand, peat,and black mould. Perhaps these are nearly all the simple soils that are found on or near the surface of the earth, in this country ; though others, distinct from them all, are found by digging deep. There is not one of these earths, in its unmixed state, that is so friendly to the growth of plants, as when mixed with some other sorts ; and it is happy for us that nature in most places has blended them. Though the original soils are so few, they are so variously compounded in different places, as to present us with an endless variety of soils, some or other of which are most suitable to nourish every different plant. But for most of the purposes of husbandry a sandy loam is as good as any.

Good earth for the general purposes of husbandry, is most commonly of a dark colour, or quite black, unctuous to the touch, easily ploughed, on a due medium betwixt dry and wet, not compact, nor too loose and open, and easily made to ferment.

Mortimer observes, " That mixed soils are best ; especially where the mixtures happen to be of the right kind, as those of the hot and dry soils, blended with the cold and the moist. All sands are hot, and all clays are cold, and, therefore, laying sand on clayey lands,or clay upon sandy lands, is the best of all manure for both. This alters and changes for the better, the very nature of the land itself, whereas dung only improves it for a time, and after that leaves it nearly as bad as it was before. It is not only the nature of the soil we are to consider, but the depth of it, and what kind of earth is underneath ; for the richest soil, if it be only eight or ten inches deep, and lies upon a cold clay, or upon a quarry of stone, will not be so fruitful, or advantageous to the farmer, as the leaner soil that lies upon better under strata."

But an under stratum of clay, not too near to the surface, and where the ground has not too much wetness, is found to be good, as the strength of manures does not escape through it. A stratum of clayey gravel, or mere clay, or almost any that is not too easily penetrated, is good: But one of loose sand or gravel must necessarily be bad, as the soil above it will not hold its manure. See *Soil*.

EFFLUVIA. The effluvia of rotten substances are supposed to breed diseases. The farmer, therefore, should be cautious that he do not breathe in the steams of his old dung-hills more than is necessary, especially when they have a disagreeable stench. And the greatest care should be taken to remove or cover up all the filth of the house, and of the hog-pen, which no doubt occasions some of those autumnal epidemic diseases, not unfrequently as fatal in country villages, in proportion to numbers, as in the thicker settled towns on the coast.

ELDER, *Sambucus nigra*, an ill smelling shrub, which grows plentifully in most parts of this country, produces a black berry, and is too well known to need describing. I mention it, because it is believed to be an excellent antidote against destructive insects. It may be well for farmers to try solutions of it, by sprinkling it over plants, infected with lice or insects of any sort.

The juice of elder, it is said, kills skippers in meat, cheese, &c.

ELFSHOT, or ELFSHOTTEN, a disease in horned cattle, the symptoms or concomitants of which are slugglishness and loss of appetite. The original of the name seems to have been a superstitious opinion, that cattle were shotten and wounded by elves, or fairies. The disease, however, is not imaginary. It is believed to be an opening in the peritonæum, or film of the belly, caused by relaxation. It resembles a hole made by a bullet, and may be felt through the skin which remains unhurt. These openings are closed, and the animals cured, by rubbing the part with salt and water. It should be repeated two or three times in the course of a day.

ELM, *Ulmus Americana*, a tree that is commonly found in our forests. It is tall and beautiful, long-lived, and grows to a large size.—The wood is not apt to split, or crack; and is very fit for the naves of wheels for carriages. Of this tree there are said to be two varieties, the white and the red. The elm is a proper tree to plant in groves. It is sightly and durable; and not apt to be broken by high winds.

EMPLOYMENT. No one that considers the condition of a farmer, can doubt of his having sufficient employment. He has so many objects to attend to, that his life must be filled up with carefulness or exercise. If he grow remiss, he will soon find that he has lost something through neglect, or failed of availing himself of some advantage.

In our climate, besides care, the farmers are necessarily hurried with their business during much the greater part of the year, that is, from April to November inclusive. But in the winter, they may be in some danger of spending some of their time idly, if they do not take care to prevent it. Feeding and tending their cattle, if they do it faithfully, will take some considerable part of each day, if the stock be large. The dressing of hemp and flax requires some time, and ought to be done in winter. Getting home fuel for maintaining fires through the year, and hauling stuff and fitting it for the building and repairing of fences; threshing and cleaning of corn and grain, and preparing farming implements,may all be done at this season. And these things ought to be done at this time of the year, to prevent hurry at a more busy season. So that, though our farmers cannot plough, or do any thing to the soil in winter, unless it be sometimes in part of December, they need not be idle. In maritime places they may employ themselves and their teams in getting manure from flats and creeks, and drawing it to their hungry high lands. This will turn to very good account, and pay them well for their labour. Holes may be dug in the ice over flats, from whence rich mud may be taken,and drawn upon sleds to the high parts of a farm. And this will be found to be a profitable employment.

ENCLOSURE, a piece of ground fenced by itself, to prevent the entrance of cattle, &c. In some places men farm in common fields.— But this method, pasturing excepted, is not eligible. Some lose more by it than enough to pay for enclosing. And it is too often the occasion of quarrels, and endless uneasiness among neighbours.

EWES, the females of sheep. That they may be profitably managed, we should keep none for breeders that have not long and fine fleeces. The rest should be killed off during the first year.— Otherwise the stock will degenerate; and a large proportion of their wool will be coarse, or too short, and of little value.

From the first of October, to the twentieth of November, the rams should be kept from them; that so their lambs may not come till the twentieth of April, when the ground is most commonly bare,and the grass begins to spring in many places.

For a few days, or weeks, before yeaning time, they should be more generously fed. Some juicy food, which they are fond of, should be given them, such as turnips, potatoes, &c. that they may have plenty of milk for their lambs: For it is the opinion of careful observers, that want of milk is the cause of the dying of so many lambs in the first stage of their existence.

From their first going to pasture to the last of June, or the middle of July, the ewes should have plenty of feed, by means of which the lambs will come forward rapidly in their growth, so as to be fit for weaning. Nor will the ewes become so lean, but that they may be fattened in autumn; which would be otherwise,were the lambs to suck them as long as they are permitted to do in this country.

EXPERIENCE. Perhaps no man ever attained to a thorough knowledge of husbandry merely by books, or by oral information. Experience is needful to fix the knowledge of the multifarious branches of it in our minds. It is needful, also, to teach us the easiest methods of performing a thousand things which depend on circumstances so minute, that they were never committed to paper, and scarcely are thought to be worth mentioning.

But experience, however necessary, is not all that is needful to make an accomplished farmer.—Observation is equally necessary. And without argumentation, none will be fit for any thing greater than going on in the most beaten tracks. None ought to conclude from their having had the longest experience, that they have the greatest degree of knowledge: For some will learn more by experience in one year, than others will in forty. Theory and practice should certainly concur, to render persons skilful in husbandry, or in any other profession. An early apprenticeship is as necessary to the attainment of this art, as any other; as some have been convinced, who have entered on farming when they were past the meridian of life.

EXPERIMENTS, trials of practice in husbandry. It is greatly to be wished, that more of these were made in this young country, where the knowledge of agriculture is yet in its infancy. Experiments made in other countries are not to be relied on, as proofs of the utility of one mode of culture in preference to another, in this country. Therefore, we should not trust to the experiments of Europeans, but make experiments for ourselves. Till this is done, we are not to look for great improvements in husbandry.

It may be true, that he who makes a new experiment is in some hazard of losing more or less by it.—Therefore, I would not press it upon farmers in indigent or low circumstances, to venture upon any thing of the kind, unless it be in very small matters, or on a small scale; for the failure of one year's crop would almost reduce them to beggary. They would do well, however, to compare the profit of one crop with another, reckoning the cost laid out upon each; and of one course of crops with another; and the success of different manures on the same, or on different soils. Thus they may find which of the old methods is to be preferred, by a small degree of attention, without any risk, which is a matter of some consequence.—For we need to learn what methods to drop, or discontinue, as well as what to adopt or bring into use.

Gentlemen of large estates, who can bear some considerable loss without feeling it, in case they fail of success, are the persons that should try new crops, or new ways of raising old ones. Love of their country should prompt them to it; for there is no reason to doubt but that our husbandry may admit of a variety of important improvements. It is wished that an enterprising spirit were more excited, that we might have reason to hope for great improvements in husbandry.—There is an extensive field for experiments; and making them might be a good and laudable amusement to persons who have leisure.—Trench ploughing, which has never yet been attempted in this country, ought to be tried, at least, by those who have deep soils, clear of rocks and other obstacles. Trials should be made of the advantage of ploughing flat land in ridges; and whether ridge ploughing will not secure grain from destruction by winter frosts. Attempts should be more extensively made to raise winter wheat, which is the most valuable of all grain. We should endeavour to find out the best steeps for grain and other seeds, to quicken their vegetation, and to secure them against insects and smut;—what are the best quantities of seed for sowing in different grounds;—whether sowing seeds with a drill be not the best method when horse hoeing is not applied; when is the best time for sowing of winter grain;—whether good peat and marle be not to be found in plenty in various parts of the country, and the advantage of marling, and sowing peat ashes;—whether drained swamps are not the most profitable of all our lands;—whether new dung or old will produce the best crop, and whether compost will not do better than either;—how lime will answer as a manure in our hot summers—on what kind of soil it is most serviceable, &c. &c.

But, in making experiments, great care should be taken that we do not draw a conclusion too hastily; certainly we must not do it from one single trial. For a thing may answer well at one time, owing to the peculiarity of a season, or to some indiscernible circumstances, which will not at another. If men allow themselves to be too sanguine and sudden in their conclusions from single experiments, they will rather embarrass and mislead, than increase agricultural knowledge.

But if improvements be wished for, experiments should be carefully recorded. If this be neglected, husbandry must be expected to remain in its present low state. For want of such records, a great deal of useful knowledge has been already lost. Though many have made experiments, by which they have satisfied themselves, but few have recorded them. The experimenters themselves have forgotten them, to such a degree, that they are apt to misrepresent them, when they attempt to relate them. And too many suffer useful discoveries to die with them. To prevent these evils, the forming of societies in various parts of the country might be of great use.

F.

FAGGOT, a bunch of bushes, or limbs of trees, bound together by a withe. Faggots for fuel are cut to the length of about two feet. In many parts of this country, the scarcity of fire wood makes it expedient that farmers should no longer go on in the practice of burning such materials on the ground. They should preserve them in faggots for fuel in their houses. They will serve to heat stoves; and for heating ovens there is no better wood.

FALL, of the year.

In a country where the springs are backward, as in the northern parts of New-England, farmers should do all they can in autumn, to diminish or lighten the labours of the following spring, when they will have much work to perform in a short time. Summer dung and composts should be carted out at this season. Fences should be built or repaired, not only to prevent having them to do in the spring, but to keep cattle from injuring the lands with their feet. All the ground should be ploughed in the fall, that is to be seeded the following spring. That which is intended for spring wheat should be ploughed twice. Though all that is ploughed in the fall, for spring tillage, must be ploughed again before seeding, the fall ploughing saves labour, as one ploughing may answer in the spring where two would be otherwise needful. It is saving labour at a time when teams are most apt to be faint and feeble, and when there is too often a scarcity of food for them. But ploughing in autumn is of great importance in a clay soil, as, by exposing it to the frost, the cohesion of its parts is much broken.

Some prefer *fall-transplanting of trees*. It would be very convenient if it could be done at this season as well, but, however it may answer in the middle and southern states, experience has shown that it will not do as well in the New-England States.

FALLOWING, is one of the words which requires a more distinct definition than it has yet had, because we are persuaded that the term is differently applied in different countries. In England, it invariably means a suspension of crops of any sort, for a greater or less period of time, accompanied by constant ploughing and harrowing of the land, on which the crops are so suspended, for the express purpose of rooting out pernicious weeds, and of dividing and pulverising the soil more perfectly.

There are also theoretical notions, perhaps true, but of which we have no evidence, that during this repose from bearing crops, and by the means of these repeated ploughings, the land collects some principles or elements from the atmosphere, which contribute to its fertility. A vast many obscure remarks, and conjectures have been made upon this subject, which if they had not been incomprehensible would have been admitted to be nonsense. It seems now to be generally acknowledged, that the principal benefit of fallows, upon the British plan, arises from the destruction of weeds, and the more perfect division of the soil—and it is a little surprising that this idea never suggested itself to the earlier writers on this subject, when they were so well acquainted with the benefit of *trenching*,—a practice, which seemed to lead to so direct and necessary a conclusion, that a minute *subdivision* of the soil, rendering it not only more permeable by water, and by the roots of plants, but also more retentive of moisture, was in truth the great secret of the advantages of fallowing, and of the frequent use of the plough. Much idle speculation might thus have been spared in agricultural inquiries, and we should not have been compelled to read so much about the collection of nitre, and other salts from the atmosphere, when there was, in fact no proof that any such deposits were made, and when it is quite as probable that these if any, which were in the land before, might escape by this exposure to the air, as that any more should be collected. We, however, mean to confine ourselves in this article, simply to *facts*, which are important to the farmer, leaving to those, who prefer to plunge into the mists of theory, the pleasure of so doing.

Though the English farmer understands by Fallowing, a season of greater or less labour on his uncropped lands, the French, and some of the American Farmers understand by it an entire abandonment of the soil to such weeds, or plants, as are useless, or nearly so to man and animals, for a geater or less period, under some vague notion, that the land will recruit itself, in a few years, and be fit again for cultivation. There are, however, many American farmers, who pursue the English mode of fallowing, and instead of a period of repose, the land is constantly moved by the plough, and the harrow.

We understand, that what we have described as the plan of some American and French farmers—that of letting the land lie untouched has been of late years entirely exploded by every man of sense in both countries; And, if it were necessary, for the recruiting of exhausted lands, to abandon the raising of grain, potatoes, or other roots, it would be better to plant such lands with any species of trees that would thrive in it, rather than to abandon it to such useless plants as might voluntarily spring up upon it.

With respect to the spirited, and expensive mode of summer fallowing adapted in England, which often extends to four, five, and even seven successive operations on the soil, in one summer, it is thought that it may be useful, in lands peculiarly troubled with pernicious weeds as the most successful mode of extirpating them. But in general, it is now thought, even in that country, that a judicious rotation of crops is much preferable on every account. It saves the loss of income in lands left in fallow; and as it is now believed on better evidence than the theories can boast, at which we have just smiled, that each plant has a preference to some species of nourishment over others; and that when this is exhausted in any given soil, and the plant will no longer thrive, another plant may flourish perfectly well on the same lands, it seems to be probable, that, by a rotation of crops, and a proper application of manures we may render fallows entirely unnecessary. Such, at least, we believe to be the prevailing opinions of intelligent men, in Great Britain, France and the United States.

WINTER FALLOWING, is only breaking up the land, or ploughing it in the fall, and leaving it exposed to the action of the frosts of winter. There is no country in the world, where this can be done more easily, or with more benefit than ours. Our winter sets in early. We cannot often plough till Christmas. Our frosts are very severe. Our ground more open than that of our easterly neighbours; and highly beneficial effects must be produced in pulverising the soil by the power of frost, one of the most irresistible agents in nature.

The following observations from English writers of great respectability will serve to corroborate some of the sentiments which we have advanced on this subject.

" The raising clean, smothering, green crops, and feeding stock with them upon the land, is not only much more profitable, as far as relates to the value of the crop substituted in lieu of a fallow, but is also a more effectual method of procuring large crops of wheat, or any other crop, which may succeed the green crop.

" The smothering and hoeing crops of tares, turnips, potatoes, cabbages, savoys, cale, hemp, and other plants, which cover the ground, and cause a stagnation of air, preserve the moisture of the soil, and promote the putrefaction and decomposition of such animal and vegetable matter, as may be contained in the soil, are more likely to prove economical and beneficial than any system of fallowing." *Code of Agriculture.*

" When weeds are buried in the soil, by their gradual decomposition they furnish a certain quantity of soluble matter; but it may be doubted whether there is as much useful manure in the land at the end of a clean fallow, as at the time the vegetables clothing the surface were first ploughed in. Carbonic acid gas is formed during the whole time by the action of the vegetable matter upon the oxygene of the air, and the greater part of it is lost to the soil in which it was formed, and dissipated in the atmosphere.

" The action of the sun upon the surface of the soil tends to disengage the gaseous and the volatile fluid matters that it contains; and heat encreases the rapidity of fermentation: and in the summer fallow, nourishment is rapidly produced at a time when no vegetables are present capable of absorbing it.

" Land, when it is not employed in preparing food for animals, should be applied to the purpose of the preparation of manure for plants; and this is effected by means of green crops, in consequence of the absorption of carbonaceous matter in the carbonic acid of the atmosphere. In a summer's fallow a period is always lost in which various vegetables may be raised, either as food for animals, or as nourishment for the next crop; and the texture of the soil is not so much improved by its exposure as in winter, when the expansive powers of ice, the gradual dissolution of snows, and the alternations from wet to dry, tend to pulverize it, and to mix the different parts together. *Agricultural Chemistry.*

FALSE QUARTER, a rift or chink in the quarter of the hoof of a horse, from top to bottom. It happens generally on the inside, that being the weakest and thinnest; and proceeds from the dryness of the hoof, but especially when a horse is ridden in dry, sandy, or stony ground, in hot weather, or in frosty weather, when the ways are flinty and hard. It is, likewise, caused by bad shoeing, and all other accidents whereby a horse becomes hoof bound: For the narrowness of the heels, and brittleness of the quarters, continually expose a horse to all the said accidents.

"This accident is both dangerous and painful; for as often as a horse sets his foot to the ground, the chink widens; and when he lifts it up, the sharp edges of the divided hoof wound the tender flesh that covers the coffin bone, which is for the most part followed with blood; and it must of course be apt to render a horse lame, as it is very difficult to form a re-union. To remedy this imperfection, First, draw the whole length of the cleft with your drawing iron, then anoint the hoof with tar, honey, and suet, molten together; for nothing can be more proper for the hoof; and lay a thin pledgit dipt in the same along the cleft. After this, take rope-yarn, such as the sailors use, which is no other than hemp moistened in melted tar, and spun loose: Apply the yarn all down the hoof, beginning at the coronet and descending downwards, one lay after another, as close as the binding of the hoops of wine casks, laying a smooth pledgit of flax behind, to keep it from fretting the heel.-- This should be opened once in three or four days, that the clefts may be drest. And to prevent any inconveniency that may happen by the opening, a thin staple may be also contrived with points like horse-shoe nails, cast off obliquely, to take a slender hold, the plate of it crossing the cleft, where part of the shoe is cut off (as it must be under the cleft) and the nails coming out on each side of the cleft, on the upper part, to be clinched as the other nails. By this method a cleft in any part of the hoof may be easily cured, if the horse be not very old, or diseased." *Gibson's Farriery.*

FARCY, a disease in horses, similar to the scurvy in men, and arising from a similar cause. The farcy is caused in horses from their being for a long time confined to dry meal. And as the scurvy in men is cured by a diet of green vegetables; so the farcy in horses may be cured by turning them into a good fresh pasture. But it is only in the beginning of the disease that it can be so easily cured.—Gibson prescribes bleeding, and moderate purging; and afterwards doses of antimony. See *his Farriery.* Mr. Mills calls it a cording of the veins, and the appearance of small tumours in several parts of the body. Mr. Bartlet deems this distemper easy of cure, when it appears on the head only. Mr. Bourgelat says, a decoction of the woods, antimony, powder of vipers, with some mercurial preparations, are looked upon as so many specifics in this disease—and that hemlock will cure it.

FARM. A writer in the Mass. Agr. Repository, vol. V. p. 320, in treating "*on the extent of land* necessary for a farm, and sufficient to support a family well and independently," has the following among other valuable remarks: "We know men, active, intelligent and industrious, possessed of thirty or forty acres of land, who are labouring for others, or taking charge of their neighbours' concerns upon the avowed reason, that they cannot support their families on so small an extent of land. But they do not realize the actual efficiency of the soil. Undoubtedly there are many honourable exceptions to the observation we are about to make; as a general rule, however, it may be asserted, that THE FARMERS OF MASSACHUSETTS ARE YET TO LEARN THE IMMENSE PRODUCTIVE POWER OF A PERFECTLY CULTIVATED ACRE. Instead of seeking riches in augmenting the number of their acres, let them be sought in better modes of husbandry. As a general truth, we believe it may be asserted that every farmer in Massachusetts, possessed of one hundred acres of land, might divide them fairly by quantity and quality, into thirds, and by a suitable cultivation, make either third more productive than his whole hundred acres are at present. This is the operation at which those interested in the agriculture of Massachusetts ought to aim—to make farmers realize what cultivation can effect, and to teach the modes, by which the productive powers of the soil can best be elicited."

FARMER. "It is indispensable for the success of every undertaking that a sufficient capital to carry it on, should be at command; and for that of farming in particular. When there is any deficiency with respect to that important particular, the farmer cannot derive sufficient profit from his exertions; for he may often be obliged to dispose of his crops at an undervalue, to procure ready money; or he may be prevented from purchasing the articles he may require, though a favourable opportunity may present itself. An industrious, frugal and intelligent farmer, who is punctual in his payments, and hence in good credit, will strive with many difficulties, and get on with less money, than a man of a different character. But if he has not sufficient stock to work his lands properly;—nor sufficiency of cattle to raise manure;—nor money to purchase the articles he ought to possess, he must, under ordinary circumstances, live in a state of penury and hard labour; and on the first unfavourable season, or other incidental misfortune, he will probably sink under the weight of his accumulated burdens. In general, farmers are apt to begin with too small a capital. They are desirous of taking large farms, without possessing the means of cultivating them. This is a great error; for it makes many a person poor, upon a large farm, who might live in comfort, and acquire property upon a small one. No tenant can be secure without a surplus at command, not only for defraying the common expenses of labour, but in case any untoward circumstance should occur. When a farmer on the other hand, farms within his capital, he is enabled to embrace every favourable opportunity of buying with advantage, while he is not compelled, if the markets are low, to sell with loss." *Code of Agriculture.*

FENCE, a hedge, wall, ditch, or other inclosing made about farms, or parts of farms, to exclude cattle, or include them. Fencing is a matter of great consequence with farmers; and, as it is managed in most parts of this country.is a great drawback upon their profits. But however costly fencing may be, it is good economy to make fences strong and fully sufficient to answer their purpose. It would be folly to save a trifle by making a fence too slightly, and be liable to lose a whole crop, by the breaking of cattle through it.

The kinds of fence, and manner of fencing, should vary according to the difference of soils; and according as one kind of materials for fencing is more plenty and cheap than another.

In the new plantations of this country, log fences are most used; as they certainly ought to be; because the wood is of little or no value. To build these fences with, the best wood that I am acquainted with is white pine. A fence built with logs of this kind will stand twenty years, with little or no repairing.

But if this kind of wood be not at hand, and other sorts be plenty and near, it may be as well to make use of some other kinds: Such, for instance, as pitch pine, Norway pine, hemlock, ash, oak, and white maple. Several, or almost any of these kinds, if they do not lie too near to the ground, will last for a considerable time. If a fence be made partly of white pine, and partly of other wood. the former should be laid nearest to the ground.

But let farmers beware of building their log fences of bass wood, poplar, birch, beach, or rock maple, unless in cases of necessity; for as they will be soon rotten, the labour of building them is in a manner lost. If logs are peeled they will last the longer in fences. The largest logs should lie lowest in a fence, both for strength and durableness. The lowest are soonest rotten, when all are of the same size; and the largest logs will last longest.

Log fences should always be braced with strong stakes across; and heavy riders add strength to a fence.

When ground is wholly subdued, and the stumps of its original growth of trees quite rotted out, if stones can be had without carrying too far, stone walls are the fences that ought to be made. Though the cost may be greater at first than that of some other fences,they will prove to be the cheapest in the end. Building stone walls is not only the way to clear ground of a bad incumbrance; but when the fence is made, it is certainly the best of all fences. On a hard,sandy, or gravelly bottom, if built with good stones, a wall will stand many years without any repairing. And it will stand well on any soil, clay and mire only excepted. On a clay soil it will stand, if the foundation be laid in a trench, near as low as the earth commonly freezes in winter. But a wall of flat or square shaped stones, will stand tolerably well on any soil, laid only on the surface.

It is true that walls will gradually settle into the ground, where the soil is at all mellow,and heaves with the frost ; so that it may be necessary, in a century or two, to dig them up and re-build them. I find some of this work has already been done in some of our oldest towns. But this is a slight objection against the utility of this kind of fence. For future generations will bless themselves, if they have materials on the spot to build fences with, when wooden materials must unavoidably be scarce in most places, and very costly.

I am aware it will be objected, that stone walls are not sufficient fences against sheep. But it is easy to make them so. A row of flat stones laid on the top, and jutting over, will make a wall sufficient for this purpose : Or any wooden poles, laid on the top of the stone wall, supported by stakes will check the passage of sheep. The limbs of trees,with their small branches upon them, laid on a stone wall, make a cheap and effectual guard against the passage of sheep. Riders with some of the limbs on them are best for this purpose.

Farmers need not fear that they shall impoverish their land by clearing it of stones. For, after all they can do to a soil that is naturally stony, there will be stones enough remaining, a little way below the surface, to render the ground moist and warm.

Rail fence is perhaps as much used as any. The timber for posts and rails should be felled in the winter. To sharpen rails before they are dried saves labour : And posts should be mortised while they are green. Rails are cut twelve feet long. Posts should be six feet and a half, or seven feet. The best timber for rails is Cedar : It is easy to split, light to carry and to handle, sufficiently strong, and the most durable of any. A rail of cedar will last an age. Next to cedar, rails of chesnut, white pine and ash are best. But, for want of better, some use rails of oak.— Cedar is also best for the post, in this and in board fence. The locust tree is said to be excellent. But posts of white oak, which in most places are more easily got, will last about fifteen or twenty years. If the lower ends of posts be scorched in a hot flame, before they are put into the ground, they will last the longer. Also soaking them in sea water will tend to keep them from rotting. Juniper, the Larch,is much used for posts in this part of this country. They will last forty years. For *Hedge Fences*, see *Hedges*. For ditch fences, see *Ditch*.

FERMENTATION, an internal motion excited in substances, by which the cohesion of their parts is destroyed, and their nature changed. But, that a fermentation may take place, it is necessary that some particles in the fermenting body be fluid; or that the body be moist. Bodies perfectly dry can have no degree of fermentation in them.

Fermentation does much towards the production and growth of plants. It is therefore a thing of much consequence to the farmer; and he ought to know by what means he may increase it in his ground.

The pasture of plants is increased by fermentation, as it loosens the soils, so that their roots do more easily find their food. All rich soils contain the principles of the food of plants in abundance: And a fermentation is produced among them by any thing that alters the arrangement of their particles. A fermentation is produced by heat from the sun, and by rain: But when the soil is too much filled with water, the fermentation is abated, or destroyed. Ploughing, and otherwise stirring the ground, is a principal cause of fermentation in the soil. The plough not only increases the pasture of plant by pulverizing the soil, but by mixing the salts and oils contained in it, so as to bring on a degree of fermentation, if the soil have neither too much,nor too little water in it at the time of ploughing.

I suspect that our severe frosts in winter may have a tendency to excite a degree of fermentation,which takes place after the ground is thawed. For the heaving and settling of the soil will make some alteration in the disposition of its particles, and conduces to its imbibing more freely, snow water and rains, which contain food of plants.

But dung, and other strong manures, are perhaps the chief causes of the fermentation of soils. Dung is no sooner mixed with the soil, when there is a proper degree of warmth in the earth, than it strongly ferments in itself, and brings on a new fermentation in the earth which is in contact with it, which is communicated to remoter earth: By all which the cohesion of the parts of the soil is broken, the soil highly pulverized, and the pasture of plants proportionably increased, so that their roots can freely extend themselves in quest of their food.

By the same fermentation, the food or nourishment of plants is increased; because the dung itself is dissolved, its salts and oils mixed, its fine earthy particles set at liberty, the vegetable substances, such as roots, weeds, &c. corrupted and dissolved: All which conspire to increase the food of plants, and prepare it to enter the minute pores of their roots.

That plants may flourish, it is thought to be needful that a fermentation of the soil be continued during their growth. Otherwise a sufficient quantity of steam will not arise to their roots; a probable consequence is, that they will be stinted in their growth. It may be for this reason that tillage, during the growing of plants, is found to be so very advantageous to them; especially when they are hoed to a good depth, by which the fermentation of the soil among the roots is increased.

FERN, OR BRAKES, *Polypo-*

dium, a well known sort of weeds, that is often troublesome to such of our cleared, or partially subdued lands, as have not been tilled. They are so full of salts, that they should be cut green, and laid in our barn yards to putrefy, and mix with dung. Perhaps there is scarcely any better method of increasing manure. Pasturing the land where they grow, especially with hungry cattle, that will eat them as fast as they come up, will help to subdue them. Folding will kill them; for there is nothing so fatal to them as urine: But not less than two or three year's tillage will subdue them. They are hardest to subdue in deep soils. Plentiful dunging, with tillage, will be effectual; but a most certain remedy is urine; this they get in plenty by folding.

"Fern, cut while the sap is in it, and left to rot on the ground, is a very great improver of land; for if burnt, when so cut, its ashes will yield double the quantity of salt that any other vegetable can do. In several places in the north parts of Europe, the inhabitants mow it green, and burning it to ashes, make those ashes up into balls, with a little water, which they dry in the sun, and make use of them to wash their linen with; looking upon it to be near as good as soap for that purpose." *Dict. of Arts.*

In the *Farmer's Calendar* you may read, under September, "Now is the proper time to cut fern, called in some places brakes. This is most profitable work, and should never be neglected. Carry it into your farm yard, and build large stacks of it for cutting down through the winter, as fast as the cattle will tread it into dung; also for littering the stables, ox houses, cow houses, hogsties, &c. By having great plenty of it, you will be able to raise immense quantities of dung, which is the foundation of all good husbandry; and it is well known that no vegetable yields such a quantity of salts as fern: from which we are to conclude, that it is best adapted to the making manure."

It is a lamentable thing that we should hitherto be so inattentive to our own welfare, as to suffer this weed to render our lands in a manner useless, when it might be turned to so great profit. It is a double advantage to cut brakes, as they not only make plenty of good manure, but every cutting helps to destroy them. The work may be done after the hurry of hay making is over; and perhaps no labour on a farm can turn to better account.

FISH, all the parts of fish, shell fish and all other, are excellent manures. They may be used, either salted or fresh; salted fish are said to be best. The offals of fish, and fish that are spoilt for eating, may be converted to this use; But I should prefer using them as an ingredient in compost. They are so strong a manure, that it has been said, one single alewife will answer as well as a shovel full of the best dung, in producing Indian corn. But they cause land to exert itself so much, that it will be apt to grow poor, unless care be taken to prevent it.

FLAIL, an instrument for threshing. A flail consists of the handstaff, the swiple or flyer, the caps or caplins, the string or band. The staff should be of the lightest timber, such as ash, and made perfectly straight; the flyer should be of a heavy kind of wood, as walnut, elm, or beetle wood. Some make the caps of wood, but stiff soal leather is better. The string or thong, which connects the cap with the flyer, may be of the neck of deerskin. But the skin of an eel will last much longer than any other string I have met with.

FLAX. The following observations on this subject are extracted from "Essays on Flax Husbandry. By S. W. Pomeroy, Esq. First Vice President of the Massachusetts Society for promoting Agriculture.

"Change of seed. Notwithstanding it is an opinion well established among experienced Flax growers in this country, that a change of seed is advantageous, it is apprehended that they are not aware of the extent of the benefit to be derived by selecting seed from a soil or climate essentially different; and it may be owing to a want of attention in this particular that the flax crops are so uncertain, and the quality inferior, however perfect in other respects the system may be conducted. Mr. Young observes that foreign flax seed was universally used in Ireland, when it could be obtained, otherwise they were careful to procure seed which grew upon soil of an opposite quality from that which was to be sown," "that American seed was to be preferred, and produced finer flax than any other. Baltic seed produced more but of a coarser quality. It is well known that American seed always bears the highest price in the Irish market."

Mr. Pomeroy cites a number of examples to show the importance attached to the culture of flax in Europe, "and to justify the conclusion that in this country, a continued, judicious change of seed will be indispensible to the successful prosecution of flax husbandry; and a further inference may be drawn, that experiments on various soils, with seed the growth of different climates, are requisite to direct the farmer to the quarter from whence his best seed may be obtained. Here opens a legitimate field for our numerous agricultural societies to labour in: on their exertions the farmer must depend in the outset; but let it once be ascertained that Riga seed is best in one section, Dutch or German in others, and mercantile interest, if not patriotism will distribute them.

"Should it be objected to importing seed on account of the expense, we reply that large quantities of linseed oil are constantly imported, and the difference of price between our own seed and that imported, will not much exceed what is now paid for good clean seed for sowing or export, and that which is sold for crushing; but if it is fifty cents per bushel, or more, it can be no object, compared with the advantages that may be reasonably expected to result, and the farmer need not to be told that "in all his operations, parsimony is never so ill judged, as when it is exercised in the selection of his seeds." "It is not pretended, however, but that from the great variety of soil and climate in the United States, the object in view might be obtained without importation; yet it may be important to have a good stock to begin with, when trials could be instituted with its produce. At any rate it cannot be expected that individuals will embark in such a course of experiments, either with foreign or domestic seed, unless encouraged by agricultural societies, or other public bodies.

In speaking of the soils most suitable for flax Mr. Pomeroy observes "The soils which rank first in this country on the flat bottoms, that are covered by the fall and spring floods, which subside early enough in the season to get in a crop; those river flats on the second banks, that have a depth of strong alluvial soil; the reclaimed marshes and swamps with a black unctuous soil not too peaty, with as much clay in the composition as will permit its being rendered soon dry and mellow, and not retain water on or near the surface, if it stands two feet below, so much the better, but it must be well guarded by ditches and dykes against sudden freshets. Such is the soil of the province of Zealand, where more flax is raised, and of better quality than in any other part of Holland. The next in estimation are the strong black loams on clay, or hard pan, that will retain moisture. Yellow loams, with a holding subsoil, may be rendered suitable for flax, by proper cultivation; and since the discovery that plaster of Paris is an excellent manure for it, a crop may be obtained with much more certainty on lighter land than formerly.—Perhaps the characteristic of best garden mould may be applied to a flax soil, viz. retaining sufficient moisture, and all that falls, without ever being saturated; but on any soils the surface should be completely pulverized, and never be worked when wet.

"No dung should be applied to the land when the flax is sown, but may be put on bountifully with the previous crop. The objection is, that dung forces the growth so rapidly that the plants draw weak, have a thin harle, and are the more liable to lodge.—Lime, marle, shells, leached ashes, &c. do not produce such effects. Top-dressings, soon after the plants appear, of plaster, ashes, soot, &c. are highly beneficial, as they not only encourage the growth, but are a protection against worms, which sometimes attack young plants, and may be considered the only enemy they have, except weeds.

"Salt has been mentioned by the late Dr. Elliot, of Connecticut, as an excellent manure to plough in with flax at the rate of five bushels to the acre;* probably more would be better. Plaster is now much used in Duchess coun-

* See Elliot's Essays on field husbandry.

ty, the best cultivated district in New York as a manure for flax, on which its good effects are as apparent as on corn.

"The late Chancellor Livingston viewed a piece of flax on the 20th of May, 1791, belonging to a poor tenant, very injudiciously sown on a dry sandy declivity, it looked so extremely sickly that the tenant thought of ploughing it up; the Chancellor gave him three bushels of plaster, which was sown the next morning before the dew was off, and had the satisfaction of seeing his tenant gather more flax from his half acre in an uncommon dry season, than was produced from any acre in the neighbourhood.

"The best preparatory crops in this country, at present appear to be potatoes, corn and roots; they will most generally repay the extra manure, and if well managed, check the production of weeds.

"The following rotations may serve as an outline subject to be varied, and hemp or other crops introduced as circumstances require, viz:

No. I. *Low, cold, or reclaimed soils.*

1st year, *Potatoes.*

2d do. FLAX with seeds.

3d do. *Herds grass* and *red-top*, or *tall meadow oat grass*, to continue three years or more, and the course repeated.

No. II. *Strong Uplands.*

1st year *Potatoes* or *Corn.*

2d do. *Corn* or *Roots.*

3d do. FLAX with seed.

4th do. *Clover.*

5th do. *Orchard-grass* or *Herds-grass*, to continue three years or more.

No. III. *Light Lands.*

1st year *Potatoes* or *Corn.*

2d do. *Corn* or *Roots.*

3d do. FLAX with seed.

4th do. *Clover*, to be mown once, the after growth to be turned in, and rye sown thick on the furrow which may be soiled, or fed in the spring by sheep or milch cows, and ploughed in for

5th year *Corn.*

6th do. *Spring Wheat* or *Barley.*

7th do. *Clover*, and the course to be pursued as before, when flax will occupy the land every seventh year. In all cases, except when hemp is substituted, the tillage crops should receive the dung.

If the land is ploughed into beds, or convex ridges like turnpike roads, about a rod wide, especially if low and level, the crop will be much more secure from injury by heavy rains, and the grass crops will be better if it remains in that form. On any soils, fall ploughing in narrow ridges will facilitate its early working in spring, and should not be dispensed with."

Mr. Pomeroy gives the following directions relative to

CHOICE OF SEED.

"That of the last year's growth should be obtained if possible. The usual marks of good seed are, that it be plump, oily and heavy, of a bright brown colour, sinking readily in the water, and when thrown into the fire to *crackle* and *blaze* quick. A very simple method of trial is to sprinkle it thin between two pieces of wet paper, which plunge into a hot-bed or dung-hill, and in less than twenty-four hours the proportion that will vegetate can be discerned, which should be ascertained in order to regulate the

QUANTITY TO BE SOWN.

"On this head no particular directions can be given, as it depends on the various qualities of soil, goodness of seed, &c. The rule for sowing small grains is *reversed*; flax requiring to be sown thickest on rich soil as not more than one stalk is wanted from a plant. In England and Scotland never less than two nor more than three bushels to the acre are sown. Two and an half is the most usual portion. In Flanders and Ireland seldom less than three bushels are sown, except when seed is an object. Thick sowing is to obtain fine flax. In this country it will be important at present, to sow at such a rate, as will insure good crops of each; and experience only can determine the exact point. It is probable that six pecks is the least, and two bushels the extent that should be sown to obtain the most profitable results, till the demand for seed is considerably lessened."

SOWING.

Mr. Pomeroy recommends sowing as early as it is possible to prepare the ground, says that it is important that the seeds should be equally distributed, and "fortunately what has long been a desideratum is now attained. A machine for sowing small seeds broadcast with perfect regularity, has lately been invented, and performs to great satisfaction.*

WEEDING.

"Weeding is considered in Europe, and by good husbandmen in this country, as necessary to secure a good crop of flax, which is a very tender plant when young, and more easily checked in its progress by weeds than any other. It is not supposed to be injured by the clover and grass sown with it; on the contrary the Flemish farmers think them beneficial, by protecting the tender roots from drought, and keeping the weeds under. It should be carefully wed when the plants are three or four inches high; they are not then injured by the labourer going barefooted over them.

PULLING.

"This should be performed as soon as the leaves begin to fall, and the stalks shew a bright yellow colour, and when the bolls are turned a little brown. The seed will continue to ripen afterwards.—When the flax is lodged it should be pulled immediately, in any stage of its growth, or it will be entirely lost; great care is requisite in sorting the different lengths, and keeping them separate till after the flax is hackled, or much waste will ensue in that process.

SAVING SEED.

"As soon as the flax is dry enough to be put under cover it should be rippled, as it is termed. A comb, resembling the head of a

* Bennett's machine for sowing broad-cast, a description and drawing of which are given in the memoirs of the Philadelphia Agricultural Society, vol. IV. with ample testimony of its usefulness. It is pushed forward by a man, like a wheel-barrow, and will sow more than one acre in an hour, unimpeded by wind or light rain. They are for sale at Harrison's & Earle's Repository for Agricultural Implements, in the City of New-York.

rake, but with teeth longer and nearer together, made of hickory or oak, is fastened upon a block, and the flax, taken in parcels no larger than the hands can firmly grasp, is drawn through, and the bolls rippled off; attention to sorting at the same time should be continued. The bolls are to be riddled and winnowed immediately; spread thin on a clean floor, or on sheets, in the sun, and when sufficiently dry, and beginning to open, threshed. By this method the foul seeds are completely separated with little trouble, and good clean seed is ready for an early market, often the best without the use of expensive machinery to make it so. Here the operations of the farmer ought to end! The process or preparation being foreign to and unconnected with his other pursuits; and which has been the greatest objection to extensive flax culture. Can there be a reason why the farmer is to prepare his flax more than the hides of his cattle, which he sends to the tanner? They are both chemical processes; and to dissolve the glutinous or resinous substances by which the fibres are attached to the stem, without impairing their strength, is perhaps as critical, and requires as much care and judgment, as to extract the animal juices from the hides, and fill the pores with tannin. In short, the flax grower, and flax preparer and dresser, should be distinct professions. They are said to be so in Flanders and Holland, and were extensively so in Scotland, where the farmer sold his flax on the ground, or in sheaves at his barn or rick.

"The preparation of flax by steeping is very general in the great flax growing countries in Europe, but it is not quite finished in the water. It remains spread some days on the grass, which is necessary to render it soft, and give that silvery appearance so desirable.—The destructive process of dew rotting, is most commonly practised in this country, and when water is resorted to, it is at an improper season, and the process imperfect; which is the cause of its being so harsh and brittle. Perhaps no part of the system requires such an allowance for difference of climate. In the humid atmosphere of Ireland, it is not very material when it is spread; but in this climate, when exposed to a July or August sun, every drop after a shower, becomes a burning-glass, and literally scorches the fibres; besides such a highly putrid fermentation as will then take place in the water, though it separates the harle more speedily, not only injures it, but communicates a stain that renders the process of bleaching much more tedious and expensive.

"The flax should not be put into the water till about the first of October, and remain from ten to fourteen days, according to the temperature of the weather, and should be taken out before the fibres will separate freely, spread on the grass, when the frost will very much assist the operation, and the flax exhibits a gloss and softness that it is impossible to give it otherwise. The following method of preparing hemp will apply with great force to the point under discussion. During the late war an experienced ship-master in Connecticut, and who was also a good farmer, raised a crop of hemp. As soon as it was dry enough to be stowed away it was put under cover and remained till October; was then put into clear soft water, till the fibres would separate with some difficulty, when it was spread on the grass; the frost completed the operation, and when dry it was immediately secured. There was no putrid fermentation to deteriorate the harle, nor was it mildewed by being exposed to the weather, and when dressed, exhibited that fine silver green hue by which the Russian hemp is distinguished;* and when worked up, was pronounced by the rope-makers to be equal to any hemp ever imported! Here is a lesson for our western brethren, that is worth more to them than mines of silver. Clear, soft stagnant water is preferred in Europe. A canal forty feet long, six broad, and four deep, is said to be sufficient for the produce of an acre of flax, at one time. It should be formed on a clay or some holding soil, where the water from a spring or brook can be conducted in with convenience; the expense would not be great, and on most farms suitable sites may be had.—May not boiling or steaming be found the most advantageous process of preparing flax? The very superior sample of thread exhibited at Brighton, in 1818, for which Mrs. Crowninshield, of Danvers, received a premium, was spun from flax prepared by boiling. It appears by the "transactions of the Swedish Academy," that a method was practised in Sweden, of preparing flax to resemble cotton, by boiling it ten hours in salt water, spreading on the grass, and frequently watering, by which it becomes soft and bleached. Boiling or steaming will not appear very formidable or expensive when we examine the subject. A box twenty feet long, six feet wide, and four deep, well constructed of stout planks, a boiler, from which a large tube extends into, and communicates with the water in the box, will boil the produce of a quarter of an acre in a day, that is, if we allow double the room to boil in, that is required for steeping. A steam pipe, instead of the tube, and having the top of the box well secured, would permit the process of steaming to go on. It is probable that by either method, spreading on the grass will be necessary to obtain soft flax. The yarns of which the sail-cloth is made at Paterson, are all steamed. The Navy Board expressly forbid their being boiled in alkaline lye, as is usual in most manufactures of linen. It is from this precaution, that their canvas has the pliable, oily feeling, which so much recommends it. It should not be lost sight of, that by boiling or steaming much time and expense will be saved in bleaching.

In dressing, says Mr. Pomeroy, "our climate gives a decided ad-

* The best Riga Hemp, supplied for the British Navy, is prepared by steeping; during which it is shifted three times.

vantage over Ireland, Flanders, or the north of Europe, where flax is dried on hurdles, over a peat fire, in ovens, or kilns, requiring great care in regulating the heat, to prevent injury. All this trouble and hazard is obviated by our dry atmosphere, and keen north-west winds. Dr. Deane estimated the expense of dressing flax by hand at one third of the product. I believe the present price does not vary much from his estimate. A respectable gentleman from Dutchess County, New-York, informed me, that mills or machines, impelled by water, have been erected there, that break, and completely dress the flax for a toll of one tenth! It is said one or more of them are in operation in the western part of this State. These mills were invented in Scotland, and are now said to be brought to great perfection. They are erected in all directions in the principal flax districts in Ireland, and notwithstanding the low price and limited demand for labour, are resorted to by the poorer classes of people, the dressing by hand being mostly abandoned. There are machines in England that dress the flax immediately from the field, without any preparation whatever. An account of them may be found in the 5th vol. of the Massachusetts Agricultural Journal. It appears by the report of a Committee of the House of Commons, that in 1817, they were in successful operation. A man and three children impelled the machines and dressed sixty pounds a day. Should they be susceptible of the application of water or steam power, in any degree proportionate, the advantage may be incalculable; but in the present inquiry, we place these machines, however desirable, entirely out of the question.

PRODUCT.

"It is not uncommon in Great Britain and Ireland, to obtain eight hundred pounds of flax from an acre! Six hundred pounds is estimated, in some districts, as an average; but it should be observed that little if any seed is obtained. The average crop in New-England, as far as our information extends, cannot be estimated at more than two hundred pounds, and six or eight bushels of seed.' (We do not include the rich bottoms on the Connecticut and some other rivers.) Dr. Deane was of opinion that four hundred pounds might be calculated on with proper management.

"We think that four hundred pounds of good clean flax, and eight or ten bushels of seed, may fairly be assumed as a medium crop on favourable soils, where the culture becomes such an object as to make other farming operations subservient to it, and due attention is paid to change of seed.

"Those who grow flax to any extent are of opinion, that the seed, at the price it has been for some years past, pays for all the labour bestowed on the crop to the time the flax is ready to be prepared or rotted.

"If we are correctly informed, flax of a fair quality cannot be imported from Ireland for less than fourteen cents per pound. And the price of the best of Russia flax delivered on ship board at St. Petersburgh, is ten and a half cents per pound. The quality called "twelve headed," costs nine and an half cents on board.

"The quality of flax raised in this country, varies more than any other product; and of course the price, which is from six to eighteen cents: The medium is about ten cents per pound.

"It must be acknowledged, that no great exertions can be expected in the pursuits of any people, till "the prospect of reward sweetens the labour." And I anticipate the question that some may be disposed to ask, "Where is the farmer to find a market, if flax is extensively cultivated?" We will ask where could the planter have found a market for his cotton, if machines had not been invented for spinning it? And how could he have supplied it, if the labour of two thousand hands had been required to clean it of the seeds, that is now performed by the Cotton Gin, invented by Whitney? We have shown that the expense of dressing flax has been reduced from one third to one tenth of its value;* and it is a fact well established, that there are now in this country, machines for spinning flax, that perform as well, and more expeditiously, except for the finer threads, than those for spinning cotton! The Paterson sail-cloth is fabricated entirely from yarns spun and twisted by machinery, assisted by as little manual labour as cotton machines. In these manufactories are six hundred spindles. In the State of New-York and Pennsylvania, about three hundred more are employed for sewing thread, shirtings, bed-ticks, shoe-thread, twine, &c. The expense of labour, after the flax is hackled, in attending a machine of twenty-four spindles for spinning common shoe-thread, is thirty-three cents per day, spinning on an average twenty-four pounds a day for each spindle; equal, it is said to the production of a cotton spindle for five or six days.

"Can any thing be wanting but the application of power looms for weaving linen, to place the manufacture nearly upon an equality with cotton? And is there any doubt but they can be so applied?

"The perfection of cotton spinning machinery, and the invention of power looms, with such improvements as are exhibited at Waltham, it is well known are about to produce an entire revolution in the India trade! If they can stop the spindle and the shuttle of the Hindoo, who is supported upon a handful of rice a day, in a climate where little is required for clothing or shelter, what must be the effect of corresponding machines in the linen manufacture, upon the Russian and the German? There is probably at this moment, a millions tons of American shipping clothed with Russian canvas! What, but the raw material of good quality is required to elicit capital, to manufacture in a few years even so as to compete with European nations in the linen market?

* *The usual toll for ginning cotton, in Alabama, we are informed, is one twelfth.*

"The exportation of linen from Germany to North and South America has been, and is at present, of vast amount. The single province of Silesia has sent in one year to Hamburgh and other ports, linens to the value of nearly five millions of dollars, to be shipped by the circuitous route of Cadiz to the Spanish colonies. These customers are at our doors. The United States possess the "Golden Gates of this Commerce," and with exertions well directed to her agriculture, Europe will be obliged to surrender the keys." See Mass. Agr. Rep. vol. VI. p. 304.

Since the above was written a machine has been constructed in New-York for dressing flax, or hemp in its unrotted state, belonging to Messrs. Anthony Dey, and James Macdonald, of that city. It is made to go by water or animal power, and on trial, was worked with facility by four men. It is estimated that when driven by the proper power, the machine, will clear one ton of flax-plant or hemp, rotted, or unrotted in a day. Flax, dressed in this machine, resembles floss silk, and it is estimated, that it may be cleaned through the machine, and brought to this state for about two cents a pound. The machine is of American invention, and said to be unlike any English machine invented or used for the same or a similar purpose. See Mass. Agr. Rep. vol. VII. p. 66.

FLAX BRAKE, a machine used in dressing flax. New improvements of it are, placing the teeth so as to converge towards the fore part, and laying the upper teeth higher at the hinder part. That this machine may last for any considerable time, care should be taken that it be not exposed to the injuries of the weather.

Brakes may be constructed to go by water. Either a mill may be built for that purpose ; or, which is attended with less expense, the machinery may be an appendage to some larger mill, and moved without a distinct water wheel.—But such brakes are attended with sundry inconveniences, besides extra cost in building them, and wasting of the flax : Though it cannot be denied that the work may be performed with much greater expedition.

Not only brakes, but scutchers, or swinging mills, have been invented, to be moved by the foot. Part of the exertion of the labourer may undoubtedly be saved by them. At least, when they are used by way of change, the work may be lightened on the whole.—They who think it expedient to have these machines, may find them described, with cuts annexed. in the *Complete Farmer*.

FLOODING, FLOATING, OR DROWNING, covering of low lands with water, when a rivulet passes through them, by making a dam at the outlet. When there is a sufficiency of water, and a short dam will answer, this is a piece of husbandry that ought not to be neglected. Oftentimes it may be of great advantage.

Sometimes it is done for the purpose of destroying the natural growth of trees, bushes, &c. The water not only makes an essential alteration in their food, but also excludes them from the free air, which is essentially necessary to vegetation. It is no wonder, therefore, that it proves their destruction.

The flowing of two summers is found sufficient to kill every plant of the woody kind, so that it will not sprout any more. But some advise to drawing off the water in August, that the ground may be, for a few days, heated by the sun. The plants thus suddenly pass from one extreme to another, which will doubtless tend to destroy them the sooner. But when the season is so dry that another pond of water could not be immediately raised, the drawing off had better be omitted.

Another intention of flooding is, to enrich the soil. Some lay their low grass lands under water during the whole of the winter. This may be a good method for lands which are so low and wet, that none of the best grasses can be made to grow on them. The poor water grasses will grow the faster ; and the crops of such hay as it is, will be the larger.

But places where clover, or herds-grass, or red-top will flourish, should not be flowed during the winter : Because the winter frosts are known to be necessary to the production of these grasses.

Flooded lands should always be laid bare early in the spring, that the growth of the grass be not prevented : Or that the ground may be dried so early as to be fit for tillage crops. And ditching of flooded lands, at least round the borders, will be necessary to lay them dry enough for tillage.

As standing water catches dust from the atmosphere, and always contains more or less of the finest particles of soil, it deposits a rich sediment ; a fat slime, therefore, will remain on the surface after the water is removed. And a time should be chosen for drawing it off, when the air is calm, and the water clearest, that as little a quantity as possible of the food of plants may pass off with it. Such land is no more liable to suffer by drought than the fertile land of Egypt, which is yearly enriched by the overflowing of the Nile.

Though winter flooding do not suit the nature of good grasses, a few days flooding in the spring and fall will not hurt them ; but will enrich the soil, and so promote their growth. The soil will have the same advantage as intervale land, which is made rich and fruitful by occasional flooding : Yea, a greater advantage, as the water may be applied and removed at pleasure.

FLOUR. "It has been generally supposed, that if wheat be much injured during a bad harvest, the flour made from it will not ferment, or bake into loaf-bread, and that it is only fit for distillation, or to be eaten by live stock. But such ideas seem to be erroneous. With the aid of soda the flour may be much improved ; and at any rate may be made into cakes, or biscuit, and consumed with safety and advantage." *Code of Agriculture.*

The carbonate of magnesia of

the shops, when well mixed with new flour, in the proportion of from 20 to 40 grains to a pound of fiour materially improves it for the purpose of making bread. Loaves, made with the addition of the carbonate of magnesia, rise well in the oven ; and after being baked, the bread is light and spungy, has a good taste, and keeps well. In cases where the new flour is of an indifferent quality, from 20 to 30 grains of the carbonate of magnesia, to a pound of flour, will considerably improve the bread. When the flour is of the worst quality, 40 grains to a pound of flour is necessary to produce the same effect. As the improvement in the bread from the new flour depends on the carbonate of magnesia, it is necessary that care should be taken to mix it intimately with the flour previous to making the dough. A pound of carbonate of magnesia would be sufficient to mix with two hundred and fifty-six pounds of new flour at the rate of 30 grains to a pound.

FOAL, a colt. " Foals are usually foaled about the beginning of summer, and it is the custom to let them run till Michaelmas with the mare, at which time they are to be weaned. When first weaned they must be kept in a convenient house, with a low rack and manger for hay and oats ; the hay must be very sweet and fine, especially at first, and a little wheat bran should be mixed with their oats, in order to keep their bodies open, and make them eat and drink freely. When the winter is spent, they should be turned into some dry ground, where the grass is sweet and short, and where there is good water, that they may drink at pleasure. The winter after this, they may be kept in the stable, without any further care than that which is taken of other horses: But after the first year, the mare foals and horse foals are not to be kept together. There is no difficulty to know the shape a foal is like to be of ; for the same shape he carries at a month, he will carry at six years old, if he be not abused in after keeping."

We often hear it lamented, that our breed of horses is so bad. But I am convinced that, as our colts are managed, if we had any other breed, we should soon make it appear to be as mean as our own, if not worse. The abusing of colts in the first winter, is the principal cause of their proving so bad. For our farmers seldom allow their weaned colts any food besides hay, and that is not always of the best kind. So that they seldom fail of being stinted in their growth, in the first winter, to such a degree, that they never get the better of it. A colt that is foaled late, should not be weaned till February or March, and should have oats during the whole of the winter. In some countries they allow a young colt fifteen bushels. We need not grudge to feed them with meal, oats and bran, besides the best of clover hay ; for they will pay for it in their growth. After the first winter, they will need no extraordinary feeding till they are grown up. Were the above directions observed, we should soon see an improvement of our breed of horses. They would be capable of doing much greater service,and be likely to hold out to a greater age.

FODDER, dry food for horses and other cattle. The term includes corn or grain,hay and straw, the stalks and leaves of Indian corn, the haulm of pease and beans, &c. Dried weeds, and leaves of trees, may also serve as fodder for hungry and hardy cattle.

Mr. Lisle recommends elm leaves, dried on the small branches, as a great relief to cattle in winter. He says the cattle will eat it before oats, and thrive exceedingly with it. Also, the chaff of all kinds of grain, in the old countries, is reserved for fodder, and made more account of than the straw. In this country it is suffered to be driven away by winds. This is an instance of our want of economy.

In such a country as ours,where the winters are long and cold, and where grass does not serve for the cattle so much as half the year, providing fodder, and preserving it are matters of high consequence. In this business, a great part of the farmer's care and strength is employed. For there are not more than two months in a year, in which farmers are not either preparing, and laying up fodder for their flock, or else dealing it out to them.

The ways to increase the quantity of fodder, will be found under other articles. The ways to preserve it, so as to make the greatest advantage from it, may be here considered.

One important caution to be observed is, that hay, which is the principal fodder, should not be so much dried as to occasion its wasting. When it has been properly made, it should not be carted in, if it can be avoided, at a time when the weather is dry and windy, nor in the hottest part of the day.— Mornings and evenings are the best times for removing it, as there is a dampness in the air which prevents it being too crispy. The leaves will not crumble, nor the seeds shatter out. The best parts of the hay are often lost by not observing this caution ; or at least much diminished.

The hay which is to be stored in small or narrow mows, and on scaffolds, will keep well with little drying. That which goes into a large mow, will need to be drier, as the air will not penetrate so near to the centre of it.

In disposing of the different kinds of hay and other fodder, some regard should be had to the places, or parts of the barn, in which the different sorts of cattle are kept. The clover hay, for instance,should be laid up near to the stable where horses are kept, as this is the most suitable fodder for them. The good hay of other kinds, should be put where it can be handily given to the calves, milch cows, and working oxen. The meanest fodder nearest to the apartment of the growing young stock, on which it is commonly bestowed, and which is more proper for them than for the rest.

In those parts of the country where salt hay cannot be had, it is a good method to apply salt to hay that has been damaged in making, and to straw, and hay of low meadows, as it is put into the mow. The salt will make it more palatable both to horses, and neat cattle. One peck of salt is enough for a ton of hay.

Some choose that a barn should have large gaps between the boards on the sides, that the hay, &c. may have air. This is surely a mistaken notion; for the hay that is nearest to the gaps will lose its sweetness. The roof of a barn should also be kept very tight; and none of the hay should be laid very near to the ground.

I do not approve of stacking any kind of fodder, excepting in case of necessity. For some inches of the outside of a stack is certainly spoilt by the weather. It is well if the rest happen to be well saved. It often proves otherwise.

When a farmer has more hay than his barn will hold, let him stack it near to the barn; and, as soon as he has made room, in some damp or calm day take it in. There will be the less danger of its getting damage.

Farmers, who mean to keep good their stocks, and to have plenty of manure, should not be fond of selling hay. If they should have some left in the spring, it will not grow worse, but some sorts will be better, by keeping. And if a short crop should happen, they will be glad they have kept it.

Straw that is reserved for fodder, may help to preserve the husks and bottom stalks of Indian corn, which commonly have too much sap in them to be mowed by themselves. If they are put in a mow together, in alternate thin layers, the straw will preserve the corn stalks, and the stalks will impregnate the straw with their sweetness, so that the cattle will eat them together with a good relish, and be well nourished by them.

Another method of managing straw, which I have found to be of singular advantage, is to mix it with salt hay which is not more than half dried. The hay is thus kept from heating, and the straw is so tinctured with the salt and sap of the hay, as to be rendered an agreeable fodder for cattle.

It is well known that cattle prefer short straw to that which is long: Therefore many farmers cut their straw as short as oats, and to tempt the horses to eat it, mix some oats or barley among it.

But the most effectual, and in the present state of agricultural science, the most approved method of increasing fodder for winter consumption, is to provide large quantities of pumpkins, potatoes, carrots, Swedish turnips and beets, by which the consumption of hay is exceedingly diminished, and the cattle, particularly milch cows, are kept in much better conditon, and it is believed at much less expense. Two or three acres, in the culture of such vegetables, furnish as much food, as eight or ten acres of the best grass lands in hay. There is no doubt, also, that a great saving may be made by cutting fine the straw and hay given to cattle. Various machines have been invented to facilitate this operation, and great ingenuity has been displayed in their construction; but the general opinion seems to be that the simplest machines, costing not more than from three to five dollars, are the most economical on the whole. There is no doubt that the subdivision of the food, and frequent feeding which our long winters afford ample time to effect, are exceedingly useful, and economical, compared with the wasteful modes heretofore in use, particularly that of throwing down large bundles of hay and cornstalks to the cattle in the yard. No man of the slightest observation can have failed to remark that at least one fourth part of the fodder so administered is given to be destroyed.

"Salting all fresh hay, when put up, is a great addition to it. A respectable farmer of Herkimer County, New York, who keeps a large stock of cattle, says, he is certain, that adding eight quarts of salt to each ton of hay will make it go as far as a ton and a quarter that has not been salted. At the same time hay may be put into the mow, when salted, in a much greener state than without it; and when taken out will be found almost as green, and apparently as fresh, as when first stowed away." *Farmer's Assistant.*

"Meadows which produce wild grass ought to be mowed very early and salted. But cattle should not be kept constantly on salted fodder.

It is observed in the Domestic Encyclopædia, (art. Cattle,) that "the first object in the article food is wholesomeness: wild cattle feed entirely on green vegetables which they find throughout the year. Similar nutriments should, if possible be procured for tame cattle in all seasons. Therefore cabbages, turnips, carrots, and the mangel wurzel are recommended as winter food."

FODDERING, feeding cattle with dry food. We have occasion to begin to fodder, most commonly, about the beginning of November; and to continue doing it till the middle of May, and sometimes later.

We should take care not to begin to fodder till it is really necessary: Because cattle that are foddered, will not graze so diligently. When it is once begun, the cattle will expect it, and it must be continued. When we first begin, we should fodder early in the morning only; for at that time of the day the frost is usually on the grass; so that the cattle will not graze. They should not yet be housed, horses excepted: But in wet weather the whole stock should be housed; for they bear cold better than wetness. Or if not put into the barn, they should have a shed in the yard, under which they may shelter themselves.

The meanest fodder should not be dealt out first of all. The husks and stalks of Indian corn are suitable for this season. The straw and the worst hay should be reserved to give them in the coldest weather; for it is then that they have the keenest appetites.

The hay of low ground, straw and haulm, if salt hay be not to be had, may be sprinkled with salted water, if salting it in the mow has been neglected. They will not only eat it heartily, but live well upon it.

Wild grass hay is not fit for horses, nor any of the water grasses. They will need some grain, if they be fed on any other hay besides clover. They should have a small window against their rack, to let in fresh air to their fodder, and at the same time give them light. If horses have not grain through the winter, they should have it at least in the fore part of winter; for the coming on of winter is the most trying season for them. If they be fed with Indian corn, it should be well soaked and swelled; it will give them the more nourishment.

Neat cattle and horses should not have so much laid before them at once, as will quite serve to fill them. The hay they have breathed on much, they will not eat up clean, unless when they are very hungry. It is best, therefore, to fodder them twice at night, and twice in the morning. Let neat cattle as well as horses have both light and fresh air let in upon their fodder, when the weather is not too cold, or stormy, to allow the windows to be open. What one sort of cattle leave, should be thrown to another sort. Those that chew the cud will eat the leavings of those that do not, and *vice versa*.

It is also well known to farmers, that what cattle leave in the barn, they will eat abroad in the open air; and most freely when it is laid upon clean snow. Not only this, but the meanest of straw should be given them in this way. What is left will help to increase the manure in the yard.

Every farm yard, where any considerable stock is kept, should be furnished with a large shed, and a rack under it. For where there is no clean snow to lay the straw, and other mean fodder upon, it should be put into the rack. A larger proportion of the dung will be dropped under the shed, than in any other part of the yard. And this dung will be better than the rest, as it will not be washed by rains, nor so much dried by the wind and sun.

Sheep, when they are under cover, should draw their hay through a rack, made so close as just to admit their noses. They should have good hay, and a cool and dry house. Beans is a sort of food they eat very greedily, and even the straw. But it is said, that ewes with young should not be allowed to eat many beans; as it will make their lambs grow too large within them. Neither should they be fed too generously, nor to the full, till near the time of lambing.

When a farmer thinks that he has too much stock for his fodder, as will sometimes be the case, it is not best to pinch them in their allowance so much in the fore part of winter as in the latter part. For the cattle are more liable to be pinched with the cold, in December and January, than afterwards.

And no man knows how favourable the latter part of winter may be. Advantage also may be made of browsing in the latter more than in the former part of winter, as the buds then begin to swell, and the twigs have more sap in them than before.

When browsing is depended on, the farmer who has salt hay, should preserve a sufficient quantity of it to the latter part of winter. It will give the cattle a high relish for browse. If they have no salt hay, they should have salt, to increase their appetite.

Cows that are near calving, should not be driven out after the browse, for fear of accidents. They should be kept on the best fodder; not be tied up with the other cattle; but each one should be fed in an apartment by herself, without tying.

FOG, FOGGE, OR FOGAGE, long grass and stumps of grass, remaining in mowing grounds and pastures till winter.

This is accounted in general a benefit to the land; especially when the grass is not of a bad and sour kind. The snow presses it down close to the surface, where it shelters the roots of the grass, corrupts it, and turns it to manure. But when mowing grounds are fed very close in the fall, the ensuing crop is poorer, the roots being more injured by the feet and teeth of cattle, and more exposed to the weather. The dung they drop, though it be considerable, will not wholly repair the damage of close feeding and trampling.

But fog is most essentially serviceable on a soil of the clay kind. It forms a cover which retains the rains and dews, in the following spring and summer, so as to give the surface a more equable and constant moisture; and prevents the binding and cracking of the surface by the heat of the sun.—Nothing can better oppose the ill effects of a dry season on such a soil.

Where fog is left very long, it checks the growth of the young grass of the next year, causes it to spring up feebly, and materially impedes the progress of the scythe. Where the field is not adjacent to houses or wood lots, so that they may not be endangered by fire, it is very useful, in a dry day, to set fire to the fog, and burn it. This is in no respect injurious to the future growth, but the ashes produced by the combustion in burning the old fog, causes the new grass to spring with increased strength and luxuriance. The experiment, however, should not be attempted, except in insulated fields, remote from houses and trees, and especially from evergreens. It should be done early in April.

FOLDING *of land*, confining sheep, or other cattle, nightly, in a small lot or yard, for the purpose of enriching the soil. The benefit arising from this is so great, that it ought not to be neglected, especially in those parts of the country, where the wolves do not come.

Some turn in their other cattle with the sheep. This is good conduct, when the soil is warm sand or gravel; and not bad when it is loamy. But it may be better to yard the black cattle without sheep, on a a very dry soil ; such as hungry sand or gravel ; and the sheep without the black cattle, on a soil that is heavy and cold. Thus both these kinds of manure will be applied to the soil which will be most helped by them.

Folding is a much better method than carrying dung from the barn-yard, when the season is suitable for doing it. One great advantage of it is, that none of the stale is wasted, but every drop of it instantly absorbed by the soil that needs it, and will make a good return for it.

Folding, or yarding, is but little attended to in this country ; and not half the advantage is made from it that might be, when it is attempted. It is said that one hundred sheep in a summer will enrich eight acres, so as to need no other manuring for six years.

This matter is certainly misconducted, when a farmer, either to save the labour of fencing, or through ignorance of the advantage of folding, makes his enclosures too small, and folds the land too much for his own profit.

Let a spot of half an acre be ploughed and fenced. Turn in, each night, a dozen head of neat cattle, and fifty sheep. Continue to do it for three weeks, harrowing the surface once in three days, to mix the excrements with the soil. The ground will be sufficiently folded to produce a fine crop of turnips, or almost any other good crop. It is reckoned by some that a sheep will fold one yard square in a night ; or rather one rod square in about a fortnight.

A yard for cabbages or turnips, may be begun about the middle of May ; or when the cattle first go to grass. About a month after will be nearly the right time to transplant cabbages ; and six weeks or about two months after, to sow turnips. And, for a general rule,it is best that a crop should succeed the manuring as soon as possible.

When a crop of wheat is wanted, the ground may be folded in July, as the seed is to be sown in August. And frequent ploughing and harrowing for this crop should not be neglected. If the land be wettish, do it in the middle of the day ; if dry, in the morning before the dew is off.

Low grass grounds, which are cold and sour, and produce bad hay, may be surprisingly meliorated by a little folding. It kills fern and mosses, and roots out the wild and watery grasses, even without breaking up the soil. At the same time it encourages the growth of better kinds of grasses. This may be done at certain seasons that are unsuitable for the folding of ploughed lands, they being too wet and dirty for the sheep to lie upon, as in October, November, March and April. Sheep are more proper for this sort of folding than larger cattle, as their excrements are hotter.

FOOD OF PLANTS. "No one principle affords the pabulum of vegetable life ; it is neither charcoal nor hydrogene, nor azote, nor oxygene alone ; but all of them together in various states and various combinations." *Agricultural Chemistry.*

FOREST, a tract of ground, producing wood. Each farm of any considerable bigness, should have a forest to afford a supply of fuel and timber. In clearing farms in a new country, due regard should be had to preserving a perpetual forest. Some have mistaken their interest so much, as not to leave a sufficient quantity of land uncleared. So that they are put to the disagreeable necessity, either of buying their fire-wood, or else of going some miles after it. That part of a farm should be set apart for this purpose, which is least adapted by nature for tillage, or grass. Land which is swampy, with a very thin soil over a sandy bottom ; land that is rocky and mountainous, or which will but poorly bear a dry season, or even the most sandy or gravelly heights, or steep declivities which cannot be ploughed, may answer well for a forest. Forest trees having long roots, some of which penetrate deeply, will find sufficient nourishment in places where corn and grass cannot be cultivated to advantage. So that it is very bad economy to suffer any such places to be destitute of growing trees. For if they do not produce wood they are in a manner useless. Or if they produce any grass, trees will not hurt them for pasturage, but in some cases make it better.

The quantity of ground that should be set apart for this use, must vary according to the largeness of the farm it belongs to, and according to the demand for wood, the quality of the soil, and the nature of the climate. If the climate be hot, the forest may be smaller. A small farm cannot so well admit of a large lot for wood as a larger one. Some intelligent farmers in this country have thought they could make a lot of ten or a dozen acres answer the purpose of supporting one constant kitchen fire. But it certainly will not, unless the soil be uncommonly fruitful, and the trees be such as are of the quickest growth. If land be poor and dry, it will require twenty acres or more, to supply one single fire, and keep the stock of trees undiminished.

To thicken a forest, or to prevent its becoming too thin, cattle should be kept out at all seasons, that all the trees which spring out of the ground may live and grow up to maturity. And when it is found needful, acorns, or other seeds, should be planted, so that none of the ground may continue unoccupied.

In our most southern climates, I find that hard wood is more rapid in its growth than in the northern. And sprouts oftener grow up from stumps of trees that are felled.— The trees that grow up quickest in general should be most cultivated. Those of these kinds should be more generally left standing than others ; such, for instance, as the red and grey oaks, ash, white maple, &c.

When a number of suckers spring up from a stump, all, excepting one or two, should be taken away as early as possible ; then the remaining ones will grow with rapidity. Those are to be left which are tallest, and most rapid in their growth.

When a farm is quite destitute of a forest, some spot, or spots, the least adapted to tillage of any part of the farm, should be converted to this use, and be planted with such trees as may be expected to thrive best.

If these spots be tillable, "cattle of all kinds, and swine should be fenced out; and the ground well ploughed and harrowed, and made mellow. Acorns may be put in, in rows four feet asunder, two inches apart, and two inches deep. The intervals may bear some hoed crops, while the trees are small. They should be hoed the first year with the hand hoe; the second with the horse hoe, and so on afterwards. When they are a year old begin to thin them. When they are, by repeated thinnings, as they grow larger, reduced to the distance of eight feet, all the rest may stand for timber, till some of them are fit for some uses. But the final distance for large timber trees, is from twenty to thirty feet." *Complete Farmer.*

But if places designed for forests cannot conveniently be tilled, the trees should be raised in a nursery, and transplanted into such places. The cost of doing it will be trifling, to compare with the advantage to be obtained by doing it, especially in those parts of the country where wood is become a scarce article. Small clumps of trees on little eminences, have an excellent effect on the beauty of a country.

In the Mass. Agr. Rep. vol. V. p. 32 to 62, were published some valuable "*Remarks on the gradual diminution of the Forests of Massachusetts and the importance of an early attention to some effectual remedy. With extracts from the work of M. Michaux, on the Forest Trees of North America.* Our intended and necessary brevity render it inexpedient to give the whole of this paper, but we will present such extracts as we think will best serve to excite attention to this very important subject.

In adverting to the great solicitude manifested by some European nations to preserve their woodlands, the author observes that "In France, great and extensive public forests were sacredly preserved, even during the period of the revolution, when almost every thing else was abandoned to destruction. By the system of management which now prevails, it is computed, that the forests of France, and the supplies which they furnish, will never diminish, and that there will be always sufficient for domestic consumption for fuel, as well as for architectural and naval purposes."

Stating the causes which induced a too general destruction of forest trees in this country, the author observes; "The cost and expense of clearing our lands compared with their value after they were cleared, and the difficulty of eradicating completely the after growth, were so great, and the forests themselves appeared so vast in proportion to the probable demand for fuel, and wood for building, and other purposes, that no man dreamt that the day would arrive when their descendants might regret the improvidence of their ancestors. Hence there seems to have been a sort of hatred, an indescribable prejudice against trees, especially round their dwellings. An exception, perhaps, ought to be made in favour of some counties, in which a little degree of mercy was exercised towards our native elms, which were permitted to retain a parsimonious possession of the waste lands of our public roads, where they exhibit a melancholy sample of the beauty of our native forests.

The above causes alone can account for the fact, that in our climate, where the summer months are so hot, compared to the climate of Europe, and where the clearness of the sky seems to render shade so much more important, we find such a general warfare waged upon trees in the vicinity of dwelling-houses.

" The Trustees of the Massachusetts Agricultural Society, impressed with these ideas, and alarmed at the constant increase of the price of fuel, have for twenty years past, by offering premiums for planting, and by remarks in their periodical publications, endeavoured to call the attention of the farmers of this State, to this important subject. It is remarkable, that but one premium for planting, during this period, has been claimed or granted. Perhaps there was one defect in the form in which the premium was offered. It was confined to those who should raise the greatest number of trees, not less than two thousand, *from seed.*— There seems to be no good reason for this limitation; on the contrary the experience of European farmers and cultivators would seem to authorize a preference for planting. The great obstacle to the latter mode, in our country, is the defect of extensive nurseries. Land must be uncommonly well prepared, and for a long succession of years, watched and cultivated, in order to raise trees from seed.-- The best mode of raising forest trees in great quantities, is unquestionably that which has been adopted in Europe. Forest trees, though so hardy and vigorous after they have attained a certain size, are remarkably tender in their early growth. They require the aid of professional men, and skilful gardeners; and it is well known that one or more transplantations, before they are finally planted out, is requisite to give them that vigour and abundance of small roots, which are necessary to their success.

" The practice of transplanting trees from our forests, of six or ten years growth, robbed, as they must be of the greater portion of their fibrous roots, and suddenly exposed to a soil and air, to which they have been unaccustomed, cannot be too much reprobated. The only thing which can be urged in its favour, is the necessity of the case, our nurseries not furnishing stocks in sufficient quantities for any considerable experiment. It is however true, that it would be cheaper for an American cultivator, who should be disposed to cover several acres with forest trees, to import plants of two or three years of age from Europe, than to attempt to effect the same object by planting acorns, walnuts, or sowing the seeds of other plants in the spot in which they are intended to grow, and watching them during their feeble state. Still more true it is that such an importation would be cheaper than to attempt to introduce the sickly and mutilated plants from the native forest into new ground prepared for this purpose. A thousand young healthy trees may be imported from Europe, and planted out at less expense than that at which one hundred could be removed from a neighbouring forest, into a new plantation.* This remark is the result of much experience. I would not be understood to recommend the use of the European nurseries to the exclusion of our own. I would rather excite an attention in cultivators of opulence to the formation of nurseries, and of all our farmers to the encouragement of nursery men. The employment ought to be distinct from that of a farmer. It requires peculiar skill and talent, and ought to be encouraged. It is a mortifying truth, that the inhabitants of Massachusetts import seven-eighths of their fruit trees, and most of their ornamental forest trees from New-York. I would not quarrel with this state of things, if I did not believe that it has been in part owing to a want of attention to a very interesting subject.

* Taking into view the proportion which would finally succeed.

What the Agricultural Society has failed to do by its premiums, has been in part effected by individual taste. The Hon. Mr. Gore led the way, which others have followed, in the formation of plantations designed principally for ornament. Farmers may, perhaps, be disposed to treat with levity, these efforts to adorn our country, but in truth if they knew the interesting and important effects of this taste in Great Britain, they would materially change their opinions."

The author takes notice of Monsieur Michaux and his Treatise on the Forest Trees of North America; from which the following is extracted.

" America is more favoured, says Monsieur Michaux, in the variety of her forest trees, than France.—The number of sorts of American forest trees, whose growth amounts to thirty feet at least, which Monsieur Michaux describes, is 137, of which ninety-five are employed in *the arts*.

" In France there are only thirty-seven which grow to that size, of which eighteen only are found in their forests, and seven only of these are employed in civil and maritime architecture.

" Monsieur Michaux says, that the object of his voyage was to acquire an intimate knowledge of the most interesting trees of America, of those which were the most useful, either for combustion or different sorts of building. He adds that he shall suggest the manner in which the American forests ought to be managed, by pointing out the sorts which ought to be favoured, and those which should be destroyed; for in his judgment a bad tree ought never to be suffered to grow in a place which might produce a better; and in no country does he think this choice more important than in America; for reasons which he does not assign. He says, he should not hesitate to allege that of two collections of forest trees, situated in the same district and of equal extent, the one from which the bad sorts should have been taken away, would, whenever the whole wood should be cut down, be worth fifty per cent more than the other. He must intend to say after a lapse of some years.

" Monsieur Michaux remarks on the singular confusion which prevails in our country, in the popular or common names of our forest trees." "The same species receives almost always different denominations in different districts; frequently also, the same name is given to species *very distinct*, and very often indeed three or four different names are given to the same tree, in the same district." This is a fact well known to every man of any observation, and is of serious inconvenience. It introduces a confusion into conversation, and even into contracts, which is very inconvenient. I shall cite one example, out of twenty, which have occurred to myself. A carpenter recommended juniper posts, as the best with which he was acquainted for durability. I at first supposed that he had got the true name of the red cedar, and asked him if he intended that tree. 'No; he meant the juniper, which was obtained in New Hampshire, and came down the Middlesex canal, a close grained, hard and heavy wood.' I presume we should never have understood each other, if he had not recollected, that he had seen it in a new plantation on my estate. On carrying me to this tree, I found it to be the larch, which I had known under the popular name of hackmatac.

The author has added to this article a List of the Forest Trees, described in the work of Monsieur Michaux, which occupies about twenty pages of the Repository, and is too long for insertion in this book. This list contains the botanical and vulgar names of ten kinds of pine; four of spruce, twenty six of oak, &c. &c.; and observes " We have given the above list, because we thought it might be of general use, as the work of M. Michaux is too large to be within the reach of intelligent and well educated farmers generally, and because we know not any catalogue of the forest trees of the United States, detached from large botanical works could be got. It is also very convenient, as exhibiting the variety of common names applied to our trees in different states."

The next number of the Mass. Agr. Repository, vol. v. p. 180, contains another valuable paper from the same hand, on " *The different Uses to which the Woods of the American Forest are* applied in the various parts of the United States.

FOUNDERING, a very painful disease in the feet of horses. A horse affected with this disease draws himself up in a heap, and is loth to move. It is said to be occasioned by bruises on the legs, by bad shoeing, by standing in cold water after being heated with exercise; or even by standing still in the stable for several days. As the disorder is in the feet, covered by the hoofs and soles, it is difficult to make application to the parts affected. But drawing out the sole Mr. Snape does not approve of, without paring the hoof. Something must be done without delay, lest imposthumations come on in the feet, by which the hoofs will be cast off: In which case, the horse must lie by useless for a number of weeks before the new hoofs will be grown. The same writer directs that the hoofs be razed from the coronet or top to the bottom, quite through the hoofs to the quick, so as to make the blood run. These channels in the hoofs may be readily made with a common marking iron.

To cure the wounds made in the hoofs, apply to them tar, turpentine and honey, melted together, with a fourth part of spirit of wine,soaking pledgits of clean flax, or tow, in this mixture, and laying them upon the chinks, not opening them till two days after the first dressing; afterwards making fresh applications every day, till the channels in the hoofs are grown up.

The same applications must be made to the sole, if that has been drawn. But similar channels in that, as I apprehend, may answer well enough, and paring the sole thin. They must, however, have the same dressings as the hoofs. A piece of leather should be laid over the sole, and the whole foot so bound up with strong bandages, that the applications may not get displaced. See *Gibson's Farriery.*

"Mason's Farrier" gives the following remedy for a founder: "Take from his neck vein at least one gallon of blood; give a drench of one quart strong sassafras tea, one tea spoonful of salt petre, and a quarter of an ounce of assafœtida, and do not permit him to eat or drink for five or six hours—at the expiration of which time should he not be evidently better, repeat the bleeding, taking half a gallon of blood, and give another drench: at night offer him some bran or oats, scalded with sassafras tea, and if it can be procured, let him have green food fresh from the field, for it has the happy effect of opening the bowels, and cooling the system: his feet should be nicely cleaned out, and stuffed with fresh cow manure: his drink should be at least one half sassafras tea, with a small handful of salt thrown therein.

"By the morning should the horse be better, nothing further is necessary, only being careful not to over-feed him. But should there be no change for the better, tie a small cord just above his knees, and with a lancet or phleme, bleed in a vein that runs round the coronet just above the hoof: take from each leg a pint of blood: give a pound of salts dissolved in three half pints of water, in form of a drench: keep his feet stuffed with fresh cow manure, and bathe his legs with equal parts of sharp vinegar, spirits, and sweet oil, or lard. By attention to these directions, in two or three days the horse will again be fit for service."

FREEZING. As a more complete subdivision and pulverization of the soil is one of the chief ends of culture by the plough and the harrow, it is very important that every farmer should be deeply impressed with the TRUTH, that frost operates as powerfully as any human means towards the same end. It is indeed generally known and faintly acknowledged, that frost breaks up the soil; but too small a proportion of our farmers apply this principle sufficiently in practice. Fall ploughing is very important for this purpose, and back furrowing, so as to lay the land in ridges, still more so.

The European gardeners, who have been introduced here within the last twenty years, annually either trench our gardens, which is nothing more than digging them two spades deep, or throw them up in high ridges to be open to frost. They learned these usages in Great Britain, where, however, the frost has much less power than with us. We have seen the soil so entirely broken and pulverized by the course just mentioned, as to require no digging in the spring, only levelling, before the very smallest seeds were sown. It cannot be questioned that frost ought not to be overlooked as a very efficient auxiliary in cultivation.

21

FRUIT TREES. The old practice of pruning fruit trees in autumn has been condemned by modern writers. Mr. Forsyth, in his *Treatise on Fruit Trees*. (p. 18, Cobbett's edition) says, "I have a great dislike to autumnal pruning of fruit trees; of all kinds of stone fruits in particular; for by pruning at that season, you seldom fail to bring on canker. Whereas, in spring, when the sap is beginning to follow the knife, the lips will quickly grow."

A writer, whose essay on the subject of apple trees, is published in the "Massachusetts Agricultural Repository," vol. V. p. 121 to 127, mentions three modes of mismanagement, which injure orchards in the vicinity of Boston. 1. Beginning to prune them in March, "when there is still much wet and frosty weather, and no activity in the sap of the tree." 2. The "old practice of *hacking* and mutilating apple trees in a manner ruinous to an orchard. It is a universal practice among the old farmers, to mount the tree with a hatchet or bill hook, and hack off any branch which is in a state of decay, or which is misplaced, about six or eight inches from its insertion, leaving a stump to rot, and to operate as a conductor of the water frost and canker into the mother branch in which it grew, or into the body of the tree according to its situation. This was done originally from an idea that if you cut close to the mother branch, or to the body of the tree, the rot or canker will seize more readily on its trunk, than if cut at a distance, and that the tree will decay the sooner. The practice has been followed without reflection, and without reason by many ; but the error is so obvious, that any man of observation may see it yearly ; and any one, who doubts, may satisfy himself in one season of the incorrectness of the practice, by making his experiments on a young tree."

This writer advises the farmer, when he has fixed upon a limb to be lopped off, if it is large and heavy, to cut it first at some distance from its insertion, to prevent its weight in falling from lacerating the bark at the shoulder, whence your final cut is to be ; because this leaves an opening for water to get under the bark,and cannot easily be healed. You may now saw the stump close to the branch from whence it proceeds with safety ; or if it be a portion of a branch which is to be lopped off, the cut should be down to a sound, healthy lateral branch, growing from the same limb ; or if the limb to be cut off, proceeds from the body or trunk of the tree, then it should be sawed close to the shaft. The wood in all cases should be smoothed over, and the edges of the bark carefully pared with your knife, or hatchet, so that the water will run off the wound. If the cut be made on a lateral branch, it should be sawed obliquely or slanting, so as to leave no dead wood, or wood to die,and in all cases the cut should be on a sound and healthy part of the tree. If the branch on which it is cut, is a healthy and vigorous one, it will heal without difficulty, if pruned the last of April, or beginning of May, but if in March, the wound should be covered with a compost ; but if the wound be large, so as to require several seasons before it can heal entirely, it will be better to apply the compost,whether it be pruned in March, or later.

The third error in managing fruit trees, according to this writer consists " in the habit of encouraging luxuriant upright branches to the great injury of the natural horizontal fruit-bearing branches ; these are very properly called *glutton* branches, because they consume the sap which would otherwise go into the lateral and fruit-bearing branches, and in the course of a few years, they leave the fruit branches decaying and decayed ; the farmer then resorts to his axe, cuts away the dead and dying wood, and leaves the glutton in full possession of all the nourishment which the roots afford ; but in return, this voracious member of the orchard gives no fruit until many years, and then it is of an inferior quality.

" To prevent this the cultivator should suppress all the stiff, upright shoots the first year they appear, by cutting them off *close* to the branch from which they issue, taking care not to leave the shoulder of the shoot, as he will in such case have the same duty to perform again ; but if the shoulder of the glutton be cut away, the sap will be distributed among the lateral fruit-bearing branches, which will be kept in vigour, and continue in a healthful bearing state.

" The compost best suited to

cover the wounds of all trees, is a composition of tar, bees-wax, and red ochre, boiled or simmered for half an hour or twenty minutes together. The proportion which I have used are, a pint of tar, into which I put a piece of bees-wax as big as an English walnut; when these are incorporated, I scatter a small quantity of pulverized red ochre, say half a gill, and stir them well together, while boiling hot or simmering. When this compost is cool, it should be stiff enough to resist the heat of the sun sufficiently to prevent its running, and yet soft enough to be applied to the wound with a small flat smooth stick; it will last two years at least, without requiring to be renewed; it yields to the sap as it issues from under the bark round the wound to cover it, while it continues to protect both the wood and the edge of the bark from water, and of course from decay."

A writer in the Massachusetts Agricultural Repository, No. IV. vol. IV. p. 133, after speaking of the decline of fruit trees in his neighbourhood, says that a gentleman "directed their trunks or bodies to be washed and well rubbed with *soft soap;* and it is not easy to imagine the early change which appeared in their bark, and foliage: the bark became smooth and glossy, and seemed sound and beautiful; and he thought the tree improved in every respect. I have tried the same experiment, and with equal advantage to Apple Trees, Pear Trees, and Peach Trees; and am persuaded they have been greatly benefitted by this process: it is used in the spring, and may be repeated in following years as frequently as the trees appear to require it." It is likewise recommended to white wash trees in the spring with a mixture of lime and water.

Fruit trees may be forced to bear in the following manner. With a sharp knife cut the bark of the branch, which you mean to force to bear, not far from the place where it is connected with the stem; or if a small branch or shoot, near where it is joined with the larger bough; the cut is to be made round the branch so as to encircle it, and penetrate the wood. A quarter of an inch from the first cut, make a second cut like the first, so that by both encircling the branch, you have marked a ring a quarter of an inch between the two cuts—then with a knife separate the bark from the wood, removing even the fine inner bark which lies immediately upon the wood, so that no connection whatever remains between the two parts of the bark, leaving the wood naked, white and smooth.

This operation must be performed when the buds are strongly swelling, or breaking out into blossoms; the same year a callous is formed on at each edge of the ring, and the bark is again restored without detriment to the tree or branch operated upon. It seems calculated to force those trees to bear, which put out a proportion of blossoms, and yet bear no fruit; or if they bear, the fruit often drops from the tree before ripe. The fruit from trees so operated upon, will be larger and fairer, and ripe earlier than the other fruit on the same tree. It is well known to botanists that the sap ascends in the wood, but descends in the bark; and the above operation prevents its return.

It is recommended in The Code of Agriculture to cover the incision with a rag, which will expedite the filling up of the ring or hollow.

For a nomenclature of the *best sorts of fruits*, known in the United States See *Massachusetts Agricultural Repository*. Vol. III. p. 92 to 99.

" The value of fruits for the manufacture of fermented liquors may be judged of from the specific gravity of their expressed juices. The best cyder and perry are made from those apples and pears that afford the densest juices; and a comparison between different fruits may be made with tolerable accuracy by plunging them together into a saturated solution of salt, or a strong solution of sugar; those that sink deepest will afford the richest juice." *Davy's Agricultural Chemistry.*

For further observations on Fruit Trees See Massachusetts Agricultural Repository. Vol. VI. p. 55.

FURROW, the trench made by a plough in going, also the earth thrown out of the trench. The European writers often use the word furrow, to signify a ploughing. They tell of sowing on one furrow, that is, after only one ploughing; on the second furrow, or on two furrows, that is, after two ploughings, &c. Though I see no need of our adopting this way of speaking, I think it not amiss to mention it, to prepare readers to understand those writers the better, when it falls in their way to peruse them.

FURROWING, in this country, is understood to mean marking ground into little squares with a horse plough, in order to plant Indian corn, or any other plant that requires the like culture. The goodness of this operation consists in making the furrows straight, equidistant, and at right angles; neither too deep nor too shallow; that the dung and seed may lie neither too low nor too high. When dung is to be laid in the furrows they should be deeper; when ground is to be seeded without putting dung in the furrows, or holes, the furrows should be very shallow. The nearer the time of planting this work is done the better. If a rain fall between furrowing and planting, it is detrimental. It soddens the ground, or makes it more heavy and compact, and causes the furrows to be less visible.

G.

GARDEN, " a piece of ground cultivated and properly ornamented with a variety of plants, flowers, fruit trees, &c. Gardens are usually distinguished into flower garden, fruit garden, and kitchen garden: The first of which, being designed for ornament, is to be placed in the most conspicuous part, that is, next to the back front of the house; and the second and third, being designed for use, should be placed less in sight." *Dict. of Arts.*

I consider the kitchen garden as of very considerable importance, as pot herbs, sallads, and roots of various kinds, are useful in housekeeping. Having a plenty of them at hand, a family will not be so likely to run into the error, which is too common in this country, of eating flesh in too great a proportion for health. Farmers, as well as others, should have kitchen gardens: And they need not grudge the labour of tending them, which may be done at odd intervals of time, which may otherwise chance to be consumed in needless loitering.

It is best that a garden should be on a declivity. If it be very steep, it may be thrown into banks, and level plats. There is commonly a variety of soils on a declivity of any considerable extent. This will give a material advantage to a garden, as a variety of different plants may have each the soil that best suits them.

A kitchen garden should not be situated at any great distance from the house, lest being too much out of sight, it should be out of mind, and the necessary culture of it too much neglected.

A garden should have a close fence, that the winds may not drive seeds of weeds into it. The fence should be at least seven feet high, and picketed, to prevent the entrance of thieves. The height and closeness of the fence, will increase the vegetation by increasing the warmth of the air in the garden, excepting perhaps in the parts which are shaded by the fences. The rage of high winds will be so opposed as to prevent the tearing and distorting of tender plants; and fowls may be the more easily kept out.

A garden should have a border of about three feet, and next to the border a walk of the same width or one foot wider. The walk through the middle may be from six to eight feet as the owner pleases. This may be crossed by one, two, or three narrower ones, if the shape of the ground requires it; or if it is half as long again the one way as the other, which is more elegant than an equilateral square. On these cross walks may be espaliers for grapes. Trees should not be in the outer border, but on the opposite sides of the outer walks; not two many of them; perhaps one of the dwarf kind in 20 or 30 feet.—Standard trees in gardens give too much shade. Dwarfs are commonly cut into espaliers.—But this torturing of trees makes them less fruitful, and shorter lived. Those who prefer it may make this sacrifice to elegance and beauty. In fruit trees which need much heat, and placed against northern walls, I object not to it.

GARDENING, a kind of agriculture, usually called Horticulture. It may be considered as farming in miniature. It is conversant in preparing ground for different kinds of seeds, and in treating them properly during their growth. The garden is the fittest place to make the first experiments in, with exotic roots and seeds, as the loss is inconsiderable, if they should not prove agreeable to the climate. If they prosper well in the garden, they should afterwards be tried in the field: And even then not at first on a very large scale.

He who would make his gardening profitable, should have his kitchen garden near to the dunghills, that the manure may be applied without too much labour. Dung that is old, and destitute of seeds, should be used, that too many weeds may not be propagated. And that a garden may be kept clean, not one weed should be suffered to have its seeds ripened in it: And every perennial weed, or weed which lasts over one year that appears in autumn, should be extirpated in such a manner that, if possible, no parts of its root may remain in the ground. The seeds of many weeds may also be destroyed, by laying the ground in high ridges during the winter. At the same time, it will help to enrich the soil; and many of certain kinds of insects, or their eggs, will be destroyed: Especially if the ridging be performed about the last of November, or the beginning of December. Ground that is so managed, will be dried the earlier in the spring, to such a degree, as to be fit for digging and seeding. It is of more advantage in land that is apt to be too wet, than in that which is sandy and dry.

GARGET, a disease in cattle. Cows sometimes have their udders greatly distended, and indurated, with this distemper; of which they will pine away and die, unless a remedy be speedily applied. The method of cure is, to make an opening in the dewlap, and insert into it a piece of the root of mechoacan, as big as a nutmeg, with a string made fast to it, that it may be drawn out when the cure is affected. The humour, in about twenty four hours, will be revulsed from the udder to the dewlap, and soon discharge itself at the orifice, which completes the cure.

From repeated experiments I should prefer to every other recipe, patient and often repeated warm fomentations, of soap suds, vinegar and spirits. It should be treated as nearly as possible like the lumps, which often appear, and are dispersed in the breasts of females. Disperse it by fomentations, if practicable, but if it must maturate, assist this by poultices.

GIGS, little tumours or bladders in the mouths of horses. To cure slit them open to discharge the matter; and wash them with salt and vinegar.

GLANDERS, a very foul and often fatal disease in horses. It is always accompanied with a copious discharge of mucus from the nostrils, and swelling of the glands under the throat and tongue. In its advanced stages the discharge becomes purulent. And when the bones become carious, the disease is attended with an intolerable stench, and may be pronounced incurable.

In the first and second stages, Gibson directs to purges, diaphoretics, and rowelling in the hinder parts by way of revulsion. To clear his nostrils, burn brimstone, feathers, and bits of leather under his nose, passing the fumes into his nostrils, through a funnel. And when much matter is discharged by sneezing, syringe the nostrils with brandy, or red wine. Afterwards a small quantity of Unguentum Egyptiacum, dissolved in oil of turpentine, may be injected through a large pipe, which will be helpful towards cleansing the ulcerated parts. See *W. Gibson on Farriery*.

The glanders is said to be a contagious disease, and care should therefore be taken lest it be communicated to other horses. A work called *The Complete Farrier*, printed in Philadelphia, in 1809, which has been highly approved of by those who have practised according to its directions, observes that "The strangles has been sometimes mistaken for the glanders or sore throat, but in this disease the inflamed glands very soon suppurate and burst, whereby all the other symptoms are generally removed, whilst in the glanders, the glands seldom or never suppurate; in order, however, to avoid all danger, it is advisable, the moment a horse is perceived to have a discharge from his nose, to put him into a stable where he can have no communication with other horses: if the glands of the throat are enlarged and inflamed, apply a large poultice to them, steam the head three or four times a day, let him be well clothed particularly about the head, and give him of powdered nitre one ounce, and an unwashed calx of antimony two drachms, mixed for a dose every day, or once in twelve hours.—Should the discharge arise from a cold, it will soon be removed by that means. When considerable ulceration is perceived in the nose, with the other concomitant symptoms of the glanders, the horse should be destroyed instantly.

"The most effectual mode of purifying stables in which glandered horses have been kept, is to remove or carefully wash every thing on which the horse may have deposited any matter, and afterwards to cover every part of the stable with a thick coat of lime and size."

GOOSE, a well known bird.—The tame kind are some of them entirely white, but they are mostly parti-coloured, grey and white.—The belly and wing feathers are white, even in those that have most of the grey colour.

Geese are more profitable than most other tame fowls, on account of the cheapness of their feeding, and the value of their flesh and their feathers, besides their grease and quills. Some strip them of most of their feathers twice a year. But this hurts the animals, and is on the whole, no profit to the owner. Moulting time is the right season for plucking them; for then the feathers are loose, and begin to fall off of themselves. Geese begin to lay their eggs in March; and begin to sit on them in March or April. The time of incubation is four weeks.

It is said that geese may be advantageously fed on turnips, cut in small pieces, similar to dice, but not so large, and put into a trough of water. Geese should be kept in places where they can have plenty of water, which will cause their feathers to be of a better quality, and they will be less subject to vermin.

GRAFTING, OR ENGRAFTING, the taking a shoot from one tree, and inserting it into another, in such a manner, that both may unite and become one tree.

Nut trees, of the same species, will take readily on each other; but those of different species will take with difficulty, and never make good trees. From curiosity, apple scions may be put into peach stocks, and the reverse, and pears will grow on thorns, but they make miserable trees. It is a well authenticated fact, that peach buds will take by inoculation on the willow, but it is a monstrous and unnatural union, and cannot serve any other purpose than the gratification of a philosophical curiosity.

The methods of grafting are various. The first, which is termed Rind or Shoulder grafting, is seldom practised but on large trees, where either the head or large branches are cut off horizontally, and two or more cions put in, according to the size of the branch, or stem; in doing this the cions are cut flat on one side, with a shoulder to rest upon the crown of the stock; then the rind of the stock must be raised up, to admit the cion between the wood and the bark of the stock, which must be inserted about two inches, so as that the shoulder of the cion may meet, and closely join the crown of the stock; and after the number of cions is inserted, the whole crown of the stock should be well clayed over, leaving two eyes of the cions unconnected therewith, which will be sufficient for shooting.

The next method is termed Cleft or Stock grafting; this is practised upon stocks or trees of a smaller size, and may be used with success where the rind of the stock is not too thick, whereby the inner bark of the cion will be prevented from joining to that of the stock. This may be performed on stocks, or branches, that are more than one inch in diameter: The head of the stock, or branch, must be cut off, with a slope, and a slit made the contrary way, in the top of the slope, deep enough to receive the cion, which should be cut sloping like a wedge, so as to fit the slit made in the stock; being careful to leave that side of the wedge which is to be placed outward much thicker than the other. And in putting the cion into the slit of the stock, there must be great care taken to join the rind of the cion to that of the stock; for if these do not unite the grafts will not succeed.

A third method which is termed Whip, or Tongue-grafting, is performed on small stocks by cutting off the head of the stocks sloping; then there must be a notch made in the slope toward the upper part downward, a little more than half an inch deep, to receive the cion, which must be cut with a slope upwards, and a slit made in the slope like a tongue, which tongue must be inserted into the slit made in the slope of the stock, so as that the two rinds of both cion and stock may be equal and join together exactly. Then there should be a ligature to fasten the cion, so that it may not be easily displaced.

Another species of grafting is called Inarching, for which see *Inarching*. See *Forsyth's Treatise on Fruit Trees*.

GRAIN. "Wheat, if not in good condition, is much improved by kiln-drying ; but it should not be used, unless in cases of necessity, until sometime after it has undergone that operation. It ought to be moderately kiln-dried, with a slow heat, and frequently turned. But if the grain be musty, it ought to pass through a previous process, which is thus described by an eminent chemist.

"The wheat must be put into any convenient vessel, capable of containing at least three times the quantity, and the vessel must be subsequently filled with boiling water ; the grain should then be occasionally stirred, and the hollow and decayed grains, (which will float,) may be removed ; when the water has become cold, or in general, when about half an hour has elapsed, it is to be drawn off. It will be proper then to rinse the corn with cold water, in order to remove any portion of the water, which had taken up the must ; after which, the corn being completely drained, it is without loss of time, to be thinly spread on the floor of a kiln, and thoroughly dried, care being taken to stir, and to turn it frequently, during this part of the process.

"By this simple operation, it is said that corn, however musty, may be completely purified, with very little expense, and without requiring previous chemical knowledge, or any expensive apparatus. Mere ventilation, however, has been recommended as a means for preparing grain for use sufficiently effectual." *Code of Agriculture.*

GRANARY, a store-house for threshed corn. A granary should be so constructed, that corn may be kept free from dampness, insects, and vermin. To avoid the last of these evils, its being mounted on blocks, capped with flat stones, like some of the houses for Indian corn, is no ill expedient. But for large granaries this will not be convenient.

The Lombardy poplar is recommended as a timber adapted for flooring granaries, which is said to prevent the destruction of corn by weovils and insects. Poplar wood will not easily take fire.

GRASS, a general name for most of those plants which are used in feeding cattle, both in their green and dry state.

"The land on which grass seed is intended to be sown, should be well ploughed and cleared from the roots of noxious weeds. Before the seed is sown, the surface of the ground should be made level and fine : Otherwise the seeds will be buried unequally. When the seed is sown, it should be gently harrowed in, and the ground rolled with a wooden roller, which will make the surface even, and prevent the seed being blown into patches. It is the common way of proceeding ; if a farmer wants to lay down his land to grass, he either takes his seeds indiscriminately from his own hay rick, or sends to his neighbour for a supply. By this means, besides a certain mixture of rubbish, which must necessarily happen, it is not unlikely but that which he intends for dry land, may have come from moist, where it grew naturally, and so on the contrary : And the consequence of this slovenly method frequently is, that the ground, instead of being covered in one year with a good seed, is filled with weeds, not natural to it, which would never have sprung up, if they had not been brought thither. This slovenly practice may have prevailed when Dr. Deane's book was first published, but we believe it is now exploded.

"Some say that if you manure your ground well, good grasses will come in of themselves. I own they will. But the question is, how long will it be before that happens ? And why will you be at the expense of sowing what you must afterwards try to kill ? Which must be the case, so long as people sow all kinds of rubbish under the name of hay seeds. Others say it will be better to have a mixture of different seeds. I suppose this to be true. But cannot a mixture be had, though the seeds be gathered and separated ? And is not a mixture by choice more likely to be proper than one by chance ? Especially after sufficient experience has been had of the particular virtues of each sort, and of the different grounds where they will thrive best ?

"It is said by some, that weeds will come up along with the grass, though what is called clean seed be sowed. No doubt of it. Can any one imagine that grass seeds should be exempted from what happens to every other kind of seed ? But I will venture to say, that not near the quantity of weeds will spring up, as they imagine, if the grass be sown thick." *Stillingfleet.*

It is undoubtedly best to sow clean seed, which is known to be suitable to the soil, when land is laid down to grass. For though grasses will gradually come in, no great crop is to be expected the first year, unless it be a crop of rank and useless weeds. And he that misses of the first year's crop, loses much, as the longer the land lies, the more compact, or bound, it will become, and produce the smaller crops.

It would require a large volume to describe all the kinds of grass which are and might be cultivated in the United States. Sir John Sinclair observes, (*Code of Agriculture*, p. 219,) that "there are in all 215 grasses properly so called, which are cultivated in Great Britain." The Duke of Bedford instituted a series of experiments, to try the comparative merits and value of a number of these grasses to the amount of 97, the result of which is annexed to Sir Humphrey Davy's Agricultural Chemistry.—According to these experiments, tall fiscue grass, (*festuca elatior*,) stands highest, as to the quantity of nutritive matter afforded by the whole crop, when cut at the time of flowering; and meadow cat's-tail grass, (*phleum pratense*, called in New-England *herd's-grass*, and *timothy grass*, in the southern States) affords most food when cut at the time the seed is ripe. Sea meadow-grass produces the greatest quantity of aftermath.

We shall add to this article a concise description of some of the most useful grasses, which have been cultivated in the United States, referring our readers to longer and more elaborate treatises for a more particular description of this class of vegetables.

1. *Tall Meadow Oat*, or Tall-oat-grass, *Avena elatior*. This kind of grass was imported from England, and has been cultivated in Massachusetts, as well as in the State of New-York, and the southern States. It is thus described by Dr. Henry Muhlenberg, of Lancaster, in Pennsylvania.

"This grass is of all others, the earliest, latest, and best grass for green fodder and hay; it blossoms about the middle of May, with red clover, and the seed ripens a month after. It grows but in a clover soil, and rises to a height of from five to seven feet—it ought to be cut for hay in blossom, about the end of May. The seed may be sown in the fall or spring, with or without grain, and must be brushed in or lightly harrowed. If mixed with clover it will make good upland meadow. Horned cattle prefer this grass to all others, but some horses do not relish it green. It must be pastured, or cut at least three times in a season. If suffered to grow old, it will become straw-like; when intended for hay it must be salted." *Willich's Domestic Encyclopædia*, American ed. vol. II. p. 194. See likewise Massachusetts Agricultural Repository, vol. III. p. 38, vol. IV. p. 111, and p. 209.

2. *Tall Fescue Grass*, *Festuca Elatior*. Dr. Muhlenberg describes this grass as "a native perennial grass of the United States, it is very luxuriant and productive, but rather coarse. Cows are fond of it, but horses do not relish it.—It grows in moist meadows, and in woods. The leaves are broad, stem very high, (sometimes 6 feet) flowers in July."

3. "*Meadow Cat's-tail*, *Timothy-grass, or Herds-grass*, of the northern States. (*Phleum pratense*) is the grass most used for hay in the northern States. It is also erroneously called *Fox-tail;* but this is another grass. The cat's-tail has a long head, somewhat resembling the tail of a cat, with very fine seeds; the *Fox-tail* has a short bushy head, more like the tail of a fox, with coarser seeds.—In other respects they have considerable resemblance.

"Cat's-tail grows best in rich *moist* soil; but it will grow well, for a few years, in a rich wet, or a rich arable soil. In the rich wet soil it gradually lessens in product; while at the same time it gives way to wild grasses. In a rich arable soil it gradually fails, by reason of the ground becoming bound and the sward thickened with other grasses. Probably, if it were well torn with the harrow in the spring, and not too closely pastured in the fall, and none in the spring, it would grow well for many years in such soil. By close pasturing in the fall, it is apt to be torn out by the roots, and by cropping it again in the spring, it suffers greatly." *Farmer's Assistant.*

4. The *Herd-grass* of the southern States, is called also White-top, Foul Meadow Grass, and Red-top. It is thus described in the second American edition of Dr. Willich's Domestic Encyclopædia, vol. II. p. 268. "Dr. Muhlenberg thinks this grass is the *Agrostis stricta* of Wildenow. It is particularly adapted to wet low grounds. It mats and consolidates the surface, continues many years, excluding every other grass and all weeds. Many worthless swampy spots in the low parts of the State of New-Jersey, have been rendered valuable grazing grounds by this grass, loaded waggons having passed over places, which, two or three years before sowing it, would scarcely permit an animal to walk through without sinking. It makes excellent hay, and cattle are said to prefer it to that made of either clover or timothy. It is more succulent than timothy, though not so coarse. The same bulk of the hay of this grass will weigh one third more than that of timothy hay. Four tons is a common crop from one acre. It yields no second crop, but affords excellent late and early pasture. It was first brought to New-Jersey from New-England by the late William Foster, who resided near Mount Holly ; and introduced into Pennsylvania about ten or twelve years since.*

The *Red-top* and *White-top*, are only varieties of the above species of grass. The latter is the larger of the two. The red-top is particularly valuable, as it will grow and sod the first year on banks, when no other grass will thrive. See an essay on artificial grasses, by Col. Taylor, of Virginia, republished in the Mass. Agr. Rep. vol. V. p. 291.

5. *Lucerne*, *Medicago Sativa*. By some experiments made by the Hon. *Robert R. Livingston*, recorded in the *Transactions of the Agricultural Society of New-York*, it appears that with good cultivation, and abundant manuring, from six to nine tons of hay may be obtained from an acre of this grass in a season. The ground must be highly pulverized to insure a good crop. Twenty pounds of seed are required for an acre if sown in broadcast, six pounds if drilled. It answers very well with red clover, and is not injured by the cold or the changes of our climate.

Mr. L. advises as the result of his experiments, 1. Never to sow on ground that is not perfectly pulverized. 2. Not to sow till the ground has acquired a degree of warmth friendly to vegetation, viz. in May. 3. To sow with no crop that will probably lodge. 4. If sown with buck-wheat, to apply no gypsum or other manure, till the buck-wheat is off. 5. When the quantity sown is small, and the farmer can afford to lose a crop, to give the ground one turn in the au-

* There is a great confusion of names in this paragraph. The foul meadow grass is not the Red-top of New-England. The former is a plant which flourishes only in a wet soil—the latter one which will flourish in the driest soils.

tumn, another in April, harrowing fine, and a third the beginning of May, and then if the weather be mild and warm sow, if the ground be in perfect tilth, otherwise give it another ploughing.

When lucerne turns yellow it should be mowed, and the plants will come up free from the disorder.

6. *Guinea Grass, Panicum Maximum.* This plant was first discovered on the coast of Guinea, whence it was brought to Jamaica.

" We will give the reader the account which Mr. Oglesby, of Kentucky, gives of its product in Wilkinson County in the State of Mississippi ; and from this, and from the certificates of others, it will be seen, that it must become productive of immense advantage to the southern part of our territory, and perhaps to every part of our country.

" I have (says Mr. Oglesby) been accustomed to both timothy and clover meadows, and have frequently assisted in cutting some of the best in the State of Kentucky. At Percyfield, near Fort Adams, I cultivated a lot of Guinea-grass, somewhat less than a quarter of an acre ; from which I fed six or eight horses, during the summer of 1812. I planted it in the second week in May, and began to cut it the 20th of June, and cut it five times before the 15th of October, and obtained from each plant (which occupied a square yard) about sixty pounds of green grass.

I have frequently observed it to grow four inches in twenty-four hours. From the astonishing growth, and from the result of all my experiments, I have no hesitation in saying, that it will yield ten times as much as any timothy or clover meadow I have ever seen." *Farmer's Assistant.*

7. Fiorin Grass. (*Agrostis Stolonifera,*) called likewise *Bent Grass.* This grass has been said to be a native of Ireland, but it has likewise been found growing spontaneously in this country. In the Orchestron meadow in England, it has yielded nine tons to the acre in a season. Sheep, neat cattle and horses are extremely fond of it. Given to cows, it increases the quantity and improves the quality of milk.

The method commonly practised in Ireland, for rearing it, is to cut the strings in short pieces, strew them evenly over the ground, and cover them with suitable earth, or with compost, as the nature of the ground may require. They are to be kept clear of weeds till the young growth has taken possession of the soil ; when it will cover the ground to the exclusion of every other plant.

It delights in a wet soil, and thrives in boggy lands, and covers them with so tough a sward that teams may go on them.

The author of the analysis of grasses, appended to Sir Humphrey Davy's Agricultural Chemistry, observes of this grass, that " From a careful examination it will doubtless appear to possess merits well worthy of attention, though perhaps not so great as has been supposed, if the natural place of its growth and habits be impartially taken into the account. From the couchant nature of this grass it is denominated couch-grass by practical men, and from the length of time that it retains the vital power after being taken out of the soil, is called squitch, quick, full of life, &c." See *Massachusetts Agricultural Repository*. vol. III. p. 32 to 37.

8. Darnel, or Rye Grass, *Lolium Perenne*, "Sheep eat this grass when it is in the early stage of its growth, in preference to most others; but after the seed approaches towards perfection, they leave it for almost any other kind. A field in the park of Woburn was laid down in two equal parts, one part with rye grass and white clover, and the other part with cock's-foot and red clover; from spring till mid-summer the sheep kept almost constantly on the rye grass; but after that time they left it, and adhered with equal constancy to the cock's-foot during the remainder of the season." *Agricultural Chemistry*.

9. *Dactylis Glomerata. Cock's-Foot.* Oxen, horses, and sheep eat this grass readily. The oxen continue to eat the straws and flowers, from the time of flowering, till the time of perfecting the seed; this was exemplified in a striking manner in the field alluded to. The oxen generally kept to the cock's-foot and red clover, and the sheep to the rye grass and white clover." *Agricultural Chemistry*.

10. "*Alopecurus Pratensis. Meadow Fox-tail.* Sheep and horses seem to have a greater relish for this grass than oxen. It delights in a soil of intermediate quality as to moisture and dryness, and is very productive. *Agricultural Chemistry*.

11. "*Green Grass. Poa Viridis*, is a native of the United States, and especially of Pennsylvania, where it grows in all meadows and on rich soils. Dr. Muhlenberg says it is not described by Linnæus, though nearly allied to his *poa angustifolia*. It may be easily known by the following description. "Culm (or haulm) erect and round (columnar) panicle diffuse, spicules five flowered and hairy at their base." Cattle are very fond of this grass, if cut when the blossom opens. It produces less than the *Avena elatior*, or tall meadow-oats, but horses prefer it. It continues green until even after frost, and when all other herbage is destroyed; and if manured will continue forever. The fine grazing farm of Mr. Wm. West, of Upper Darby, Delaware County, consists entirely of this grass. Mr. W. finds it necessary to sow clover thinly on the green grass sod every three or four years, to correct a slight tendency which green grass has to bind the soil. When the *Green Grass* appears upon meadows made by banking out rivers, care must be taken to secure a supply of water; otherwise, according to Mr. J. Cooper, the ends of the seeds will become affected with a black spear, about one fourth or one half an inch in length, similar to the smut on rye, and cause a loss of the hoofs of cattle that eat the grass.

12. "*Blue Grass, Poa compressa*, a native grass of the United States, having "a compressed ob-

lique culm or haulm pancile squeezed, spicules round, (columnar) and eight flowered." This is eaten tolerably well by cattle when young, and remains green until after frost. It binds the soil, in the course of three or four years, so as to require ploughing up. This is often mistaken for the green grass, and both are called occasionally *Spear Grass*, and wire grass." *Domestic Encyclopædia.*

13. Col. Taylor, of Virginia, in a paper on artificial grasses, communicated to the Agricultural Society of Virginia, says, "The best grass which I have tried in many respects is one commonly called "the highland meadow-oat." I have had no means of ascertaining whether it is a species of rye grass, or of the avena pratensis, or neither, nor whence it derived the appellation "Peruvian," by which I have heard it distinguished. With its qualities I am better acquainted, having carefully observed them for many years. It ripens as early as the red clover, and is easily made into fine hay, if cut in proper time. Its earliness is of vast importance in our climate. Thence it happens that it produces heavy spring crops, like red clover, as it commonly perfects its growth, before a drought occurs. It is the hardiest grass I ever saw, and bears drought and frost, and heat and cold much better than any I have tried. It keeps possession of the land in spite of severe grazing. It flourishes but on soils suitable for red clover, but it will live on and improve lands, whereon red clover will perish. It furnishes better grazing early in the spring, late in the fall, in droughts and in winter, than any grass known to me. Ripening with the red clover, it is peculiarly fitted for being sown with it, because it greatly facilitates its conversion into hay, retains possession of the ground for years after the clover has disappeared. Alone, cut before the seed ripens, its hay is as nutritive and pleasant to stock of all kinds, as any I have ever used; and it will yield both seed and tolerably good hay at one cutting, as it ripens soonest at top. Mixed with favourite grasses of grazing animals it is partially rejected, but eaten as they fail. Alone, it is greedily fed on. After being cut or grazed, if left to grow it rises anew, almost with the rapidity of lucerne, with a vigour but little diminished. Fifteen years experience has not enabled me to decide as to its capacity for improving the soil, because the small quantity of seed first obtained, by confining experiments to small patches, long concealed its qualities; and the large lots first sown have remained too flourishing to require manure, and too valuable to be ploughed up. It produces, (after it has come to perfection, in doing which it is one year slower than red clover,) ungrazed and uncut, a warmer, and more lasting cover than clover, which has recently induced me to mix and sow it with wheat on a large scale. for the end of improving the soil; in which anticipation I have yet discerned no cause for apprehension. For this purpose it possesses one recommendation

beyond clover. It does not grow and spread so rapidly in the fall or spring as to injure the crop of wheat with which it is sown, as is sometimes the case with clover.—Upon the whole, whatever doubt remains as to its fertilizing power, the conclusions that it eminently possesses the qualities for lasting, grazing, and making highland hay, may, I think, be confided. But it does not, so far as my trials have extended, succeed in lands, originally wet, however well they are drained."*

GRAVEL, earth of the same nature with sand, only more coarse and harsh. Both seem to consist wholly of minute pebbles. Gravel is useful in mending roads, in making dams, and for walks in gardens, &c.

A soil of mere gravel is the meanest of all soils; and will produce next to nothing, till it be mended with something mixed with it; and even then it will need a wet season, unless it be in a wet situation, as at the foot of a hill, or watered with springs.

The best manures for this sort of land in general, are marl, clay, the mud of swamps, ponds, rivers and creeks. If applied in large quantities, they will meliorate it for a long time. The best yearly dressings are the dung of cows and swine, sea weeds, straw partly rotted, bits of leather, woollen rags, and almost any spungy substances which retain moisture for some time.

* This grass as far as we can judge from Col. Taylor's description, is the Avena elatior, or Tall meadow oat grass mentioned in the first instance.

This kind of soil, well manured, sometimes produces good crops of such plants as require much heat, as Indian corn and tobacco. And it does well in a good season, for rye, clover, beans, pease and potatoes.

GREASE, a distemper so denominated, is a swelling and gourdiness of the legs of horses, which frequently happens to them after a journey. Most people have believed their grease to be melted by hard riding, and fallen into their legs: And that which may have given encouragement to this opinion, is the colour of the matter issuing from the chinks and sores in those parts, when they come to break, somewhat resembling grease. The distemper may arise from various causes. If the grease be an attendant on some other distemper, the cure will be the more difficult, and it will be in vain to expect a recovery, until the disease is removed which occasioned it.—Therefore, methods for the cure of those distempers must be followed, and applications made outwardly for this. But if it be an original disorder, and if the horse have been pampered, or well fed, the cure ought to be begun by bleeding and purging, to lessen the redundancy of humours. Neither should these be too often repeated: But what is wanting in that way had much better be effectuated by a more spare diet, with daily exercise. After moderate evacuations, a rowel may be made on the inside of the thigh, or on the belly; which may be continued for a month, or longer if necessary. In the mean time the cinnabar or antimonial balls ought to be constantly given. And while these things are doing internally, the legs should be frequently rubbed, not with hard instruments, but with a good wisp of hay, or a brush. Baths and fomentations, such as may cause the humours to go off by perspiration, or render them fit to return in the circulations, are also to be made use of. For this purpose the following is recommended.

Take wormwood eight handfuls, John's wort, centaury, camomile, of each four handfuls, elder flowers two handfuls, bayberries half a pound: Boil them in two gallons of water till one third is consumed, and make a fomentation.

The horse's legs are to be bathed three or four times a day, with woollen cloths wrung out of the liquor, and applied as hot as he can bear them, adding a little of the spirit of wine or brandy. And if they be much inflamed, as happens when the sinews are affected, a good quantity of the ashes of the green twigs of vines, walnut or oak, may be boiled in the dedoction, adding more water, when the other ingredients are easily to be had.

The lees of wine, with a mixture of soap, are also very proper to be applied warm: As also cow's dung boiled in vinegar.

Suitable cataplasms in bad cases are proper. The camphorated spirit of wine alone is good, viz. an ounce of camphor to a pint of spirit. Frequently used, it will answer well when the swelling is new. See *Gibson's Farriery*.

GREEN-DRESSING, turning a crop of green plants into the ground in summer, to enrich the soil, and fit it to produce a good crop of wheat. By repeating this culture, poor or worn out land may be brought to any degree of richness that is desired, without any other manure. Buck wheat, rye, pease or oats, may be sowed in the spring, and in June ploughed in, when they are fullest of sap, and most easily rotted. The ground should be again ploughed in the fall, sowed with winter grain, and well harrowed. The cost of ploughing and seed, is not so much as that of dung, when it can be had, and carting it. This management, therefore, may often appear eligible, especially in places where manures are not plenty. On account of the cheapness of the seed, Mr. Elliot recommends millet as a most suitable crop for green dressing: and some have used clover and rye grass. In Britain, buck wheat is much used, as the stalks, when green, are very large and juicy, and as they require but a short time to rot. It is asserted, that about ten days are sufficient for it to lie under the furrows.

The chief difficulty I can think of, which tends to discourage this practice, is, the choaking of the plough in going among a tall growth of plants. It may be needful for a boy to tend it. But in Britain, to prevent choking, they recommend to pass a roller over the crop to be turned in, which lays it flat, and in the same direction that the plough is to pass.

GREENS, the general name of those pot herbs which are boiled for food when they are young and tender. Some of the most useful of them known in this country, are spinnage, kale, French turnips, dandelion, purslain, white and black mustard. There is a Scotch kale which may be reared earlier than almost any other greens, and is equal in goodness to any. To have greens early, let kale and French turnips be sowed in October, and the young plant covered closely with eel grass, or straw, during the winter, and till the influence of the sun be sufficient to renew their vegetation.

GREEN SCOURING, "a disease to which sheep and bullocks are often subject. The best remedy for this distemper is verjuice: A wine glass full is enough for a sheep, and a pint for a bullock." *Complete Farmer.*

GRIPES, *or cholic pains.*—Horses are very subject to griping, or cholic pains. They may proceed from flatulencies, or wind pent up in the stomach and bowels, from inflammation of the coats of the stomach and intestines, or from worms, spasms, &c. In such case it is very wrong to give him heating things by the mouth, as is too commonly practised. Bleeding should be the first thing in these cases, if the disorder be violent, which may be known by the creature's motions, frequency of lying down, and starting up again, &c. As horses are costive in these cases, the rectum should be cleared of the hard dung, by *back racking*, as it is called, that is, it should be taken out by a hand, which gives a horse great relief. For the pressure on the neck of the bladder being thus removed, he will be able to stale. Emollient clysters are then of great advantage, as they not only bring away the excrements, which affords a passage for the wind backwards; but they act as an internal fomentation, to remove spasms from the bowels. They may be frequently repeated, till the confined air finds a passage backwards. If it should be found necessary, a spoonful of laudanum may be given in a pint of watergruel, either by the anus in a clyster, or by the mouth. See *Clark's Farriery.* Nearly the same treatment is proper for horned cattle under the same disorders.

GROVE, a row or walk of trees, planted close or a little open, for ornament and shade.

Groves in gardens are both ornamental and useful, if the trees be not too large. They shade the walks in the borders; so that we may walk in gardens with pleasure, in the hottest part of the day. It is scarcely needful to say that these garden groves should consist of fruit trees; and they should be of the smaller kinds, if in a garden of a small or middling size. A double row has the best effect, as it respects shade, one near the wall, the other on the opposite side of the walk. But this on the whole I do not recommend, unless it be in gardens uncommonly large.

In other situations groves of larger trees are preferred. Lanes and avenues, leading to mansion houses and other buildings, may be ornamented with rows of trees, either on one, or on both sides: If only on one, it should be the southernmost, on account of the advantage of shade in the lane. Such trees are best, the limbs of which are not apt to be low; such as elm, ash, maple, poplar, &c.

Lots and enclosures should be bordered with rows of trees, either fruit trees or timber trees, in close order. They will do better a yard or more from the fence, than in hedges according to the English method, as recommended by Mortimer. But such trees should be chosen, as are not apt to propagate and multiply, lest the borders be soon filled with shrubs.

It would be advantageous to the public, as well as to the owners of adjoining farms, if all our roads were lined with groves, of timber trees. They might be either within or without the fences. In the latter case, government might interpose, and secure to the planters those which stood in the roads against their lands; and oblige farmers to plant in the roads against their own lands. I should prefer this to planting within the fences, especially where the roads have a good width. But the trees should be so tall when planted, as to be above the reach of cattle; and be staked, or otherwise secured, till they arrive to a certain bigness. The expense of thus securing them need not amount to much, when compared to the advantages arising from such groves.

Or if they were planted along the southerly sides of roads only, the advantage to the public would be great. Besides providing a stock of wood and timber for future generations, the present would receive the benefit of their shadow, cast into the roads in the hottest part of our summer days. This would be extremely refreshing to travellers, to teams that pass under them, and to many tame animals that live in the roads. In this case, the adjoining lots would not be injured with the shade; but for the beauty of their appearance, trees on both sides of the road would be best.

If the country were well stocked with these groves, their perspiration would help to abate the scorching heat of the sun, in a dry season, by moistening the atmosphere. They would serve to impede the force of high, driving winds and storms in summer, which often tear our tender vegetables, or lay our crops flat to the ground.—Our buildings would be also in less danger from them. The winds in winter would not be so keen and violent. The force of sea winds on our fruit trees would be abated. The snows that fall would be laid more even on the ground. Roads would be less blocked up, and seldomer rendered impassable by them. But for these last purposes, groves of evergreens will have the greatest effect.

Groves should be planted thick at first, that the above advantages may be had from them while young. When the trees become so large as to be crowded, they should be thinned. And thus a considerable quantity of fuel and timber may be soon realized by the proprietors.

The increasing scarcity and dearness of wood, especially in the older settlements in this country, affords an unanswerable argument in favour of such a piece of good husbandry.

GUINEA CORN. "Guinea Corn, or *Holcus Sorghum L.* an exotic vegetable, growing on the coast of Africa: its stalks are large, compact, generally attaining the height of 7 or 8 feet, and producing abundance of grain. It may be easily raised in sheltered situations, especially in exhausted hot-beds and other loose soils, where its seeds should be sown early in the spring. This plant is cultivated in South-Carolina, and yields from sixty to eighty bushels of seed to an acre. It is often mowed green for the purpose of soiling cattle.

"The seed is used for feeding poultry, and sometimes hulled by beating in a mortar, boiled and eaten with milk, and is said to be equal to rice; but it is not valued highly for making bread, which made of this material is black and heavy."

H.

HARROW, a kind of drag used in tillage. By drawing a harrow over ploughed ground, the clods which remain after ploughing, are broken, and the ground made mellow and fine. It serves also to destroy weeds, by pulling out their roots, and exposing them to the sun and wind. And it is used to cover seeds newly sown. The wood of a harrow should be the strongest and best seasoned white oak.

There are two kinds of harrows commonly used; the square harrow, and the bifurcate harrow; the former is for old and clear ground, the latter for land that abounds with stumps of trees and other obstacles. The square harrow is armed with sixteen, or with twenty-five tushes, or teeth. The sharper these teeth are, the more they will pulverize the soil. If they be steeled at the points, they will hold their sharpness the longer, and stir the ground more effectually. And the cost of doing it so little, that it is surprising to see that it is so generally neglected by our farmers.

It has been the common practice in this country to place the teeth in the joints of the square harrow. But this has a tendency to weaken the joints, and the teeth are more apt to become loose.—They should be placed in the solid parts between the joints. The best way to fasten them is, with shoulders under the harrow, and nuts screwed on above.

Some use harrows with wooden teeth, but they are of so little advantage to the land, unless it be merely for covering seeds, that they may be considered as unfit to be used at all. The treading of the cattle that draw them, will harden the soil more, perhaps, than these harrows will soften it.

The bifurcate, or triangular harrow, is either a fork of natural growth, or else made artificially.—The artificial one is commonly strongest, when well made, as timber may be chosen which is sufficiently tough and strong. The two legs may either be lapped together at the angle, or else framed together like a pair of rafters, excepting that the butt ends, being toughest, and strongest must be put together. But the joint must be strengthened by a good iron hoop smartly driven on to the nose, after the wood is thoroughly dry, and fastened with strong nails; and further strengthened with a brace from one leg to the other, framed in, about two feet from the juncture of the legs.

The angle may be more or less acute, according to the state of the land in which the harrow is mostly to be used. For the roughest ground the angle must be more acute; but for well cleared ground, the angle may be of 45 degrees, or more. The more obtuse the angle is, the more near together the teeth must be placed. In this kind of harrow some put 9, some 11, and some 13 teeth, or even 15. The rougher and harder the land, the fewer the teeth; and the fewer they are, the longer and stronger they should be. Twelve inches clear of the wood is not too long, nor three pounds too heavy for a tooth in the strongest harrows.

To prevent this machine from fastening itself often in immoveable stumps and roots, the teeth may be set leaning a little backwards. But where there are no obstacles, they should rather incline the contrary way, or at least they should be perpendicular.

Some make use of a horse harrow of the forked kind, and very narrow, to mellow the ground and kill weeds, betwixt rows of Indian corn. But the horse plough answers the purpose better in general, unless it be upon green sward ground, in which the horse plough will not answer at all. The stiffness of the old furrows will prevent its regular going. Lord Kaimes recommends what he calls a cleaning harrow with no less than 56 teeth, which teeth are no more than six inches apart. The use of it is to clear land of roots, in an expeditious and effectual manner.—The weight of a tooth is one pound only. If they are set raking forward they will penetrate the deeper, and have a better effect.

The harrows most generally used are of an oblong shape, each containing 20 tines, 5 or 6 inches long beneath the bars in which they are inserted. It is still common for every harrow to work separately; and though always two, and sometimes three harrows are placed together, each of them is drawn by its own horse. The great objection to this method is, that it is scarcely possible, especially upon rough ground, to prevent the harrows from starting out of their places, and *riding* on one another.—To obviate this inconvenience the exterior bars of each are usually surmounted by a frame of wood, raised so high as to protect it from the irregular motions of its neighbour; but in many instances, they are connected by chains or hinges, and cross bars, which is a preferable plan. Another objection, which has been made to the common harrow is, that the ruts, made by the tines, are sometimes too near, and sometimes too distant from one another; but this is not a great fault;

when the soil requires to be pulverized as well as the seed covered, especially when they are permitted to move irregularly, and in a lateral direction. Where the soil is already fine, as it ought always to be before grass seeds are sown, lighter harrows are used, which are so constructed that all the ruts are equidistant. *Supplement to the Encyclopædia Britannica*, p. 115.

Sir John Sinclair gives the following principal rules regarding the formation of harrows: 1. That not any two of the teeth should move in one tract. 2. That the tracts should be at equal distances from each other. And, 3, That the teeth, should either be round, or perhaps with a sharp edge bent forward, like so many coulters, as they make themselves cleaner than when they are square, or of any other shape, and work easier after the horses. The teeth of harrows are generally of unequal lengths, the front row being about half an inch shorter than the second, and the third row being about one inch shorter than the first; so that each row backwards, is about one half inch shorter than the one which precedes it. Where crops are drilled, an instrument called "the dr ll harrow," is found a very great improvement. It thoroughly cuts the weeds in the intervals, and harrows them up to the top in a very complete manner.

HARROWING, working the soil with a harrow. A team that travels quick, is best for harrowing, unless the land be too full of obstacles. Horses, therefore, are better for this work than oxen, because their motion is quicker. The faster the harrow moves, and the more it jumps, the more the hard clods are broken, and the turfs torn.—The teeth will also keep cleaner and go deeper; so that the land will be more mixed and mellowed. But clayey land is so apt to be cloddy, that it is often necessary to follow the harrow with a maul, or a hoe, to break the remaining clods.

Besides pulverizing the soil, covering seeds, and drawing out the roots of weeds, the designs of harrowing are to make the land level, or smooth; and, on fallows, to cause the seeds of some weeds to vegetate by exposing them to the air, in order that they may be destroyed by after operations, either with the plough or the harrow.

When land is wet and poachy, or at all muddy, it can be of no service to harrow it. It will rather do damage, as it will make it more compact and stiff.

Land that is too light and puffy, as drained swamps often are, cannot easily be too much harrowed. The more it is harrowed, the more compact it will be; and this is what it wants.

The harrowing of new ground for feeding, without ploughing, may be performed in almost any weather, if the ground be only dry enough to be mellowed by the harrow. And the sooner, after burning, this work is done, so much the better, as it will prevent the ashes being blown away by high winds, and as it will spread it more equally, and more effectually mix it with the soil. Here the strongest harrow must be used; and it ought to be heavy, in order to make any considerable impression on the soil. It is often necessary that the harrow pass several times in the same place,in order to raise a sufficient quantity of mould. There is no reason to fear its being lost labour. The more such ground is harrowed, the better crop may be rationally expected.

On furrows of green sward newly ploughed, the harrow should pass the same way that the plough did: Otherwise, some of the furrows, which lie a little higher than the rest, will be turned back again, grass upwards. This sort of land requires a heavy harrow, or one made so by loading it. A light one will sink into the furrows but little, and be of little service.

On old ground, ploughed plain, the harrow should pass the first time, across the furrows, as the teeth will better take hold of the roots of weeds, and more deeply penetrate the soil. It will also do more towards levelling the ground. Afterwards it should be harrowed the other way, lengthwise of the furrows.

Harrowing commonly does the most service immediately after ground is ploughed, as the teeth go deeper and raise the more mould. If it be neglected at this juncture, a time should be chosen when the soil is not too dry. After a gentle rain the clods will crumble the more easily; and the soil underneath being drier, will not be hardened by the treading of cattle.

In light sandy, or gravelly soils, or where there is occasion for harrowing land which is excessively dry, or in danger of soon becoming so, it should be done when the dew is on the ground, early in a morning. This will increase, rather than diminish the moistness of the soil. And on the contrary, land which is apt to be too wet, should be harrowed at a time when it is driest, as in the middle of a fair day. The first scratching will cause it to dry fast, and so prepare it to be made fine and mellow by the second.

The European farmers recommend harrowing ground once over before corn is sowed, and then to harrow in the seed the contrary way. The grain will be the more even, and not appear so much in rows, as if it were sowed upon the furrows; but it will not be so deeply covered. Perhaps sowing upon furrows, both winter and summer grain, may be generally the better method in this country, which is so much more liable to suffer by severe frosts and droughts. Some of our farmers even think it best to plough in the seed with a shallow furrow. The roots will lie the deeper, and be less exposed to suffer by frost and drought.

Harrowing fallows is doubtless a beneficial operation. If it be done two or three times between ploughings, the seeds of weeds will be encouraged to vegetate, and consequently will be killed at the next ploughing or harrowing. Thus the land will become very clean after a year of fallow; and the food and pasture of plants will be more increased than it could be by ploughing only. For every weed that consumes in the soil is of some advantage.

Some have found their account in harrowing mowing grounds, when they have become bound and stiff. Though the roots of the grass are much torn and mangled by harrowing, the soil will be loosened at the surface, and the vegetation of the grass so much increased, that the excess of the next crop will more than compensate the labour of harrowing. It should be done in autumn, and before heavy rains fall, but after a gentle one, when the surface is a little moistened. It would be best, before harrowing, to afford the land a sprinkling of old dung, or compost: Or else immediately after, and bush it in. Its fruitfulness will thus be greatly increased.

The harrowing of land that is ploughed in ridges, should be performed lengthwise, and by two harrows abreast, or three, if the breadth of the ridges require them, that the trenches may not be too much filled. The second harrowing may be across, if the land needs to be laid even for mowing. But then the trenches should be cleared out with a shovel or plough, if the land be so flat and wet as to make it proper or necessary to lay it down in ridges.

Harrowing of winter grain, in the spring, is approved of beyond the Atlantic. When the roots are well set, and in sufficient plenty, I think this may be a laudable piece of husbandry. The harrow will destroy but a few of the plants; and the loss of them will be more than made up in the increased growth of the rest. But, in order to make the loosened plants take rooting, Mr. Lisle advises to drive a flock of sheep about over the field. Others advise to rolling the ground, which appears more rational.

" There are wo modes of driving the harrows, either leading the horses, or driving them by whip-reins. The latter ought to be preferred, for as young horses are frequently employed in harrowing, before they are broken to regular work, the person, who *leads*, is exposed to considerable danger: whereas, when he directs the horses by long whip-reins, he runs no risk, and is always at hand, to remove any obstructions that may occur, from the accumulation of weeds, and other extraneous substances among the tines, or by the harrows riding on each other.

" Harrowing is usually given in different directions: first in length, then across, and finally at length, as at first. An excess of harrowing is prejudicial to wheat, it is better for that crop to have the land rough and cloddy; but the process ought to be effectually done for barley, (more especially if accompanied by grass-seeds), and for turnips. *Code of Agriculture.*

HARVEST, the season when corn is cut down, and secured.

Wheat and rye are harvested in much the same manner. Both are reaped and bound in sheaves. It is usual to cut rye rather greener than wheat, that the flour may be the whiter.

When a severe blight or rust has struck the stems of wheat, or rye, it answers no purpose to let it stand longer to ripen, or grow hard. It is agreed that it should be cut though full in milk. And afterwards it may lie on the ground, exposed to the sun and weather, till the grain is hardened. But the heads should lie so as not to touch the ground; which may be easily done, if the reapers will only take care to lay the top end of each handful on the lower end of the preceding one. Some say it will answer to cut it three weeks before the usual time, and before the stems are turned yellow.

If grass or weeds grow among grain, it should be cut high, that so the less quantity of trash may be bound up in the sheaves. And when taking weeds with the grain cannot be avoided, it should be reaped a little the earlier, that it may have time to lie in the field, till the weeds are well dried, without danger of scattering the corn by its being over dried.

The bands should be made in a morning early, when the dew is greatest, and the straw most supple. But the best time to bind the sheaves, is when the air begins to be damp towards evening, as the least degree of moisture will toughen the straw and prevent the scattering of the grain: And there is some degree of dampness in the air, for an hour or two before sunset.

A late writer advises to make the sheaves with only one length of straw.

After binding, it should be made up into shocks without delay, or after standing in sheaves one day, if the weather be settled and dry; where it is to stand in the field till not only the straw, but the grain, be thoroughly dried; and till a suitable opportunity present for carting it in. It should be done when the air has a small degree of dampness, to prevent the scattering of the grain.

It would be best on some accounts, that grain should be thrashed as soon as it is carried in. But as it is usually a hurrying season, it is but seldom that the farmer can spare time for it. It must, therefore, be stored most commonly.

The best method of storing it, is, to lay the sheaves up in the barn. But if want of room require them to be stacked, care should be taken that the grain may not draw moisture from the ground, by laying boards, straw, or rubbish under the stack. A better way still is to have a tight floor of boards mounted on four blocks, set in the ground, and so high from the ground as to prevent the entering of vermin.

In building a stack, care should be taken to keep the seed ends of the sheaves in the middle, and a little higher than the outer ends. No fowls nor birds can then come at the grain; and the rain that falls on the straw ends will run off, and not pass towards the centre. The stack should be well topped with straw, that the rain may be completely turned off. As to the harvesting of barley, oats and pease, see those articles.

With respect to harvesting Indian corn, I would observe, that many do it much too early, to their own damage and loss. As long as there is any greenness, or sap, remaining in the whole length of the stalk, below the ear, or even in the cob; so long the corn improves by standing. For the sap will continue to discharge itself into the grain. Though a crop harvested earlier may measure as much in ears, or more, when it is newly husked, it will shrink a great deal, sometimes so much that not two corns on an ear will touch each other. Besides, there will be the greater difficulty in drying and keeping it. Corn that is harvested early, will not be fit to store in out door cribs, nor in our common corn houses, unless it be first spread thin on floors, and dried. And this is troublesome, at least, if not impracticable.

Squirrels, and rapacious birds, disorderly cattle and bad fences, drive persons to harvesting early. But there is commonly more lost than saved by it. When the corn stands tolerably secure, and is in no danger from frost, nor from thieves, harvesting early is an error. I should not think the beginning of November at all too late.

It is not safe to let it lie long in the husks after it is gathered, lest it should heat, or contract dampness. One unripe ear or green stalk, in a heap, may damage many. The common practice of collecting large companies to husk the corn as soon as it is gathered, is a laudable one. And after it is husked, it should have a dry place, and so much benefit of the air, that it may be sure not to grow warm, let the air prove to be ever so moist.

Sometimes a severe early frost drives the farmer to harvesting, as he knows the frostbitten corn is apt to rot in the husks. But in such a case, or when corn holds its greenness uncommonly late, an approved method is, to cut it up close to the ground, bind it in small bundles, and set it up in small shocks in the field. It will ripen kindly, and take no damage. By this method the grain has the benefit of all the sap contained in the stalks, to bring it nearer to maturity.

I have heard of some persons in the county of Lincoln, who, finding their Indian corn very green at harvesting, have boiled it in the ears after husking; By which expedient they were able to dry it in the ears, without its rotting, or moulding. This may be no ill method at a pinch. But rather than be obliged to do it yearly, I should think they had better lay aside the culture of this plant, or else use no seed but from the northward, which will ripen in season.

HATCHEL, an instrument called sometimes, a Comb, full of long pins of iron or steel for teeth, with which flax and hemp are combed. They who manufacture these articles, as perhaps all the families of farmers should, ought to be provided with several hatchels of different finenesses. Where only flax is manufactured, two combs, one coarse, and the other fine, will be sufficient.

HAY. It has been found that the greatest quantity of nutriment is found in the leaves and stalks of plants, just at the moment when the seed is forming; and that the grass loses rapidly its value, as the seed advances to maturity. It is the better opinion also that the plant itself is the more exhausted by suffering it to mature its seed, and in crops that are to last several years this consideration deserves weight. The rule applied to clover is applicable to all plants, that they should be cut for hay, when one half the plants of a field are in flower.

HAY HOOK, an instrument to pull hay out of a mow, or stack.—This instrument is often made of wood; but an iron one is far preferable. It should be sharp pointed, armed with a fluke, and have a socket to receive the wooden handle. The handle should have a turn at the end for the ease of pulling. There can be no better handle than the half of an old ox bow: Or a little more than half. But this instrument will waste the hay, and divest it of much of the seed. A better way is, to cut off slices of two or three feet in thickness, from a mow or stack, as it is wanted for use.

HAY-MAKING, the curing, or drying of grass for fodder. The first thing to be considered about hay-making, is the time of cutting the grass. It should not be cut too early, or before it has got its growth: For this will cause it to shrink too much in drying. On the contrary, it should not stand too late, or till the seed be quite ripe. It is not only harder to cut, but the ripeness of the seed will cause it to shatter out while drying, which will be a considerable loss, as the seed is the most rich and nourishing part; and the soil will be the more exhausted by nourishing the seed till it come to maturity, and the next succeeding crop will be the poorer. There never can be any advantage in mowing late, unless it be thickening the grass roots, by scattering some of the seed, where they were before too thin. He that mows early has the advantage of longer days for drying his hay; and of shorter nights, when the dews are less detrimental to hay-making.

But the farmer who has many acres of the same kind of grass, cannot always expect to cut the whole of it in exactly the right season. That he may approach as near to right as possible, he should cut the thickest grass first of all; especially if it be in danger of lodging, or so thick that the lowest leaves perish, or the bottoms of the stalks turn yellow. The thinnest of his grass should be cut next, which is apt to be ripe soonest: And last of all, the middling sized grass, or that which is on a medium between thick and thin.

Where a second crop is expected the same year, thick grass should be cut a little the earlier, that the roots may not be injured so much as to prevent their speedy recovery, by being closely covered too long by the first crop.

Some regard should be had to the weather, when the time of cutting is in contemplation. Those, especially, should regard it, who are able to call in as much assistance as they please in hay-making.

Grass, which has not been washed by rain for several days, has a kind of gum on it, which is known by its adhering to the scythe. This gum is thought to be a benefit to the hay; and farmers are fond of mowing their grass when this gum appears, rather than just after the grass has been washed by rain.

As to the drying of hay, or the manner of making it, I know there are a variety of opinions. The right way is to do it in such a manner that as much of the sap as possible may be retained, and in the best state that is possible. In this I should think all would agree.–All persons will allow that too much drying is hurtful. It is certainly a loss to rake it, or stir it at all, when it is so dry that the leaves will crumble. And doubtless as much of the sap should be retained as is consistent with its being kept in good order for fodder, and for long keeping.

Some grasses will keep well with less drying than is needful for others. The Rhode-Island bent, as it is called, or red-top grass, will do with less drying than some other grasses. It has been much practised to put it up with so little drying that it heats in the mow to so great a degree, as to make it turn brown like tobacco; and it is known that cattle will eat it well, and thrive on it. But the mow will certainly send out part of the virtue of the hay in steams. I cannot but think that all grasses should be so much dried, that mows and stacks, though they have a degree of heat, should not emit any sensible steam; and I would not wish to have hay made brown by mow-burning. It surely does not appear to so good advantage at market.

Were it not for the labour and cost, a good way of hay-making would be, for the hay-makers to follow at the heels of the mowers, at least as soon as the dew is off, and spread the swarths evenly; turn the grass about the middle of the same day; make it up into cocks before night; open the hay, and turn it the next day; and so on till it be sufficiently dried, doubling the cocks if signs of rain appear. It will not commonly take more than two or three days to dry it, unless it be very green, or uncommonly thick and rank. A person who has but little hay to make, need not be much blamed, if he do in this way; especially if the weather do not appear to be settled.

The practice of the best English, Flemish, and French farmers, is to expose the hay as little as possible to the sun. It is carried in dry, but it preserves its green colour; and you see hay of one or two years old in their market, of so bright a green colour, that we should scarcely conceive it to be cured. Yet they are in the practice of preserving it for years, and value it more for its age. If such a course be best in climates so cool and cloudy, how much more important would it be under our scorching summer suns?

But if the weather be unsettled, or if showers be frequent, it may be better to spread grass well, as soon as it is mowed, stir it often, cock it the same day it is mowed, open it the next fair day when the dew is off, let it sweat a little in cock, and house it as soon as it is dry enough. It will bear to be laid greener on a scaffold, than in a ground mow; and in a narrow mow greener than in a broad one. And that which is least of all made, should be put upon a scaffold.

When grass is very thin, and not full of sap, having stood beyond the right time of cutting; it may be cut in the forenoon, and raked in the afternoon of the same day; and then dry sufficiently in cocks, in two or three days. But if a heavy rain fall, it will need to be opened, and exposed to the sun for a few hours. If there be only a small quantity of rain, it may be sufficient to pull out some of the hay round the bottoms of the cocks, or only on that side which was windward when the rain fell, and lay it on the tops. If the cocks are so situated that the water has run much under their bottoms, they should be turned bottom upwards, and trimmed at least; but it will most commonly be necessary to spread them abroad.

Sometimes hay will become too dry, notwithstanding every precaution to prevent it: For it will dry twice as fast in some fair days as in others, because of the different dryness of the air. When this is the case, it should be removed to the barn only in the evening, or morning, when the air is damp. And it is good to have some greener hay to mix with it.

Some think that mown grass should never be exposed to the full influence of the sun, lest it be robbed of too much of its sap, while it is in its most fluid state. A very ingenious gentleman, of my acquaintance, does not permit his grass to lie in swarth, but for an hour or two after it is cut; or no longer than till its wetness be gone, and it just begins to appear withered: He then gathers it into very small parcels, which he calls grass cocks, not more than a good forkful in each: Turns them over once in a while, about sun-set is the best time: Doubles them as they grow drier: And when the hay is almost dried enough, makes up the whole into large cocks. Grass that is thus dried, will not waste at all by crumbling; nor will much of its juices evaporate. I have seen his hay, the flavour of which excelled almost any other that I have met with. The colour of it, indeed, was rather yellowish than green: But that is a matter of no consequence to the farmer who does not send his hay to market. I cannot but think that, in dry settled weather, this is an excellent method of hay-making. But in catching weather, perhaps a method which takes less time is to be preferred. From the above Mr. Anderson's method is not much different. "Instead," says he, "of allowing the hay to lie, as usual in most places, for some days in the swarth, after it is cut, and afterwards putting it up into cocks, and spreading it out, and tending it in the sun, which tends greatly to bleach the hay, exhales its natural juices, and subjects it very much to the danger of getting rain, and thus runs a great risk of being good for little, I make it a general rule, if possible, never to cut my hay but when the grass is quite dry ; and then make the gatherers follow close upon the cutters, putting it up immediately into small cocks, about three feet high each when new put up ; always giving each of them a slight kind of thatching, by drawing a few handfuls of the hay from the bottom of the cock all around, and laying it lightly on the top, with one of the ends hanging downwards. This is done with the utmost ease and expedition ; and when it is once in that state, I consider my hay as in a great measure out of danger; for unless a violent wind should arise, immediately after the cocks are put up, so as to overturn them, nothing else can hurt the hay ; as I have often experienced that no rain, however violent, ever penetrates into these cocks but for a very little way.—And, if they are dry put up, they never sit together so closely as to heat ; although they acquire in a day or two, such a degree of firmness, as to be in no danger of being overturned by wind after that time, unless it blows a hurricane.

"In these cocks I allow the hay to remain, until, upon inspection, I judge that it will keep in pretty large tramp cocks, &c. The advantages that attend this method are, that it greatly abridges the labour, that it allows the hay to continue almost as green as when it is cut, and preserves its natural juices in the greatest perfection. For it is dried in the most slow and equal manner that can be desired. Lastly, that it is thus in a great measure secured from almost the possibility of being damaged by rain." *Essays on Agriculture.*

Clover is a sort of hay that requires a critical attention in curing: Because, though the stalks need much drying, the leaves and heads will bear but little without wasting. It is best to rake it towards night, when the dampness of evening begins to come on; open it the next day, and never stir it much when there is danger of its crumbling.

Clover, requiring so long a time to dry, and its stalks being so succulent that the leaves, which are the best part, will crumble before the hay is sufficiently dried, it has been found expedient to cart it in before the stalks are dry, and either to put it up with alternate layers of straw, or to salt it at the rate of from one half to one bushel of salt to the ton.

Salt hay, in this country, has usually been hurt by lying too long in the swarths. The method in which I have treated it for several years, is, to cock it the next day after it is cut, and carry it in, without delaying more than one day, and put a layer of some kind of dry straw between load and load of it, in the mow, to prevent its taking damage by over-heating. The straw contracts so much of its moisture and saltness, that the cattle will eat it very freely ; and the hay is far better than that made in the common way.

If this hay be permitted to lie out in rains, the saltness of it will be diminished, which they who have but little other fodder may be apt to consider as an advantage.-- But it will contract no virtue, while it loses its saltness. The fresh water will damage it ; especially for those who have plenty of other fodder, or even straw to mix with it.

Salt hay should not be cut when the full or change of the moon is approaching, lest the tides should be high, before it can be got off from the marsh.

"Gen. Smith, of Suffolk, makes use of a *horse-rake*, for raking on his smooth mowing grounds, which with one man, a horse, and a boy to ride the horse, will gather hay as fast as six men in the ordinary way. The rake is about ten feet long; the teeth about two feet; and at right angles, from these are some upright slats of the same length, set, at the lower end into the piece into which the teeth are morticed, and into another light slender piece at the top.

"The teeth, when in operation, run along the ground nearly horizontally, with the points a little the lowest, so as to run under the hay, and as they take it up, the upright slats retain it till the rake is full, when the man who follows it behind turns it over, and thus empties it in a row; then lifts it over the hay thus emptied, and sets it in beyond it; and so it proceeds on, till it is again filled, and the same process is again repeated.

"When one strip across the piece is thus raked up, the horse is turned round, and another strip is raked in the same manner, emptying the hay at the ends of the last heaps raked up, so that in this way winrows are formed. When it is thus raked into winrows, it is dragged up by the rake into bundles, large enough for making into cocks." *Farmer's Assistant.*

HEDGES. A variety of shrubs and trees have been made use of for hedges, but that which appears to be in highest repute is the American Hedge Thorn, (*cratægus cordatæ*) Mr. Quincy, of Massachusetts, gives a statement of his mode of making hedges of this plant, which was published in the third volume of the *Massachusetts Agricultural Repository*, p. 27, and from which we extract the following particulars. The seedling thorns 10,000, were imported in March 1808, from the nursery of Thomas Main, near Georgetown, in the District of Columbia, and planted in a hedge course of two hundred and fifty-five rods, so far as was necessary to fill that extent in one line, each plant being five inches apart. The residue were planted in a nursery for the purpose of filling vacancies.

The hedge course was made in sandy land, ploughed, of the width of four feet, and manured and prepared, precisely as for Indian corn; except only that after ploughing, the centre for two feet wide was turned over with the spade. Without other preparation, the hedge was planted in April, 1808, on a level, without either bank or ditch.

Mr. Quincy's communication is dated 25th of June, 1813, at which time he remarks "I consider the experiment now as completed, so far, at least, as is necessary for a satisfactory ascertainment of the cost of this species of fence. For the expence of the hedge, this year, has been nothing except the annual trimming. It is, upon an average nearly five feet high; and a sufficient security against cattle, for almost the whole extent; and is every day strengthening, without any application of attention or labour.

Mr. Quincy then gives a particular account of every item of the expence in preparing the ground, purchasing, setting out and trimming the trees, and makes the whole expense of making two hundred and fifty-five rods of hedge fence $167,93 cents, *less than sixty six cents a rod*! He concludes the article with the following remarks.

"The course best to be adopted, having reference to the economy of labour, is thought to be the following:

"Plough the hedge-course *six feet wide*. Plant the whole course one year to potatoes. This pays for the labour as much as other land thus planted. Set the thorns *eight* inches apart. This is near enough in a country like this, where hogs are not permitted to run at large, and makes a considerable saving in labour as well as the cost of the plants. Keep both sides of the hedge planted with potatoes, during the whole six years that the hedge is coming to perfection. The potatoes will nearly pay the cost of the labour. The manure for the potatoes benefits the hedge; and while having the potatoes, keeping the hedge clean of weeds is easy.

"*To keep the hedge clear of weeds, and to fill up the vacancies regularly in the spring of every year, with plants of the same age with those of the original hedge*, are the two essential objects of attention after the hedge course is prepared and the plants are set. Younger plants may answer, but whoever would make a hedge, in the most speedy and perfect manner, ought to procure at the time of obtaining the plants for the original hedge, a sufficient extra number to supply all deficiencies likely to occur during the whole time the hedge is forming; to be kept in a nursery thriving, if possible a little better than those in the hedge course. Experience has satisfied me that *two* for every *ten* planted in the hedge course, is a number more than sufficient for this object.*

In the statement of expense, I have made no allowance for *protecting fences*. Where these are necessary, their expense must be added. In my experiment, by excluding cattle, the necessity for them was obviated. Whatever these may cost, the economy of this species of fence, when its durability is taken into view, (to say nothing of its ornamental nature,) must be sufficiently apparent."†

It is often found necessary to *plash*, or interweave the branches

* This calculation is too large, one in fifty, is quite sufficient.

† It ought to be stated, that the Virginia Thorn, of which Mr. Quincy's hedge was made, though rapid in growth, and though in high reputation at the date of his letter, is found to be less adapted to the purpose, than the English white Thorn.

of hedges. Some advise to cut off the tops or head them down to about three feet, or three feet and an half from the ground, when the plants are about four years old. Others direct to saw off the plants when the hedge has completed its second year, and they are about the size of a man's thumb to within about an inch or an inch and an half of the surface, and rely on the shoots. It is likewise advised to place poles or rails along the top of the plants, and fasten the ends together by withes or some kind of pliant bark. See *Farmer's Assistant*, Transactions of the Philadelphia Agricultural Society, and *Dr. Willich's Domestic Encyclopædia.*

HEMP, a plant with a tough fibrous coat, which answers the same purposes as flax, but is coarser and stronger.

The plant is tap rooted, and therefore does best in a deep and free soil. It is luxuriant, and quick in its growth, and therefore requires a rich, and well prepared soil.—The soils which have been found to suit it best, are a rich gravelly loam, or a loose black mould, which is dry and deep. It is an error to think that it needs a wet soil, for it bears drought almost equally with any plant that we cultivate.

Mr. Eliot found by experiment, that it answered very well on a drained swamp: And he tells of a man in the Jerseys, who raised as much hemp yearly, on half an acre of such land, as brought him fifty pounds York money. It is not uncommon for one acre to yield half a ton, which will sell for twenty pounds in cash, at the lowest. And I am told by one who is much acquainted with it, that it is more easily broken and swingled than flax; and that, oftentimes, the brake will do all that is necessary in cleaning it.

To prepare land for a crop of hemp, the land should be ploughed to a good depth in the fall of the year preceding. If it be greensward land, it should be ploughed as early as August or September, that the sward may be perfectly rotten. And if it were ploughed in ridges it would be the better, and fit for sowing the earlier. And by cross ploughing and harrowing in the spring, it should be made extremely fine and mellow. A little dung should be applied, if the land be not in the best heart; and the fall is the best time to apply it.—But if composts are used, they should be laid on just before sowing.

The time of sowing the seed is as early in the spring as the soil can be got into good order, as it is a plant that is not easily injured by frost; but the middle of May will not be too late.

The seed for sowing should be of the last year's growth, as older seed is not wont to come up at all. I once sowed seed which was brought from England. It looked as well as any I ever saw; but not one in ten thousand ever sprouted. The quantity of seed for an acre, in the broad cast way, is three bushels; but half that quantity, in the drill method, will be enough. If the land be poor, a smaller quantity of seed will serve. The ground should be watched after sowing, that birds do not take away the seeds,

The drill method is on some accounts preferable to the other.—For though in the first crop it will fall short, it exhausts the land less; and, therefore, in the long run, it may be more profitable. But in this way it produces more seed, and this method is certainly advantageous on account of the more convenient pulling of the hemp. If sown on narrow ridges, or beds, and the trenches shoveled out after sowing and harrowing, I suspect the broad cast way would have the preference. But of this I have had no experience.

As the correspondent parts of generation are on different plants, they are of two distinct sexes, male and female, and require different treatment. I will venture to assert, contrary to M. Mercandier, that the male is the plant which bears the flowers, and the female that which bears the fruit, or seed.

That which bears the flowers, will be fit for pulling about the end of July. Its ripeness is known by its growing yellow at the top, and white at the root, by the falling of the flowers, and the withering of the leaves. If care be taken in pulling, not to hurt those plants which are left, they will thrive the better after it, as they will have more room, and as the earth will be stirred about their roots. And the drill method is favourable to this work, as the pullers need not tread among the thickest of the hemp. And sowing in beds has the same advantage.

After pulling, it must be put into the water without delay, to steep. Ponds and still waters are best. It will not take more than four or five days to water it enough. But it must be watched, lest it should be overdone. After watering, it must be spread and dried in the sun.

The fruitful kind does not ripen till about five or six weeks later. Its ripeness is known by the seed's turning brown. After it is well dried, and the seed taken off by a kind of coarse comb, it must be watered. It will take almost three times as much watering as the first kind. The one kind is more fit to be manufactured into thread and cloth, the other more suitable for rigging of ships, and ropes. But the latest kind may be made pliable and fine, if labour enough be bestowed upon it. Instead of steeping, spreading hemp in the dew will answer, as I have found by experience; and this method is practised in England.

The dressing of hemp may be performed in the same manner as that of flax if it be not uncommonly large and long. A person, who is well acquainted with the culture and manufacture of hemp, assured me, that when his neighbour raised it on a drained swamp, he had it twelve feet long; and, that he might manage it easily in dressing, he cut it in the middle. It was then as long as ordinary hemp, and as strong for every purpose.

If some of the stalks of hemp should be too large and stubborn for the brake, they may be put by themselves to be peeled by hand. The doing of it may be an amusement for children and invalids.

But to facilitate the dressing of hemp, mills should be erected for doing it. Or the machinery may be an appendage to some other mill. Two brakes should be moved together, a coarser and a finer, placed head to head, that the handfuls may be easily shifted from one to the other. It is light work for two boys to tend them. But the breaking of large hemp by hand, is severe labour for the strongest men.

If no convenient stream be at hand, a mill may be constructed to be worked by a horse.

It was formerly the custom to beat hemp abundantly with mallets, or with pestles in large mortars, or in fulling mills, to make it soft and fit for spinning. But Mr. Mercandier has shewn how it may be more easily done, by steeping it in warm water, or in lie, and washing it. See his *Treatise on Hemp*.

The great profit on a crop of hemp, and its being an article that will readily command cash, should recommend the culture of it to all our farmers. Besides the hemp itself, of the value of twenty pounds per acre, after it is dressed, the seed of an acre must be allowed to be of considerable value. Persons need not fear their crops will lie upon their hands, when they consider the vast sums of money which are yearly sent to other countries for this article, almost enough to deprive the country of a medium, and how naturally the demand for it will increase as it becomes more plenty. There is no reason to doubt of success in raising hemp, if the soil be suitable, and well prepared; for it is liable to no distemper; cattle will not destroy it, unless it be with their feet; and it is an antidote to all sorts of devouring insects. Neither is the plant difficult as to climate. Though the hottest climates do not suit it, temperate and cool ones do; and it has been found, by the small trials that have been made, to thrive well in the various parts of New-England. The most northern parts are very suitable for the growing of hemp. The southern are equally so.

A new method of rotting hemp was communicated by M. Bralle, and published in a foreign Journal, and is, in substance as follows. To 25 gallons of water boiling hot, add 12 ounces of green soap, and when the soap is dissolved, 22 pounds of hemp are to be immersed, so as to be entirely covered with the liquor, the vessel closed, the fire put out, and the hemp left to macerate for two hours. Several steepings may be made in succession, care being taken to add soap, each time, to replace what has been absorbed, and to heat the water to the former temperature. The same water may be employed for fifteen days continually.

When the bundles of hemp are taken out, they are covered with straw, that they may cool gradually, without losing their humidity. Next day they are to be spread on a floor, the hands shifted, and a heavy roller passed over them, after which the hemp separates easily from the reed by beating. The hemp thus separated, is spread on the grass, and turned, and after five days removed to the ware-house. In steeping the hemp, the bundles should be kept in a vertical position, as the operation is found to succeed better so than when they are horizontal.

The advantages of this method are, 1st. The superior speed of the process to that in common use: 2d, Its being practicable at all seasons: 3d. Its not being injurious to health, or producing any bad smell: 4th. A saving of expense, when a proper apparatus is used: 5th. A superior quality of the hemp so prepared, and less waste, so that nearly a fourth more hemp is obtained from the same raw materials: 6th. The extending the culture of hemp to all situations, which can now be carried on only in the vicinity of running water.

A very good apparatus, for the process, is formed by a boiler, with covers for steeping vessels. See *Steam Boiler*.

A process similar to the above described, would, probably, save the trouble and expense of water rotting flax. Lie made of wood ashes, would, perhaps, answer the purpose of water impregnated with soap.

The Farmer's Assistant states that in the bog meadows in Orange County, N. Y. the hemp is cut close to the ground with an instrument made for the purpose. That when dried it is gathered in bundles, bound with straw and stacked in the field till about Christmas. It is then spread on the snow, and when the snow dissolves in March, it is generally found sufficiently rotted.

For further directions relative to the culture of hemp, see a "letter from Hon. Justin Ely," published in the Massachusetts Agricultural Repository, vol. III. p. 105. And for the mode of dressing it without rotting, see *Flax*.

HERD'S GRASS, or Cat's-tail, *Phleum Pratense*, called Timothy Grass in the southern States. This grass is a native of New-England. Mr. Eliot says it was first found at Piscataqua in New-Hampshire, by one Herd, who propagated it, whence the name. It is cultivated in our improved fields for hay. It requires about ten or a dozen quarts of the seed for an acre. It does best in rich and moist land.—More needs not to be said of a grass, the great value of which is so well known in this country; especially in the northern parts, where it prospers more than in the southern.—It is of more importance to our farmers than any other grass that they cultivate.

HIDE BOUND, a distemper into which horses fall when they are poorly fed and neglected. "A horse that is hide bound grows lean, has a feverish heat, his skin sticks to his ribs, the spine becomes harder than usual, small biles break out on his back, and yet his appetite sometimes continues good. As this disorder seldom is an original complaint, but generally arises from some former cause, regard must be had to that cause, in the method of cure. But as to the disorder itself, Vegelius directs the anointing the whole body with oil and wine mixed together, rubbing them strongly against the hair, in a warm sun, in order that the skin may be relaxed, and a sweat break out; after which the horse should be well curried, and placed in a warm stable, with plenty of litter.

"The authors of the *Maison Rustique* advise that the next day after bleeding the horse, a fomentation be made of emollient and aromatic strengthening plants, boiled in lees of wine, or beer, and that the whole body of the horse be rubbed with these plants, whilst they are warm, till it is thoroughly wet; and that the loins, belly, and neck, as well as the rest of the body, be anointed with a mixture of one part honey and three parts of ointment of elder, rubbing it strongly in with the hand, that it may penetrate the skin. This done, the horse should be covered with a cloth dipt in the warm fomentation, and doubled, and another covering should be put over this, tying it on with one or two surcingles. The horse should remain in this condition 24 hours, and then be fomented, rubbed, &c. twice as before. These fomentations being finished, a warm covering must be continued, lest the horse catch cold; and he should then have an opening clyster, and the next morning a purging medicine; continuing to wash his head and neck, and also to rinse his mouth with the decoction.

"For food, put into a pail or two of water about half a bushel of barley meal carefully ground, stir it well about, and let it settle. When the heaviest parts have subsided, pour the thin part off for the horse to drink, and give him what remained at the bottom, at three different times in the day, mixing with it a due quantity of crude antimony.—The horse must have rest for some time, and be fed with the best hay, or grass according to the season of the year. In spring, there is nothing better than new grass. In about three weeks, he will begin to mend remarkably." *Mills on Cattle.*

HOE, a well known instrument used in tillage. It is called by some writers the hand hoe, to distinguish it from the horse hoe.

Hoes are chiefly of two kinds, narrow and broad. The use of the narrow hoe is to break up spots of hard, or tough ground, as the balks left by the plough in swarded land, or the corners of lots where the plough cannot conveniently reach; or to take up strong roots, such as those of the shrub oak, &c. Therefore, this tool must be made thick and strong, with a large eye, that it may admit a strong helve.

It has also the name of a breaking up hoe; but it is seldom made to do the work of a plough in this country of late, unless by the poorest people, and in new places where teams cannot be easily had.

The broad hoe is a very important implement among farmers, as it is much used, though not so much as it should be. The more mellow the land is, the larger the hoe should be, that work may be done more expeditiously. The tough and hard soil requires a narrower hoe, to render the labour more easy.

Where land is not stony, hoes should be kept sharp by grinding. They will enter the ground the more easily, and destroy weeds and their roots more effectually.

For the ease of the labourer, hoes should be made as light as is consistent with the needful degree of strength: Their handles especially should be made of some light kind of wood, as ash, or white maple, or a young tree of spruce. Great and important use is made in weeding between carrots, beets, and other crops, that do not require hilling, of the Dutch hoe, which is used for cutting up weeds, the labourer passing backwards, and pushing the hoe forwards. It is expeditious, and effectual, for weeding. A small hand hoe, of steel, not more than two inches wide, and six inches long is excellent for light weeding. For the *Horse Hoe*, see that article.

HOEING, either burying seeds in the earth with the hoe, or breaking and stirring the soil, chiefly when plants are growing in it.

This, after tillage, as I may call it, has been found to be of great advantage to almost every kind of plants, and to some it is so necessary that no crop is to be expected without it. The deeper land is hoed, the greater advantage do plants receive from hoeing, if due care be taken that their roots be not disturbed, or too much cut to pieces.

The ends to be answered by hoeing are chiefly these:—1. To destroy weeds, which are always ready to spring up in every soil, and which would rob the cultivated plants of most of their food. Scraping of the surface, if it be done frequently, may answer this purpose; but to destroy the roots of weeds, deeper hoeing is necessary. 2. To keep the soil from becoming too compact, which prevents the roots extending themselves freely in search of their food, at the same time keeping up a fermentation, by which the vegetable food is concocted, and brought into contact with the roots. For this purpose, the deeper land is hoed the better. But hoeing should cease, or be only superficial, when the roots are so far extended as to be much injured by hoeing. They will bear a little cutting without injury. For where a root is cut off, several new branches will come in its place. 3. To render the soil more open and porous, so that it shall greedily drink in the nightly dews, and that rain may not run off, but readily soak in as it falls, and be retained. Accordingly, the more and oftener land is hoed, the more moisture it retains, the better it bears drought, and the more its plants are nourished. 4. Another design of hoeing, and which has not been enough attended to, is to nourish plants by drawing fresh soil near to them, the effluvium of which enters their pores above ground, and increases their growth. 5. At the same time, earthing of plants makes them stand more firmly, and increases their pasture in the spots where the roots most abound. At the same time it prevents the drying of the earth down to the roots.

But earthing, or hilling of plants, should be done with caution.

Hilling excessively is hurtful, as it does not permit the roots to have so much benefit from the rains, and too much hinders the influence of the sun upon the lowermost roots. Whatever hilling is done, should be done by little and little, at several hoeings, that the roots may gradually and easily accommodate themselves to the alteration of their condition. Lastly, frequent hoeing serves to prevent the standing of water on the surface, so as to chill the ground, and check all fermentation in it.

When all the hoeing between rows of plants is performed with the hand hoe, the labour is severe, and more expensive to the owner; and the plants will, on the whole, receive far less advantage from hoeing. Therefore, where land is tolerably free from obstacles, I would earnestly recommend that the hoe plough, or the common horse plough, which answers nearly the same end, be much used; and the earth stirred with it to a good depth, and frequently, during the proper season of hoeing, which is the former part of summer, but varies with respect to different crops.

A plough, called a cultivator, has been constructed, with two mouldboards, which turns the mould both ways at once, towards each of the two rows between which it passes. But, as it requires more than one horse to draw it in stiff ground, two furrows made with a hoe plough, or horse plough, according to the customary practice, may answer full as well. When the soil is light and mellow, it will be a saving of time to use this cultivator; and the work will be done with more regularity and neatness, if guided with skill and due care.

The usual method of horse hoeing is as follows: At the first hoeing, turn the furrows from the rows, so that they form a veering, or ridge, in the intervals between the rows. The plough should pass as near to the rows as may be without danger of eradicating or disturbing the plants; for it is best that the soil be loosened as near to the roots as possible: Because when they are tender and weak, they will extend their roots but little; and there will be no opportunity afterwards of ploughing and stirring the earth so near to them, without too much danger of tearing and injuring their roots. After ploughing, the rows are to be cleared of weeds with the hand hoe, and a little fresh earth brought into contract with them.

At the next hoeing, and all after hoeings, in our common husbandry, the furrows are to be turned towards the rows, so as to form a henting, or trench, in the middle of each interval; and cross the furrows last made, that the land may be the more thoroughly pulverized. This operation carries the share of the plough farther from the roots, and at the same time affords plenty of fresh earth about the plants; which must be finished with the hand hoe. But if, in ploughing, any of the plants should chance to be covered, they must be set free without delay.

At the last hoeing, either of Indian corn, or of any thing that is planted in hills, as it is vulgarly called, it is best to make but one furrow in an interval, and to pass the plough both ways, or cut the ground into squares with the plough, or rather with the cultivator. This leaves the roots the more room, and less work will remain to be done with the hand hoe.

If the horse be weak, or the ground hard and stiff, it may be needful to let the plough go twice in a place, which makes four times in an interval. For the plough should go as deep for hoeing, as in any other ploughing, or else the intention of it will be partly defeated; which is to keep that quantity of soil light and mellow from which the plants are to draw the most of their nourishment.

We apply horse hoeing to Indian corn, when the ground is well cleared from obstacles, and could not be easily persuaded to neglect it. Every farmer knows how much it saves labour, and that the crop is increased by it. Why then will they not be persuaded, by all that has been experienced, and written, by some of the wisest farmers, to apply this method of culture to many other plants? I have no doubt it might be done with equal advantage. Indeed, we cultivate but few plants in tillage, for which this kind of culture would be improper. In Europe, they horse-hoe all kinds of grain, and even some kinds of grasses.

In a dry season, or in land that is in no danger of ever being too wet, it is advisable to hoe only in the morning and evening. And if farmers will work as early and late as they can, they may afford to desist, and rest themselves from nine till four, when the air is hottest. The ground will get and retain the more moisture which is thus hoed early and late. And in the middle of some of our hottest days, there is danger of hurting tender plants, by drawing the scalding hot earth close to their stems. But the opinion entertained by many, that no hoeing at all should be done in a dry season, is irrational and ridiculous. They deprive their land of the benefit of the dew, by neglecting to hoe it, suffer it to be overrun with destructive weeds, which rob the plants of most of their nourishment, and allow the ground to be so compacted and hard, that the rain when it comes will not penetrate it. This strange opinion will occasion much loss to those whose conduct is influenced by it.

HOGSTY, a kind of building in which hogs are confined and fed. The ways of constructing these houses are various: But the best are those which are framed and boarded. The boards, that the swine may not gnaw them to pieces, should be of some harder wood than white pine, and they should be fastened with ribbings and spikes. Whatever be the constructure of sties, they should always have one part close and warm, with a tight roof over it; and the other part open, in which the trough is placed. Swine will not well bear to be wholly secluded from the weather and sunshine; and it is hurtful to them to have a cold and wet lodging; more hurtful than many people are ready to imagine.

Although there should be a part of the sty, planked and boarded and warm, covered from rain and sun, and amply supplied with litter, yet the greater part of it should have no wooden floor. The trough in which pigs are fed should be firmly fixed to the floor, so that they may not overset it, and it is very important that the divisions of a pig-stye be so numerous, that pigs of nearly equal size and strength only should be permitted to feed together, otherwise the more powerful will exclude the others, and materially stint their growth.

The more comfortably hogs are kept, the less nourishment they require. The trough should be on the upper side covered with one or more lids, and upright pieces set before it at such distances that one hog only can put his head between any two of them. If sixteen hogs are to be kept in the same sty, it should be thirty-two feet long and twelve wide, and the apartments should be so divided, that too many of the animals may not be forced to lie together. And it would, probably, be well to divide likewise the feeding apartment. Posts should be set up in the sty for the hogs to rub themselves. If thirty-two hogs are to be kept or fatted, perhaps the better way is to have two hog-stys of the dimensions last described, placed together with a roof over the whole, and a passage between them for the purpose of carrying food to the troughs.

The upper part of the sty, or some part of it may be appropriated to storing the different articles of food, which are wanted for feeding the animals, and it would be well to have a steam boiler beneath the same roof. If a part of the roof be made to extend considerably beyond the sty, it will afford a cover for forming a heap of compost with the dung of the swine. See *Farmer's Assistant*, Art. *Hog-sty*.

HOP, *Humulus*, a narcotic plant of the reptile kind, the flower of which is an ingredient in beer,ale, &c. As I have not had much experience in hops, I shall give an account of the management of them, chiefly abstracted from the *Complete Farmer*, and abridged.

A rich, deep, mellow, dry soil, rather inclining to sand than clay, is best adapted to the cultivation of hops. A black garden mould is excellent.

The ground should be ploughed very deep, or dug with a spade,reduced to a fine mould by repeated ploughing and harrowing, and laid even.

When the ground is in proper readiness for planting, let a line be stretched on a straight side of a field, with knots or rags in it,as far asunder as you design your hills shall be; and stick in the ground a sharp pointed stick at every knot, as marks for the places where the hills are to be made. Remove the line to such a distance as to make the hills equidistant both ways; and so on through the whole ground.

The distance of the hills should

be regulated by the strength of the soil. But in every case they should be far enough asunder to admit the hoe plough at all times. If the soil be dry and shallow, six or seven feet will be a convenient distance: But if it be rich, moist, and apt to bear large hops, it may be right to allow eight or nine feet.

The time to plant hops is when they begin to shoot in the spring. The sets are cuttings from the roots, or branches which grow from the main root. They should be from five to seven inches long, with three or more joints or buds on each, all the old and hollow part being cut off. Make holes twelve or sixteen inches wide, and of a depth proportioned to the nature of the ground. If shallow, with hard clay or gravel under, dig not into it, lest you make a basin to retain water; but raise a small hill of good mould. If there is a good depth of rich mellow mould, dig the hole a foot and a half, or two feet deep; the hops will thrive the better.

When all things are ready for planting, fill up the holes with the mould before thrown out, if it be good; but if the same earth be not rich enough, make use of fine fresh mould, or of a compost provided for the purpose, a peck in each hill, but no dung on any account.

Then with a setting stick make five or six holes, one in the middle perpendicular, and the rest round it sloping, and meeting at the top near the centre. Put in the sets so that they may stand even with the surface, press the mould close to them, and cover them with fine mould two or three inches thick. A stick should be placed on each side of the hill to secure it.

The ground being thus planted, all that is to be done in the following summer is to keep the hills and alleys clear of weeds by frequent hoeings, to dig the ground in May, and carry off the stones raised by digging; to raise a small hill about the plants, and throw some mould on the roots, and in May or June to twist all the vines and branches together in a loose knot, and lay them thus twisted on the top of the hill.

Early in the following spring, when the weather is fine, open the hills, and cut off the shoots of the first year, within an inch of the stock, together with the younger suckers that have sprung from the sets, and cover the stock with fine earth.

In the third and following years, when you dig your hop ground, let the earth be taken away with a spade or hoe, round about the hills, very near them, that you may more conveniently come at the stock, to cut it. Then in fair weather, if your hops be weak, begin to dress them: But if strong, do it later; for dressing late restrains their too early springing, which hurts the hop.

After dressing in the second year, the next thing is to pole them.—Poles ten or twelve feet long will do then: But in the third year, when they come to their full bearing state, they will require poles of full size: This, if the ground be rich, and the hop vigorous, will be from sixteen to twenty feet; or there will be danger of losing great part of the crop.

The hop will soon run itself out of heart, if it be over poled. Neither can a good crop be expected from over poled ground ; because the branches which bear the hops grow very little, till the buds have overreached the poles, which they cannot do when the pole is long. Two small poles are sufficient for a hill in a young ground.

A hop garden, Mr. Young says, will last almost forever, by renewing the hills that fail to the amount of about a score annually : But it is reckoned better to grub up and new plant it every 20 or 25 years.

In forward years hops are ripe at the beginning of September.—When they begin to change colour, or are easily pulled to pieces ; when they emit a fragrant smell, and when their seeds begin to look brown and grow hard, you may conclude that they are ripe. Then pick them with all expedition ; for a storm of wind will do them great mischief at this time.

When the poles are drawn up in order to be picked, the vines around should be cut asunder at the height of three or four feet from the ground : For cutting them lower, especially while the hops are green, would occasion so great a flow of sap, as would weaken the root.

It has been remarked by one who had much experience, that hops which are late picked bear more plentifully the following year than such as are picked early : For which reason he recommends late picking. But the hops which are picked early look better, and are undoubtedly stronger.

The best way of drying hops is on kilns. Four pounds of undried hops, will make one pound after they are dried.

Before hops are bagged, they should be laid in a heap, that they may sweat and grow tough : And if they are covered for a while with blankets, they will be the better. The bags are coarse linen cloth. They are commonly about eleven feet long, and near two yards and a half in circumference, and contain about 250 weight of hops. The small bags, called pockets, contain about half as much.

The manner of bagging is thus. Make a round or square hole about 26 or 30 inches over, in the floor of the chamber where the hops are laid in heaps after sweating. Tie with a piece of pack thread,a handful of hops in each lower corner of the bag, to serve as handles for the more easy lifting or removing the bag ; and fasten the mouth of the bag to a frame, or hoop, somewhat larger than the hole, that the hoop may rest on its edges. The upper part thus fixed, the rest of the bag hangs down through the hole, but not so far as to touch the lower floor. Then throw into it a bushel or two of hops, and let a man go into the bag, and tread the hops down till they lie close ; then throw in more and tread ; and so on till the bag is full. Loose it from the hoop, and sew up the mouth as close as possible, tying hops in the upper, as was done in the lower corners. The harder the hops are pressed,and the closer and thicker the bag is, the longer and better the hops will keep.

A small manuring of hop ground every second year is sufficient.—Dung was formerly more in use than at present, experience having shewn that lime, sea sand, marle, ashes, &c. answer the end better, and last longer. But hog dung prevents mildew from taking hops.

Each pole, according to Dr. Hales, has three vines, which makes six vines to a hill. All the sprouts above this number, should be broken off in the spring.

" *Method of extracting the virtue of Hops in Brewing*. The usual method is to put in hops without any preparation into the strong beer, or alewort; the consequence is, the richer and better the wort is, the less it will partake of the essence of hops. The rich fast wort sheathes up the pores of the hop, and as it were embalms the leaves, so that the beer or alewort can extract scarcely any part of the necessary quality of the hop; but when it is put into the small beer wort, a fluid of a more thin nature, there the pores are unsheathed, and the small beer is rendered too bitter: therefore the hops, before they are put into the strong drink, should be previously soaked in a pail of hot water." *Domestic Encyclopædia.*

HORN DISTEMPER, a disease of neat cattle, the seat of which is in their horns. Cows are more subject to it than oxen. It does not attack bulls; and steers and heifers, under three years old, have not been known to have it. The distemper gradually consumes the pith of the horn. Sometimes it is in both horns at once, but more usually in one only.

The disease is discoverable by the coldness, or loss of the natural warmth of the horn; by dulness of the eyes, sluggishness, loss of appetite, and a disposition to lie down. When the brain is affected, cattle will toss their heads and groan much as if in great pain.

To effect the cure, the horn should be perforated with a nail gimblet, through which the corrupted thin matter will be discharged, if care be taken to keep it open. By this boring, which should be nearly horizontal, or in the depending part of the horn, and two or three inches from the head of the animal, the cure sometimes is completed. When it proves otherwise, a mixture of rum and honey with myrrh and aloes, should be thrown into the horn with a syringe; and be several times repeated, if the disease continue. For a more particular account, see a letter from the Hon. C. Tufts, Esq. in the 1st vol. of the *Memoirs of the Academy of Arts and Sciences.*

HORSE, one of the most useful of tame quadrupeds. The marks or evidences of a good one are these, a high neck, a full breast, a lively eye, a strong back, a stiff dock, full buttocks, ribs reaching near to the hips, well made hoofs, rather large, and a good gait.

The size of a horse should be in proportion to the work in which he is chiefly to be employed. Small sized ones often prove good in the saddle. They are apt to be hardy, and in proportion to their size, and the quantity of their eating, usually are the most profitable. Plough horses, and all draught horses, should be large, as their weight is of importance in drawing; and as it is often inconvenient to put two horses to one plough, especially in horse hoeing. Largeness is also of importance, when they are used single, in journeying, as they most usually are, in a chaise or sleigh.

A horse's manner of going is a matter of no small importance.— The ambling gate, or what in this country is vulgarly called pacing, is not good, neither for the horse nor the rider. It is tiresome to both. It habituates a horse to carry his feet too near to the ground, so that he is the more liable to trip and stumble.

The method so much practised formerly in this country, of teaching horses to pace swiftly, and racing in that gait, is highly pernicious. It puts them to a much greater strain than running; and numbers have been thus ruined.— Some colts naturally amble, and others trot. But all may be made to trot, if due care and pains be taken with them while they are young, or as soon as they are first ridden. In a carriage an amble is tiresome to a horse, appears highly improper, and is disgusting to every one. And I do not see why it should appear at all more tolerable in the saddle.

When any change of gait is wanted for the ease of the rider, the canter is to be preferred, than which none can be more easy.

The way of breaking a young horse that is mostly used in this country, is highly absurd, hurtful, and dangerous. He is mounted and ridden before he has been used to the bridle or to bearing any weight on his back. If he will not go forward, he is most unmercifully beaten; by which his spirits are broken, and his strength impaired. If he rears up, he is pulled backwards, with the risk of hurting both horse and man. If he runs and starts, as he probably will under such management, he flings the rider, perhaps is frightened, gains his liberty, and is encouraged to do just so the next opportunity; and the unfortunate rider blesses himself, as he has reason to do, if he escape without broken limbs. Or if the horse should chance to go kindly, the rider continues the exercise till the horse is fatigued, discouraged, and injured.

Instead of this mad management, the way practised in the older countries should be adopted. Let a horse first of all be tamed with the bridle, by leading him again and again; in the first place, after, or by the side of another horse; and after he walks well, bring him to trot after his leader. In the next place, put on the saddle, and lead him in that, time after time. Then lay a small weight on the saddle, and if he be apt to start, fasten it, that it may not be flung off, increasing the weight from time to time, till he learns to carry what is equal to a man's weight. Lastly, let a man gently mount him, while another holds him by the bridle, and fix himself firmly in the saddle.— The place of riding is recommended to be a ploughed field. Let him thus be ridden with a horse going before him, till he learn the use of the bit, and will stop, or go forward, at the pleasure of the rider, and without the application of much force. Being exercised in this manner a few times, and treated with all possible gentleness, there will be no more occasion for leading him. He will go well of himself; and be thoroughly broken, without so much as giving him one blow, and without danger or fatigue, to the horse or his rider. And, what is much to be regarded, the horse's spirits will be preserved, though he be sufficiently tamed. In teaching a horse to draw, gentleness must be used. He should be tried first in company with other horses, whether in carting or ploughing; and the drought should not be so heavy as to fret him or put him to great exertion till he has learned to draw steadily. After this he may be put to draw light loads by himself. Lastly he may be put to a pleasure carriage, but coupled with another rather than alone, and to a sleigh rather than a chaise.

It may be taken for a general rule, that the gait which is easiest to a horse, will be the easiest to his rider. For jaded horses, it has always been observed, are apt to go hard, and to tire their riders.

The feeding of horses, as I conceive, has not been sufficiently attended to in this country; which is, doubtless, one reason why they are in general so mean and despicable. Too many keep horses who cannot well afford to feed them. They should neither run upon the roads and commons, nor in pastures that are filled with wild and water grasses. They love a dry pasture, not too much shaded, and short grasses of the best kinds. Clover and white honey-suckle, both green and dry, are excellent food for them. It nourishes them well, and prevents costiveness, which is very hurtful to them.—The best of clover hay will keep them as well as most other kinds of hay with oats.

To fit a horse for a journey he should not be suffered to grow too fat and gross. He should for some time be kept in the stable rather than in the pasture, and fed mostly with hay and provender: But rather sparingly if he incline to be fat. He should have exercise daily to harden his flesh, and keep him in the habit of travelling. He should be shod some days before he begins a journey, that the shoes may be well settled to his feet, and the nails a little rusted at the points, that they may hold the faster. And the pads of the saddle should be well fitted to his back, so as to fill the hollows, and bear equally on every part. And while he is on the journey, he should be stabled every night. It is destructive to expose a horse to the dampness and cold of the night after severe exercise. But it would be best, if neither horses, nor any of our cattle, were wholly confined to dry meat in winter. Horses indicate this, by their eating snow with their hay. Set a basket of snow within reach of a horse, when he is at his manger, and he will take a mouthful from each alternately. Of all juicy food for horses in winter, writers on husbandry seem to give carrots the preference. They have been found by experience to answer well instead of oats for labouring horses ; and to fatten those which are lean.

He that would be sure to keep his horse in good order, must beware whom he suffers to ride him, and must see that he is never abused. Profuse sweating should always be avoided. And when a horse is much warmed by exercise, he should not be exposed to cold air, or night dew, and much less to rain and snow. If he cannot be instantly rubbed down and housed when warm, he should be covered with a blanket ; and he should always have a dry stable, and be well littered. The neglect of these precautions may bring on incurable disorders.

Horses should not be too much deprived of the liberty of motion, as they too often are. Close confinement after hard labour, will be apt to abate their circulations too suddenly, make them chilly, and stiffen their joints. To be deprived of motion, is bad for man and beast. Horses therefore should not be straitened for room in their stables. Stables should not be so low as to prevent their tossing up their heads as high as they please. Some stables have so little room over head as to bring horses into a habit of carrying their heads too low. They become afraid to lift them up. They should also have room in their stables to turn their heads to any part of their bodies, that they may defend themselves from the biting of insects, allay itching, &c. And their halters should always be so long, and their stable so wide, that they may lie down conveniently. Nor should horses be so placed as to be able to deprive each other of his fodder.

When horses are kept in stables, as they generally are in the coldest half of the year, they should be daily dressed, as it is called. The curry comb, and the brush, should be well used on all parts of their skin, which are covered with hair. This increases perspiration through the pores of the skin, which is necessary to health ; and causes the blood to move faster in the veins. This treatment will not only cause them to look better, but they will have better health, and more activity and courage. They will digest their food better, and be better for service. But if rubbing and friction be wholly neglected, or slightly performed, the hair will appear dry and rough ; the perspirable matter hardens in the pores of the skin, or remains lodged at the roots of the hair, and has the appearance of a dirty white dust : And sometimes like small scales attended with itching. More especially is rubbing necessary for horses, when they are growing cold after being sweated by labour. In such cases it should never be omitted.

Columella observes " that the bodies of cattle ought to be rubbed down daily, as well as the bodies of men ; and says it often does them more good to have their backs well rubbed down, than their bellies well filled with provender."

But in warm weather it would be best for them, that they should not have the confinement of the halter, nor even of the stable. A small spot of feeding ground, if it were only a few rods, adjoining to the stable, and the door left open, that a horse may go in and out alternately as he pleases, would greatly conduce to the health of the animal. This degree of liberty will be most needful, when the flies are troublesome; and be better for him than confinement to a stable that is perfectly dark. In fly time it gives a horse much ease and comfort to smear his limbs, neck and head, with rancid fish oil, or something else that will keep the flies from attacking him. And in all seasons, when horses have been heated with exercise, they should be well rubbed, or curried.

When a horse runs in a pasture during the grass season, he should have some shelter, not only a shade to defend him from the intense beat of the sun, but a shed, or a clump of trees, that he may retreat from the inclemencies of the atmosphere.

But horses that are daily worked, in summer, should be mostly kept upon green fodder in stables, rather than grazed in pastures. The tendance of them will not be so burdensome, with a spot of high and thick grass at hand, as leading them to and from a pasture, at the distance of a quarter of a mile. This will prevent their being often chilled by feeding in wet nights. A large quantity of manure will thus be saved. And a very small quantity of land will answer, in comparison with what it takes for the pasturing of a horse. Keeping a sithe and a basket at hand, a horse may be foddered in this way, in two or three minutes; and by the time that the whole spot has been once mowed over, that which is first cut will be grown up again. Where a number of horses are soiled, a pair of poles, or a hand cart, will be better than a basket to carry the hay to them. This practice, called soiling, answers well near to cities and large towns, where lands for pasturage are not plenty; and where, by means of the plenty of manure, lands may be made to yield the greatest crops of grass. For very thick grass should not be fed off; because the greater part of it will be wasted by the trampling, and the excrements of animals.

When grain is given to horses it is an economical practice to have it either ground or boiled. When horses are soiled, or fed in a stable on green grass, it should be cut and carried in during the morning, while the dew is on.

A disorder, called *Ptyalism*, has for some years past been gaining ground among horses in various parts of the United States, which is an excessive watering or slavering at the mouth. Various causes have been assigned for this disorder, but none of them satisfactory. Soiling them, is, however, a certain remedy.

HORSE HOE, a small light plough, drawn by a single horse,

either with one or two mould boards. It answers in extensive culture, all the purposes of the hoe, though it may be well to have it followed with the hand hoe in every species of culture. In the cultivation of potatoes and Indian corn it seems to be indispensable that it should be so followed. But in drill husbandry, with potatoes planted in rows, or drills, with Indian corn planted in the same manner, and with carrots, Swedish turnips, and grains of all sorts, it need not be followed with the hand hoe, but a man must follow to pull up the weeds, which the horse hoe had not fully destroyed.

HORSE SHOEING. The following directions for shoeing horses are taken from an English publication. "Let nothing be cut from the sole-binder, or frog, except the loose rotten scale. No opening of heels on any occasion,—it infallibly causes in time the disease called hoof-bound.—No shoes to be fitted on red hot. Shoes always to be made of the best hard and well wrought iron, with not a convex, but a flat and even surface next to the ground. So that the horse may stand in a natural and easy position. No caulks for either heel or toe of fore or hind feet. The web of the shoe not so wide as usual, nor so thick nor strong at the heel, and never to project beyond it, in order that the foot may stand perfectly level, and the frog be not prevented from touching the ground. It is as absurd to pare down the frog as is usually done, as it would be to pare away the thick skin, which nature has placed over the human heel! All the horses in England are now shod according to the above directions."

"When a horse is well shod, if water is poured upon the bottom of his foot, it will not pass between the hoof and the shoe. Shoes for draft horses, that have seldom occasion to go out of walk, should be heavy, strong, and with high heels, and pointed at the toe with steel." *See Mason's Farrier, published by Peter Cottom, Richmond, Virginia*, 1821.

HURDLE. The hurdles used in husbandry, for fences, and frames of wood, consist of two poles, four feet apart, connected with small sticks across from the one to the other. Spruce poles are good for this use, being light and tough. The sticks may be of split timber, such as does not rot too soon; or round sticks of natural growth, such as thrifty suckers from the stumps of oak trees. If they are wattled, or have twigs wove into them, the sticks may be a foot, or eighteen inches apart; and they will resemble the hurdles on which fish are dried. If they are not wattled, the sticks must be so near together, that neither sheep nor hogs can pass between them. Cheap gates may be conveniently made in this way. A hurdle is often wanted, to make a good fence across a run of water, being most suitable for this purpose, as it may be fastened by strong stakes at the ends, and as it resists the current of water but little. They are useful to fence small pens and yards on any sudden occasion.

And as they are easily removed, they are useful in England, in eating off a crop of turnips with sheep. If there should be need of preventing the climbing of boys over them, the ends of the cross sticks may rise a few inches above the upper pole, and be made sharp at the points.

HURTS, *and Bruises in the withers*. Horses are very often hurt, or wrung in the withers, by the biting of other horses, or by unfit saddles, especially when the bows are too wide; for by that means they bruise the flesh against the spines of the second and third vertebræ of the back, which form that prominence which rises above their shoulders. When the swelling is moderate, the usual method is to wash the part with salt and water, or to apply horse dung, or salt and black soap mixed together, which very often succeeds.—Any restringent charge, as bole and vinegar with whites of eggs, has the same effect; as also the whites of eggs beat up into a foam with a piece of alum. This is very much commended.

"Sometimes the hair is rubbed off, and the part becomes galled, in which case nothing is preferable to the rectified spirit of wine or brandy, which ought to be used often, covering the part with a flaxen cloth dipped in bees-wax, and a little oil melted together, to keep the dirt from it and defend it from the air." *Gibson's Far.*

I.

IMPROVEMENT. This is not the mere use, and occupying of lands as the word is very improperly used, but it implies *such* an employment of them as shall constantly render them profitable to the occupant. If lands are so inferior in their natural qualities and productiveness, as *not* to return to the cultivator a reasonable profit for his labour, rent, and interest, it would be better to abandon them either altogether, or to bestow more labour and manure on a part of them, leaving the residue for a scanty pasture for their cattle and sheep. Perhaps the greatest error into which our farmers too often fall is the attempt to cultivate too much land even when it is good—but the error is still greater when the lands are of inferior quality.—The practice of cropping their lands, till they will no longer yield sufficient to pay the expense of culture, is perhaps the source of the poverty of many farmers. Nothing can be more pernicious than the custom of exhausting lands, and then leaving them to recruit by neglect, and permitting them to bear such weeds, and exhausting plants, as any soil, however impoverished, will furnish. The European mode of fallowing, which consists of repeated turnings of the soil, enriched by the weeds, which spring up, spontaneously, upon the ground so abandoned is perhaps the most judicious course; but it is doubted, whether in a country like our own, in which the lands of first quality, are not yet exhausted, and are more than competent to supply all the wants of our own population, and all that other nations will take at a price, which will return the expense of labour and capital can be advantageous. It would seem, therefore to be the evident policy of our farmers to cultivate no more land than what they can attend to thoroughly, and instead of spreading their labours over large tracts, no one acre of which on computation yields a fair remuneration for their toil and capital, that they should confine themselves to smaller portions, and cultivate these with spirit and intelligence.

To apply these general remarks, suppose a farmer, possessed of one hundred acres of sandy or gravelly land, and to own 10 or 20 head of cattle. It is scarcely to be believed that on a farm of such a size, there should not be 10 or 15 acres of meadow ground, or of richer soil. If he would devote all his manure from his cattle to his best lands—if instead of reaping 10 or 15 bushels of rye, or 25 or 30 bushels of Indian corn to the acre, he would apply all his manure carefully preserved, and intelligently increased by all the substances, calculated to make a compost heap to 10 acres of the best part of his land, he would be able to gather 1000 bushels of potatoes, 1000 bushels of carrots, as many of Swedish turnips, and 150 bushels of Indian corn on his 10 acres of cultivated land. Can it be doubted that his stock would be better fed, his family better supplied, his net income from articles sold from his farm much more increased, than if he should persevere in the old system of raising rye at the rate of 15 or 20, or corn at the rate of 35 bushels to the acre? We only ask that the experiment should be tried—but above all that the culture of roots, and the increase of manure should be attempted for only a few successive years. We have no fears that this system would be ever afterwards abandoned.

In ascertaining the composition of sterile soils with a view to their improvement, any particular ingredient which is the cause of their unproductiveness, should be particularly attended to; if possible they should be compared with fertile soils in the same neighbourhood, and in similar situations. If a barren soil contains salts of iron, or any acid matter it may be improved by quick lime. If there be an excess of lime or chalky matter, sand or clay should be applied.—Where there is too much sand, clay, marle, or vegetable matter is required. Peat makes a good manure for a sandy soil. The improvement of peats, bogs, or marsh lands must be commenced by draining.

The materials necessary for the purpose of improving soils are seldom far distant: Sand is generally found beneath clay, and clay often beneath sand. Peat and bog earth are commonly to be obtained in the neighbourhood of gravel and sand.

Swamp-land after being drained, may often be improved by quick lime, which is sometimes better than paring and burning, as by the latter process much vegetable matter, capable of being converted into manure is dissipated and lost. See *Soil*, *Farm and Soil*.

I would entreat farmers to consider that the cost of raising a poor crop, one time with another, is nearly as much as that of raising a large one. There is the same expended in fencing—the same tax paid—the same quantity of seed sown—the same almost expended in ploughing, as rich land ploughs so much more easily than poor, as to make up for the extra number of ploughings in a course of tillage. I may add, there is the same or more labour in thrashing. An attention to these things is enough to convince any one of the great importance of endeavouring to improve crops by a more spirited and rational husbandry.

If a farmer think he cannot afford to lay out a farthing more on the tillage of an acre, than he has been accustomed to do, let him be entreated to save a little in fencing, and so enable himself to do it, leaving out some of his lands that bring little or no profit, and pay taxes for a less quantity of land in tillage; or let him turn some of his tillage land to grass; and lay out the same quantities of labour and manure on a third less land in tillage. Lands in tillage might thus be made profitable; and more so than many are ready to imagine.

It has often been observed, that those farmers in this country who have the fewest acres, commonly get the best living from their farms. It is, doubtless, because their lands are under better cultivation. And some have taken occasion to remark, that our farmers are ruined by the great plenty of land in their possession. Though this remark is just, I can see no reason why it should continue to be so; any more than, that being rich should necessarily make a man poor.—What need has the man who possesses three hundred acres, to destroy the wood, or clear the land, as they call it, any faster than he can make use of the soil to the best advantage? What need has he to be at the expense of enclosing more than his neighbour does, who has only one hundred acres, while he has no more ability, or occasion, for doing it? Or to pay taxes for more acres in grass or tillage? It is a foolish and ruinating ambition in any one, to desire to have a wide farm, that he may appear to be rich, when he is able to give it only a partial and slovenly culture.

If such improvements as are possible, and even easy, were made in the husbandry of this country, many and great advantages would be found to arise. As twice the number of people might be supported on the same quantity of land, all our farming towns would become twice as populous as they are likely to be in the present state of husbandry. There would be, in general, but half the distance to travel to visit our friends and acquaintance. Friends might oftener see, and converse with each other.—Half the labour would be saved in carrying corn to mill, and produce to market; half the journeying saved in attending courts; and half the expense in supporting government, and in making and repairing roads; half the distance saved, in going to the smith, the weaver, clothier, &c.; half the distance saved in going to public worship, and most other meetings; for where steeples are four miles apart, they would be only two or three. Much time, expense, and labour, would on these accounts be saved; and civilization, with all the social virtues, would, perhaps, be proportionably promoted and increased.

Nothing is wanting to produce these, and other agreeable effects, but a better knowledge of, and closer attention to, matters of husbandry, with their necessary consequences, which would be a more perfect culture, a judicious choice of crops, and change of seeds, and making every advantage of manures.

Improvements of vast importance, might also be made in the management of meadows and pastures. See those articles.

INARCHING, "a method of grafting, commonly called grafting by approach, and is used when the stock intended to graft on, and the tree from which the graft is to be taken, stand so near, or can be brought so near, that they may be joined together. The method of performing it is as follows: Take the branch you would inarch, and having fitted it to that part of the stock where you intend to join it, pare away the rind and wood on one side, about three inches in length. After the same manner, cut the stock or branch in the place where the graft is to be united, so that the rind of both may join equally together: Then cut a little tongue upwards in the graft, and make a notch in the stock to admit it; so that when they are joined, the tongue will prevent their slipping, and the graft will more closely unite with the stock. Having thus placed them exactly together, tie them with some soft tying; then cover the place with grafting clay, to prevent the air from entering to dry the wound, or the wet from getting in to rot the stock. You should also fix a stake in the ground, to which that part of the stock, together with the graft, should be fastened, to prevent the wind from breaking them asunder, which is often the case, when this precaution is not observed. In this manner they are to remain about four months, in which time they will be sufficiently united, and the graft may then be cut from the mother tree, observing to slope it off close to the stock. And if at this time you cover the joined parts with fresh grafting clay, it will be of great service to the graft.

"This operation is always performed in April or May, and is commonly practised upon myrtles, jasmines, walnuts, firs, pines, and several other trees that will not succeed by common grafting, or budding." *Dictionary of Arts.*

INCREASE, a word commonly used in husbandry, to express the proportion in which a crop exceeds the seed from which it is raised. It is generally true that the smaller the quantity of seed the greater is the increase; because a plant that stands by itself, has all the food that the earth is adapted to give it. But plants that are so near together that their roots intermingle, do more or less rob each other of their food. But we must not conclude from hence, that the less quantity of seed we sow, the better. Because, in getting a crop, other things beside the increase from the seed, are to be taken into consideration.

Other things being equal, those crops are most to be coveted, which require the smallest proportion of seed. But the greatest profit, on the whole, is to direct the choice of crops. The cheapness of seed sometimes misleads the farmer. To this cause may be ascribed, not seldom, the cultivation of maize on soils that are more suitable for other kinds of corn; or on soils that will produce no crop of maize worth cultivating. In a suitable soil, well dunged, it is not uncommon for one quart of maize to yield ten bushels, which is an increase of 320 fold. The expense of seed, therefore, for producing a bushel of corn, at 67 cents, is but 2 cents. But an increase of 20 fold is a good crop of wheat; the seed to produce a bushel of wheat at one dollar and 17 cents will be more than 6 cents: So that the expense of seed for wheat, is much greater than for maize. One consequence of this difference in seed is, that many of the poor can obtain seed for the former crop, who cannot obtain it for the other. And I suspect that the greater expense for seed of English grain, as we call it, has gradually brought the people of this country into a habit of sowing it too thin, and made them establish rules of doing so. It is certaih we sow much thinner than Europeans do.

It is not easy to determine what quantities of seed will answer best for given quantities of ground. But it is observable, that, in kindness to man, the beneficient Governor of nature has made most plants of the farinaceous kind, capable of getting their full growth when they stand near together. The greatest increase from the seed, is not to be accounted the most profitable crop.

A yield of thirteen for one may be of more advantage than twenty for one. If one bushel of wheat sowed on an acre produce 20, and two bushels sowed on an acre produce 26, it is manifestly more profitable to sow the two bushels. The farmer may consider one of the two bushels as yielding 20, and the other 6. But as the labour in both cases is precisely the same, it is clearly more profitable in this case to have thirteen for one increase than 20 for one. Now, in Europe, generally, they sow double, and some times treble the quantity of seed that we do. Can we be certain, till we try the experiment that one of the causes of the superiority may not be their liberality in saving seeds?

Another matter in which *increase* is to be considered, is the breeding of cattle, and other animals. The farmer may reckon increase in neat cattle as follows: He that has one cow may expect, in one year, to possess a cow and calf;—in two years, a cow, a yearling and a calf;—in three years, a cow, a two year old steer or heifer, a yearling and a calf. The two year old steer or heifer may be worth 3*l.* the yearling 40*s.* and the calf 20s. So that the increase from a cow worth 4*l.* in three years may be worth 6*l.* Consequently, he that lets out a cow for half her increase, as is the practice in some places, gets 25 per cent. simple interest on the money he buys her with. No man therefore that has a due regard to his own interest, will choose to hire cows at this rate; or take them to the halves as it is called, engaging to return the cow and half her increase at the end of three years. When cows are thus let the owner ought to risk the cow and her offspring.

The increase of sheep is a matter of greater uncertainty, as they are liable to more fatal diseases and accidents than black cattle are. But as they often bring two at a yeaning, it many times happens that ewes increase as fast as cows, or faster. But as a lamb grows to maturity in one year, and a she calf not in less than three years, ewes may be said to increase three times as fast as cows, even when they bear single.

INDIAN CORN, *Zea*, a well known and useful plant of the grain kind. It is called maize in most countries, zea in some.

The parts of generation are on different parts of the same plant. The panicles, or tossels, contain the *farina fœcundans*, which falling on the silk, or the green threads at the end of the ear, impregnate the ear, and render it fruitful. If the tossels, or spindles, were cut off before the grain in the ear is formed, the crop would be spoiled. This has been proved by experiment. But this effect will not take place, unless all the tossels be removed; because one of them will be sufficient to impregnate twenty plants. The silks, or threads, must be undisturbed to the time of impregnation. They are as necessary as the sowing itself. If part of them are taken away or pulled out as soon as they appear, part of the corn will be wanting on the ear: For every single grain has one of these threads. It is therefore a bad practice to suffer weaned calves to go among the corn, as some do, at the season of impregnation.

Maize is considered, in this country, as a most important crop. It is preferred to wheat and rye, because it is not subject to blasting, nor to any other distemper that is apt, in any great degree, to cut short the crop. A good soil, well tilled and manured, seldom fails of giving a good produce.

Though it be not so light and easy to digest as most other sorts of corn, it is found, that people who are fed on it from their infancy, grow large and strong, and enjoy very good health. There are a variety of ways of preparing it for food. The Indians parch it in embers, then reduce it to meal, and carry it with them, when they go forth to war, or hunting. When they eat it they reduce it to a paste with water, for it needs no other cooking. It is called *nocake*.

The green ears, either roasted

or boiled, are delicate food; especially some of the more tender sorts, which are cultivated for this purpose. Ripe corn, the hulls being taken off with a weak lie, and boiled till it is soft, is an excellent food; and not inferior to it is pounded corn, known by the name of samp. In either way, many account it equal to rice. But the most common use of it is in meal sifted from the bran, made into bread or puddings. For the latter, it is allowed to excel all other sorts of flour: For the former it does not answer well by itself; but is excellent when mixed with an equal quantity of rye meal.

The cheapness of seed, being next to nothing, greatly recommends, to the poorer sort of people, the culture of this corn. For it is often the case, that they are scarcely able to procure other seed for their ground. But this they can often have gratis.

In our new settlements, bordering on the wilderness, it seems to be of more importance than in other places; because the stalks, leaves and husks, being good fodder, supply the new beginners with winter food fo• their cattle, before hay can be raised.

Of all soils a clayey one may justly be accounted the worst kind for this crop. A loamy soil is best; or even sand, if it be not destitute of vegetable food, and manure sufficient is added. In the northern parts of New-England, it is not worth while to plant this corn on clay. On any soil it requires much tillage and manure in this country; if either be scanty, a good crop is not to be expected.

I think it is not the best method to plant it on what we call green sward ground, at least in the northern parts. It is apt to be too backward in its growth, and not to ripen so well. But if we do it on such land, the holes should be made quite through the furrows, and dung put into the holes. If this caution be not observed, the crop will be uneven, as the roots in some places where the furrows are thickest, will have but little benefit from the rotting of the sward.— But if the holes be made through, the roots will be fed with both fixed and putrid air, supplied by the fermentation in the grass roots of the turf. In this way, I have known great crops raised on green sward ground, where the soil was a sandy loam, but mostly sand.

But in the course of my experience, I have found pease and potatoes the most suitable crops for the first year. In the second, it will be in good order for Indian corn. This case, however, may be peculiar to the northern parts of New-England.

For this crop, it is certainly best to plough in the fall preceding; and again in the spring, just before planting. If the land be flat, and inclining to cold, it should lie in narrow ridges during the winter; and if it is naturally moist, the corn should be planted on ridges; otherwise it should be ploughed plain in the spring.

Some recommend gathering

seed corn before the time of harvest, being the ears that first ripen. But I think it would be better to mark them, and let them remain on the stalks, till they become sapless. Whenever they are taken in, they should be hung up by the husks, in a dry place, secure from early frost; and they will be so hardened as to be in no danger of injury from the frost in winter.

I would not advise the farmer to plant constantly his own seed; but once in two or three years, to exchange seed with somebody at the distance of a few miles. Change of seed is doubtless a matter of importance in most kinds of vegetables; though it has not yet been so plainly discovered in this as in some others. But let the farmer beware of taking his seed from too great a distance. If he should bring it, for instance, a hundred miles from the southward, his corn would fail of ripening; if as far from the north, he must expect a lighter crop; and in case of drought, the latter will be more apt to suffer, as it has been proved by experiment. A farmer in the county of Bristol, took seed from the county of Cumberland. It came on well at first. But the summer being pretty hot and dry, it parched up, and produced next to nothing, though the seed he had taken from his own field turned out very well.

If the farmer cannot conveniently obtain new seed; or if he be loth to part with a sort that has served him well, and choose rather to use it than seed he has not tried; let him, at least, shift seed from one field to another, and especially from one kind of soil to another.

And in the choosing of seed, some regard should be had to the state of the soil on which it is intended to grow. If it be poor, or wanting in warmth, the yellow sort with eight rows will be most suitable, as it ripens early. A better soil should have a larger kind of seed, that the crop may be greater, as it undoubtedly will.

If twenty loads of good manure can be afforded for an acre, it should be spread on the land and ploughed in: If no more than half of that quantity, it will be best to put it in holes. In the former case, the corn usually comes up better, suffers less by drought, and worms; and the land is left in better order after the crop. In the latter case, the plants are more assisted in their growth, in proportion to the quantity of manure. If the manure be new dung, burying it under the furrows is by far the better method. None but old dung should be put in the holes.

Let the ground be cut into exact squares, by shoal furrows made with a horse plough, from three to four feet apart, according to the largeness or smallness of the sort of corn to be planted. This furrowing is easily done with one horse, and is by no means lost labour, as the more the ground is stirred, the more luxuriantly the corn will grow. If dung is to be put in the angles where the furrows cross each other, the furrowing should be the deeper, that the dung may not lie too light.

The right time of seeding the ground may be from the first to the third week in May; or a little sooner or later according to the dryness of the soil, and the forwardness of the spring. The farmers have a rule in this case, said to be borrowed from the aboriginals, which is, to plant corn when the leaves of white oak begin to appear. But so much time is commonly taken up in planting this corn, it being tedious work to dung it in holes, that it will be necessary to begin in the driest part of the field a little earlier than this rule directs.

Shell the seed gently by hand, that it may not be torn or bruised at all, rejecting about an inch at each end of the ear. And, if any corns appear with black eyes, let them also be rejected, not because they will not grow at all the contrary being true; but because the blackness indicates, either some defect in drying, or want of perfection in the grain. Put five corns in what is called a hill, and let them not be very near together; for the more the roots crowd each other, the more they will prevent the growth of each other. Four corns would perhaps be a better number, if it were certain they would all prosper. The true reasons for putting more than one in a place I take to be, that by means of it, the rows may be so far apart as to admit of ploughing between them; and that some labour in hand hoeing is saved, it being no more work to hoe a hill with five plants, than with one in it.

Some steep their seed. But in general it had better be omitted; for it will occasion it to perish in the ground, if the weather should not prove warm enough to bring it up speedily. If planting a second time should become necessary, by means of the destruction of the first seed; or if planting be delayed on any account till the beginning of June, then it will be proper that the seed should have boiling water poured on it. Let it not soak more than half a minute, and be cooled speedily, and planted before it dries. The corn will be forwarder in its growth by several days.—The seed should be covered with about two inches of earth.

To prevent birds and vermin from pulling up the corn, steep some corn in a strong infusion of Indian poke, or refuse tobacco, and scatter it over the ground before the corn is up. White threads stretched over a field of corn, will prevent crows from alighting upon it: But I doubt whether this will deter any other birds.

A handful of ashes on each hill, will nourish the plants, and have a tendency to prevent their being annoyed by worms. Some lay it on just before the first, or second hoeing. It will have a better effect in preventing worms, if laid on before the corn is up. But it is commonly designed to answer chiefly as a top dressing; and for this purpose it would answer better near the third hoeing; for then the plants want the greatest degree of nourishment, as they begin to grow very rapidly. Two dressings with ashes, to answer the two purposes, would not be amiss.

When the plants are three or four inches high, the plough must pass in the intervals, making two furrows in each, turned from the rows; and then the weeds killed with the hand hoe, and a little fresh earth drawn about the plants. This operation we call weeding.

In about half a month after, plough again, but across the former furrows, and turn the furrows towards the rows. Then with the hand hoe earth the corn as much as it will well bear. This is called moulding, or half-hilling.

When the plants are about knee high, and before they send out their panicles, or spindles, give them the third and last hoeing. The best way at this hoeing is to plough one furrow in an interval, both ways. The cultivator with two mouldboards would be better for this work, than the common horse plough, as it would throw the mould equally towards each row, and save labour in hand hoeing. The ground would thus be cut into squares, and the hills almost completely formed. In finishing them, care should be taken that they be not made too high, or steep; that so they may not divert the water, which falls in rains, from the roots. When hills are too much raised, they also prevent the warm influence of the sun upon the lowermost roots, by too great a thickness of earth; in consequence of which, the plants are put to the exertion of sending out a new set of roots, at a suitable distance from the surface.

Some think high hills are needful to make the corn stand upright. I never could perceive the advantage of it. But I am confident it is oftener broken by winds when the hills are uncommonly high, which is a greater evil than its leaning half way to the ground, if indeed that be any evil at all, which I think may be doubted.

The farmer, who wishes for a large crop of this corn, should not annoy it with running beans, or pumpions; the former, by winding round the stalks and ears, cramp them in their growth, and sometimes bend them down to the ground by their weight; the latter, by their luxuriant growth, rob the hills of much vegetable food, and by their thick shade, shut out the influence of the sun from the roots of the corn. So that they must needs be very detrimental to its growth, and ripening.

Instead of the common method of planting, if your land be rich and easy to till, and free from obstacles, I should think it would be best to plant the corn in the drill method, the rows being of the same distance as in the common way, placing the corns about five or six inches asunder. I have found by experiment, that a greater quantity of corn may be produced in this method, than in hills; and the labour is but little, if at all increased. In a small field, where the dung had been evenly spread, and ploughed in, I planted one row thus, the rest being in the common way; and it yielded, at harvest, one eighth part more corn by measure than either of the two nearest rows, the corn being equally ripe and good.

Within a few years, and since Mr. Deane's death his hints as to planting corn in drills have been tried and with great success. Mr. Stevens, of New-York, raised 118 bushels on an acre. Mr. Ludlow, of the same State, 98, and Mr. Hunnewell, of Massachusetts, 111 in the drill method. Mr. Stevens planted double rows, only 8 inches apart, and the seed in rows also 8 inches asunder, but planted diagonally so as that no one stalk should be opposite to another. The double rows were distant from each other 5½ feet. Mr. Ludlow planted in single rows, the grains at 8 inches apart in each row, and the rows four feet apart. Mr. Steven's land was much more highly manured.

When there is reason to apprehend that the ground will prove too moist for this crop, it will be advisable to plough it into narrow ridges, and seed each ridge with one or two rows, as shall be found most convenient. Some of the finest crops that I have known, have been raised in this method.

When a season is at all wet, this would be the best culture in almost any soil, unless the very driest be excepted.

There is a kind of ridging, which would be very proper for this plant, not only on account of drying the soil, but that the land may have an alternate resting, or fallowing, between the rows. In the common method of plain ploughing, it commonly happens that a hill stands precisely in the place of a hill of the preceding year. When this is the case, the plants will receive less nourishment than if the hill had had a new situation. That each hill may always have this advantage, let a ridge be formed by two furrows, turning part of a row of hills on each side, so as to meet each other, in the last year's interval: Thus small ridges will be formed, on which the rows should be planted. If dung be first spread over the ground, the most of it will be buried where it should be, in the bottom of these ridges. At the time of weeding, or at the second hoeing, the remainders of the old hills may be turned towards the new rows. With such a mode of culture, land could not soon be exhausted, even by a successive cropping with maize. Land which has before been planted on ridges is as proper for this management, as if it had been planted in hills, or even more proper. For the success of a method not very dissimilar to this, see *Experiment for raising Indian corn*, in the *Memoirs of the American Academy*, by Joseph Greenleaf, Esq.

Care should be taken that the top stalks of the corn should not be cut too early. Perhaps the best possible rule is to examine whether the ears are pretty generally filled out, and whether they are so firm as to resist a light impression of the finger nail. In that case, they may be cut without injury, but while in a growing state it is inexpedient to cut them.

We are certainly guilty of an error when we harvest this corn too early. The difference of early and late harvested corn may be seen by the shrinking of corn in the former case. In drying, large spaces may be left between the kernels on the cob; but that which is well ripened on the stalk, will shew no such interstices. The corn will undoubtedly be growing better till the stalk below the ear is perfectly sapless, and the cob dry; receiving continual nourishment from the sap, unless the frost or some accident should happen to prevent it. Squirrels and other animals drive people to early harvesting; but there is commonly more lost than saved by it. When corn stands tolerably safe from the attacks of tame and wild animals, harvesting early is an unpardonable error. See *Harvest.*

This plant is so luxuriant in its growth that it impoverishes the soil faster than almost any other crop. Therefore it is not good husbandry to plant it more than two years in succession. It would be better still to grow it but one year in the same place.

European writers say, the land should be ploughed as soon as the crop is off, to prevent the stalks from drawing the moisture out of the ground. But the reason of this is not so evident as to carry conviction, unless the stems are quite in a green state. It is, however, a good method to plough all fields in tillage, as soon as the crop is off.

The following remarks on the culture of this valuable vegetable are extracted from a letter, written by S. W. Pomeroy, Esq. to John Lowell, Esq. Corresponding Secretary of Massachusetts Agricultural Society, and published in the Agricultural Repository, vol. VI. No. 2.

"Indian corn forms the basis of our bread stuff, and gives deserved reputation to two great staples, *Beef* and *Pork;* without corn you get none of the latter;—and a farmer who has no pork in his tubs, may be considered as fairly on the road to ruin, as a Bank with its vaults destitute of *specie*, for it is well known that Ruta Baga, potatoes or carrots, *alone*, will not fatten such pork as our farmers or fishermen have been accustomed to, or that will be found profitable for consumption."

"It is admitted, that on most farms near sea-ports, where corn and *manure* can be purchased, the system of potatoe and root culture, to the *exclusion* of corn, may be found profitable; nor would I be understood other than an advocate for such a system, in a regular rotation, upon an extensive scale; but I wish at the same time to hold up to view the GOLDEN FLEECE found by our *Pilgrim Fathers* on their first landing; and which, had it not existed, or continued with their descendants nearly a century after, the *fair inheritance* we now possess, in the opinion of many sound political economists, could not have been transmitted to us."

"On suitable soils, well tilled and manured, an average crop may be estimated at *forty bushels*, weighing 2400 pounds of the most nutritious substance to be found in the vegetable kingdom, north of the latitude of the *sugar-cane*—and which can be preserved with ease, for a number of years. You have the *fodder*, which, if seasonably and well cured, is, in the opinion of judicious farmers, equal to half a ton of good *hay*—then comes three or four tons of *pumpkins*, should the season favour, fifty or an hundred bushels of *turnips*—and, not unfrequently, a comfortable supply of *white beans!* No wonder that plenty of the necessaries of life are as sure to follow large crops of corn as effect succeeds cause in any department of physics!"

The writer combats the general opinion that corn is an exhausting crop, queries whether it is more so than potatoes, and observes that "Experience demonstrates that the *larger the crop of corn, the better the succeeding crop;* this was asserted by Dr. Eliot, the father of New-England husbandry, in his *Essays on Field Husbandry*, published in 1747; at which period it appears that *oats* usually succeeded corn, and possibly, in some districts, such a murderous course has continued; and it is probable that Corn, after supporting its *allies*, the *pumpkins*, the *beans* and the *turnips*, from the provisions charged to its account, has to answer for the deteriorating effects of oats; the most inimical to grass of any plant that can be named. Moreover, it will be recollected that formerly, the rich alluvial bottoms or intervals, were planted with corn, without a particle of manure, for a number of years in succession, till the product was considerably reduced;—would *potatoes*, or any *root* crop, with such management have continued more productive? and hence has not the reputation of corn materially suffered?

"We will next inquire what return does corn make to the soil? I cannot answer so well, as by quoting from Arator.* "Indian corn may be correctly called meal, meadow and manure; it produces more food for man, beast and the earth, than any other farinaceous plant. If the food it produces for the two first was wasted, and men and beasts should thence become poor and perish, ought their poverty or death to be ascribed to the plant which produced the food, or to those who wasted it? Is Indian corn justly changeable with the impoverishment of the earth, if the food it provides for that is not applied?

"Let us compare it with wheat. Suppose that the same land will produce as much grain of the one as of the other, which in its use will make equal returns to the earth. Here the equality ends, if indeed it exists even in this point. The corn stalks infinitely exceed the wheat straw in bulk, weight, and a capacity for making food for the earth. If any attentive man who converts both his stalks and straw into manure, will compare the product in April, when he may distinguish one from the other, he will find in the former a vast superiority in quantity. The English farmers consider wheat straw as their most abundant resource for manure, and corn stalks are far more abundant; corn therefore is a less impoverishing, because a more compensating crop to the earth, credited only for its stalks than any in England. In comparing crops to ascertain their relative

* A series of Agricultural essays entitled Arator by Col. John Taylor, of Carolina County, Virginia. This work, though adapted to the agriculture of that, and the adjoining states, will be found to contain valuable practical information to the New-England Farmer.

product, and operation on the earth, we must contrast farinaceous crops with each other; and consider the litter or offal they produce, not as wasted, but as judiciously applied to the compensation of the land. At the threshold of the comparison, corn exhibits a return from the same land of more offal, or litter in its stalks alone, than wheat does altogether. But to the stalks of corn its blades, tops, husks and cobs remain to be added, each of which will nearly balance the litter bestowed on the land by wheat. " The author concludes his encomium upon Indian corn, with observing, that " as a fallow crop it is unrivalled, if, as fallow crops aught constantly to do, it receives the manure."

" Arthur Young, who has given such an impetus to rural economy, and to root culture particularly, in Great Britain; in his travels through France in 1789, makes the following remarks. " The line of maize (corn) may be said to be the division between the good husbandry of the south, and the bad husbandry of the north of the kingdom, till you meet with maize very rich soils are fallowed, but never after; perhaps it is the most important plant that can be introduced into the agriculture of any country, whose climate will suit it. The only good husbandry in the kingdom, (some small rich districts excepted,) arises from the possession and management of this plant. For the inhabitants of a country to live upon that *plant*, which is the preparation for wheat, and at the same time keep their cattle fat upon the leaves of it, is to possess a treasure, for which they are indebted to their climate." " Planted in squares or rows, so far asunder, that all imaginable tillage may be given between them; and the ground thus cleaned and prepared at the will of the farmer, is an invaluable circumstance; and finally it is succeeded by wheat. Thus a country, where soil and climate admit the course of, 1st maize, 2nd. wheat, is under a cultivation that, perhaps, yields the most food for man and beast, that is possible to be drawn from the land."

" The perfect culture and copious manuring demanded by corn. " Mr. Pomeroy maintains, instead of being an objection to the cultivation is an argument in its favour; and he contends " that should Indian corn be made the basis of an extensive rotation system, with a pointed attention to the manures, which it has the capacity to increase in a compound ratio, the agriculture of Massachusetts may not only approximate to that of Great Britain, but with the energies such a system might be expected to elicit, equal it."

In speaking of his method of cultivating corn Mr. Pomeroy says, " Having a large bed of *Beets* planted in narrow rows or drills, in the usual way, and a severe drought ensuing, the leaves were observed, in the middle of the day to wilt and fall down, in all but the out-side rows, which remained erect and flourishing; and when the crop was taken up were nearly double the size of those from the centre of the bed! That a greater exposure to the atmosphere was a principal cause could not be doubted; and the idea immediately presented that if they had been cultivated on *ridges* so far apart as to permit a plough to pass, it would in effect, be making the whole crop *out-side-rows:* and for upwards of fifteen years since, all the arable crops upon my farm have been cultivated on ridges with manifest advantage,—as to product, labour and the *unquestionable* improvement of the soil. The ridges of beets, carrots, parsnips, turnips and rutabaga are about two feet, cabbages four, and corn and potatoes five feet and an half apart."

"But to return to the culture of corn: in describing which that of potatoes must necessarily be embraced; as I deem it wrong for two crops of the same kind to succeed each other, when it can well be avoided; nevertheless, as potatoes should be employed as the *pioneers*, of a farm, and it is frequently the case that cold moist or rough soils, not suitable for corn, are found, sufficient to be occupied by as many potatoes as the farmer can well manage, or may want under such circumstances, the permitting two crops of corn to follow *on ridges* may be justifiable; whereas in the common method it should be considered as unpardonable.

"If we begin the system with *sward* land, it is broken up in the autumn; harrowed fine in the spring, and light furrows run out five feet and an half apart: into these furrows potatoes are dropped from six to eight inches asunder, according to size, the carts follow and cover them with manure; a furrow is then turned from each side so as to meet over the manure, a little labour with the hoe may be required, to make all level and complete the planting; other furrows are turned up, and at the first hoeing the baulk is cleared and the ridge completed. In the subsequent culture the plough approaches only the sides of the ridges, and continues deepening the furrow between them. The following spring the ridges are reversed and placed directly over those deep furrows. The manure is turned in while the ridges are forming; or by opening a furrow on the top to receive it, with a pair of oxen walking on each side, covered with hoes and the crop planted. If the land is in good heart, and manure abundant, the kernels may be six inches apart, or as many dropped together as to insure four stalks to remain two feet apart, in which case the manure is deposited conformably; either method will give the same number of plants to the acre; I think four stalks together afford support to each other against winds, and are not so apt to send up suckers as when single, and there may be some advantage by concentrating the manure, in forwarding the young plants during the cold seasons which we frequently have in June; still I have not had sufficient experience to determine which method is preferable; if the land is *stony*, the last will be found most convenient. As soon as the plants appear, the earth is stirred about them with the hand, and ashes at the same time strewed on, in the proportion of a handful to four plants, the cultivation proceeds similar to that described for the potatoes—the ploughing continues as often and as long as you please, without disturbing the roots of the corn, or breaking it down, deepening the soil and turning up the dead earth to the sides of the ridges, which checks the growth of weeds greatly; and if the crop is at first thoroughly hand-weeded, they will give but little trouble afterwards.

It has not been in my power to make a comparison with a crop cultivated in the common method, but I have had several foremen since ridging has been practised on the farm, who were at first prejudiced against it, but were soon convinced of its utility; and none of them have estimated the increase of product, with the same quantity of manure, at less than one fifth, and the saving of labour full as much.

"It is probable that on rich soils with plenty of manure, larger crops may be obtained by making the ridges much nearer, but the labour will be more, and the advantage of uncommon deep culture without extra expense must be abandoned; the importance of such culture to the succeeding crops may be illustrated by the following relation.

"About fifteen months since, I was applied to severally, by two respectable farmers, from an adjoining town, who are considered as judicious and as *thrifty*, as any in the county of Norfolk, to be informed how I had managed a piece, the former state of which they had well known to make it produce such a *second* crop of clover, in such a season of drought that theirs was *dried up*, and on land naturally more moist, which they considered stronger, and had been well cultivated and highly manured? They were told in reply, that 'the land was under the sixth course of rotation, viz.:—1. potatoes; 2. corn; 3. carrots and beets; 4. corn; 5. spring wheat, of which it carried thirty bushels of 59 lbs.; 6. clover, first crop, two tons and an half; second the same season, and which had attracted their notice, about one ton to the acre.' That it had from 16 to 20 lbs. of clover, and half a bushel of herds'-grass seed to an acre; one half ploughed in with the wheat, and the remainder sown on the furrow, harrowed in and well rolled: and further, that no stock were suffered to graze upon it. It should also be stated, that a heavy crop of herds'-grass was produced the last season, and that it gives promise of another;—that the soil is a deep sandy loam, known to many farmers by the name *fox-coloured earth*, and which some of them are too fearful of turning up; that when first cultivated in the common method, which was some years previous to the present rotation, it did not shew two inches of brown soil, or vegetable mould, and had been considered as poor pasture. And I would observe, that had the land been longer under previous cultivation, the third and fourth courses might have been omitted, and the rotation thereby shortened two years."

Mr. Pomeroy recommends the following preparation of SEED CORN for "protecting it against squirrels and other vermin."

"Take equal parts of tar and train oil, simmer them together and turn over the corn; then sift on ashes, lime, or plaster, stirring it, till each kernel has taken up as much as will permit its being conveniently handled." He likewise thinks it probable that *walnuts*, *acorns*, *chesnuts*, &c. thus prepared for planting would be protected; "the destruction of which by squirrels, has very much retarded plantations of those valuable trees. The Farmer's Assistant say, that when seed corn is prepared with tar, "it is necessary first to soak it sufficient to make it vegetate; as without this the coat of tar will keep out the moisture, and prevent the seed from sprouting."

The Massachusetts Agricultural Repository, p. 241, contains a statement of the production of one acre of land cultivated by Jonathan Hunnewell, Esq. and the manner of its cultivation. By this it appears that the field was an old pasture; that it was ploughed in the fall of 1818, cross ploughed in the spring of 1819, and planted with corn in hills in the common form, but well manured in the *hill* with a mixture of horse hung, lime and ashes; when the corn was fit for weeding, it was ashed by putting about half a pint of unleached ashes to each hill; a part, however, was left without any ashes. The difference was very visible between the corn, which had ashes, and that which had none. No plough was suffered among it after an early half-hilling; nor had it any hoeing except to destroy the worst of the weeds, and to stir in the turnip seed, which was sown amongst it. One acre produced 78¼ bushels of corn when shelled. The seed of this corn was presented to Mr. Hunnewell by Samuel Blagge, Esq. of Boston, who had it of William Jarvis, Esq. of Weathersfield, Vermont, and is the same sort, we believe, which has been advertised in the newspapers as the Dutton-corn.

In 1820 the same ground was planted again, after a fall and spring ploughing. It was furrowed out at a distance of four feet, leaving each furrow one foot wide. The furrows were well manured with a mixture of *horse-dung*, *lime*, *ashes*, *and dock-mud*. The seed raised the last year was planted in the drill on each furrow, making *three rows* to each furrow. Care was taken to drop the corn as near six inches apart as possible without waiting time. The cultivation similar to that of the preceding year. The corn when ripe, appeared on the 13th of October as though some pains had been to strip it of the husks; but it is the nature of this corn to husk itself in the field. In good seasons it requires but ninety days to ripen. The product of one acre of this corn was one hundred and eleven bushels. Mr. Hunnewell received a premium from the Massachusetts Agricultural Society for this last mentioned crop.

Oliver Fiske, Esq. of Worcester, Massachusetts, has ascertained by an experiment detailed in the Massachusetts Agricultural Repository, vol. VI. p. 244, that corn planted in rows produces more than when planted in hills, and received a premium for his experiment.

INOCULATING, or BUDDING, inserting a bud so that it will live and grow, in the side of the trunk, or limb of a tree. It answers the same end as grafting. Mr. Miller says, "This is commonly practised upon all sorts of stone fruit in particular, such as peaches, nectarines, cherries, plums, &c. as also upon oranges and jasmines, and is preferable to any sort of grafting. The method of performing it is as follows: You must be provided with a sharp penknife, having a flat haft (the use of which is to raise the bark of the stalk to admit the bud) and some sound bass mat, which should be soaked in water, to increase its strength, and make it more pliable; then having taken off the cuttings of the trees you are to propagate, you should choose a smooth part of the stock about five or six inches above the surface of the ground, if designed for dwarfs; but if for standards, they should be budded six feet above ground; then with your knife make a horizontal cut cross the rind of the stock, and from the middle of that cut make a slit downwards about two inches in length, so that it may be in the form of T; but you must be careful not to cut too deep, lest you wound the stock. Then having cut off the leaf from the bud, leaving the foot stock remaining, you should make a cross cut about half an inch below the eye, and with your knife slit off the bud, with part of the wood to it. This done, you must with your knife pull off that part of the wood which was taken with the bud, observing whether the eye of the bud be left to it or not (for all those buds which lose their eyes in stripping should be thrown away, being good for nothing.) Then having gently raised the bark of the stock where the cross incision was made, with the flat haft of your penknife, cleave the bark from the wood, and thrust the bud therein, observing to place it smooth between the rind and the wood of the stock, cutting off any part of the rind belonging to the bud, which may be too long for the slit made in the stock: And so having exactly fitted the bud to the stock, you must tie them closely round with bass mat, beginning at the under part of the slit, and so proceed to the top, taking care that you do not bind round the eye of the bud, which should be left open.

"When your buds have been inoculated three weeks or a month, you will see which of them have taken; those of them which appear shrivelled and black being dead, but those which remain fresh and plump you may depend are joined. At this time you should loosen the bandage, which, if not done in time, will pinch the stock, and greatly injure, if not destroy, the bud.

"The March following" (perhaps April in this country) "you must cut off the stock close to the bud, sloping it that the wet may pass off, and not enter the stock. To this part of the stock, left above the bud, it is very proper to fasten the shoot which the bud makes in summer, to secure it from being blown out; but this part of the stock must continue on no longer than until the bud has acquired strength to support itself, after which it must be cut off close above the bud that the stock may be covered thereby.

"The time for inoculating is from the middle of June to the middle of September, according to the forwardness of the season, and the particular sorts of trees to be inoculated, which may be easily known by trying the buds, whether they will come off well from the wood. But the most general rule is, when you observe the buds formed at the extremity of the same year's shoots, which is a sign of their having finished their spring growth." *Gardener's Dict.*

INSECT. We shall only be able to notice a few of those insects, which are more injurious than others to the labours, and often fatal to the hopes of the husbandman, with the intimation of such remedies as have been found most effectual.

The *Grasshopper* is one of the most constant annoyers of the farmer. In all years it is found in greater or less numbers. In hot and dry seasons it seems to multiply more rapidly, and occasionally it lays waste whole tracts of country. The causes of the varieties in their numbers is unknown. Some insects, for instance one of the locusts, it is ascertained, lie in the chrysalis state, exactly sixteen years—but why such insects as the grasshopper, which never quits us, should sometimes over-abound so as to endanger the subsistence of man, is not understood. It is matter of regret, that no method has yet been discovered of checking this insect. The letting in of troops of turkies and chickens to the field is, to be sure, a palliative: But, it is obvious that it cannot be extensively practised. The poultry, themselves, do great mischief to the crop, and the support of them, in great numbers throughout the year, might be fairly set against the benefit they might do.

Curculio, a small bug, which perforates the young fruit of the pear, apple. and all stone fruits, and deposits its eggs within them. There are several sorts of them. The eggs soon hatch, and a small maggot is produced, which feeds either on the pulp, or on the kernel of the seed, for the tastes and habits of the various species are very different. In the stone fruits this injury destroys their growth, and they fall, with their little enemy within them, who soon seeks shelter in the ground, where he passes the winter in the chrysalis state, and comes forth, just as the young fruit is forming, or as the petals of the flowers are falling, to renew the work of destruction. As this is a winged insect, and probably flies with ease, for he is never caught, like the canker-worm, ascending the trunk, it would seem that most of the remedies proposed in agricultural works must be ineffectual, as they are irrational. Tying therefore, a bag of salt on the tree, or surrounding it with grease and blubber, can have no effect, except so far as the effluvia might operate to deter the insect. In that view pieces of shingles, covered with moist tar, suspended in many places, on the exterior limbs might be more effectual. We think, however, very lightly of all these remedies. If the Curculio is to be extirpated or checked, we think it must be in the ground. It is ascertained that pavements will check them. It is said that tanners' spent bark will do it, and this is now in a course of experiments. It certainly accords better with the history of the insect, and is more rational. Any substance, which will check the *descent*, or *ascent*, will occasion their destruction.

We must enter our protest against all projects of boring holes into trees, and filling them with mercury, or mercurial preparations, or any other substance, as preventatives against the Curculio, the Aphis, and as some say, all insects. This is too empyrical, and there is too much nonsense already in the world to attempt to increase it.

Aphis. Plant or Tree Lice. Almost every plant has a species of this insect peculiar to itself. Its powers of reproduction are almost infinite, and more rapid than can well be imagined. It is pernicious to cabbages, Swedish turnips, and to many trees and shrubs. In greenhouses they are readily destroyed by the smoke of tobacco, or of sulphur. But in the open air, fumigation, though much in vogue many years since, is of no avail. The best remedy is the simplest. Soap suds, forcibly applied, will, after one or two applications effectually destroy them, without apparent injury to the plants.

The cut worm, an insect so called, is an ash coloured worm, with a stripe almost black upon its back. At its full growth, it is about the bigness of a goose quill, and an inch and a quarter in length. The greatest mischief that they commonly do, is to young cabbages, cauliflowers, &c. They never choose to appear on the surface in the day time; but keep themselves buried about an inch or two beneath it. In the night they come up, eat off the stems of the young plants, and again bury themselves in the soil, often attempting to draw in the plants after them.

They sometimes destroy other vegetables. I have known them to cut off great part of a field of Indian corn, before the first hoeing: But this is not a common case.

They begin to devour in May, and cease in June.

I once prevented their depredations in my garden, by manuring the soil with sea mud, newly taken from the flats. The plants generally escaped, though every one was cut off in a spot of ground that lies contiguous. From the success of this experiment I conclude, that salt is very offensive, or pernicious to them. Lime and ashes in some measure prevent their doing mischief; but sea water, salt, or brine, would be more effectual antidotes. The most effectual and not a laborious remedy even in field culture is to go round every morning and open the earth at the foot of the plant, and you will never fail to find the worm at the root within four inches. Kill him and you will save not only the other plants of your field, but probably many thousands in future years.

Top worms, or *spindle worms*, a white worm, resembling a grub, found in the hose, or socket, of a plant of maize, which eats off the stem of the plant, and renders it unfruitful. When its excrements appear on the leaves, it may be known that a worm is in the socket. They are most commonly found in places that are rich and dungy, particularly in corn that grows near to barns; but they will sometimes prevail through whole fields. Sprinkling the corn, when they begin to eat, with a weak lie of wood ashes will effectually destroy them. So I suppose, would almost any bitter infusion; but of this I have made no trial.

The striped bug, or yellow fly, is a small four winged insect, the outward wings of which are striped with yellow and black. They eat and destroy the young plants of cucumbers, melons, squashes and pumpions. They begin to eat while the plants are in seed leaf; and, unless they are opposed, will totally destroy them, especially in a dry season.

These insects may be considerably thinned, by killing them in a dewy morning, when they have not the free use of their wings, and cannot well escape.

I have sometimes defended the plants in some measure, by encircling them with rock weed. But nothing that I have tried has proved so effectual, as sifting, or sprinkling powdered soot upon the plants, when the morning dew remains on them. This forms a bitter covering for the plants, which the bugs cannot endure the taste of. Perhaps watering the plants with some bitter infusion might equally preserve them, if it were often repeated. I prefer soot, as I know by experience that once sprinkling with it will answer the end, unless it happen to be washed off by rain. When this happens, the sooting should be repeated; lime and washes will answer as well.

The turnip fly, a well known winged insect, which eats the seed leaves of turnips, before the first rough leaf appears. Their ravages are so general, and of such consequence, that the ingenious have attended to the matter, and explored many methods, both preventive and remedial, to oppose them.

One of the preventive methods is, making the ground so rich that the plants will grow rapidly, and continue but for a short time in the seed leaf; for, after the evolution of rough leaves, the plants are almost or quite out of danger of this insect.

It is also recommended, to pass a roller over the ground, as soon as the seed is sown. This not only prevents the too sudden escape of the moisture in the surface, and causes the plants to rise sooner and more vigorously; but fills up or closes ten thousand little interstices in the surface, which serve the insects as places of retreat. The consequence is, either that they are destroyed by rains, driven away by winds and storms, or stiffened with the dews of the coldest nights.

Mr. Tull thought it best that the seed should be buried at different depths in the soil, and says, as they will come up at different times, either the first or the last will probably escape the fly. He accordingly constructed his turnip drill in such a manner as to bury the seed at different depths.

The same thing in effect may be done in the broad cast way of sowing. The ground may be harrowed with a common harrow with iron teeth; then half the seed sowed, and the ground smoothed with a bush harrow and rolled; then the other half sowed, and bushed in, or raked. After which the roller should be again passed over the surface.

Some writers on this subject are confident that the best method is, to sow the seed very thick, equal to double the usual quantity of seed, that when the flies have eaten all they can, there may be a sufficient number of plants remaining to insure a good crop. Another project is sowing a mixture of old and new seed, as the latter is known to come up sooner than the former, one or other of which may happen to escape.

After the turnips are up, if the flies appear in plenty, it is advisable to pass a smooth roller over them. If the roller be drawn carefully by hand, or even by a horse, turning the roller about on the head lands only, the operation may be performed without hurting the turnips; and the flies will mostly be crushed by the roller. This operation should be performed in a dewy morning, when the flies are so stiff that they cannot make their escape.

Or, instead of this, I am confident that the sifting of soot over the turnip ground in a dewy morning will be effectual; at the same time that it will answer as a slight top dressing, and increase the growth of the plants.

Some writers assert, that only drawing a green bush of elder over the young plants will save them from the fly. I think it may have some tendency towards it; but I have never made the experiment. An infusion of elder, applied by sprinkling, would probably have a greater effect. But I should expect more from an infusion of tobacco.

Some set plants of tobacco thinly in their turnip ground, thinking that the scent of them does something towards repelling the fly. I have no objection to this, excepting that a much richer ground is requisite for tobacco than for turnips, in our climate. See *Turnip*.

The red worm is another enemy to the farmer. This insect is slender, and usually about an inch long, with a hard coat, and a pointed head. It eats off wheat, barley and oats, above the crown of the roots. It perforates, or bores quite through bulbous roots, turnips, potatoes, &c. My turnips, for several years, which were sown in the spring, have been thus almost ruined, though on a soil that suited them. When a turnip is once wounded by them, it grows no bigger, unless it be in ill shapes, and hard excrescences, and becomes totally unfit for the table. As to potatoes, I have seldom known them do much hurt, unless when they were planted in a soil that did not suit them particularly in a clay. It is easier to say what will not stop those borers, than what will do it. I have manured with sea mud; applied dried salt to the soil after the plants were up; mingled dry salt with the seed when it was sowed; steeped the seeds in brine before sowing, and coated them with sulphur; but all in vain.

I suppose the burning of a stubble as it stands would destroy all the worms that happened to be very near to the surface. A certain English writer thinks that a perfect summer fallow would destroy them, partly by exposing some of them to the heat of the sun at each ploughing, and partly by depriving them of food.

I should think ploughing late in autumn might destroy many of them, by exposing them to the most violent action of the frost. Or in a garden throwing up the soil in ridges with the spade, so to lie during the winter, would have a good effect. Liming plentifully, if it could be afforded, I should rely upon as a most effectual antidote to this, and several other kinds of insects. The *Complete Farmer* mentions lime and soot as good antidotes to this insect in particular.

Red Worm, or Wire Worm. Mr. William Moody of Saco, (Maine) in a communication to Hon. Josiah Quincy, published in the *Massachusetts Agricultural Repository*, vol. IV. p. 353, observes, "I am persuaded, from experience, that sea sand, put under corn or potatoes with manure or spread on the land, will go far, if not wholly to the total destruction of those destructive worms, on which nothing else seems to have any effect. It has a beneficial effect spread on land before ploughing, or even after land is planted with corn or potatoes, not only to destroy the wire worm and other insects, but to increase the crop. With my neighbours a load of sea-sand is considered preferable to a load of the best manure, to mix in with their common barn manure, or to spread on their gardens and low flat land."

Probably sea-mud, or sea-water would have good effects as preservatives against these and other insects.

The garden flea is a minute fly that eats cabbages, and other plants of the *brassica* kind, while they are in seed leaf. They are of a very dark colour, or nearly black.

I once applied some clefts of the stems of green elder to some drills of young cabbages, which this fly had begun to eat, and could not find that they eat any afterwards. But as I made this trial but once, I dare not positively assert its efficacy. I would heartily recommend the trial of bitter steeps to gardeners who are troubled with this insect. They are earlier in gardens than any other insect; and I have never known them fail to appear in a dry spring.

Maggots. I have often found

a white maggot, of the shape and size of those in cheese, preying upon the roots of young cabbages, turnips, and radishes. My radishes, when sown early, seldom escape; those that are sown in June mostly prosper.

A person in my neighbourhood, who has often been defeated by these insects, in his attempts to raise cabbages, declares, that last spring, as usual, the maggots attacked his cabbages before he transplanted them; and that, having a scarcity of plants, he transplanted, on the same spot where they used to fail, some which had maggots in their roots among sound plants: That as soon as he had done transplanting, he watered them plentifully with sea-water: That the watering was not repeated; but the maggots did no damage at all; and that his crop was very large and good. Some that were left without watering were destroyed, as before, by the maggots.

I have since tried this experiment, with a good effect. Very few of the cabbages were touched by the maggots. But I find there is danger in applying the sea-water plentifully, unless it be in a wet season. The plants are in danger, when the ground is dry, of imbibing too much of the salt. In this case their growth is greatly obstructed.

Though I have conceived that it would kill all sorts of plants, to pour sea-water upon them, the cabbage, having an oily surface to which water does not easily adhere, is perhaps an exception. The farmers who are remote from the sea cannot apply the sea-water to their cabbages, without too much expense of carriage. But they can afford to water their plants with a brine of equal saltness. I wish them to make the experiment, not only on cabbages, but on radishes, &c. It is safe to apply salt-water in a wet season, or just after a rain.

The Hessian fly, so called, is an insect that is pernicious to wheat, while it is growing. It made its appearance in the time of the late war, in the vicinity of New-York, and is supposed to have been imported with the German troops. From thence it has spread into Connecticut and New-Jersey, laying waste whole fields in its course. A more formidable insect has scarcely ever appeared in the country.

But against this enemy it seems an easy antidote has been already discovered. A letter, signed *D. Wadsworth*, which has lately been published in the newspapers, communicates a method of preventing its depredations, which the writer says has seen used with effect. It is only steeping the seed before sowing for twelve hours in a strong infusion of the leaves of elder.

In the latest edition of Dr. Morse's Geography, there is an account, that yellow-bearded wheat, sown late in autumn, escaped the Hessian fly.

The Hessian fly, although formerly supposed to be of German origin, is, we believe, not known in Germany, nor any other part of Europe.

The palmer worm, a wanderer, as its name signifies. This is a small worm, about half an inch in length, with many legs, and extremely nimble. It appears at different times in different parts of the country. I have seen them only on apple trees and oak trees, in any great abundance. They give trees the same appearance that the canker worm does. They appeared in the county of Cumberland in the year 1791, about the middle of June, eating off the covering of the leaves on both sides, and leaving the membranous part entire. The following year there was none to be seen; and I have not known them in any place two years in succession. The seeds of them may be constant, wanting only a particular state of the weather to produce them. The spring which preceded their appearance had been remarkably dry, both in April and May. The history of this insect is so little known, that I will not undertake to say how they may be successfully opposed. I made smokes under the fruit trees, without any apparent effect. As they let themselves down by threads, they may be thinned by shaking the trees, and striking off the threads. Their ravages had not any lasting effect: For the orchards that had been visited by them bore plentifully the following year.

Weevil, an insect injurious to corn in granaries. Shutting up an apartment and filling it with the smoke of burning sulphur will destroy them. But the smoke should be continued as much as twelve hours. Grain may be cleared of them by sifting, in a sieve so made that the insects will pass through, and the grain stay behind. Sprinkling wheat with lime in the bin, which may be afterwards winnowed out; making bins of Lombardy poplar, and sprinkling salt among the sheaves before threshing, have been recommended.

The timber worms should also be mentioned. These are of two kinds. The smaller kind eats only the sappy parts of the wood, turning it to what is vulgarly called *powder post*. To prevent damage from this insect, nothing more is necessary than to fell the timber in December or January, in which month it is sure to be freest from sap. When it is necessary to fell trees that are full of sap, something should be done to divest it of the sap, or alter the quality of this juice. Soaking it, even in fresh water, will be of some service. But in salt-water, soaking will be quite effectual, against most kinds of worms.

The large boring worm is far more mischievous than the one I have mentioned; and no season of felling secures timber wholly from this insect. They make the greatest havock in pine. They are hatched in the cavities of the bark, and being small when they enter the wood, they grow larger as they proceed, till their boring may be heard, like the cutting of an augur, to a considerable distance. They proceed to eat the wood in every direction, till they become as large as one's finger, or till the juice of the wood, being altered, is unfit to nourish them any longer.

Steeping the wood seasonably in salt-water destroys the worms, or prevents their entering the wood. If the trees be scorched in a light flame, before they have entered too far, the effect will be the same.

To prevent and cure worms in timber, Mr. Evelyn recommends the following, as much approved. "Put common sulphur into a cucurbit, with as much aquafortis as will cover it three fingers deep; distil it to a dryness, which is performed by two or three rectifications. Lay the sulphur that remains at bottom on a marble, or put it in a glass, and it will dissolve into an oil; with this oil anoint the timber which is infected with worms."

Besides the destructive insects which appear more or less every year, there appear sometimes formidable swarms, or armies of worms, which suffer scarcely any green thing to escape them. They overran many parts of the county of Cumberland, in the year 1770, rather before the middle of July, to the extreme consternation, as well as the great injury of the inhabitants. They stripped the corn and grass of the leaves, leaving only the bare stems, and those deprived of their sap. They were extremely voracious; and appearing to be in the utmost haste, they all moved in the same direction. They suffered nothing that they could climb upon to stop their course. They crawled over houses, and all other buildings, unless when they found a door, window, or chink in their course, where they could enter. Whether they passed in this manner over the plants they destroyed I did not take notice.

Between twenty and thirty years ago the same dreadful insect appeared in the county of Essex; and between 1770 and 1780, in some places in the territory of Vermont.

The only ways of opposing their ravages that have been used, are, either to mow a field of grass, whether it were fully grown, and fit to cut, or not; or, to fence, against them with narrow trenches, made perpendicular, or rather hanging over, on the side next to the field. Many fields of corn have been thus saved; and bushels of the worms being unable to climb such crumbling walls died in the trenches.

If their history were attended to, perhaps it would be found they have stated periods.

It is not sufficient for the farmer to defend his vegetables against insects. There are insects also that annoy and hurt his animals.

Lice are often found on colts, and on neat cattle, especially on yearlings in the spring. When these animals become poor, they most commonly grow lousy, which makes them still poorer. Possibly it may be owing to an obstruction of perspiration. For there are doubtless many oily particles in the effluvium of healthy cattle, and oil is an antidote to this insect. Oiling their skins will clear them of lice; so will a strong infusion of tobacco. But when they are cured, better feeding is the best preservative from the return of the insects.

The tick, or tike, is the sheep louse. When these insects become numerous, they are very hurtful to the sheep. In England the farmer smears the sheep, after shearing, with a mixture of butter and tar. This fortifies them against being injured, either by the weather, or by insects. But at any time, oil, or tobacco, will destroy the ticks. A worm has of late years, near Boston, proved fatal to the apple tree, quince tree, and mountain ash. It is a borer which subsists in the sap-wood or alburnum. It can only be killed, and the trees saved, by cutting them out in the month of May by a gouging chissel. Thousands of trees have been saved by this expedient.

INTERVAL, the space between two places, or things. The word is used in husbandry to denote the space between rows of corn, or other vegetables; especially in the horse-hoeing husbandry.

By interval, also, and more usually in this country, is understood land on the border of a river. Interval land is commonly so high and dry as to be fit for tillage; and yet always so low as to be frequently overflowed by the swelling of rivers, especially in the spring. On some of these lands the water often continues so late in the spring that they cannot be seeded till June. But the increased fruitfulness of the soil seems to more than make up for this delay. For when the waters subside, they leave a fat slime upon the soil, most friendly to vegetation.

The soil on these intervals is most commonly sand, with a large mixture of the finest vegetable mould; and much of it is made, from time to time, by the shifting of the channels of rivers.

This sort of land has generally been prized highly in this country. But in some places it has become less fruitful of late than formerly. The reason of this alteration most probably is, that the floods are not so great as formerly, or that they subside quicker; owing to the more cultivated state of the country, and a quicker evaporation of the waters.

K.

KALE, Colewort, an excellent potherb, early, and of quick growth, which ought to be cultivated in this country.

Sea Kale. A plant introduced within forty years in Great Britain. It is the Crambe Maritima. Its leaves blanched by having the earth heaped round them, are as delicious and much resemble the Cauliflower. For further description, see the Massachusetts Agricultural Society's Journal of June, 1822.

KALENDAR, an account of time. That great naturalist, Linnæus, did not approve of farmers' confining themselves to certain set days, or weeks, for committing their seeds to the earth. The seasons are much forwarder in some years than in others. Therefore, he, who thus governs himself, will assuredly sow his annual seeds sometimes too early, and sometimes too late.

That a better practice might be introduced, he recommended it to his countrymen to take notice at what times the trees unfold their leaves. Nature is so uniform in her operations, that the forwardness of trees is an unfailing indication of the forwardness of the spring. And the genial warmth, which causes trees and shrubs to put forth their leaves, will be sufficient to cause seeds to vegetate.

In order to reduce to practice so ingenious a hint, an account should be made out of the first leafing, and I may add, the blossoming of a variety of trees and shrubs. I suppose trees and shrubs to be most suitable for this purpose, as they are more deeply rooted, and therefore more steady and uniform in their appearances, than any plants which are perennial only in their roots. They are especially much more so than annuals.

It is certain that such an account taken in one place will not answer alike for every part of the country; because the vegetation in every part is not equally forward. Therefore, I would earnestly recommend, that in each degree of latitude, throughout New-England at least, some attentive naturalist would make a list of a considerable number of trees and shrubs, which are common, and near at hand; carefully watch their appearances, and minute the times of the first opening of their leaves, and also of their blossoming. By comparing the accounts, the absurdity will immediately appear, of sowing the same kind of seeds at the same time of the month or year, in the 42d, 43d, 44th, and 45th degrees of latitude. This is a matter that farmers ought to attend to; that so those who remove from one degree of latitude to another, may not be confounded concerning the true times of sowing, on supposition that they have been once in the right practice. The right in one place will be wrong in another.

When these accounts are obtained, let trials be made by sowing a certain kind of seed before, at, and after the foliation, or the flowering, of some particular plant, and the produce compared. Let accurate experiments of this kind be yearly repeated, with all the most useful spring plants; by this, in a few years, complete kalendars may be obtained for every degree of latitude in this country. The consequence will be, that the farmer will be able infallibly to read the true times of sowing, by casting his eye upon the trees and shrubs that are about him. We have already such a rule as this, with respect to Indian corn; but it perhaps ought to undergo a further examination.

But such rules, after all that can be done, must not govern us invariably. The right times of seeding admit of some latitude, on account of the degree of dryness of the soil, and of its exposure to the solar warmth. Land should have the right degree of moisture when seeds are sown on it; and a southern exposure will afford an earlier vegetation than a northern.

That I may set an example of what I have been recommending, and begin the needful work, here follows an account of the leafing and blossoming of trees and shrubs in that part of New-England which lies in the 44th degree of latitude, in the spring of the year 1789.

	Leafing	Blossoming.
Gooseberry,	April 16	May 12
English Willow,	28	
Wild red Cherry,	29	19
Lilac,	30	
Currant,	May 1	9
Alder,	5	
Apple Tree,	6	25
Thorn Bush,	7	
White Birch,	8	
White Maple,	9	
Beech,	10	
Plum Tree,	12	
Hazle,	14	
Elm,	15	
Summer Pear,	17	31
Wheat Plum,		19
Common red Cherry,	19	20
Damascene Plum,		22
Grey Oak,	20	
White Oak,	23	

This plan of keeping a kalendar of the flowering of plants, has been followed up for many years by a Roxbury farmer, and it is much to be desired that in all parts of our country,gentlemen would keep and regularly publish such a record.

KALI, *Salicornia*, glass wort, or rock weed, a sea plant which grows upon rocks near the shore. By burning of this weed a hard fixed salt is obtained, which is a principal ingredient in the composition of glass. Rock weed is also an important manure.

KALMIA, *angustifolia*, a shrub commonly called laurel, or lamb poison. It is an evergreen, with narrow leaves of a dirty green colour. The flowers are red, growing round the upper part of the stem. It grows plentifully in low flat land, which has never been ploughed. It indicates a cold soil.

But I mention it in a work of this kind, on account of its poisonous quality. Sheep and goats, especially young lambs and kids, will eat it, when compelled by hunger, by which they sicken and die. The way to cure them of this sickness, is drenching them repeatedly with milk, mixed with oil, or fresh butter. Or, a tea of rue, given in season, may have the same good effect.

KID, the young of a goat. See *Goat*.

KILN, a fabric for admitting heat, to dry or burn various things. Malt is dried on a kiln. Another sort of kilns is used for the burning of lime stone. A lime kiln should be constructed of a sort of stones which will endure the fire. But if such cannot be easily obtained, hard burnt bricks will answer, and last a good while. The shape of a lime kiln should be like that of a pitcher, widest in the middle, and gradually narrower to the top and bottom. The fire will be the more confined, and act the more powerfully. In countries where limestone is plenty, each considerable farmer is furnished with a lime kiln, in which he makes lime to manure his soil. This practice might doubtless be imitated with advantage, in a few places in this country, where this sort of stone is at hand.

KITCHEN GARDEN. The species of manure for a Kitchen Garden, must depend upon the soil. Rotten dung is preferred by gardeners, as, in the course of fermentation the seeds of weeds, and the larvæ of insects are destroyed; and the more putrid the dung, the greater effect it has in promoting the rapid growth of plants. But too much manure, especially if it be of a fetid kind, gives vegetables a strong and disgusting flavour. Sea weed is said by Sir John Sinclair, to be an excellent manure for garden crops, particularly for onions. "Soaper's waste, says the same author," used in moderation by itself or in a compost with earth, is an excellent manure for garden soils. This substance not only destroys insects and their larvæ, but, consisting principally of calcareous matter, every species of vegetable is greatly improved in quality, where it is applied. Cow dung mixed with water, is a good manure, when frequently applied in a liquid state.

"Slugs or snails are very destructive in a garden. Slacked lime or sifted coal ashes, spread on the surface, or laid in rows, in various places, are useful in destroying them.

"Skilful gardeners recommend attention to a rotation of crops, and occasionally fallow portions of their gardens, or to lay them down with clover, which seldom fails to restore them to their former fertility. The most experienced horticulturists are now agreed, that even the currant, gooseberry and raspberry quarters should be changed every seven or eight years, and the strawberry ground every four or five years." *Code of Agriculture.*

L.

LAMBS, the young of sheep. The first care of them is to see whether they can come at the teat; and if not, to clip away the wool of the ewes which hinders them, as also all tags of wool on the udders of the ewes, which the lambs are liable to take hold of instead of the teats.

If a ewe refuse to let her lamb suck, she and her lamb should be shut up together in a close place, till she grow fond of him. For this purpose, some say that surprising a sheep with a dog will be effectual.

Care should be taken to feed the ewes plentifully after yeaning, and with some juicy kind of food, that so the lambs may not fail of having plenty of milk. The rams may be gelded at any time from one to three weeks old, if they appear to be well and strong.

They should not be weaned till they are six weeks, or two months old. At this age they should be taken from the ewes, and have the best of pasture during the first fortnight; by the end of which time they will be so naturalized to living wholly upon grass, that they may be turned into a poorer pasture.

The worst wooled lambs, and bad coloured ones, and those that are very small, should be destined to the knife, and not weaned. So great is the need of increasing the manufacture of woollen in this country, that I must earnestly recommend it to the farmers, not to kill, or sell for killing, any lamb, till it is near half a year old, or till the wool become to such fulness of growth, as to be valuable for spinning. To kill them earlier is so wasteful a practice as to be inexcusable.

Those ewe lambs which are kept for stock, should not come at the rams: For if they have lambs at a year old, it stints them in their growth; and they have so little milk, that their lambs commonly die for want of nourishment. Or if they chance to live, they will be apt to be always small. This practice is one reason why our breed of sheep in this country is so poor. See the article *Sheep*.

The largest lambs should be sheared at the time of the new moon in July. Their fleeces will yield as much the next year, and the wool will be better, and as cold storms rarely happen at that time of the year, the lambs will do better without their fleeces than with them.

LAMPAS, "an excrescence in the roof of the mouth, which hinders a horse from feeding, and happens usually to young horses. It is cured by applying a hot iron made for that purpose. It is successfully performed in all parts; so that there is no need of any caution, but only that the farrier do not penetrate too deep, so as to scale the thin bone that lies under the upper bars; for that would be attended with very troublesome and dangerous symptoms." *Gibson's Farriery*.

Judge Peters, of Pennsylvania says, that lampas is caused by fever in the horse that the swelling should be allayed by reducing the fever—that he would never suffer a horse to be burned in the mouth, as he had known them ruined by the operation. Sometimes a mess or two of hard Indian corn he observes will cure the swelling.

LAND, a general name applied to the earth, or to the ground.

"If land be unproductive, and a system of ameliorating it is to be attempted, the sure method of obtaining the object is by determining the cause of its sterility, which must necessarily depend upon some defect in the constitution of the soil, which may be easily discovered by chemical analysis.

"Some lands of good apparent texture are yet sterile in a high degree; and common observation and common practice afford no means of ascertaining the cause, or of removing the effect. The application of chemical tests in such cases is obvious; for the soil must contain some noxious principle which may be easily discovered, and probly as easily destroyed.

"Are any of the salts of iron present? They may be decomposed by lime. Is there an excess of siliceous sand? the system of improvement must depend on the application of clay and calcareous matter. Is there a defect of calcareous matter? the remedy is obvious. Is an excess of vegetable matter indicated? it may be removed by liming, paring, and burning. Is there a deficiency of vegetable matter? It is to be supplied by manure." *Agricultural Chemistry*. See *Soil*.

LARCH, *Hackmatack*, *Juniper*, *Pinus larix*, "a genus of trees, whose leaves are long and narrow, produced out of little tubercles, in the form of a painter's pencil. The cones are produced at remote distances from the male flowers, on the same tree : The male flowers are very like small cones at their first appearance, but afterwards stretch out in length. In autumn they cast their leaves. From the wounded bark of this tree exudes the purest Venice turpentine." *Complete Farmer.*

A sort of trees which grow naturally, and in great plenty, in the northern parts of New-England, called Juniper, I take to be the true larch, as it answers to the above description, as well as to that given by Mr. Miller. They thrive best in poor, wet and cold soils, and should by all means be cultivated. But their peculiar excellence is, that they will also thrive admirably in the driest and poorest soils. This is true of the whole race of pines, of which this is one, though it differs from all the others in being deciduous, or losing its leaves in winter. We may add that they are far superior to the Cedar for posts, or for any use where they are subjected to moisture. It is known that Venice in Italy is built on poles of the larch, and there is evidence that they have endured a thousand years.

LAYERS, tender twigs buried in earth, which having stuck root, are afterwards cut off, and become distinct plants.

Potatoes, and many other herbaceous plants, may be in this manner propagated. But there is little advantage to be gained by doing it.

As to those trees and shrubs which yield no seed in this climate, neither can be propagated by cuttings, there may be often occasion for laying them. The manner of doing it is as follows : Take shoots of the last year's growth, bend them to the earth, and bury them in good mellow soil half a foot under the surface, and fasten them with hooks to prevent their rising, bending the tops so as to bring them above the surface. A slit upwards in the twig should be made in that part that lies deepest in the soil, or a wire drawn fast round it, to prevent the sap mounting too fast; and moss should be laid on the surface, to prevent the sudden drying of the mould. Afterwards they should be watered as there may be occasion. If they form roots, they may be cut off, and transplanted the next spring into the nursery.

The time for laying evergreens is July or August; for laying deciduous trees, October.

LEAVES, the most extreme parts of the branches of trees, shrubs, &c.—" Their office is to subtilize the nourishing sap, and convey it to the little buds, and to cover and defend the flowers and fruit.

" Dr. Grew observes, that the fibres of leaves consist of two general kinds of vessels, viz. for sap, and for air ; and are ramified out of greater into less, as veins and arteries are in animals.

" If the surface of the leaves is altered, by reversing the branches of trees on which they grow, the plants are stopped in their growth, until the foot stalks are turned, and the leaves recover their former position. If leaves are eaten, or cut off the enclosed buds will not grow, and the plants will be weakened. The winter feeding of wheat, therefore, is hurtful; and it has been found so by experience.

"Another principal use of the leaves, is to throw off by transpiration what is unnecessary to the growth of plants, answering to the discharge made by sweat in animal bodies. As plants receive and transpire much more, in equal times, than large animals, so it appears how necessary the leaves are to preserve the plants in perfect health: For it has been found by the most exact calculation, made from repeated experiments, that a plant of the sun-flower receives and perspires, in twenty-four hours, seventeen times more than a man." *Complete Farmer.*

Mr. Bonnet made many experiments, which proved that leaves imbibe the moisture of the atmosphere on their under surface; excepting such as have the upper surface covered with hairs, or down. The leaves undoubtedly serve for inspiration, as well as for transpiration; and plants draw through their leaves, some considerable part of their nourishment.

Leaves also serve for ornament, and to screen vegetables, and their fruits, from the too intense heat of the sun in summer.

Leaves of trees are useful as a manure, excepting those of the resinous kinds. They should be collected into farm yards, trampled by the cattle, and mixed with their excrements. Some recommend leaves of oak for hot beds, instead of tanner's bark, as, by fermenting more slowly, they afford a more regular and permanent heat. Dr. Hunter proved the advantage of them by his continued practice. See *Georgical Essays*, by A. Hunter.

"A correspondent of the Bath Society in England warmly recommends a species of manure for potatoes, which I think peculiarly applicable to our country, because easily attainable. It is the employment of mould and *fallen leaves* taken from the woods. This the writer observes, he has found an excellent substitute for other manure. He found the potatoes, raised in this way, much more mealy and of a flavour much finer than when produced by the application of ashes and dung; and he considers it of great importance to poor people who have not the means of procuring much dung. This he observes can always be procured in woody countries, and in those which are not so, it may be obtained under hedges and ditches, and in old ponds.

"If this be a fact, and we have little doubt of it, since it is known that few substances are more favourable to vegetation than rotten leaves, and the soil formed by their decomposition, there is scarcely a farmer in Massachusetts, who may not by two or three days labour collect enough to plant all his potatoes, and save his corn and grass land.

Extracts from the Bath Society Papers, with remarks by John Lowell, Esq. Agricultural Repository, vol. IV. p. 60.

LEES, the gross sediment in fermented liquors. Most kinds of lees contain much of the food of plants. But they should not be applied to the soil as a manure till their acidity is destroyed, by mixing and fermenting them with large proportions of alkaline substances, such as marle, lime, ashes, soot, &c. Even the pomace at cyder mills, which has hitherto been considered by our farmers as good for nothing, might be thus changed into a good manure, It is nearly the same substance as the lees of cyder. Cyder lees will also produce brandy by distillation.

LIME, a crumbly soft substance, made by burning stones, and the shells of shell fish, and slacking them with water.

Lime has been proved, by the long experience of European farmers, to be one of the most efficacious manures. This may be thought strange by those who know it to be a mere alkali, containing neither oil nor salt, which are certainly the principal ingredients in the food of plants. Oil is an indispensably necessary part of this food.

But, by experiments made of late, it has been clearly proved that plants are greatly nourished by fixed air, of which it is known that lime contains a large quantity. It has been proved by the experiments of Mr. Lavoisier, that one third part of calcareous earths, and particularly of lime-stone, consists of fixed air.

But besides affording to plants this nourishment, which is known to be in plants, lime acts as a manure, by attracting and imbibing the oils and acids which are contained in the earth and atmosphere. It not only collects these ingredients of vegetable food, but so alters them as to fit them to enter the roots of plants. With the acids it forms a salt, which, by mixing with the oils, becomes a saponaceous mucilage, which is the true pabulum for the nourishment of plants.

These changes cannot be made in the ingredients of which vegetable food is composed, without a considerable degree of fermentation. This fermentation breaks and mellows the soil, and so increases the pasture of plants, that the roots can more freely extend themselves in quest of their food. Accordingly it is found that liming renders a soil very soft and open.

And as lime, when it is slacked, is a very soft substance, I can see no reason to doubt of its containing a very considerable quantity of those impalpably small particles of earth which enter into plants, and become part of their substance. If so, it must be allowed that lime is fit to answer every intention of manure. It either has all the ingredients of vegetable food, or produces and prepares them, though not in the same proportion as dung, which is allowed to be the most valuable of all manures.

Lime has been complained of, as impoverishing the soil; and it has been often remarked, that though one dressing will produce several good crops, the land is less fruitful for some time after, than before it has been limed; and that a second dressing with lime, will not have such an effect as the first, in increasing the fertility of the soil. But the farmer should consider how far he has been recompensed by extraordinary crops, for the exhausting of his soil; and that if lime will not, other manures will recruit it. So will fallowing, rest, or using it as a pasture.

It is granted that lime may have an ill effect, when it is injudiciously applied, as in too great quantities, or to an improper soil. Three cart loads, or 120 bushels, are allowed to be a sufficient dressing for an acre. But in Ireland, where they plough extremely deep, they lay on twice as much. This dressing enriches cold, stiff and clayey soils, for many years after; and in such soils it may be safely repeated. If it force any soils too much, it can be only those which are weak and sandy.

The best time for applying lime as a manure is, when land is newly broken up, or after laying a long time in grass. This may be ascribed to the plenty of roots in the soil, which the lime soon dissolves, and changes into food for plants.

Mr. Evelyn advised to the mixing of lime with turf in alternate layers, to lie in heaps for months; in which time it will become so rich and mellow as to run like ashes. He thought it would nourish the soil more than if used alone in a greater quantity, and without any danger of exhausting the vegetative virtue of the earth, which should be preserved. If it were mixed with a large proportion of clay, or with mud from the bottom of ponds or rivers, it might be applied even to sandy and gravelly soils without danger, and to great advantage.

Lime is a very important ingredient in composts, as by raising a strong fermentation it dissolves and prepares the other materials. There should be some layers of it, where it can be easily obtained, in every heap of compost. It will be the sooner fit for use, as well as prove to be a more fertilizing composition.

When lime is laid on land which has a quick descent, it should always be mixed with dung, and laid on the highest part; because it so loosens the soil, as to dispose it to be plentifully washed downwards by rains, soil and manure together.

Lime is an excellent manure for soils that are mossy, as it speedily dissolves the oil which is contained in moss, which is not soon dissolved by other manures, and changes it to vegetable food. It destroys all aquatic weeds, and dissolves the remainders of decayed vegetables in the soil. Therefore it does well in moory and peaty swamps that are drained.

While I am treating on this excellent manure, I have the disagreeable reflection, that it will be to little purpose; as lime is so scarce and dear in most parts of the country, that it must not be used as manure. Most people can scarcely obtain a sufficient quantity of it for building. But those farmers who know they have limestone or shells in plenty near them, should not neglect to make use of them as manures, after reducing them to lime.

"*Advantages of lime.*—Though there are exceptions to the rule, yet in general, it may be confidently asserted, that unless where a soil has by nature, enough of calcareous matter in its composition for the purposes of vegetation, it can neither be brought into its most fertile state, nor will other manures be so useful as they ought. if lime or some other calcareous earth, be not previously applied. By lime spread upon a moory soil, good herbage is produced, where nothing but heath, and unpalatable grasses grew before. By the same means, grass lands, instead of yield ing nothing but bent, and other inferior grasses, have been covered with those of a more valuable description. The utility of lime to turnips is so great, that though in the same field, where no lime had been applied, the crop died away; yet in the limed part, the turnips flourished with unabated vigour.

"*The principles on which Lime operates as a manure.*—Quicklime in powder, or dissolved, is injurious to plants; hence grass watered with lime-water is destroyed. But lime freshly burnt, or slacked, forms a compost with vegetable matter, which is soluble in water, and nutritive to plants. Mild lime, (as chalk, or quicklime again impregnated with carbonic acid,) chiefly operates by improving the texture of the soil, and its relation to absorption.

LIME-STONE, a stone of a calcareous nature, which, by calcination, or burning in the fire, becomes lime. There are many kinds of lime-stone; the hardest kinds make the best lime, and require the most burning. Chalk will burn into lime, of the nature of stone-lime, but a great deal weaker; lime may be made of marble and alabaster, &c. But the stones used for lime are mostly of a bluish colour or inclining to grey. They are sometimes purely calcarious, but often mixed with undissolvable stones, which lessen their value.

Some countries are very plentifully furnished with these stones, Great Britain and Ireland in particular. It is strange they have been found in so few places in New-England. It has probably been owing to want of attention. An infallible way to distinguish them is, by dropping upon them a few drops of aqua fortis, spirit of sea-salt, or oil of vitriol. All those stones, on which these, or any other strong acids, effervesce, or rise into bubbles, are limestones, and will burn into lime.

It is greatly to be wished, that some persons in the various parts of this country, would be furnished with one or other of these acids, and make frequent trials with them. They who are not furnished with the proper acids, may prove stones, by burning them for some days in a smith's fire, and then throwing them into water. Possibly we may find the benevolent Author of nature has not left us so unfurnished with these valuable stone, as we have been ready to imagine.

The various sorts of Limestone.—Sometimes limestone is almost perfectly pure, as in the case with marble, which frequently contains scarcely any other substance but calcareous matter. Several sorts of limestone, however, have mixtures of clay and sand, in various proportions, by which the efficacy of the manure, in proportion to the quantity of the substances is considerably diminished. It is necessary, therefore, to analyze limestone, to ascertain the proportion of pure lime, before it is advisable to use so expensive an article in great quantities, more especially if it must be conveyed from a distance. Bituminous limestone makes good manure. But the magnesian is the species which requires the greatest attention. Limestone contains from 20·3 to 23·5 of magnesia, in which case it would be injurious to weak soils, to apply more than 25 to 30 bushels per statute acre, though in rich soils, double that quantity may be used, and still more with peat, on which soil it would have a most powerful effect in producing fertility."—*Code of Agriculture*.

Sir John Sinclair recommends to slack lime, in order to prepare it for manure with sea-water or urine. The time "for applying lime is when the land is under summer fallow, in the months of June or July, that it may be completely mixed with the soil before the crop is sown." "For a turnip crop it should be laid on early in the spring, before the turnips are drilled, in order that the lime may be thoroughly incorporated with the soil, by the ploughings and harrowings it will receive. For potatoes lime is not to be recommended, as it is apt to burn and blister their skins. The quantity of 240 to 300 bushels of unslacked lime, may be applied on strong lands with advantage. Even 600 bushels have been laid on at once on strong clays with great success. On light soils a much smaller quantity will answer, say from 150 to 200 bushels; but these small doses ought to be frequently repeated. When applied on the surface of bogs or moors, the quantity used is considerable." Upon land in a proper state for calcarious application, lime is much superior to dung. Its effects continue for a longer period, while the crops produced are of superior quality, and less susceptible of injury from excesses of drought and moisture. The ground, likewise, if it be of a strong nature is more easily wrought, and in some instances the *saving of labour alone* would be an inducement to a farmer to lime his land. Lime having a tendency to sink in the soil, cannot be kept too near the surface. The drier and finer lime is powdered when applied to the soil, the better.

LOAM, one of the principal kinds of earth. Some suppose it to be not one of the natural soils; but gradually made since the creation, by the putrefied vegetables which have fallen upon the earth. This does not appear probable; for, if so, why do we meet with any other kind of soil? This soil consists of very fine particles, without grit, almost as fine as those of clay, but do not cohere like them. If it lie long under water, it is apt to have the appearance of clay.

It receives water readily, and retains it long; on which accounts it is preferable to clay or sand. It is better adapted by nature to nourish vegetables than either the one or the other. But its needs manure, and will commonly pay well for it by the increase of its crops.

Loams are of various kinds. Some is stiff, approaching to the nature of clay, and is apt to be adhesive in wet weather. This is not fit for the nourishing of those vegetables which require much heat. It needs to be dressed with hot and opening manures for any kind of crop. Other loam is more light, soft and mellow, and does not so much need the most heating manures. Some loam is of a dark red, hazely, or brown colour. This is commonly a most excellent soil. Other loam is of a light yellow, or whitish colour, and requires abundance of manuring to render it fruitful.

All kinds of loam are apt to be too wet, and to be covered with a short green moss, if they lie flat. In this case, ridge ploughing is best, and hollow drains often necessary. Loam that has a mixture of gravel, or sand, is warmer, and fitter for tillage; but all loams are good for the growing of grasses.

LOCUST TREE, *Robinia*, a well known tree, which grows in great plenty in the vicinity of Boston, and is a native of this country, but does not flourish so well in the state of Maine, as the frost of winter is apt to kill the extremities of the limbs. There are particular places, however, in this state, where the growth of this tree is considerably rapid.

This tree would be more prized for its beauty, were not its limbs often broken by high winds. Its leaves put out late in the spring, and fall off early in the autumn. It blossoms about the beginning of June, at which time it makes a beautiful appearance, and perfumes the circumambient air with an agreeable odour. The branches are armed with hooked spines; and the leaves composed of ten pair of oval lobes, terminated with an odd one.

The wood is not only good fuel, but excellent timber, very durable in any situation, and particularly when used as posts in fences.

This tree grows best in a sandy soil, and will propagate itself in the most barren places, where the soil is so light as to be blown away by winds. By sheltering such places, and dropping its leaves on them, it causes a sward to grow over them, and grass to grow upon them. It is not advisable to plant groves of the locust tree on the borders of fields, on account of their spreading too much by scattering their seeds, unless on those which are most barren. But those who possess hills of barren sand, and in a climate that suits them, should not delay to make forests of these trees on such spots. It may be easily done by sowing the seeds in a nursery, and transplanting them. A plenty of wood may be thus speedily produced, without the least injury to the land, yea, with advantage to it.

For "some notice of the Insect which destroys the Locust Tree," we would refer our readers to the Massachusetts Agricultural Repository, vol. V. p. 67. The insect is there called Cossus Robiniæ; and drawings are given descriptive of its different appearances, in different stages of its existence. The writer suggests no remedy against its ravages.

The Massachusetts Agricultural Repository, vol. VI. page 272, contains a paper detailing a number of experiments by H. A. S. Dearborn, instituted for the purpose of preserving locust trees from the depredations of insects. For this purpose Mr. Dearborn stopped up the holes, which they had bored into the trees, with lime mortar, and by white washing, &c. He succeeded, however, but partially, and proposed to persevere in the spring of 1822. Should his experiment fail, he proposes to head down all his locust trees in the latter part of April, and burn the wood to destroy the larvæ. "The trees," he observes, "are of rapid growth, and the stools throw up luxuriant suckers, and I shall soon have young thrifty trees, which can be more certainly preserved from the attacks of the insect, should they again visit them; for trees which are so large as to have very rough bark, supply so many fastnesses for the deposition of the eggs, and to cover the depredations of the worm that it is is very difficult to extirpate them."

M.

MADDER, *Rubia Tinctoria*. The following directions for raising this plant, are copied from "*The Emporium of Arts*."

"This plant may be propagated either by offsetts or seeds. If the latter method is preferred, the seed should be of the true Turkish kind, which is called *Lizari* in the Levant. On a light thin soil the culture cannot be carried on to any great profit. The soil in which the plant delights is a rich sandy loam, being three feet in depth or more.

"The ground being first made smooth, is divided into beds four feet wide, with alternate alleys half as wide again at the beds. The reason of this extraordinary breadth of the alleys will presently appear. In each alley is to be a shallow channel for the convenience of irrigating the whole fields, &c. That part of the alley which is not occupied may be sown with legumes.

"The madder seed is sown broad cast,in the proportion of from twenty-five to thirty pounds per acre, about the end of April. In a fortnight or three weeks, the young plants begin to appear; and from this time to the month of September, care must be taken to keep the ground well watered, and free from weeds.

"If the plants are examined in autumn, they will be found surrounded with small yellow offsetts, at the depth of two inches; and early in September, the earth from the alleys is to be dug out, and laid over the plants of madder to the height of two or three inches. With this the first year's operation ceases.

"The second year's work begins in May, with giving the beds a thorough weeding; and care must be taken to supply them with plenty of water during the summer. In September the first crop of seed will be ripe; at which time the stems of the plants may be mown down, and the roots covered a few inches with earth, taken as before out of the alleys.

"The weeding should take place as early as possible in the spring of the third year; and the crop, instead of being left for seed, may be cut three times, during summer for green fodder; all kinds of cattle being remarkably fond of it.

"In October the roots are taken up, the offsetts carefully separated. and immediately used to form a new plantation; and the roots, after being dried, are sold, either without further preparation, or ground to a coarse powder, and sprinkled with an alkaline lye.

"The roots lose four-fifths of their weight in drying; and the produce of an acre is about two thousand pounds weight of dry saleable madder."

Madder usually sells for about thirty two dollars per hundred; so that the produce of an acre as above stated, would amount to six hundred and forty dollars. *Farmer's Assistant.*

MALANDERS, a horse disease, caused by corrupt blood, or over hard labour, &c. It consists of chops, or cracks, on the inside of the fore legs against the knee, discharging a red sharp humour.

To cure this disease, wash the cracks with warm soap suds or old urine; then rub them twice a day with an ointment of hog's lard mixed with two drachms of sublimate of mercury. Or apply a poultice of the roots of marsh mallows and flax seed, softened with linseed oil, tying it on with a roller. Continue that till the seeds fall off and the sores become clean. Afterwards a mixture of turpentine and quicksilver will be a proper application.

MALT, barley, or other corn, prepared for making beer or ale. As it is of great importance that the people of this country should make a greater use of malt than they do at present, I will here give the process of making it, from the *Dictionary of Arts and Sciences.*

"In making malt from barley, the usual method is to steep the grain in a sufficient quantity of water, for two or three days, till it swells, becomes plump, somewhat tender, and tinges the water of a bright brown, or reddish colour. Then, this water being drained away, the barley is removed from the steeping cistern to the floor, where it is thrown into what is called the wet couch; that is, an even heap, rising to the height of about two feet. In this wet couch, the capital part of the operation is performed; for here the barley spontaneously heats, and begins to grow, shooting out first the radicle, then the plume, spire or blade. But the process is to be stopped short at the irruption of the radicle, otherwise the malt would be spoiled. In order to stop it, they spread the wet couch thin over a large floor, and keep turning it once in four or five hours, for the space of two days, laying it somewhat thicker each time. After this it is again thrown into a large heap, and there suffered to grow sensibly hot to the hand, as it usually will in twenty or thirty hours: Then being spread again, and cooled, it is thrown upon the kiln, to be dried crisp without scorching. If these directions be followed, the malt will always be good.

"The method of malting Indian corn, or Virginia wheat, is much less laborious. For, if this corn be buried two or three inches deep in the earth, and covered with the loose mould, in ten or twelve days time the corn will sprout, and appear like a green field; at which time being taken up, and washed or fanned from the dirt, it is immediately committed to the kiln, and by this means becomes good malt."

MALT DUST, the dust which falls from the kiln, while malt is drying. Repeated experiments made by Europeans, have established the credit of this dust as a manure for stiff loams and clays. A good dressing of it has been found to increase a crop of barley as much as fifty per cent. and wheat still more. The quantity used is from thirty to sixty bushels per acre, according to circumstances. It is used mostly, or only, as a top dressing. It exerts its strength so suddenly as to be nearly exhausted with one crop. It should not be sown together with winter wheat, but upon it in December or January following: For if it be sown early, it will exert its strength too soon, and bring the wheat forward too fast, as has been proved by experiments. For barley, this dressing should be sown with the seed and harrowed in. A small dressing of this manure on grass land, mightily increases the vegetation, and the sweetness of the grass.

Maltsters should carefully preserve this precious manure in some place where it will not contract dampness. It may be of use to farmers in their neighbourhood: But it cannot become a manure of general use, the whole quantity that is made being so small.

MANGEL WURTZELL. See *Beet*.

MANURE, any kind of substance suitable to be laid on land to increase its fertility.

Manures contribute several ways to the producing of this effect: Either by increasing the quantity of vegetable food in the soil--or by preparing the nourishment already contained in the soil to enter the roots of plants—or by enlarging the vegetable pasture in which roots spread and seek their food—or by attracting the food of vegetables from the air. Some of the manures increase fruitfulness in all these ways, particularly the dung of animals, rotted vegetables, &c. Other manures perform each office, excepting the first: And some have no other immediate effect besides opening and loosening the soil: But even these last kinds may sometimes be used to great advantage.

There are different ways of ordering and managing manures, according to their different natures. Some are to be applied to land without alteration, or mixing; the rest to be prepared by compounding and fermentation: Some are suitable for stiff and some for light soils: Some to be mixed in the soil by the plough and harrow; other kinds to be used only as top dressings.

Farmers and gardeners should not be so inattentive to their own interest, or that of their employers, as to suffer a variety of valuable manures to lie useless, while they are suffering for want of them. I have drawn up the following list for their benefit, hoping that such a variety, all of which can be had by one or other, in this country, and by most farmers in plenty, might excite the ambition of some to make use of their advantages, and suffer no manures to escape their attention.

The substances fit to be used as manures, are either animal, vegetable, fossil, or mixed.

Animal manures are such as these that follow:

Putrefied flesh, such as the carcases of animals, or meat not well saved. This may be an ingredient in compost, or buried at the foot of fruit trees to increase their fruitfulness. Dead horses, dogs, cats, rats, and uneatable birds, should, instead of putrefying the air by rotting above ground, be thus converted to an economical purpose. When the carcases of animals are buried in dung-hills, it may be proper to lay over them some bushes of thorn, to prevent ravenous dogs from taking them away.

Blood, mixed with saw dust, and used as a top dressing, &c. See the article *Blood*.

Hair, a top dressing for grass land; under the surface of a dry soil in tillage; or used in compost. In either way it is an excellent fertilizer.

Feathers, such as have been worn out in beds, or are unfit to go into them—in compost.

Refuse wool, such coarse dag locks as are not fit for carding—covered with the plough in a dry soil. They will serve as spunges to retain moisture, and be a rich food for plants when they are dissolved. So will

Woollen rags, chopped to pieces, for a light soil. They should be cut as small as an inch square. Twenty-four bushels are said to be a sufficient quantity for the dressing of an acre. These should be under the surface.

Hoofs of cattle, *sheep*, &c. If large hoofs were set in holes with the points downwards in a dry soil, so low as not to be disturbed by the plough, they would cause the land to retain moisture, and hold the manure, not only by the spunginess of their substance, but also more especially by their hollowness.

Bones, of all kinds, pounded or broken into small pieces, with hammers or mallets. This is an incomparable manure, if they have not been burnt, nor boiled in soap. But in either way they should be saved for manure. Sixty bushels are a sufficient dressing for an acre.

Raw skins of all kinds of animals. These should be cut into small pieces, and used for light soils, ploughed in.

Leather, new or old, in small bits, for dry soils, ploughed in.

Curriers' shavings, cut small, for a soil of sand or gravel, ploughed in.

Oil, of all sorts, used in composts, not applied to the soil till a year after it is mixed, that it may be dissolved and altered.

Fish, of all kinds, from the whale to the muscle; they are best used in composts; and should lie a year, that their oil may be dissolved, and fitted for the nourishing of plants.

Offal of fish, in composts, fit for one soil or another, according to the predominant ingredients of the mixture.

The vegetable manures are good, though not so strong as animal ones. They can be had in greater plenty in most places; and ought to be laid on in larger quantities.

Green vegetables, such as all useless weeds in fields and gardens. These should be collected and rotted in heaps. They are a good manure for all soils, and to nourish all sorts of plants.

Aquatic weeds, such as grow in the borders of ponds and rivers. These should be collected in large heaps on the higher ground, and covered with turfs, the grass side outwards. These heaps will be easily made in some places, and will be a valuable manure. Some say, care should be taken to prevent their taking fire by fermenting, as their heat will be very great.

Straw, and other offal of corn of all kinds, rotted in farm yards, or dung pits.

Refuse hay, both fresh and salt, rotted in yards, and trampled on by cattle, and mixed with their excrements.

Thatch, that grows by the sides of salt creeks, or the parts of it which cattle will not eat, should be thrown into the farm yard, to putrefy. Thus a great increase of good manure may be made.

The haulm of all dry vegetables, such as the stalks of potatoes, beans, peas, &c. Even the offal of flax, if it have sufficient time to rot, will be a good manure.

Fern, a vegetable peculiarly adapted to the purpose of making manure. See *Fern*.

Lees of fermented liquors, rotten fruit, and pomace, in compost.

Oil cakes, which may be got at the mills where linseed oil is expressed for top-dressing, being first pulverized.

Tanners' bark, from the oak tree, *Leaves* of deciduous trees, *Rotten wood*, *Saw dust*,	Fermented with other manures, to be laid on clayey and stiff soils.

Wood ashes, a good top dressing for almost any kind of soils, but best for a moist one.

Coal ashes, top-dressing for cold damp soils.

Coal dust, top-dressing for low meadows.

Malt dust. See that article.

Sea plants, rock-weed, eel-grass, &c. are the most valuable of green vegetables for manure. They should be either ploughed into the soil, or mellowed in compost dunghills. It is a wrong practice to use them as top-dressings. Much of their virtue in this way is lost.

Moss, mixed with dung in holes for a dry soil. Good for potatoes.

Linen rags; these will be a manure worth saving, but they take a long time to putrefy—in compost.

The fossil or earthy manures are these:

Lime, mixed with the soil, or in composts, for stiff soils. See the article *Lime*.

Marle, most suitable in general for light soil. See the article *Marle*.

Sand, in roads, washed down from hills, to open a stiff clayey soil. See the article *Sand*.

Plaister of Paris, and *Dust of hewn stones*, } Absorbent manures for cold wet soils, for top-dressing.

Gravel, for a wet puffy swamp.

Clay, to mix with the plough and harrow in a sandy or gravelly soil. It should be exposed to the action of the frost one winter before it is ploughed in. Otherwise it will remain a long time undissolved.

Swamp mud, *River mud*, *Pond mud*, *Sea mud*, } To be mixed with a sandy or gravelly soil; but best in composts, with dung. See the article *Mud*.

Ashes of sea-coal—for cold stiff land.

Peat, when reduced to ashes, top dressing for all soils, best for a cold one. See *Peat*.

Turfs, either in composts, or dried and burnt. They may be taken from the sides of highways without damage. These places are the walks of cattle and swine, where much dung is dropped; the turf is therefore a rich ingredient in manure.

Shells of shell fish, ploughed in whole, are a good manure for dry soils; and ground or pounded small for stiff land.

Brick dust, *Burnt clay*, } To open a clayey, or warm a cold soil.

Beach sand, to open a stiff, and warm a cold soil. That which has a fine grain is the best.

Pit sand of any colour, to meliorate a soil of stiff clay. It should be laid on plentifully.

The mixed solid manures are these.

Dung of all kinds. Though it chiefly consists of rotten vegetables, there is a mixture of animal juices in it, and some of the finest particles of the earth. Most dungs should be mixed with the soil, by the plough or harrow. See the article *Dung*.

Composts of every kind, fit for light or stiff soils, according to the difference of their predominant ingredients; or a general manure for all soils.

The scrapings of back yards, for all kinds of soil, but when containing chips, shavings of wood, or much saw dust, for stiff soils.

Rubbish of old houses, for cold and stiff soils. This contains much nitre—in composts it is of most advantage.

Earth that has been long under cover. This commonly collects much nitre. Best in composts.

Scrapings of streets, a general manure, fit for all soils. Farmers who live in the vicinity of cities, and great towns, should always avail themselves of this kind of manure.

Mixed liquid manures.

Old brine of salted meat or fish, which contains, besides salt, some blood, oil, &c.—in composts.

Sea water, which contains other things besides water and salt, fit to nourish vegetables. It may be sprinkled on land, or used in composts.

Soap suds—replete with a prepared food for plants ; excellent for watering gardens in dry weather. None of this should be lost. If the garden be distant, or wet, it may enrich the dunghill.

Urine of all animals. This contains earth and animal juices, salts and oil ; and is, next to dung, perhaps the most valuable and important of all manures. See the article *Urine.*

Water in the hollows of farm yards. Instead of suffering this rich liquor to soak into the bowels of the earth, it should be taken up by mulch, or some absorbent substance thrown into it, or else carried out in a water cart, and sprinkled over a soil that needs it.

Water that runs from compost dunghills. This should be thrown back upon dunghills, or else used as the preceding article.

Liquors from dye-houses. This should be used in composts.

After all, I may add *Salt*, being distinct from all other manures, an important ingredient in the food of plants, and adapted to prepare other ingredients. Some apply it as it is, but it has a better effect when used in composts.*

If our farmers in general would be persuaded to avail themselves of so many of these manures as fall in their way, or can be easily obtained, we should no longer hear of so many dismal complaints as we do, of short crops, and worn out lands. The face of the country would soon be surprisingly improved.

But that manures may fully answer their intention, they must be judiciously applied. We should not only apply each manure to the soil for which it is most suitable, but at seasons when it will produce the most valuable effect. For a general rule, it is best to apply those rich fermenting manures, which are to be mixed in the soil, as near as may be to the time when the ground is seeded. Dung should be ploughed in with the seed furrow, as it is called. Composts may be harrowed in with the seed. The reason for applying these manures at this time is obvious. They will begin to raise a fermentation in the soil, almost as soon as they are applied ; so that if there be no seed, nor plants to be nourished by them, some part of the good effects of the manure will be lost. As part of the fermentation will be past, before the plants begin to grow ; so there may be danger of its being over, before they have attained to their full growth. If so, the soil will harden, and the plants will receive the least quantity of nourishment at the time when they need the greatest.

As to those matters which raise little or no fermentation, they may be laid on at any time when the farmer has leisure for it, as sand on a clayey, gravel on a boggy and puffy soil ; or clay, marle, or mud, on a light soil.

* See the article "*Salt*" in this work for further notice of it.

It has been too much practised in this country, to apply scanty dressings to lands in tillage, hardly sufficient to have a perceptible effect, and to repeat it year after year. But this, I think, is a wrong practice. A sufficient dressing once in two years, I have always found to do better than a half dressing each year. This last method does not so well agree with a succession of crops; because some crops require a much greater degree of strength in the soil, than others do. Let us then rather follow the example of the European farmers, who commonly manure very plentifully once in a course of crops, and no more; and the year the manure is laid on, take a crop that requires the greatest assistance from manure, or that bears high manuring best, or makes the best returns for manure: Afterwards, crops that need less manure, till the end of the course. Perhaps the year of manuring in this country should be chiefly for Indian corn. This crop is not easily overdone with manure, and it pays well for high manuring. And this happens well for us, as a hoed crop, when the dung is used, will prevent the increase of weeds, which a plentiful dunging will greatly promote in every kind of soil.

The following observations on this important subject are extracted from Sir H. Davy's "Elements of Agricultural Chemistry."

"All *green succulent plants* contain saccharine or mucilaginous matter, with woody fibre, and readily ferments. They cannot therefore if intended for manure, be used too soon after their death.

"When *green crops* are to be employed for enriching a soil, they should be ploughed in, if it be possible, when in flower, or at the time the flower is beginning to appear; for it is at this period that they contain the largest quantity of easily soluble matter, and that their leaves are most active in forming nutritive matter. Green crops, pond weeds, the paring of hedges or ditches, or any kind of fresh vegetable matter, require no preparation to fit them for manure. The decomposition slowly proceeds beneath the soil; the soluble matters are gradually dissolved, and the slight fermentation that goes on, checked by the want of a free communication of air, tends to render the woody fibre soluble, without occasioning the rapid dissipation of elastic matters.

"When old pastures are broken up and made arable, not only has the soil been enriched by the death and slow decay of the plants which have left soluble matters in the soil; but the leaves and roots of the grasses living at the time and occupying so large a part of the surface, afford saccharine, mucilaginous and extractive matters, which become the food of the crop, and the gradual decomposition affords a supply for successive years.

"*Sea weeds*, consisting of different species of fuci, algæ and confervæ are much used as manure on the coasts of Britain and Ireland. By digesting the common fuci, which is the sea weed usually most abundant on the sea coast, in boiling water, I obtained from it one eighth of a gelatinous substance, which had characters similar to mucilage. A quantity distilled gave nearly four-fifths of its weight of water, but no ammonia; the water had an empyreumatic and slightly sour taste; the ashes contained sea salt, carbonate of soda and carbonaceous matter. The gaseous matter afforded was small in quantity, principally carbonic acid and gaseous oxid of carbon, with a little hydro-carbonate. This manure is transient in its effects, and does not last for more than a single crop, which is easily accounted for from the large quantity or the elements of water it contains. It decays without producing heat, when exposed to the atmosphere, and seems as it were to melt down and dissolve away. I have seen a large heap entirely destroyed in less than two years, nothing remaining but a little black fibrous matter.

"Sea weed is sometimes suffered to ferment before it is used; but this process seems wholly unnecessary, for there is no fibrous matter rendered soluble in the process, and a part of the manure is lost.

"*Dry straw* of wheat, oats, barley, beans and peas, and spoiled hay, or any other similar kind of dry vegetable matter is in all cases, useful manure. In general such substances are made to ferment before they are employed, though it may be doubted whether the practice should indiscriminately be adopted.

"When straw is made to ferment it becomes a more manageable manure; but there is likewise on the whole a great loss of nutritive matter. More manure is perhaps supplied for a single crop; but the land is less improved than it would be, supposing the whole of the vegetable matter could be finely divided and mixed with the soil.

"It is usual to carry straw that can be employed for no other purpose to the dung-hill, to ferment and decompose; but it is worth experiment, whether it may not be more economically applied when chopped small by a proper machine and kept dry till it is ploughed in for the use of a crop. In this case, though it would decompose much more slowly and produce less effect at first, yet its influence would be much more lasting.*

"*Wood ashes* imperfectly formed, that is wood-ashes containing much charcoal, are said to have been used with success as a manure. A part of their effect may be owing to the slow and gradual consumption of the charcoal, which seems capable, under other circumstances than those of actual combustion, of absorbing oxygene so as to become carbonic acid.

"Manures from animal substances in general require no chemical preparations to fit them for the soil. The great object of the farmer is to blend them with earthy constituents in a proper state of division, and to prevent their too rapid decomposition.

"The entire parts of the muscles of land animals are not commonly used as a manure, though there are

* Some agriculturists are of a different opinion. See *Dung*. Probably the question of *long* and *short* manure depends on the nature of the soil, and of the crop.

many cases in which such an application might be easily made. Horses, dogs, sheep, deer, and other quadrupeds that have died accidentally, or of disease, after their skins are separated, are often suffered to remain exposed to the air or immersed in water till they are destroyed by birds or beasts of prey, or entirely decomposed; and in this case most of their organized matter is lost for the land on which they lie, and a considerable portion of it employed in giving off noxious gasses to the atmosphere.

"By covering dead animals with five or six times their bulk of soil, mixed with one part of lime, and suffering them to remain for a few months, their decomposition would impregnate the soil with soluble matters,so as to render it an excellent manure; and by mixing a little fresh quick lime with it at the time of its removal, the disagreeable effluvia would be in a great measure destroyed; and it might be applied in the same way as any other manure to crops.

"Fish forms a powerful manure in whatever state it is applied; but it cannot be ploughed in too fresh, though the quantity be limited. Mr. Young records an experiment, in which herrings spread over a field and ploughed in for wheat produced so rank a crop, that it was entirely laid before harvest.

"It is easy to explain the operations of fish as a manure. The skin is principally gelatine; which from its slight state of cohesion is readily soluble in water, fat or oil; is always found in fishes either under the skin or in some of the viscera; and their fibrous matter contains all the essential elements of vegetable substances.

"Among oily substances, *graves* and *blubber* are employed as manure. They are both most useful when mixed with soil, so as to expose a large surface to the air, the oxygene of which produces soluble matter from them. Lord Somerville used blubber with great success at his farm in Surrey. It was made into a heap with soil, and retained its powers of fertilizing for several successive years.

"The carbon and hydrogene abounding in oily substances fully account for their effects; and their durability is easily explained from the gradual manner in which they change by the action of air and water.

"*Bones* are much used as a manure in the neighbourhood of London. After being broken and boiled for grease they are sold to the farmer. The more divided they are the more powerful are their effects. The expense of grinding them in a mill would probably be repaid by the increase of their fertilizing powers; and in the state of powder they might be used in the drill husbandry.

"*Bone dust*, and *bone shavings, the refuse of the turning manufacture*, may be advantageously employed in the same way.

"The basis of bone is constituted by earthy salts, principally phosphate of lime, with some carbonate of lime, and phosphate of magnesia; the easily decomposable substances in bone are fat, gelatine, and cartilage, which seem of the same nature as coagulated albumen.

"*Hair, woollen rags*, and *feathers*, are all analogous in composition, and principally consist of a substance similar to albumen, united to gelatine. This is shewn by the ingenious researches of Mr. Hatchel. The theory of their operation is similar to that of bone and horn shavings.

"The *refuse* of the different manufactures of skin and leather form very useful manure; such as the shavings of the currier, furrier's clippings, and the offals of the tan-yard and of the glue maker. The gelatine contained in every kind of skin is in a state for its gradual solution or decomposition; and when buried in the soil it lasts for a considerable time, and constantly affords a supply of nutritive matter to the plants in its neighbourhood.

"*Blood* contains certain quantities of all the principles found in other animal substances, and is consequently very good manure.

"Among the *excrementitious* animal substances used as manure, urine is the one upon which the greatest number of chemical experiments have been made, and the nature of which is best understood.

"Urine is very liable to change and to undergo the putrefactive process; and that of carniverous animals, more rapidly than that of graminiverous animals. In proportion as there is more gelatine and albumen in urine, so in proportion does it putrefy more quickly.

"The species of urine that contain most albumen gelatine and uria are the best manures; and all urine contains the essential elements of vegetables in a state of solution.

"During the putrefaction of urine the greatest part of the soluble animal matter that it contains is destroyed; it should consequently be used as fresh as possible; but if not mixed with solid matter it should be diluted with water, as when pure it contains too large a quantity of animal matter to form a proper fluid nourishment for absorption by the roots of plants.

"Amongst the excrementitious solid substances used as manures, one of the most powerful is the *dung of birds* that feed on animal food, particularly the dung of sea birds. It is easy to explain its fertilizing properties; from its composition it might be supposed to be a very powerful manure. It requires water for the solution of its soluble matter to enable it to produce its full beneficial effect on the crops.

"Night soil, it is well known, is a very powerful manure, and very liable to decompose. It differs in its composition, but always abounds in substances composed of carbon, hydrogene, azote, and oxygene. From the analysis of Berzelius, it appears that a part of it is always soluble in water; and in whatever state it is used, whether recent or fermented, it supplies abundance of food to plants.

"The disagreeable smell of night soil may be destroyed by mixing it with quick lime; and if exposed to the atmosphere in thin layers strewed over with quick lime, in fine weather it speedily dries, is easily pulverised, and in this state may be used.

"The Chinese, who have more practical knowledge of the use and application of manures than any other people existing, mix their night soil with one third of its weight of a fat marle, make it into thin cakes, and dry it in the sun. These cakes, we are informed by the French missionaries, have no disagreeable smell, and form a common article of commerce of the empire.

"The earth by its absorbent powers probably prevents to a certain extent, the action of moisture upon the dung, and likewise defends it from the effects of air.

"After night soil *pigeon dung* comes next in order, as to fertilizing power. It is evident that this manure should be applied as new as possible, and when dry it may be employed in the same manner as the other manures capable of being pulverized.

"The dung of *domestic fowls* approaches very nearly in its nature to pigeons' dung.

"The dung of *cattle*, *oxen* and *cows* has been chemically examined by M. M. Einhoff and Thaer. They found that it contained matter soluble in water, and that it gave in fermentation nearly the same products as vegetable substances, absorbing oxygene and producing carbonic acid gas.

"If the pure dung of cattle is to be used as manure like the other species of dung which have been mentioned, there seems no reason, why it should be made to ferment except in the soil; or if suffered to ferment it should be only in a very slight degree. The grass in the neighbourhood of recently voided dung is always coarse and dark green. Some persons have attributed this to a noxious quality in unfermented dung; but it seems to be rather the result of an excess of food furnished to the plants.

"The question of the proper mode of application of the dung of horses and cattle, however, properly belongs to the subject of *composite manures*, for it is usually mixed in the farm yard with straw, offal, chaff, and various kinds of litter; and itself contains a large proportion of fibrous vegetable matter.

"A slight incipient fermentation is undoubtedly of use in the dung-hill; for by means of it a disposition is brought on in the woody fibre to decay and dissolve when it is carried to the land, or ploughed into the soil; and woody fibre is always in great excess in the refuse of the farm.

"Too great a degree of fermentation is however, very prejudicial to the composite manure in the dung-hill; it is better that there should be no fermention at all before the manure is used, than that it should be carried too far. The excess of fermentation tends to the destruction and dissipation of the most useful part of the manure; and the ultimate results of this process are like those of combustion. It is a common practice among farmers to suffer the farm yard dung to ferment till the fibrous texture of the vegetable matter is completely broken down; and till the manure becomes perfectly cold, and so soft as to be easily handled or managed with the spade. Independent of the general theoretical views unfavourable to this practice, founded upon the nature and composition of vegetable substances, there are many arguments and facts which shew it is prejudicial to the interests of the farmer.

"During the violent fermentation which is necessary for reducing farm yard manure to the state in which it is called *short muck*, not only a large quantity of fluid but also of a gaseous matter is lost, so much so that the dung is reduced one half or two thirds in weight; and the principal elastic matter disengaged is carbonic acid with some ammonia; and both these if retained by the moisture in the soil are capable of becoming an useful nourishment of plants. Besides the dissipation of gaseous matter, when fermentation is pushed to the extreme, there is another disadvantage in the loss of heat, which if excited in the soil is useful in promoting the germination of the seed, and in assisting the plant in the first stage of its growth, when it is most feeble and liable to disease; and the fermentation of manure in the soil must be particularly favourable to the wheat crop in pre serving a genial temperature beneath the surface late in autumn, and during winter.

"Again it is a general principle in chemistry, that in all cases of decomposition, substances combine much more readily at the moment of their disengagement, than after they have been perfectly formed. And in fermentation beneath the soil the fluid matter produced is applied instantly even while it is warm to the organs of the plant, and consequently is more likely to be efficient than in manure that has gone through the process, and of which all the principles have entered into new combinations.

"A great objection against slightly fermented dung is, that weeds spring up too luxuriantly where it is applied. If there are seeds carried out in the dung they will certainly germinate; but it is seldom that this can be the case to any extent: if the land is not cleansed of weeds, any kind of manure fermented or unfermented will occasion their rapid growth. If slightly fermented farm yard dung is used as a top dressing for pastures, the long straws and unfermented vegetable matter remaining on the surface, should be removed as soon as the grass begins to rise vigorously, by raking, and carried back to the dung hill; in this case no manure will be lost, and the husbandry will be at once clean and economical.

It is the language of a lazy farmer, to say, that any species of manure is objectionable because it produces weeds. It is his business to extirpate them, and the oftener he is obliged to do it, the better for him, and for the crop. Some particular weeds may be excepted, such as the couch grass and other perennial ones.

"In cases when farm-yard dung cannot be immediately applied to crops, the destructive fermentation of it should be prevented as much as possible. The principles on which this may be effected have been alluded to.

"Watering dung-hills is sometimes recommended for checking the progress of fermentation; but this practice is inconsistent with just chemical views. It may cool the dung for a short time, but moisture is a principal agent in all processes of decomposition. Dry fibrous matter will never ferment. Water is as necessary as air to the process; and to supply it to fermenting dung, is to supply an agent which will hasten its decay.

"In all cases where dung is fermenting, there are simple tests by which the rapidity of the process and consequently the injury done may be discovered.

"If a thermometer plunged into the dung does not rise to above 100 degrees of Fahrenheit, there is little danger of much æriform matter flying off. If the temperature is higher the dung should immediately be spread abroad.

"When dung is to be preserved for any time, the situation in which it is kept is of importance. It should if possible be defended from the sun. To preserve it under sheds would be of great use; or to make the site of a dung hill on the north side of a wall. The floor on which the dung is heaped, should if possible be paved with flat stones; and there should be a little inclination from each side towards the centre, in which there should be drains, connected with a small well, furnished with a pump, by which any fluid may be collected for the use of the land. It too often happens that a dense mucilaginous and extractive fluid is suffered to drain away from the dung-hill, so as to be entirely lost to the farm.

"*Street* and *road dung* and the *sweepings of houses* may be all regarded as composite manures; the constitution of them is necessarily various, as they are derived from a number of different substances. These manures are usually applied in a proper manner, without being fermented.

"*Soot*, principally formed from the combustion of pit coals, generally contains substances derived from animal matters. This is a very powerful manure. It affords ammoniacal salts by distillation, and yields a brown extract to hot water of a bitter taste. It likewise contains an empyreumatic oil. Its great basis is charcoal in a state in which it is capable of being rendered soluble by the action of oxygene and water.

"This manure is well fitted to be used in the dry state, thrown into the ground with the seed, and requires no preparation.

"The doctrine of the proper application of manures from organized substances offers an illustration of an important part of the economy of nature, and of the happy order in which it is arranged.

"The death and decay of animal substances tend to resolve organized forms into chemical constituents; and the pernicious effluvia disengaged in the process seems to point out the propriety of burying them in the soil, where they are fitted to become the food of vegetables. The fermentation and putrefaction of organized substances in the free atmosphere are noxious processes; beneath the surface of the ground they are salutary operations. In this case the food of plants is prepared where it can be used; and that which would offend the senses and injure the health, if exposed, is converted by gradual processes into forms of beauty and usefulness; the fœtid gas is rendered a constituent of the aroma of the flower, and what might be poison becomes nourishment to animals and to men.

MAPLE, *acer*. Though Mr. Miller reckons nine distinct kinds of maple, I know of but two that are usually to be found in this part of the country.

One of these kinds is vulgarly called red maple, *Acer rubrum*. It is a very quick growing tree, and therefore ought to be encouraged in forests, especially where a quick profit is desired. But the wood is soft and white, not inclined to burn well till it is dried. As timber, it is valued chiefly by turners, by farmers for ox yokes, and for cabinet-work.

The other sort, called rock-maple, *Acer saccharinum*, is much harder and heavier, and an excellent wood for fuel, being inflammable in its green state, and durable in the fire. Both sorts will quickly decay, when exposed to the weather; the latter is less durable than the former. The sap of the white maple is drawn by many for sugar. But it yields little in comparison with the other.

It is of the sap of the rock-maple that an excellent sugar is made, which is no small advantage to the planters in the wilderness, where the trees are plenty, and the wounding and injuring them is not considered as any loss. But a farmer that wishes his trees to live and grow should not tap them for their sap; because it stints them in their growth, and often causes them to decay and rot. The best method of tapping has been lately found to be by boring the trees: So that the discharge of the sap may be stopped at pleasure with a peg, as there may be occasion. "When a plenty of sap is collected, you should have three kettles of different sizes. Fill the largest kettle with sap. To six gallons of sap put in one heaped table-spoonful of slacked lime, which will cause the sugar to granulate. Boil the sap in the large kettle, taking off the scum as it rises, till the quantity is so diminished that the second kettle will hold it. Shift it into the second kettle, and fill the large kettle with fresh sap. Let both boil till the third or smallest kettle will hold the sap contained in the second kettle. Shift it into that, and the sap in the first into the second, and fill the first with fresh sap. Boil the sap in the smallest kettle, till it becomes ropy, which you will know by taking out a little with a stick, and trying it between your thumb and finger. Put it into the cooler, and keep it stirring till the next parcel is done, and put that into the cooler, and continue the stirring. When the third parcel is ready, put that also into the cooler, with the other, and stir the whole smartly till it granulates. Put it into moulds. Earthen ones are best. Wooden ones are made by nailing or pinning four boards together, so shaped as to make the mould one inch diameter at the bottom, and ten or twelve inches at the top. The length may be two feet, or two and a half. The moulds must be stopped at the small ends. The sugar must then be put into the moulds. Next morning the stoppers must be taken out, and the moulds put on troughs to drain their molasses. In the evening the loaves must be pierced at the small ends, to make them run their sirrup freely. This may be done by driving in a wooden pin, shaped like a marling-spike, three or four inches up the loaf: After which they must be left to drain their molasses, which will be done in a shorter or longer time, according as the sugar has been boiled."—*American Museum.*

It is practised in England, to plant a large sort of maple on the margins of plantations against the sea, as they thrive well in such situations, and serve to screen the plantations of other kinds.

Mr. Miller says, "All sorts of maple may be propagated by cuttings. And that if they be cut from the trees before the buds begin to swell, and before the ground be fit to receive them, they may be wrapped in moss, and put in a cool place, where they may be kept a month or five weeks without injury." The trees may also be propagated by sowing the seeds, commonly called keys.

MARE. Breeding mares should be free from diseases; and have good eyes; because the colts are apt to inherit their distempers. They should be the strongest, best spirited, and well shaped; not of any bad colour. If any defects are dispensed with, the mare and the stallion should by no means have the same defects. In such case there can be but little prospect that the issue will be good. Some say they should not breed with stallions of the same blood. Crossing the breed is said to be of great consequence. Mares should not be suffered to breed till after four years old; and the best time for them to take horse is about the latter end of June, then they will not foal till the same part of the month of the following May, when the grass will be grown, which is better to make mares give milk than dry food is.

Mares with foal should be housed the earlier in the fall, and fed well till foaling. For the last month or two before foaling, they should not be ridden swiftly, nor be put to draw at all nor to carry heavy burdens on their backs.

MARKING *of cattle.* As one man's cattle, horses, and sheep, have very often such a resemblance to those of another, that they cannot easily be distinguished; and as they often graze together on commons, or in common pastures, marks for these different animals have been found necessary.

I have known no other marking used for horses than branding with a hot iron, on the shoulder or thigh. As these marks are not ornamental, most persons choose that their horses should have no marks, but natural ones, as they are called, such as particular spots on them of different colours, &c. In this case, these natural discriminations should be registered; because, in cases of dispute in law, no owner's word, who is a party will be taken as evidence.

The marking of neat cattle on the horn, with the branding iron, is so easily done, and without giving them pain, and is so permanent, that it should never be neglected. The brand should be made nearer the point than the root of the horn, on the outside which is most exposed to view, and not very deep, especially on young cattle, which have thinner horns than the older ones. Burning a horn through to the pith will hurt a creature, and will spoil the horn for certain uses afterwards.

The same kind of mark would be preferable for sheep, if they all had horns; as they have not, some other mark, alike suitable for all, should be used. Marking them on the wool is a bad practice. Some of the wool is spoiled and lost by it; and, at longest, it can last only to the next shearing; oftentimes not so long; and an uncertain mark is worse than none. The ear mark must be used, though the operation gives some pain to the animals. These marks may be distinct for a great number of flocks. And these marks should be matter of record.

MARLE. Although we do not know that marle has been discovered, or at any rate ever applied in New-England, yet we feel it a duty in a work like the present to introduce a manure, which has had so great a reputation, and produced so great effects in Europe.

Marle effervesces with acids; but this effervescence does not distinguish it from other calcarious fossils.

It has been said that a most infallible way to distinguish marle from other earths, is, to drop a piece of dry marle, as big as a nutmeg, into a glass of clear water, where it will send up many sparkles to the surface of the water, and soon dissolve into a soft pap. But I have found that some clays exhibit nearly the same appearances.

Sometimes the beds of marle are near the surface, but they are oftener found deep in the earth.

It is sometimes found on the banks of ditches, by means of the rank growth of weeds and grass on it. Boring with a long auger, or the screw borer, may discover where it is. Two kinds of marle were lately found at Penobscot in digging a well. Sometimes it is very dry and compact in the earth, but in some places almost liquid. Earths, thrown out of wells, if they have a clayey appearance, should always be examined.

Marles have been known to fertilize all kinds of soil, but light sandy ones more than any other. But as Dr. A. Hunter, by decompounding, has proved that marle consists of particles of lime-stone, mixed with clay or sand, or both; according as either of these ingredients is more predominant in it, the soil will be indicated for which it is most suitable. That which contains the least proportion of clay will be proper manure for a stiff soil, being of the most absorbent kind; that which has the largest proportion of clay should be applied to a sandy soil. To discover the proportion of these substances in marles, the same ingenious writer advises as follows:

"Having dried and powdered the marle to be examined, pour upon any given weight of it a small quantity of water. To this mixture, well shaken, add a little of the acid of sea salt, and when the consequent effervescence is over, add a little more. Repeat this addition at proper intervals, till no more effervescence ensues. Then throw the whole, with an equal or greater proportion of water, into a filter of grey paper, whose weight is known. When all the fluid parts have passed through, fill up the filter again and again, with warm water. By this means the dissolved particles of calcareous earth, adhering to the residue, or entangled in the pores of the paper, will be washed away, and nothing but what is really unsoluble will remain in the filter. This *residuum*, with the filter, must be completely dried and weighed. Then the difference betwixt it weight and the original weight of the filter, gives you the weight of unsoluble parts contained in the marle under examination. This being known, the proportion of calcareous earth in the same marle is evident. The proportions of clay and sand in it are discovered by subjecting the *residuum* to a proper elutriation. This operation is very simple, and performed thus: Having weighed the dry residue, mix and shake it well with a sufficient quantity of water. After allowing a little time for the subsidence of the grosser parts, let the water, with the finest particles of clay suspended in it, be gently poured off. When this is done, add more water to the remainder, and after sufficient mixture and subsidence, pour off that likewise. In the same manner repeat the operation, again and again, till the water comes over perfectly pure. The substance which then remains is sand, mixed perhaps with some flakes of talc; and whatever this substance wants of the weight of the residue employed, is the weight of pure clay carried away by the water in the process of elutriation."—*Georgical Essays.*

If five parts in six prove to be calcareous in a piece of marle, the lime is predominant, and it is fit for the stiffer soils; if two-thirds only be calcareous, and the rest clay, it is fit for a sandy soil, &c.

The calcareous part of marle does not produce so quick an effect as lime, when used as manure; because the latter is burnt, and flakes suddenly. This seems to be the true difference, which is not essential; because the calcareous part of marle gradually flakes in the earth without burning. Like lime, it attracts and imbibes the acids of the earth and air, forming a salt, which dissolves the oils, increases the pasture of plants, and prepares the food of plants to enter their roots.

The quantity of marle to be applied to an acre is about sixty loads. Some sandy soils may bear more of the clay marle; rich soils need not near so much, of the kind of marle which suits them.

Marle should be mellowed by the frost of one winter before it is buried in the soil; even in this case, it will not fertilize the soil so much the first year as afterwards. Some marles do not produce their full effect till the third year, as they dissolve slowly. Some say the good effect of one full dressing with marle will last thirty years.

As good soils may be overdone with this manure, it is better to err at first in laying on too little than too much. More may be added at any time. As the principal effects of marle are like those of lime, it is not to be expected that marling a second time will have so good an effect as the first. This observation is said to be confirmed by experience.

There is another sort of marle no less valuable than the former kind; and much used in old countries. It is composed chiefly of broken shells, which were undoubtedly once the shells of marine animals, mixed with a proportion of sand. It sometimes also contains a mixture of moss and decayed wood.

This marle is usually found under moss, or peat, in low sunken parts of the earth; and especially those which are nigh to the sea, or considerable rivers. Mr. Mills says, "Whoever finds this marle finds a mine of great value. It is one of the best and most general manures in nature; proper for all soils, and particularly so for clay." This sort of marle, as well as the other, may be easily found by boring. It has been sometimes discovered by ant hills, as these insects bring up some small pieces of shells from their holes. One would think that this country must be furnished as plentifully as any other with this kind of marle; whether we suppose the beds to have been formed by the general deluge, by the raging of the sea and inundations since that great event, or by the shifting of the beds of rivers.

The goodness of this marle depends upon the shells, which are the principal, and sometimes almost the whole that it contains. It is much of the nature of lime, and will go further than other marle. It effervesces strongly with all acids.

MARSH, according to Dr. Johnson, a fen, bog, or swamp. In this country the word is used only to signify flat land, bordering on the sea, and lying so low as to be often overflowed by the tides, when they are fullest.

Marshes are distinguished into high marsh and low marsh. The former bears a very short grass, but in many places very thick; the latter produces a tall rank grass, called thatch. Both these sorts of grass are too highly impregnated with salt to be a constant food for cattle; but the long grass is salter than the short, as it is oftener wetted with sea-water during its growth.

It is esteemed healthy for horses, cattle, and sheep, to have some of this sort of land in their pasture; or to be turned, now and then for a few days, into a marsh. At least it saves the trouble and expense of giving them salt. In England, it is thought to save sheep from that fatal distemper, the rot.

Marshes are certainly the richest of our lands, as appears by the astonishing degree of fruitfulness, apparent in those pieces from which the sea has been excluded by dikes. Marsh may be so far improved by diking and tillage, without manuring, that instead of producing less than one ton of salt hay per acre, it shall produce three tons of the best kinds of hay. The value of this soil must needs be great, as it is not exhausted by cropping and needs no manure, unless it be sand, or some other cheap substance, to dry and harden it.

Some marshes require a dike to exclude the sea, long in proportion to the land it contains; others a short one, as where the marsh is narrowest towards the sea. He that possesses a marsh of the latter kind, can undertake no business that will be more profitable than diking it. Two men can easily build a rod of dike upon high marsh in a day. Through the hollows and creeks, more work will be required.

If a marsh, after it is diked, should be rather too wet for tillage, a ditch should be made round by the upland to cut off the fresh water, both above and below the surface, and lead it to the outlet or sluice. See *Dike*, and *Sluice*.

MATTOCK, a pickaxe. This is a useful instrument in sinking wells, digging trenches, ditches, &c.

MEADOW, grass land for mowing. In this country the word is seldom used to signify upland mowing ground, but that which is low and moist, and seldom or never ploughed. In other countries it is the name of all mowing grounds.

Too much or too little moisture is hurtful to these meadows. Those that are apt to be too wet should be made drier by ditching or by draining, if it be practicable. They may be made drier also by spreading sand, gravel, or coal dust, upon them: At the same time, their fruitfulness will be increased, and better kinds of grass may be introduced.

When they are become dry, they should be ploughed and tilled, if the soil be not a tough clay with only an inch or two of black mould above it. In this case, I think a low meadow should not be ploughed at all. Instead of ploughing, perhaps, it would be better to cut away the hillocks and unevennesses; which by rotting in heaps, or burning, may be converted into good manure for the soil. And to increase the thickness of good soil, let sand and other earths, with dung, be spread over it.

When the soil is a loose crumbly clay, such as is found under some meadows, such a meadow may be converted to tillage land with great advantage.

Flooding in the spring not only enriches the soil of meadows, but makes them bear a sharp drought better. It causes the grass to grow so rapidly that the soil is sooner screened from the scorching heat of the sun.

Particular care should be always taken to keep cattle out of meadows in the spring and fall, when they are very wet and soft. For they will so break and spoil the sward with their feet, that it will not be fit for mowing, nor bear more than half a crop. All the fall feeding of such land should be over, before the heaviest rains of autumn. In the spring, no hoof should, by any means, be suffered to go upon a soft meadow. It occasions so much loss and damage, that a farmer had better give treble price for hay to feed his cattle, or buy corn for them, than to turn them in, as some do, to eat the grass that first springs, and which has but little more nourishment in it than water. No husbandry can be worse, if husbandry it may be called.

Meadows that bear poor water-grasses should be mown rather before the grass is grown to its full size. The hay will be so much sweeter and better, that what it wants in quantity will be more than made up in its quality. And the loss of quantity may perhaps be made up in fall feeding; or else a second crop may be taken.

I have long observed that heavy rains commonly fall before the end of August, by which low meadows are often flooded. Therefore, there is danger in delaying to mow them till it is so late. The crop may be either totally lost, or men must work in the water to save it in a damaged condition.

MEASLES, a disease in swine. The eyes are red and inflamed, and the skin rises in pimples, and runs into scabs. To cure a swine of this disease, take half a spoonful of spirit of hartshorn, and two ounces of bole armoniac, mix it with meal and water, and give it him in the morning when he is hungry. Repeat the dose every day, till he is cured, which will be in four or five days.

MEAT. *Preservation of.* Meat may be preserved fresh many months, by keeping it immersed in molasses. A joint of meat, or any provision, suspended in a flannel bag will keep sweet much longer than by most of the modes commonly practised. The cooler and dryer the meat is, when the flannel is put round it, the better, and the flannel should be perfectly clean. Fresh meat put in a close vessel, containing vinegar will be preserved a considerable time. Tainted meat may be rendered good, by pickling it in pearl-ash water some time. Before it is cooked, however, it should be dipped in vinegar a short time, and then salted in brine. When meat has become in some degree putrid it may be cured by putting it into a vessel and boiling it, and skimming off the scum. Then throw into the saucepan a burning coal, very compact and destitute of smoke; leave it there for two minutes, and it will have contracted all the smell of the meat.

When meat has become tainted, after having been pickled or salted down, the following process, it is said, will cure it: Take a sufficient quantity of charcoal, and after taking out the meat, and throwing away the offensive pickle, repack it in the barrel, lay pieces of charcoal between the pieces of meat, and make a new pickle, adding a little salt petre. In about five or six days the meat will become as sweet as when it was first packed. See *Salting of Meat*.

MELON, a pleasant tasted, cooling fruit. It grows best in a warm climate; and is large and excellent in the southern States. But they will ripen in New-England, in the common way of planting; but are not so large, nor so early in the most northern parts. Some improvement has lately been made in this fruit, by bringing seeds from the southward. Whether this will be a lasting advantage time will shew.

Melons grow best on a sandy loam, which has a warm exposure to the south or south-east. The vines should be sheltered against cold winds which stop their growth; and against boisterous winds from any quarter which will hurt them, by disturbing and displacing their vines.

A good manure to be put under melons, is an old compost of good loam. with the dung of neat cattle or swine. The ends of the runners, and the fruit latest formed, should be taken off, that the fruit first formed may have more nourishment, grow larger, and arrive to the greater perfection. To raise melons on hot beds, under frames, or under hand glasses, see *Gardener's Dictionary*.

METHEGLIN, a pleasant fermented liquor, made of honey and water. It is made thus: Put so much new honey into spring water, that when the honey is dissolved, an egg will not sink to the bottom. Boil the liquor for an hour. When cool, barrel it up, adding a spoonful of yeast to ferment it. Some add ginger half an ounce to a barrel, and as much cloves and mace; but I have it very good without any spices. One hundred weight of honey will make a barrel of metheglin, as strong as good wine. I once had a barrel made with 90 pounds of honey. After fermenting and fining, it was an excellent liquor; some part of which I kept bottled several years; it loses the honey tast by age, and grows lighter coloured: But on the whole it does not improve by age, like some liquors.

MICE, a well known genus of quadrupeds, troublesome to all housekeepers, but more especially to farmers, and those who keep quantities of grain in their houses, or in granaries. Farmers should know the best ways of opposing their depredations, and of destroying them. The field mouse eats the bark of trees in nurseries and young orchards, when snow is on the ground, and mostly when it is deep. A good way to prevent this mischief is to tread down the snow, and make it very compact, about the stems of the trees. And though laying mulch about the roots of trees be good for the trees, it occasions the mice to increase; therefore I do not go into that practice, while the trees are small, and have a smooth bark. It is only while the trees are young that mice eat the bark.

It has been said of late that the application of tanner's spent bark is an admirable substitute for the mulch so much recommended by early writers, to be laid about the roots of trees to keep the ground open. The evidence in favour of it is very strong, and it is certain that it will afford a much less favourable shelter to field mice. From its structure, it is impossible for them to burrow into it, and to form nests, and it is probable that it will produce all the desired advantages of keeping the roots free from plants which will exhaust the soil; at the same time that it will secure our trees from the effects of our severe droughts. We think that sufficient attention has not been paid to the difference between *our* summers, and those of any of the northern parts of Europe. Our sky is more unclouded—our droughts of longer continuance, and therefore the rule applicable in Europe, will not answer for this country. In winter the same distinction ought to be kept in mind. Our frosts are so penetrating, and so deep, that animals, accustomed to live on vegetable matter, are driven from the roots of grass, to subsist on the barks of trees, and the injuries thus inflicted are more lasting, and more severe than any, which the European farmer encounters. If the application of tanner's bark should prove a successful remedy against the attacks of field mice, and of drought, we shall owe much to the intelligent farmer of Massachusetts, who first introduced it into notice. There is scarcely any town without a tan-yard, but in defect of tanner's bark, it has been suggested, that the breakings of flax, or hemp, would answer the same purpose. We have not so much confidence in this, except against insects, which make a lodgment in the earth. Against them it would seem to be an effectual preservative, if it were possible to procure it in sufficient quantities. See *Orchard*.

Take a spoonful of flour, mixed with some scrapings of old cheese, and seeds of hemlock, made as fine as possible. Set it where the mice haunt. If it be set in a house, let it not be in the same apartment with any thing that is to be used as the food of man. This mixture will destroy all the mice that eat it.

But since many fear to use poison, they may take them alive in wire cages. However, instead of the round ones which are commonly used, I would recommend square ones, enclosed in thin wooden boxes, with a hole in the box against the entrance of the cage; because a mouse will not so readily enter into a place where he sees another confined. The bait may be a rind of cheese scorched, made fast to the centre of the bottom of the cage, and so far from the hole that a mouse cannot reach it till he has got quite into the cage. For if he should stick in the passage, he will prevent the entrance of others.

MILDEW. It has been ascertained by discoveries since the date of Dr. Deane's writings on agriculture, that this disease is occasioned by a minute parasitic fungus or mushroom called by botanists Uredo, and Puccinca, on the leaves, stems, and glumes or chaff of the living plant. The roots of the fungus, intercepting the sap, intended by nature for the nutriment of the grain, render it lean and shrivelled, and in some cases rob it completely of its flour; and the straw becomes black and rotten, unfit for fodder.

The same fungus is generated on many other vegetable substances besides wheat. Those receiving the infection at different seasons of the year, form, as it were, conductors from one to the other,in which fungi germinate, effloresce,disseminate and die, during the revolutions of the seasons. The fungus having arrived at maturity in the spring on a few shrubs, bushes or plants, its seeds are taken up the next humid atmosphere, (hence the erroneous idea that the rust or mildew is caused by fog alone) wafted into the adjoining fields, and the nearest wheat is sure to suffer the most from it. In damp weather also, its seed is more immediately received into the leaves of trees and shrubs, or into the barks and fruits, or the stems of plants, through the medium of those valves or mouths, with which nature has supplied them,for the admission of moisture.

The remedies against Rust or Mildew are, 1. Cultivating hardy sorts of wheat; 2. Early sowing; 3. Raising early varieties; 4. Thick sowing; 5. Changes of seed; 6. Consolidating the soil after sowing; 7. Using saline manures; 8. Improving the course of crops; 9. Extirpating all plants that are receptacles of rust; and 10. Protecting the ears and roots of wheat by rye, tares and other crops. The above remedies are enlarged upon by Sir John Sinclair in "*The Code of Agriculture*," but his observations are too voluminous to quote in this place. We will give, however,his 10th remedy, as follows:

"A curious and most important circumstance, connected with the rust in wheat remains to be stated. In the northern counties of England, where it is the practice to sow what they call *meslin* (blind corn) or a mixture of rye and wheat, it has been there remarked, that wheat, thus raised *is rarely infected by the rust.* It is singular that the same circumstance has been observed in Italy. In an account drawn up by Professor Symonds,on the climate of that country, it is recorded that "wheat mixed with rye or tares escapes unhurt." It would appear from tares being so useful that the seed of the fungus must be *taken up by the root*, and that if the root be protected it is sufficient. This seems to be so countenanced by other circumstances, as that by treading the ground, and thick sowing of crops of wheat, the crop is less liable to be affected by this disease; the access of the seeds of the fungi to the root being rendered more difficult. Mr. Knight is decidedly of opinion that the disease is taken up by the root, and indeed if it were introduced at the ear of the plant, how could it descend, and infect solely the stem, which is the case unless when the disease is inveterate.

"If a field be evidently affected, and the progress of vegetation stopped, the only way to preserve the straw and the grain, if any has been formed from being entirely lost, is to cut it down immediately, even though the crop should not be ripe. The straw is thus preserved either for food or litter; and it is maintained that any nourishment in the stem, will pass into and feed the grain, and make a greater return than could well be expected.

It has been recommended to sprinkle wheat while growing, which appears to be in any degree affected with this disease, with a solution of salt and water, which may be applied by means of a mop. The sprinkling should be several times repeated, so that every part of the plants may be wetted, and it is said that wherever the brine touches the rust disappears. Sir John Sinclair observes that " In the course of a most extensive inquiry into the causes of mildew in wheat, and the means of its prevention, it appeared, that a farmer in Cornwall was accustomed to manure his turnip land with the refuse salt from the pilchard fishery; and that ground, thus treated was never liable to the mildew, though it infested the neighbourhood.

A gentleman, who is at once a scientific and practical farmer, and who has had great experience on this as well as most other subjects connected with the agriculture of New-England, observes that "Mildew is the principal obstacle to the growth of wheat and rye in New-England, and it is probable that it is connected with an exposure to the desolating winds, which blow from the ocean. It is probable that nature has pointed out to us the true limits, within which cereal or graminiferous plants can be advantageously raised, and that those to whom this blessing is denied, should seek to obtain an equality, by confining their attention to those roots, and plants, which are not so frequently subject to such evils. We can say with confidence that no project or system has yet been discovered to correct, or even mitigate the evil."

Perhaps, however, some of the remedies of Sir John Sinclair, above mentioned may be worth the trial, particularly that of early cutting.

MILK, a nutritious liquor, which nature prepares in the breasts of female animals, for the nourishment of their young. The milk of cows is that with which the farmer is mostly concerned.

That the greatest quantity of milk may be obtained from cows, they should not calve out of the right season. April is a good time of the year, if the calves are to be reared; if not, perhaps May is better, being rather more favourable to the diary. But that cows may give plenty of milk to nourish their calves at this season, they should not be wholly confined to hay, or any other dry meat: But be daily fed with some kind of juicy food, such as potatoes, turnips, carrots, &c. until they have plenty of grass.

In feeding milch cows, the flavour of the milk should be attended to, unless it be when their calves suck all their milk. Feeding them with turnips is said to give an ill taste to the butter made of the milk. The decayed leaves of cabbages will undoubtedly give a bad taste to the milk, though the sound heads will not. There is no fear of potatoes and carrots having any bad effect upon the milk in this way. The quantity of milk is greatly increased by potatoes, but it becomes thinner. Some think carrots have a tendency to dry up the milk in cows; but I have assured myself of the contrary by much experience.

The milk of cows in summer is sometimes made very bitter by their feeding on ragweed, which they will do, when they are very hungry. To prevent this evil it is only necessary that they should not be forced to eat it by the want of other food.

" Milk consists of three parts, namely *caseous*, *butyrous*, and *serous*. The first comprehends the grosser earthy particles, which serve to suspend the butyrous part; and which, when coagulated by art, are formed into Cheese. The second ingredient comprises the *butyraceous* or oily particles, or cream, which float on the surface of milk, and are by agitation converted into Butter. The *serous* are the more watery parts, constituting what is called whey, and serving as a vehicle for the two before stated. The most wholesome milk is that which contains a due proportion of the three constituent parts.

" *To preserve milk*, add five grains of magnesia to a quart. To ascertain the quantity of cream, which will be afforded by various specimens of milk, procure half a dozen glass tubes, closed at one end, 9 inches long, and half an inch in diameter; fix them in a wooden frame; mark on each 4 several inches from the top as a scale. The cream settling at the top, after they have been filled with milk, will shew the quantity." *Domestic Encyclopedia.*

MILLET, *Panicum*, a round yellowish white grain, which grows in panicles at the top of the stalk. The stalks and leaves are like those of Indian corn, but smaller. It grows to the height of three or four feet. A sandy warm soils suits it best. It should be sown about the middle of May, in drills three feet apart. The plants should be so thinned at the first hoeing as to be about six inches apart in the rows. It will produce as large crops as Indian corn, and bears drought admirably well. Cattle are fond of eating it green, preferring it to clover. A crop of it sown thick, and mowed green, would be excellent fodder.

Some say a crop may be obtained by sowing it at about mid-summer. Perhaps it may be so in hotter climates. I tried the experiment in the 44th degree of latitude, and the crop was little better than mere chaff, for want of continuance of heat to fill the grain.

This grain appears to be subject to no distemper; but when it is nearly ripe, the birds are apt to get a great deal of it, if it be not watched carefully.

The way to harvest it is, to cut off the panicles with a knife, near the uppermost joint of the stalk, put them into sacks or sheets carry them to the barn floor, and empty them into heaps, covering them with cloths. After lying five or six days, it must be thrashed and cleaned. It should be dried well in the sun, before it is stowed away in the granary; for it will not keep well with any moisture in it.

Millet is an excellent food for

fowls and swine; for the latter it should be ground into meal. Some mix it with flour in bread; but it is better for puddings. There is also a red sort of millet; but this I have never seen.

MOLES. The Hon. Jacob Rush of Philadelphia, from some experiments was of opinion, that dried cod-fish, cut into small pieces, and put into the earth, will drive away moles from gardens. Take a small round stick of about an inch in diameter, sharpened at the point, and perforate the ground in the roads of the moles; drop in a small piece of fish, and cover the hole with a lump of dirt, and your garden will be rid of these mischievous animals.

MOSS, *Lichen*, a sort of plant that is injurious to the growth of other plants in general. It was formerly thought to be an excrescence; but even the minutest kinds are now known to be propagated by seeds.

Low meadows are often infested with moss, which prevents the flourishing of the grass, and indicates the coldness and sourness of the soil. To cure meadows of moss they should be top dressed with lime, ashes, and other absorbent manures; as well as laid drier by ditching or draining. After which fire should be put to it at a time when it will burn freely.

Tillage lands, when they are laid down to grass, often become mossy, especially when they are too long in grass. Cold loamy soils are most subject to this evil. The moss on such land is often so small, as to appear only as a green mouldiness of the surface. But this mould consists of distinct minute plants, as well as all other mouldiness, as may be seen by the help of microscopes. If dressings of warm manures do not prove sufficient to clear the ground of this moss, it should be scarified, or harrowed, or else broken up and tilled. For if it be permitted to continue, it will rob the grass of most of its food.

A very long white or yellowish moss grows in wet swamps. Draining the swamps, and setting fire to the moss in a dry season, will commonly be sufficient to subdue it.

Moss on fruit trees is detrimental to their fruitfulness. "The remedy is scraping it off from the body and large branches, with a kind of wooden knife, that will not hurt the branches; or with a rough hair cloth, which does very well after a soaking rain. But the most effectual cure, is taking away the cause. This is to be done by draining off all superfluous moisture from about the roots of the trees. And it may be guarded against in planting the trees, by not setting them too deep in the soil.

"If trees stand too thick in a cool ground, they will always be covered with moss; and the best way to remedy the fault is to thin them. When the young branches of trees are covered with a long and shaggy moss, it will utterly ruin them; and there is no way to prevent it, but to cut off the branches near the trunk, and even to take off the head of the tree, if necessary, for it will sprout again. And if the cause be in the mean time removed by thinning the plantation, or draining the land, the young shoots will continue clear after this.

"If the trees are covered with moss in consequence of the ground's being too dry (as this will happen from either extreme in the soil) then the proper remedy is, the laying mud from the bottom of a pond, or river, pretty thick about the roots, opening the ground to some distance and depth to let it in. This will not only cool it, and prevent its giving growth to any quantity of moss; but it will prevent the other great mischief which fruit trees are liable to in dry grounds, which is the falling of the fruit too early." *Mortimer's Husbandry.*

"One of the greatest difficulties in the management of old pasture lands, is to prevent that immense growth of mosses, by which the finer species of grasses are apt to be overwhelmed. Drainage, and the use of rich composts, are in this case necessary, and tillage itself must be sometimes resorted to. Harrowing, and cross-harrowing, with a common harrow loaded with a weight, so as to go from one to two inches deep, with a sprinkling of grass seeds afterwards, and some lime, or well prepared compost are the most likely means of destroying the moss, and improving the pasture. Feeding sheep with oil cake, and allowing them to pasture on the land, has also been found effectual for the destruction of moss, and bringing up abundance of grass. But the radical remedy is to plough up such grass lands, upon the first appearance of moss, or before it has made any considerable progress." *Code of Agriculture.*

MOULD. "Mould, which contains a mixture of animal and vegetable remains, particularly from putrefaction, is an essential ingredient in all fertile soils. It arises from the decomposition of the roots and leaves and vegetables on grass land, and the stubble and roots of grain crops, on arable. This process on grass lands is very slow; for it is calculated that it proceeds at the rate of only one inch in a century. Considerable accumulations of this mould, are likewise produced from the leaves of trees, amassed for a number of years, and rotted on the surface. When in addition to vegetable, there are animal remains, in a state of decomposition, the soil is distinguished for its fertility." *Code of Agriculture.*

MOULDBOARD, that part of a plough which turns over the furrow. For ploughing green sward, an iron mouldboard is best: If it be wood it ought to be plated with iron to prevent its being soon worn through. See *Plough.*

MOW, a quantity of hay, or grain in the straw, piled in a barn for keeping. Ground mows are more liable to take damage by moisture, than mows upon scaffolds. Mows of grain should be laid upon the latter. The larger the mow, the drier the hay or sheaves should be of which it consists. See *Fodder.*

The following modes of preventing hay from heating in the bay or mow are recommended in the "Farmer's Manual" by Frederick Butler.

"Let your bays be open under the bottom for a free circulation of air; fill several large bags with hay, set them erect upon the floor of your bays, mow the clover around them with as little treading as possible; raise up your bags with the rise of your mow, and when your mow is finished, remove the bags; these openings will serve as ventilators, and secure your mows from heating. If you reserve your wheat or rye straw for this purpose, and cover your clover occasionally as you mow it with straw; your straw will not only prevent your mow from heating, but imbibe the moisture of your clover, and become valuable feed for your horses and cattle, and thus become a double saving. One bushel of salt sprinkled on your clover, as you mow it, will preserve it against heating, and doubly pay you in the value it will give to your hay."

MOWING, the operation, or art of cutting down grass, corn, &c. with a scythe.

They who have not been in their youth accustomed to do this work, are seldom found to be able to do it with ease or expedition. But when the art is once learnt, it will not be lost.

As this is one of the most laborious parts of the husbandman's calling, and the more fatiguing as it must be performed in the hottest season of the year, every precaution ought to be used which tends to lighten the labour. To this it will conduce not a little, for the mower to rise very early, and be at his work before the rising of the sun. He may easily perform half the usual day's work before nine in the morning. His work will not only be made easier by the coolness of the morning air, but also by the dew on the grass, which is cut the more easily for being wet. By this means he may lie still and rest himself during all the hottest of the day, while others who begun late are sweating themselves excessively; and hurting their health, probably, by taking down large draughts of cold drink to slake their raging thirst. The other half of his work may be performed after three or four o'clock; and at night he will find himself free from fatigue.

If the mower would husband his strength to advantage, he should take care to have his scythe, and all the apparatus for mowing, in the best order. His scythe ought to be adapted to the surface on which he mows. If the surface be level and free from obstacles, the scythe may be long and almost straight; and he will perform his work with less labour, and greater expedition. But if the surface be uneven, cradley, or chequered with stones, or stumps of trees, his scythe must be short and crooked. Otherwise he will be obliged to leave much of the grass uncut, or use more labour in cutting it. A long and straight scythe will only cut off the tops of the grass in hollows.

A mower should not have a snead that is too slender; for this will keep the scythe in a continual tremor, and do much to hinder its cutting. He must see that it keeps perfectly fast on the snead; for the least degree of looseness will oblige him to use the more violence at every stroke. Many worry themselves needlessly by not attending to this circumstance.

Mowing with a company ought to be avoided by those who are not very strong, or who are little used to the business, or who have not their tools in the best order. Young lads, who are ambitious to be thought good mowers, often find themselves much hurt by mowing in company.

Mowers should not follow too closely after each other: For this has been the occasion of fatal wounds. And when the dangerous tool is carried from place to place, it should be bound up with a rope of grass, or otherwise equally secured.

"Mr. de Lisle introduced in England, the mowing of wheat. The method is this: The scythe he uses is at least six inches shorter in the blade than the common scythe; and instead of a cradle, has two twigs of osier put semi-circular wise into holes made in the handle of the scythe, near the blade, in such a manner that one semi-circle intersects the other.

"By this method of mowing wheat, the standing corn is always at the left hand. The mower mows it inward, bearing the corn he cuts on his scythe, till it come to that which is standing, against which it gently leans. After every mower follows a gatherer, who, being provided with a hook or stick, about two feet long, gathers up the corn, makes it into a gavel, and lays it gently on the ground. This must be done with spirit, as another mower immediately follows." *Complete Farmer.*

As reaping is slow and laborious work, it would be right for our countrymen to learn this method of mowing their wheat; which will undoubtedly answer also for other sorts of grain.

MOWING GROUND, a name commonly given in this country to land that is mowed for hay; which being fit for either mowing or tillage, is occasionally used for the latter.

The generality of farmers, in this country, lamentably mistake their interest, by having too large a proportion of their lands in grass for mowing. Half the usual quantity with the best management, would produce as much hay as they need, a great deal more than they commonly get; besides saving them expense and much hard labour; and allow them to convert half their mowing land to tillage or pasture; especially to the latter, which is most wanted.

A New England farmer is not contented, unless he yearly mows over the greater part of his cleared land; because he supposes that if he does not, he shall be able to winter but a small stock. His grass on the most of his acres must needs be very thin, even when the seasons are most favourable; therefore, if a summer happen to be dry, the soil, which is so poorly covered as to retain neither dews nor rains, is unavoidably parched and bound. The grass, thus deprived of its nourishment, does not get half its usual growth in a dry season; and the crop turns out to be almost nothing. The distressed farmer, not knowing how to get fodder for his cattle in the ensuing winter, with severe labour or cost, mows his dead grass, and gets perhaps four or five cocks from an acre. He cannot sell off many of his stock, because of the general scarcity of hay; nor fatten them to kill, for want of grass; therefore he keeps them along poorly and pinchingly, till the ground is bare in the following spring; then, to save their lives, he turns them into the mowing ground, as soon as there is the least appearance of green grass. They potch the soil to the depth of six or eight inches, which is sufficient to prevent the growth of a good crop that year; as it sinks a great part of the surface to such a depth that it can produce nothing; tears and maims the roots which remain in their places; and leaves the surface so uneven, that if a crop of grass should grow, it could not be mown closely, if at all. Therefore, through want of hay, the soil and sward must be mangled in the same way the spring following; and so on from year to year perpetually. How absurd and ruinous this practice!

If our farmers would resolve they will mow but half the quantity of ground which they have mowed hitherto, I should think they might soon find their account in it. But it will be necessary that they should adopt a new kind of management, with respect to their mowing grounds.

In the first place, let them not lay down to grass for mowing, any lands that are quite exhausted by severe cropping; nor without manuring them well. Good crops of grass are not to be expected when there is no strength, or next to none, in the soil. Therefore the lands should be dunged when the grass is sown, unless we except clover and other biennial grasses. And even for these it is often quite necessary, always advantageous.

Also, the surface should be rolled after the seed is sown, to close the mould about the seeds, to prevent their being removed by strong winds, to prevent the surface from being irregularly torn by the frost of winter, and to make the soil smoother for mowing.

Grass land, by lying, is apt to become uneven, and knobby. For this reason the good farmers in England pass a roller over their grass-land every spring and fall. It gives the roots of grass a more equal advantage for nourishment and growth, and facilitates the mowing of the grass, and the raking of the hay.

When land becomes bound, or mossy, so as to diminish the growth of the grass, if it be not convenient for the farmer to break it up, it should be cut, or scarified by a spiked roller, or if the farmer does not possess this, by a heavy loaded harrow, when the ground is softened by rains, or by the coming out of the frosts. Then dressed with some short rotten manure suited to the soil; bushed, and a roller passed over it. There is no danger of destroying the roots of the grass by this operation. Though they are broken they will be speedily renewed; new offsets will be more plentifully formed, and the crops will rise with renewed vigour.

Let farmers keep their mowing land so completely fenced, that cattle and swine may be effectually prevented from breaking in at any time of the year. I think every one must be sensible of the necessity of this.

It is ridiculous to think of taking many crops of hay from any piece of upland, in uninterrupted succession, without affording it any manure. For it does not imbibe the richness of the atmosphere so plentifully as land in tillage. Grass land should, therefore, once in two or three years at least, have a dressing of good rotten dung, or of a compost suitable for the soil. But the best way is to do it every year. Autumn is the time for applying the manure, according to long approved practice. But a writer in the Georgical Essays recommends doing it immediately after the first mowing, when a second crop is expected, which will undoubtedly be the larger. Whenever it is done, a bush harrow should be drawn over the surface, which will break the small lumps remaining in the manure, and bring it closer to the roots of the grass. By this management, four or five tons of hay may be the annual produce of an acre. Or if the surface be not dunged, the crop should be fed off once in three years; that the excrements of the cattle may recruit the soil.

No cattle should, on any account, be turned into a mowing ground in the spring. The mischief they will do, will be ten times more than the advantage they can get. In the fall, neat cattle may get the aftermath: But sheep and horses may be apt to bite so close as to injure some of the roots. Therefore I think they should be kept out, especially after the grass comes to be short. Whatever dung is dropped by the cattle, should be carefully beat to pieces, and spread, before winter, or early in the spring.

These lands should never be fed so bare, but that some quantity of fog may remain on them through the winter. The snow presses it down to the surface, where it rots; it holds the rain water from passing off suddenly; and the virtue of the rotten grass is carried into the soil, where it nourishes the roots.

Grass lands, with such a management as is here recommended, would produce crops surprisingly large; especially in the northern parts of New-England, which are extremely natural to grass. The surface would be covered early in the spring with a fine verdure. The crops would cover the ground so soon as to prevent most of the ill effect of drought in summer. It would, by forming a close cover to the soil, retain most of the moisture that falls in dews and rains. So that a dry summer would make but little difference in the crop; and the rich lands would often produce two crops in a year.

On this plan of management, much labour might be saved in hay making; and the grass might all be cut in due season; not only because the farmer has more leisure, by having so much less mowing to do; but also because a good crop is not apt to dry up so suddenly, as a poor and thin one. The grass in our mowing grounds is often said to be winter killed. It is observable that this happens only in the little hollow places, where the melting snow towards spring forms little ponds of water. A cold night or two turns these ponds to cakes of ice, which lying long upon the roots chills them so much that they cannot soon recover. Or the ponds made by the thawing of the ice destroy the roots by drowning them; so winter flooding destroys all the best grasses. The grass, however, only of one crop is destroyed in the hollows; for it rises again by the midsummer, or autumn following.

Laying lands very smooth and level, according to the above direction, will do much towards preventing this evil. But if a field be perfectly flat, and apt to retain too much wet when it is in tillage, it should be laid down to grass in broad ridges or beds. I am acquainted with some farmers who have found advantage from this method. The trenches, or furrows between the beds, should be the breadth of two or three swarths asunder, that the grass may be mowed with the less inconvenience. It is near as much work to mow a half swarth as a whole one; which is a good reason why the beds should not be very narrow. Ten or twelve feet is a good breadth, as it is equal to two swarths.

MUD, a black or dark-coloured sediment, found at the bottom of ponds, rivers, creeks, ditches, and wet sunken places. It is mostly composed of a fine vegetable mould, mixed with the substance of perished vegetables, &c. and therefore it contains much of the natural food of plants.

In ponds and rivers, this sediment is made up of fine dust, together with a rich variety of other substances, which have been wafted in the air, and have fallen into the water; together with the subtilest particles of the neighbouring soils washed down into them by rains. That is supposed to be the richest mud, which is near to the borders, and which has been alternately flooded and fermented; as it will ferment when it lies bare, in some degree.

In rivers, and in long ditches that have currents, there is a greater proportion of soil in the mud. It has been brought down from soft, mellow lands, through which the rivers pass; and some of it doubtless from beds of marle, which are often found in the banks of rivers, and which readily dissolve in the water.

Some ponds are totally dried up in a hot and dry summer; and all ponds and rivers are so diminished by a copious evaporation, as to leave part, and the richest part, of their beds uncovered. And these beds, where there has been no rapid current, are always found to contain a rich mud. In some places it reaches to a considerable depth. This mud, though taken from fresh waters, has been found to be a valuable manure; more especially for dry, sandy and gravelly soils. I have known it to have as good an effect as barn dung, in the culture of Indian corn, upon such soils. The advantage of it is not found to be only for one season; it meliorates the land for several years. It restores to a high piece of ground what vegetable mould the rains, in a long course of years, have been washing away from it.

It is happy for the farmer that Providence has prepared for him these magazines of manure in all parts of the country. None but the stupid will let them lie unnoticed, or unremoved. When a dry autumn happens, the prudent farmers will be very industrious in carting mud up from evaporated ponds, and other sunken places in their farms, and laying it upon their light soils, especially upon high gravelly knolls; or into their barn yards, if the distance be not too great.

But with respect to using mud as a manure, the maritime farmers have the advantage of all others. For the sea oose, which appears on the flats, and in creeks and harbours, along the shores of the sea, has all the virtues of fresh water mud, with that of sea salt superadded, which is one of the most important ingredients in the composition of the best manures. I might add, that it abounds, more than any other mud, with putrefied animal substances. Much of these are contained in the sea itself: And innumerable are the fowls and fish that have perished upon flats since time began; and the component parts of their bodies have been inclosed by the supervenient slime.

Mud taken from flats where there are shell fish, or even where they have formerly lived, is better for manure, than that which appears to be more unmixed. The shells among it are a valuable part of its composition. If it abound much with shells, it becomes a general manure, fit to be laid upon almost every kind of soil.

That mud, however, which is a richer manure than any other, is taken from docks, and from the sides of wharves in populous towns. For it has been greatly enriched by the scouring of foul streets, and from common sewers; as well as from an unknown quantity of animal and vegetable substances, accidentally fallen, or designedly thrown into such places.

Sea mud may be taken up at any season, whenever the farmer has most leisure. It is a good method to draw it up on sleds from the flats in March, when the border is covered with firm ice. I have thus obtained mud from flats, with great expedition and little expense.

Mud that is newly taken up, may be laid upon grass land. But if it is to be ploughed into the soil, it should first lie exposed to the frost of one winter. The frost will destroy its tenacity, and reduce it to a fine powder; after which it may be spread like ashes. But if it be ploughed into the soil, before it has been mellowed, it will remain in lumps for several years, and be of less advantage.

A layer of mud will be no bad ingredient in a heap of compost. But it should be contiguous to a stratum of lime, if that can be obtained. But where this is wanting, new horse dung is the best substitute, to excite a strong fermentation.

The best method of managing all sorts of mud, were it not for increasing the labour, would be to lay it in farm yards, and let it be thoroughly mixed with the dung and stale of animals. When it is so managed, the compost is excellent, and fit for almost any soil, though best for light ones. Perhaps the advantage of it is so great as to pay for the increased expense of twice carting. For it will absorb the stale of cattle, and retain it better than straw, and other light substances.

MULBERRY, *Morus*, a well known tree, the leaves of which are the proper food of silk worms. The only mulberry, in any degree proper for the food, or successful culture of silk worms, is the white mulberry. All attempts to raise them on the leaves of the common black mulberry will be unsuccessful.

It would be right for us to propagate these trees, as it might be done with the greatest ease. We may do it by their seeds, or by layers, cuttings or slips. If we are not disposed to make use of them for the feeding of silk worms, they would pay for the trouble of rearing them, by their fruit and their timber. They suit our climate, and grow rapidly, at least in Connecticut, and in the western part of Massachusetts.

Possibly the time may come when we may be glad to make silk for our own use in this country. If this should happen, it will be regretted if there be no trees in the country from which the worms can be fed. They will grow well in a deep dry soil which is moderately rich.

MULCH, rubbish of decayed vegetables. Litter is a word of the same import.

MURRAIN. "Murrain, a contagious disease incident to cattle; it is known by the animals hanging down their heads, which are swollen; by short and hot breathing; palpitation of the heart; staggering; an abundant secretion of viscid matter in the eyes; rattling in the throat and a shining tongue.

"The murrain is occasioned by various causes, but principally by a hot, dry season, or a general corruption of the air. It raged about the middle of the last century in various parts of Europe, and carried off great numbers of cattle. The remedy then employed both for prevention and cure, consisted in a mixture of equal parts of gun-powder, salt, soot, and brimstone; one spoonful of this composition was given for a dose, and was hid down with warm water.

"In the 36th volume of *Annals of Agriculture*, the following recipe is inserted for the *murrain in hogs*:—A handful of nettles is to be previously boiled in a gallon of small beer, when half a pound of flour of sulphur, quarter of a pound of elecampane, three ounces of liquorice, and a quarter of a pound of anniseeds are to be added in a pulverized state. This preparation should be administered in milk, and the quantity here stated is said to be sufficient for six doses.

"But the most effectual preventive of this contagion is to keep the cattle cool during the summer, and to allow them a sufficiency of water: all carrion should be speedily buried: and as the feeding of those useful animals in wet places, or on rotten grass or hay frequently causes this malady, their food ought to consist of dry and sweet fodder."—*Domestic Encyclopedia.*

N.

NAVE, the middle part of a wheel, through which the axle passes. See *Wheels.*

NAVEL GALL, "a disorder on the top of the spine, opposite to the navel, whence the name. It is most commonly caused by an ill formed saddle, or want of good pads, and being neglected turns to a foul fungous excrescence; and sometimes, after long continuance, to a fistulous ulcer. While there is moisture and sensibility in the part, an ointment may be applied of quicksilver and turpentine; an ounce of the former or two ounces of the latter, rubbed in a mortar till they be well incorporated; and then spread upon tow. On each side of the spine, over the swelling, may be laid smooth dry pledgits, or bolsters, which may be girt round with a sursingle. But if the sore be dead and lifeless, a good sharp knife must be used to cut it to the quick; then let it be dressed according to the directions for the cure of wounds.

"*A sit fast* also proceeds from a saddle gall, and is another of the accidents that happen to the spine. It is dry and horney, and may be cured by anointing it first with oil of bays, until it turns soft; then by dressing it with quicksilver and turpentine, as above directed. This will make a cure, especially if the hard horney substance be gently scarified in some places." *Gibson's Farriery.*

NETTLE, *Urtica.* This genus of plants is said to comprise fifty-seven species. We shall mention only the common nettle, which has a fibre, said to be finer and stronger than that of hemp. Dr. Willich says, "The roots of the common nettle, when boiled, communicate a yellow tinge to yarn. But the most valuable part is its fibrous stalk or stem; which on being dressed in a manner similar to flax or hemp, has in some parts of Europe been advantageously manufactured into cloth. As, however, this plant requires a rich soil to obtain it any great quantities, and as a much greater degree of attention and accuracy is necessary in the operation of *retting*, than is requisite either for flax or hemp. Dr. Anderson is of opinion, that the cultivation of the nettle will be attended with difficulty."—*Domestic Encyclopedia.*

The cultivation of this plant has been patented in the United States, and the exclusive right given to a Mr. Whitlow. But if it was cultivated in Europe, described in scientific works, and applied to the uses, which Mr. Whitlow proposes, we do not perceive how he can secure any exclusive right to its culture or application. For a further account of this plant and the uses to which it is proposed to apply it. See *Massachusetts Agricultural Repository*, vol. III. page 41 to 46.

The species for the manufacture of which Whitlow obtained a patent, was, as he says, a new one—which he called Urtica Whitlowii.

NEW HUSBANDRY, *drill husbandry*, or *horse-hoeing husbandry*. It chiefly differs from the old husbandry, in this, that the soil is tilled while the plants to be nourished are growing in it. This mode of culture was introduced into England, by the ingenious Jethro Tull, Esq. who wrote largely and repeatedly on the subject. His volume in folio, entitled, *New horse-hoeing Husbandry*, was published in the year 1731. An Essay on the same subject, in the year 1733. A Supplement to the Essay, in 1735. Addenda, and Conclusion, in 1738, and 1739. This gentleman expended as it were his whole life, in zealous and benevolent exertions to convince mankind of the great utility of his new system, and directing them in the practice of it. But he had the mortification of finding, that only here and there an enterprising genius adopted it in practice. And though more than sixty years have now elapsed, since he made it public, it is so far from having become the general practice of farmers in that country, that there is no reason to suppose that it ever will: Although it has been recommended, and further explained and improved, by writers of note in several nations.

The author of this husbandry meant to apply it chiefly to wheat, as being the most important kind of corn. The new husbandry differs from the old in the manner of preparing the ground for a crop, and in the manner of sowing the seed. The ground is ploughed into ridges, or beds, five or six feet wide, and smoothed with harrows. Instead of sowing at random with the hand, or broad-cast, as it is called, the seed is dropped by a drill, in straight lines, in little furrows about two inches deep. Either two or three such rows are on one bed, eight or nine inches apart; and the seeds are closely covered in the furrows, by a small harrow annexed to the drill.

Mr. Tull invented a drill, or drill plough, on a new construction. With this machine one may sow such a quantity of seeds, and as many rows as may be thought necessary, lay the seeds at a convenient depth, and cover them nicely, only by drawing the machine once along the ridges.

As soon as the plants are a few inches high, the horse-hoe is introduced, which differs but little from a horse-plough, excepting in the manner of connecting it to the horse that draws it. With this plough, passing it within three or four inches of the rows, the earth is turned from the rows into the intervals or alleys, so that the furrows meet each other, and form a sharp ridge. This is the first hoeing, and is performed late in autumn, just before winter. It lays the young plants so dry, that it is thought they are in no danger of being killed by the frosts of winter. But some improvers on this system have recommended omitting one of these furrows, or if both be ploughed, to turn back one of them towards the row before the hard frosts of winter; lest the ridges should be too much in danger of being washed away by rains, and the young plants removed. This seems to be a real improvement upon Mr. Tull's method.

Early in the following spring (they say in March, but it must be April in this country) the earth is turned toward the rows; then in May, from them; and lastly, in June, it is turned back to the rows, and partly against the stems, when the grain is just out of blossom; which last ploughing is thought to do more service than any other, as it greatly helps to fill out the grain; and must not, therefore, on any account, be omitted.

Each of the ploughings must be very deep, so as to keep the ground very loose and open. But care must be also taken to uncover plants that chance to be buried by the plough; to weed the grain once or twice in the rows, and to stir the earth between the rows, with a prong-hoe or hand-hoe, as often as the intervals are ploughed, or horse-hoed.

The advantages of this method of culture are said to be these: That indifferent land will produce a good crop, which would produce little or nothing in the old way; that a good crop of wheat may be raised each year from the same piece of ground, without impoverishing the soil, as the intervals are always followed; that there is no need of manuring the land at all, as the extraordinary tillage will answer the same end as manure, and at less expense; that there will be no crop missed or prevented by a year of fallow, which must take place every second year in the old way of cultivating wheat, to prevent exhausting the soil; that the crops will be larger, better and fuller grain by far, and entirely free from the seeds of weeds.

The editors of the last edition of Mr. *Tull's horse-hoeing Husbandry*, by a computation of the expense and profit of the old husbandry and the new, and comparing the accounts, make the clear profit of the latter appear to be more than double to that of the former. This may be seen at large in the *Complete Farmer*, under the article *Husbandry*. Other ingenious writers in Great Britain, since have written in confirmation of this opinion. See *Encyclopedia*, article *Agriculture*.

I do not at all scruple the fairness of the computations; nor the accounts of writers in other countries to the same purpose. But there is no arguing with any certainty from the advantage of the new husbandry in England, or other parts of Europe, to the advantage of it in this country. Because, in the first place, labour is more than twice as dear in this country; and that there is a greater quantity of labour required in the new husbandry than in the old, is very obviously true. There are at least two or three ploughings extraordinary to a crop, besides weeding and hand-hoeing; and weeders will not accept of the weeds they pull as sufficient pay for pulling them, as poor women sometimes do in the old countries.

Another reason for suspecting that the new husbandry may not answer so much better than the old in this country, when applied to wheat and rye, is, that these grains are here very subject to blasting; and the later they ripen, the more they are in danger of this distemper. Hoeing of grain will cause it to ripen later, as may be seen in the border of a field that is contiguous to hoed ground. The plants that stand nearest to the hoed ground retain their greenness much longer than the rest of the grain, because they are more plentifully fed. Hence there appears to be some reason to doubt of the advantage of hoeing wheat and rye in this country.

But if there were no weight in this, nor in the foregoing argument, yet the difference of climate must be taken into consideration. Our lands are hoven and mellowed by the frost of every winter, to a greater depth than the hoe-plough can ever stir them, by which the roots of winter grain are often hoven out of the soil; but in England the ground seldom freezes to half the depth that a plough goes. Therefore, the most forcible argument in favour of the new husbandry, which is used by its advocates, will not so well apply in this country; which is, that the ground settles and becomes very compact, during the long continuance of a crop of grain upon it. I see no reason to doubt but that our extraordinary degree of frost may, on the whole, have nearly as much effect towards loosening and breaking the soil in tillage ground, as one ploughing has. But this bye the bye.

Not only is the success of the new husbandry in this country for the above reasons uncertain; but there are several disadvantages and inconveniences, attending this husbandry, which are common to all countries. One of these disadvantages is the cost of the drill-plough. This is every where a material objection to the new husbandry in the minds of common farmers. And the curious and complicated structure of this machine, which renders it liable to get out of order, is no small inconvenience; for common labourers are not expected to have skill enough to rectify, or repair it. Besides, the accuracy of the work of drill sowing requires so much thought and attention, that the ignorant and careless, who are apt to despise new inventions, will not perform it in the best manner. So that a gentleman must always do his own sowing himself, if he wishes to have it done well. And not every gentleman who has a farm will be disposed to submit to this employment. Neither does the drill plough perform well on sidling situations and declivities. To which it may be added, that there are many kinds of seed which it is next to impossible to sow well with this machine. Such are all the hooked, winged, flat, long shaped, and extremely light seeds ; such as those of carrots, parsnips, lettuce, &c. It will not well deliver any but those which are ponderous, smooth, and so round, or regular shaped, as to be easily put in motion.

These difficulties are complained of in the old countries ; but there is a more material one to conflict with in many parts of this. In many of our fields, stumps of trees, roots, rocks and stones, are so frequently met with, that the drill plough could not be used. It is necessary that the ground should be perfectly clear of every thing that can obstruct or hinder the going of the drill. These obstacles, I confess, are not insuperable ; but in process of time may be removed. And in future generations the drill may be more conveniently used.

These objections apply chiefly to the newly settled parts of our country.

I have not mentioned these things with any view to deter my countrymen from attempting to apply the new mode of culture to winter grain. There is nothing that I more sincerely wish, than to see careful exments made with it. But I think this caution ought to be observed, never to attempt to raise spring wheat, or spring rye, in this manner. Though I have never read, nor heard, of horse hoeing spring wheat in England, I have known it tried by several persons to their mortification and loss, in this country. The crops were so entirely blasted as to be scarcely worth reaping. This has been the case, when the culture has been conducted by some of the most judicious persons, with great attention, and with the proper apparatus. The true reason of their miscarriage I take to be this, that as spring grain ripens later than winter grain, and hoed later than unhoed grain, it could not be ripe till some time in August, when some of the nights are so cold as to blast the grain, by stopping the ascent of the sap.

But let the new husbandry be tried on winter wheat, sown in August, or September, on a warm soil with a southern exposure, and where there are no stones, nor any other obstacles ; and let the seed be brought from some place at least a hundred miles northward. If with these advantages for ripening early, and in favourable seasons, a good crop of wheat cannot be obtained, it will not be worth while to make any further trials. But it should be tried on rye also ; for as that is known to be a hardier grain than wheat, it is possible it may answer better in this husbandry.

We need not be at the expense of procuring drill ploughs, and horse hoes, to make experiments of these kinds. After the ground is ploughed into ridges and well harrowed, the channels may be expeditiously made two inches deep with the head of a common rake, and the seed may be scattered in them by hand, and covered with the rake. The horse hoeing may be well enough performed with a common horse plough, passing it twice in a furrow, if it be found necessary, that the ground may be stirred to a sufficient depth.

If, after a fair trial or two, the new culture of winter wheat and rye should prove unsuccessful, it need not discourage any from sowing their grain with a drill plough. In land that is fit for it, the sowing may be performed with great expedition. If the seed were to be drilled in rows about nine inches apart, leaving no wider intervals, it would be attended with several advantages. Half the seed may be saved by it, which is a matter of some importance, especially in a time of scarcity of grain.

If the seed be good, it will undoubtedly all come up well and prosper: Because it will all be buried at the most suitable depth in the soil. But in the common way of sowing, some of the seeds are buried at such a depth, that they scarcely come up at all. Some are so near the surface, that the least drying of the soil prevents their vegetating, or alternate moisture and dryness turns them to malt. And some will be uncovered, which will be taken away by birds. Many stinted plants will appear; the crop will be uneven, some part of it being better, and ripening sooner, than the rest. Another advantage of drilling will be, that weeders may pass through a field to weed it, if there should be occasion for it, without any danger of hurting the plants. And all fields of wheat that produce weeds, ought to be carefully weeded. Sowed in this way the ground might also be stirred in the narrow intervals with a small hoe, which would encourage the growth of the plant, and keep it cleaner from weeds.

Instead of the drill husbandry, Dr. Hunter recommends a new scheme of his own, which partakes partly of the new, and partly of the old husbandry. He calls it alternate husbandry. The scheme is as follows: He ploughs his ground in flat ridges, or in lands, nine feet wide. When seed time arrives, he sows one land in the broad cast way, and leaves the next, sowing the third, and so on alternately through the field. The lands which are not sown he fallows, allowing them three or four ploughings in the fallow year; sows them the next year, and fallows the other.

He finds this to be a good mode of culture for land that is weak, and which lies remote from manure. A mean soil will thus bear pretty good crops without dressings, or with very small ones. The grain has greater advantage of a free air than in the old husbandry. No new implements are needed, nor any greater accuracy in the culture required than any ploughman is capable of. Perhaps a row or two of potatoes, or carrots, in the middle of the fallow ridges, might not be amiss in this husbandry; but rather an improvement.

But, to return to my subject: Every one must be easily convinced, that plants in general receive a greater degree of nourishment, if the ground about their roots be frequently stirred during their growth. We find the benefit of this in our gardens. We see that bare weeding does not answer so well as hoeing, among the plants we cultivate in them.

The great advantage of horse hoeing husbandry must appear, if we only attend to our ordinary method of cultivating Indian corn, which differs but little from that husbandry. If ploughing and hoeing were to be totally neglected, while the plants are growing, we should have no good crops. On the contrary, the deeper we plough the intervals, and the oftener we stir the mould with the hand hoe, the better is our crop. And why should not the advantage of the same culture be equally great, when appli ed to most of the plants which we cultivate? The more the ground is opened by frequent stirrings, the more vegetable nourishment it will receive from the atmosphere; and the roots will find a freer passage in extending themselves after their food. They will, therefore, receive a greater quantity; and their growth and perfection will be answerable.

I have not the least suspicion that barley and oats will fail of receiving great advantage from this culture; in both of which I have had some experience. Several years of late I have applied this culture to barley, in single rows or ridges three feet apart; and have never once failed of gaining at the rate of 40 bushels per acre. The grain has been perfectly clear from seeds of weeds, and more full and large than when cultivated in the common way. After ploughing the ground, and harrowing it, I form the ridges with the cultivator. I sow the seeds with a most simple drill of my own inventing. The weeds are killed, and the plants earthed, by passing the cultivator between the rows, with the addition of but little hand hoeing. That it does well for hemp, has been proved by trials in this country. None will doubt the advantage of it in raising potatoes, our common culture of which is so similar to that of Indian corn. But if they were set in drills, instead of hillocks, the produce would be greater, in both corn and potatoes, as I have found by several trials.

The new husbandry may as well be applied to all siliquose plants, as pease, beans, &c. and to all esculent roots, as parsnips, carrots, beets, and the like. The same may be said with regard to cabbages, asparagus, and most kinds of pot herbs. The trials that have hitherto been made upon such plants, in this country, have been so successful, that I trust the practice will soon become general. See the Rev. Mr. Eliot's *Essays*, p. 111.

These kinds of plants require so much less labour, in the drill way, than is usually bestowed on them in gardens, that when they are cultivated for the market, or for feeding of cattle, they should by all means be sown in drills, and horse hoed. The above writer from his own experience concluded, that five bushels of carrots might be as easily raised, as one bushel in the common method. My own experiments have fully justified this opinion.

Although, in a work like this, some notice of the New Husbandry is proper, some have doubted whether it can be successfully applied on an extensive scale in this country. See *Drill Husbandry*.

NURSERY, a garden, or plantation of young trees, to be transplanted. In a nursery for fruit trees, the land should not be quite so rich as that into which they are to be transplanted; because it will be better for them to have their nourishment increased than diminished, as they increase in age. Therefore, a nursery will need but a little manure, unless the soil be uncommonly poor.

A nursery should not be on a spot where fruit trees have lately grown, or indeed any other deep rooted plants. It should be on a medium between the two extremes of wet and dry.

To prepare the ground for sowing, it should either be trench ploughed, or dug with a spade to a considerable depth. From a foot to fifteen inches is not too deep. This should be done in the latter part of summer, and the ground well cleared of the roots of all perennial weeds and grasses. The season for planting either seeds or stones, is about the month of October. If it were done in the spring, none of the plants would be up in less than a year: And a considerable proportion of the seeds would perish. The seeds may be sown promiscuously; and they should be pretty thick, because they will not all come up. Some think it necessary to sow the pomace with the seeds of apples. I have sown them with and without it, and do not see that sowing seeds with the pomace is to be preferred.

When you transplant trees of one or two years growth in the nursery, mark the ground in lines three feet apart. Then open a trench a foot wide on the first line, and of a depth proportionable to the length of the roots: Take the stocks out of the seed bed, with a spade, preserving the roots as entire as possible: Cut off all the very small fibrous roots; and if a root tends directly downward, it must be shortened: Plant them in the trench twelve inches asunder. Then dig a trench and plant it in the next line, and so on, till the business is completed.

The main branch for the top should not be cut off, but carefully preserved. Several of the lateral branches should be taken off more or fewer in proportion as the root is more or less diminished. In this situation they are to grow till they are transplanted into orchards, &c. And they must be carefully tended, or they will not become good trees. Every spring and fall the ground between the rows must be well digged, and so carefully as not to injure or disturb the roots; or else the intervals must be horse hoed, If the latter be intended, the rows should be planted at least three feet and a half apart. But the plough must not go so near the rows as not to leave some ground to be dug with the spade, or stirred with a dung-fork ; and in using the plough, great care should be taken to avoid galling and injuring the trees.

A nursery should always be kept clear of weeds by frequent hoeing. No suckers that spring up from the roots should be suffered to remain. They will need a little pruning each year, to prevent their becoming mishapen; and all buds should be speedily rubbed off, which would make branches too low on the stems. A nursery requires so much attention, that it should be in a situation where the owner cannot avoid seeing it often ; otherwise it will be in danger of suffering through neglect.

The fruit trees should be allowed to grow to the height of five or six feet, before they are budded or grafted. *See those articles, Inoculation, and Grafting.*

Trees, to be transplanted into forests, may be cultivated in a nursery in the same manner as fruit trees. But, as Mr. Miller advises, it would be best to have a nursery of these in the place where the forest is designed to be planted; where a sufficient number of the trees may be left standing, after the rest have been removed.

If a nursery be in such a situation that the young trees are in danger of being broken down by deep snows ; either the fence on the windward side should be made so open, that the wind may have a free passage through it, and drive away the snow : Or else the trees may be defended by staking. A stake a little taller than the tree, made of a slip of board, should be set close on the windward side, and the top of the tree fastened close to it with a soft string. Or two such stakes may be so set, that the upper ends may meet over the top of the tree.

NUT TREE, *Juglans*. According to Michaux there are ten sorts of Nut Trees found in the United States, viz. the *Juglans Nigra*, or Black Walnut ; *Juglans Cathartica*, Butternut, White Walnut, or Oil Nut ; *Juglans Olivæ Formis*, Pacane Nut, or Pacanier ; *Juglans Amara*, Bitter Nut Hickory, or White Hickory ; *Juglans Aquatica* ; Water Bitter Nut ; *Juglans Tomentosa*, Mockernut Hickory, White heart Hickory, or Common Hickory ; *Juglans Squamosa*, Shell bark, or Shag Bark Hickory ; or Shagbark Walnut ; *Juglans Laciniosa*, Thick-shell-bark Hickory, Gloucester Nut Hickory, or Springfield Hickory ; *Juglans Porcina*, Pignut Hickory, or Hognut Hickory ; and *Juglans Myristicæformis*, or Nutmeg Hickory.

There are two sorts that are most common in New-England ; the common walnut, and the shagbark, so called. The first of these is a very hard and tough wood, which our farmers find useful for many purposes. It will bend into almost any form without breaking, especially the lower part of the body of a young tree. It is white and smooth ; it is therefore much used for ox bows, goads, and axe helves. But it soon decays when it is exposed to the weather. The fruit of this tree has a thin smooth shell, and is of very little value. The inner bark is useful for making a yellow dye.

The shagbark tree is so called, on account of the roughness of its scaly bark, which hangs in slips on the bodies of old trees. This has a small rich nut, enclosed in a very thick shell; but it is not so much esteemed for its timber as the other sort. The nuts naturally adhere strongly to the trees, but the first hard frost causes them to drop.

The black walnut tree is said to grow naturally in Virginia, and particularly on the banks of the Ohio. Though it be rather brittle, it receives a good polish; is hard and heavy, and is much prized for its beautiful brown colour, and used in all sorts of cabinet work.

We have another sort, not indigenous, but the only one that is cultivated with care in this country. It goes by the name of the English walnut. The fruit is larger and better than that of either of the other sorts, except the oil nut. In its tender state, it is used in pickles for sauce. But the nuts are too solid for this use when they are come to their full growth. The tree is so tender that it will only grow in sheltered situations, in vallies, or on the south side of houses and barns; and there it will amply reward the cultivator by its crops affording a pleasant addition to his table in the winter, and finding a good, and certain market in the great towns, both when gathered in July, as a pickle or preserve, or in winter, when in maturity.

A moist loamy soil seems to be the best situation for walnut trees; but they will grow on almost any upland. They are not well adapted to be cultivated in nurseries. They bear transplanting but poorly, unless when they are very young. The roots should not be wounded, but it is not easy to avoid it in taking them up, as they naturally run deep. Though the transplanted trees are best for fruit, they grow short and bushy, and are not fit for timber. Therefore, he who wishes to cultivate a grove of them for timber, should plant the nuts in the places where he wishes the trees to remain.

As there is a considerable pith in the limbs of walnut trees, they do not admit of much pruning. The water is apt to enter at a wounded limb and cause it to rot.

NYMPHA, "the state of winged insects, between their living in the form of a worm, and their appearing in the winged or most perfect state. The eggs of these insects are first hatched into worms, or maggots; which afterwards pass into the nympha state, surrounded with shells, or cases, of their own skins: So that, in reality, these nymphs are only the embryo insects, wrapped up in this covering; from whence they at last get loose, though not without great difficulty.

"During the nympha state, the creature loses its motion. Swammerdam calls it *nympha aurelia*, or simply aurelia; and others give it the name of chrysalis, a term of the like import." *Dict. of Arts*.

It is in their winged state only, that they copulate. The female lays eggs; and their offspring go through the same changes. The state of these animals may serve to remind mankind of the manner of their existence, first in mortal bodies, then in a state of death, afterwards possessed of glorious bodies. In their aurelian state, these animals have no vital motion, but are to all appearance dead. So that in their last state of existence, insects have,as it were,resurrection bodies.

O.

OAK, *Quercus*, a well known tree, the timber of which is of great use and importance in ship building, and architecture, and is valuable for fuel and many other purposes. The timber is both strong and durable.

Mr. Miller reckons eighteen species of the quercus, or oak. I know of but five that grow in this country, unless the swamp white oak, so called on account of its growing in wet swamps, may be a distinct species from that which grows on the upland.

The first and best is the white oak, *Quercus alba*, which bears a long shaped, small and pleasant tasted acorn. The bark is of a very light ash colour. The timber is more strong, and far more durable than the other kinds. Staves for casks, made of this tree, bear a higher price than any other. As it does not soon decay, the farmers find it convenient to have their wheels, carts, ploughs, and several other implements of husbandry, made of this timber. The butends of the trees which have grown in pastures, are commonly found to be extremely tough,and are most fit for the naves and spokes of cart wheels.

The black oak, *Quercus nigra*, has a very dark coloured, hard and rough outer bark. The inner bark is of a bright yellow colour, and may be used to advantage in dies. Little or none of this oak is found in the District of Maine. Of all the kinds of oak produced in our country, this is esteemed the best for fuel, as it will burn freely in its green state : But it is not so much prized for timber as some other sorts.

The grey oak is next in quality to the white for building. The red, *Quercus rubra*, which is so called from the colour of its wood,answers well for staves,especially for molasses hogsheads. But as it is not a lasting timber, it is more proper for fuel; and for the last purpose, it does not answer well in its green state. The acorns of the grey and red oak, are much larger than those of the white. The leaves are also larger, and very deeply sinuated. They are probably not different species of the oak,but only varieties.

The last kind, and the meanest of all, is the dwarf, or shrub oak, it being fit for neither fuel nor timber. It is always crooked and small, and seldom rises to the height of ten feet. It delights in a poor soil,and overruns many of our sandy and gravelly plains. It has a strong root, which will continue to send up new shoots, though they are cut off yearly; so that there is no effectual way to subdue them, but by grubbing them, or pasturing goats upon them.

As all the kinds of oak bear fruit,

the shrub oak as plentifully as any, these trees are of some advantage in feeding swine and poultry. They are fondest of the acorns that grow on the white oak, as the other kinds have a bitter taste.

The bark is of great use in tanning hides, and a good ingredient in dies.

The oak produces a fungous ball, or apple, of a loose, soft contexture, which soon dries and falls off, and is of no use.

But besides, it has little round hard kind of excrescences, called galls, which are of great use in dying and making the best writing ink. Though they grow as large as nutmegs in other countries, those which I have found in this, have been much smaller. Perhaps trees must stand single many years, before they will be apt to produce galls of a large size. I have not found them but upon the white oak, and those not larger than peas.

I beg leave here to give the reader the history of galls, from the *Dictionary of Arts*. "An insect of the fly kind is instructed by nature to take care of the safety of her young, by lodging her eggs in a woody substance, where they will be defended from all injuries. She, for this purpose, wounds the branches of a tree; and the lacerated vessels, discharging their contents, soon orm tumours about the holes thus made. The hole in each of the tumours, through which the fly has made its way, may for the most part be found; and when it is not, the maggot inhabitant, or its remains, are sure to be found within, on breaking the gall. However, it is to be observed, that in those galls which contain several cells, there may be insects found in some of them, though there be a hole by which the inhabitant of another cell has escaped."

It is to be wished, that persons in the oldest parts of the country, when an oak is felled, would search for galls. If they are produced here in plenty, it will not be right to persist in sending our money for them to foreign countries.

As trees, both for timber and fuel, are become scarce in some parts of the country, it is high time to begin to make plantations of trees for these purposes. And I know of no kind that will answer, all things considered, better than the oak. The trees are so hardy as never to be damaged by the severe coldness of our winters: Neither have they been known to suffer much by any kind of insects. The red and grey kinds are very rapid in their growth, and will soon repay the cost and trouble of rearing them: And the white is of so essential importance for timber, that a scarcity of it is to be dreaded.

Some of our pasture lands, which are high and quite bare, would be much improved, if every hundred feet square were shaded by a lofty oak: besides gaining a beautiful appearance, especially if they were placed in regular order. Barren heights, in some pastures, are in great want of trees to shade them. Copses, or clumps, in such places, would have excellent effects. There would be more grass, the appearance would be beautiful, and the profit considerable. But the question is, in what manner shall oaks be propagated? They may undoubtedly be raised in nurseries, and transplanted, as well as other trees. But this method is not universally approved.

Mr. Miller says, oaks are best produced from the acorns in the places where the trees are to remain; because those which are transplanted, will not grow to so large a size, nor remain sound so long. He advises to planting the acorns as soon as they are ripe in October, which will come up in the following April; because if they are attempted to be kept, they will sprout, although spread thin. He directs that the ground designed for a plantation, should have a good and durable fence; that it be prepared by three or four ploughings and harrowings; that the acorns be taken from the largest and most thrifty trees; that they may be sowed in drills about four feet asunder, two inches deep, and two inches apart; that the ground should be ploughed and hoed among them, during the first eight or ten years; that after two years some of them should be drawn out where they are too close; and so from time to time as they grow larger, till they come to be eight feet distance, each way, when they will want no further thinning for a long time. But after the trees come to be large, he thinks 25 or 30 feet apart will be the right distance.

Another writer directs that the acorns be gathered as soon as they fall in autumn, and kept in a box or boxes of sand till the following spring. Then open them, and carefully plant those of them which are sprouted, which he says will not fail to come up. But no time should be allowed for the sprouts to dry. I incline to prefer this method, especially since I have tried that which is recommended by Mr. Miller without success. Not one in a hundred ever came up.

A rich deep soil suits the oak best, and in such land they will grow to a large size. The timber is apt to be tough and pliable: But in a gravelly soil, or one that is dry and sandy, the wood is more hard and brittle. The oak, however, will grow in almost any soil that is not too wet.

Many are apt to object against attempting to raise timber trees, that they shall not live to receive any advantage from them. But do they think they were born for themselves only? Have they no great regard for the welfare of their own children? Do they not care how future generations fare after they are gone? The more growing trees they leave upon their farms, the better will their children be endowed; and does this appear as a matter of indifference? Or if they should providentially be under the necessity of selling their farms while they live, will they not be prized higher, by any rational purchaser, for having a few hundreds of thrifty young trees growing upon them? But it is possible that while they hold their farms, they may receive actual advantage from their trees themselves. Possibly trees may grow faster than they apprehend. The Marquis of Lansdown planted with trees a swampy meadow, with a gravelly bottom, in the year 1765, and in the year 1786 the dimensions of the trees were as represented in the following table.

	Height in Feet.		Circumf. Ft.	In.
Lombardy poplar	60 *to*	70	4	8
Abele poplar -	50	70	4	6
Elm - - -	40	60	3	6
Chesnut - - -	30	50	2	9
White or Weymouth pine -	30	50	2	5
Scotch fir - -	30	50	2	10
Spruce - -	30	50	2	2
Larch - - -	50	60	3	10

The measures were taken five feet above the ground. It appears that if trees can be waited for 21 years they will repay the cost, by becoming fit for many important uses. And I am persuaded that some of the species of oak will grow as fast as some of the trees in the foregoing table.

One acre will bear 160 oaks, at the distance of 15 feet from each other: If each tree will grow in 30 years to half a cord of wood, worth 12s. per cord, the whole produce will be 90 cords of wood, worth 160 dollars, which is four dollars and a third per acre per annum, for the use of the land, a greater profit than we expect from other acres in general. It ought to be considered that intermediate trees taken out young, may pay the cost of planting and culture; and that the land may serve most of the time for tillage or pasture; for tillage while the trees are small, which will hasten their growth. The increasing dearness of fuel and of timber should put the holder of land, in old settlements, upon thinking of the cultivation of all trees that are useful for either of these purposes. The day is at hand, if not already arrived, when this will be one of the most profitable, as well as important, branshes of husbandry.

"The natural history of the American Oaks was first partially given by John Philip Du Roi, who published his observations in two volumes 8vo. at Brunswick in Germany, in 1771: and in 1801 more extensively, by Andrew Michaux, the celebrated botanist, who resided many years at the *Botanical Garden* which was established by the late Louis XVI. king of France, near Charleston, South-Carolina. The species and varieties described by him are twenty-nine. This truly superb work *on American Forest Trees*, ought to be in the hands of every gentleman farmer in the United States. It may be had of Wells and Lilly, Booksellers, Boston."—*See Forest Trees.*

OATS, *Avena*, a well known grain, very pleasant and nourishing to horses, and conducive to keep them in health. Though other sorts of grain are too binding, oats have a contrary effect; and even too much so, unless they be sweated in a mow before they be thrashed. The flour of this grain is no bad ingredient in table provisions. It is highly approved for gruels and puddings: And would be more used, were it not for the difficulty of divesting the grain of its husk.

There are varieties of this grain, distinguished by their different colours, the white, the black, the grey, and the brown oats; but as these differ only in colour, they are not considered as distinct species.

The white oats which are most commonly cultivated in this country, are generally preferred in other countries, as producing the best crops. But I suspect that sufficient trials have not yet been made here, in the culture of the black oats. The produce of them from a few corns sown in a garden, has been astonishing. But this might be owing to the newness of the seed in our climate, or to some circumstance less considerable, or less obvious.

There is also a species of the naked oats. This, one would think, must have the advantage of other oats, as it is threshed clean out of the husk, fit for grinding. But with this grain we are yet unacquainted.

I have lately met with the Tartary oats, which resemble our white oats, but differ in their manner of growing. They bear very plentifully: But are rather apt to lodge.

Oats cannot be sown too early in the spring, after the ground is thawed, and become dry enough for sowing. The English farmers sow them some time in February. But in a wet soil they sometimes answer very well, though sowed in June.

Three bushels of seed is the usual quantity sown on an acre. This quantity say some will be rather more than enough on a rich soil. If the soil be poor, the quantity of seed should be the greater, say they, as the plants will be smaller, will not tiller; and so may stand the nearer each other without crowding. But this is a matter of opinion only, and may be a mistake.

Oats have strong piercing roots, and are called hearty feeders, so that they can find their nourishment in stiff soils; and for the same reason they sometimes produce great crops when sown after one ploughing. But two ploughing, are generally better for them than one.

When they are cultivated according to the new husbandry, they should be sown in double rows, fifteen inches apart, on beds six feet wide. For they will grow taller than wheat, and therefore require more room. One bushel of seed will be sufficient for an acre in this way. Some advise to brining and liming the seed; but this may as well be omitted, unless when they are sowed late. It may serve in this case to quicken their growth.

Oats should be harvested in a greener state than other grain. The straw should not be wholly turned yellow. It will be the better fodder, if it do not stand till it be quite ripe and sapless. Mr. Cook, an English writer, recommends cutting them about four or five days before the state of ripeness; and says they will improve by lying on the ground. But if they be quite ripe when they are cut, they will be apt to shed out by lying.

Though they should be well dried on the ground after cutting, they should not be raked, nor handled at all, when they are in the driest state. It should rather be done in mornings and evenings, when the straw is made limber and pliable by the moisture of the air. If they should be got in when they are somewhat damp, there will be no danger, having been before thoroughly dried; for the straw and chaff are of a very dry nature.

Some choose to reap them: But the straw is so valuable a fodder, that it is better to cradle or mow them. And that the ground may be well prepared for mowing and raking, a roller should be passed over it after sowing and harrowing: But some prefer rolling the ground after the grain is some inches high; it is said to close the soil to the roots, and make the grain grow with fresh vigor.

Oats are so apt to rob the land of its richness that they should not be sown on the same spot twice in succession, unless the soil be very plentifully manured. In a succession of crops, oats may sometimes be sown to advantage the first year after the breaking up, before the land can be made mellow enough for other grain: Or they may follow wheat or barley. In the latter case, the wheat or barley stubble should be ploughed in as soon as the crop is off.

The Farmer's Assistant says, that oats are not so great exhausters of the ground as is generally imagined, and that a piece of intervale land was sowed eighteen years successively with this grain without manure, and without any sensible diminution of the crop.

Gypsum is said to afford an excellent manure for oats. A change of seed is as necessary in this plant as in most others.

A writer in the Massachusetts Agricultural Repository, vol. V. pp. 331, 2., says:—" It appears to us best, all things considered, that the first crop after turning over sward, should be oats. The reason, why an oat crop should precede a potatoe crop is, that it not only pays well by its product for the year's labour, but chiefly because it enables the husbandman to deepen his ploughing, preparatory to the second year's series in the rotation."

ONION, *Allium*, a well known esculent root. The common sort have purple bulbs. The white, or silver skinned, which are supposed to have come from Egypt, are by some preferred to the other. They have not so strong a taste.

This plant flourishes so well in the southern parts of New-England, that it has long been a considerable article of exportation; in the northern parts, it requires the very best culture; but even there, onions may be raised in sufficient plenty for home consumption.

A spot of ground should be chosen for them, which is moist and sandy; because they require much heat, and a considerable degree of moisture. A low situation, where the sand has been washed down from a neighbouring hill, is very proper for them. And if it be the wash of a sandy road, so much the better. The most suitable manures are old rotten cow and horse dung mixed, ashes, but especially soot. A small quantity of ashes or sand. or both, should be spread over them after sowing, especially if the soil be not sandy. And it is not amiss to roll the ground after sowing; or harden the surface with the back of a shovel.

I have many years cultivated them on the same spot; and have never found the land at all impoverished by them. But on the contrary, my crops are better than formerly. But the manuring is yearly repeated; and must not be laid far below the surface.

The ground should be dug or ploughed in autumn, not very deep; and then made very fine in the spring, and all the grass roots, and roots of weeds, taken out; then laid in beds four feet wide. Four rows of holes are made in a bed, the rows ten inches apart, and the holes in the rows ten. About half a dozen seeds are put in a hole, or more if there be any danger of their not coming up well, and buried an inch under the surface. This is allowed by the experienced cultivators in Connecticut, to be the best way of setting the seeds. For they will grow very well in bunches. I have lately found that they grow full as well in drill-rows a foot asunder. They crowd each other up out of the soil, and lie in heaps as they grow upon the surface. Though the largest onions are those that grow singly, some inches apart, those that are more crowded produce larger crops. And the middle sized onions are better for eating than the largest.

The last week in April is the right season for sowing the seeds, if the ground be capable of being got into proper order so early. In wet ground it is often necessary to sow them later.

Last year I sowed my onions in drills, twelve inches apart, across the beds: And I found my crop was near double to what it used to be, when they were sown in bunches. Perhaps this will prove to be the better method. But I gave them also a slight top-dressing of soot, just before they began to form bulbs, which might be the true reason of the great increase: So that I dare not yet absolutely prefer the drill method to the other; though I am much inclined to give it a decided preference.

Onions should be hoed three or four times, and kept quite clear of weeds, before the tops arrive to their full height. At this time the bulbs will begin to swell; hoeing should therefore be laid aside, and the weeds pulled up by hand as often as they appear. Weeds not only rob the plants of their food, but injure them much with their shade; for they have occasion for all the warmth of the sun that they can get.

To promote the growth of the bulbous roots, I have found it advantageous to trample the ground hard between the rows or bunches, and to draw the soil away from the bulbous roots, laying them bare to the sun. They are the more warmed, and grow faster.

Some think it proper, and even necessary, to pass a roller over beds of onions, or cripple down their tops by hand. But I have never been able to find the least advantage from either of these methods: Nor do I think they ought to be practised; for I cannot easily conceive how the crushing and wounding any plant, while it is growing, should conduce to its improvement. Though some may have good crops, who treat them in this manner, I am persuaded that if they neglected it, they would have much better crops. For, besides the mischief already mentioned, the sun is shut out from the bulbs by crushing the tops down upon them; but the more upright the tops are, the more the sun will shine upon the roots. I would sooner cut off part of the tops than go to crushing them.

Others shake and twist the tops, to loosen the bulbs in the soil, which I cannot approve of: For if it do not snap off some of the fibrous roots, it gives too free a passage of the air to them, by which, if dry weather follow, they will be injured, rather than assisted in their growth.

When onions are thick-necked, do not incline to bottom, but rather to be what are vulgarly called scallions, the more care should be taken to harden the ground about them, and to lay the bulbs bare to the sun. And it may be proper to let them touch the soil only in that part which sends out the fibrous roots.

At the worst, if they fail to have good bottoms the first year, and chance to escape rotting till spring; they may perhaps get them by being transplanted. Even an onion which is partly rotten will produce two, three, or four good ones, if the seed stems be taken off as soon as they appear. They ripen earlier than young ones, have the name of rare-ripes, and will sell at a higher price.

When onions are so ripened that the greenness is entirely gone out of their tops, it is time to take them up: For from this time the fibrous roots decay, and no longer convey any nourishment to the bulbs, as appears by their becoming quite loose in the soil, and easy to take up.

After they are pulled up they should lie on the ground for ten days or a fortnight, to dry and harden in the sun, if the weather be fair. Then, in fair dry weather, be moved into a garret, and laid thin. The scallions should not be mixed with the good onions, lest they should cause them to rot; but be hung up in some dry place in small bunches, where they will not be too much exposed to frost.

That onions may keep well through the winter, they should not be trusted in a warm and moist cellar; but have a situation that is dry and cool. Moisture soon rots them, and warmth causes them to vegetate. A degree of cold which would ruin most other esculent roots, will not injure them at all. The spirit that is in them is sufficient to enable them to resist a considerable degree of frost. Accordingly, in the southern parts of this country, as I am informed, they are usually kept through the winter in dry casks placed in chambers, or garrets. But they should not be removed, or touched, while the weather is very frosty.

Those which are shipped for market, are usually made into long bunches, by tying them to whisps of straw.

When onions are kept long, they are apt to sprout, which hurts them for eating. To prevent this, nothing more is necessary than to sear the fibrous roots with a hot iron. The pores of the roots will thus be stopped, through which the air enters and causes them to vegetate.

To obtain seed from onions they should be planted early in beds, about nine inches apart. The largest and soundest are best. In a month the tops will appear; and each one will send up several stems for seed. They should be kept free from weeds; and when the heads of the flowers begin to appear, each plant must have a stake about four feet long, and its stems be loosely tied to the stake by a soft string of sufficient strength. If this be neglected, the heavy tops will lay the stalks on the ground, or the winds will break them. In either case, the seeds will fail of coming to perfection.

ORCHARD, an enclosed plantation of fruit trees, not again to be removed.

An orchard may consist wholly of pear trees; or of quince, peach, plum, &c. or it may be a mixture of various kinds of trees. But orchards of apple trees are the most important, and are almost the only ones in this country. Other fruit trees are commonly planted in the borders of fields, or gardens; because only a small number of them is desired, or considered as advantageous, by farmers.

The soil for an orchard should be suited to the nature of the trees planted in it. Though a clay soil will do well for pear trees, it is not at all suitable for apple trees. Dry sand and gravel are not good; but a deep hazel loam is preferred to any other soil; and it is the better if it be somewhat rocky and moist.

Plains, hollows, or high summits, are not so good situations for orchards, as land gently stopping: And a southeastern exposure is generally the best. But when this exposes the trees to sea winds, a southwestern exposure may be accounted better.

If the land be swarded, it should be broken up and tilled one year before the trees are planted; and if it be dunged it will be better for the trees. The rocks should also be taken out; because it cannot be done so conveniently afterwards. And if there be any large stumps of trees, which would last long in the ground, they should be taken out. Otherwise they will render the operations of tillage in the young orchard very difficult.

Trees which are ungrafted are supposed to bear as good fruit as any for cyder. They commonly bear more fruit, and will last longer.

But when grafted trees are to be transplanted, those should be chosen that have not been grafted more than two years. Old stunted trees, the refuse of a nursery, are to be avoided, which will grow very slowly, if at all. For directions concerning the time and manner of planting an orchard, see *Fruit trees and Transplanting*.

Concerning the right distance of the trees in an orchard, there are a variety of opinions. But the coldness and wetness of the climate, an argument used in England for placing them far asunder, does not so well apply in this country. Trees in that cold and cloudy region need every possible advantage of exposure to the sun and air. It should be considered at the time of planting, to what size the trees are likely to grow: And they should be set so far asunder, that their limbs will not be likely to interfere with each other, when they arrive to their full growth. In a soil that suits them best, they will become largest. Twenty-five feet may be the right distance in some soils; but thirty-five feet will not be too much in the best, or even forty. If, contrary to expectation, they should be too close when they are grown up, they may be easily thinned: And it will be better to take away here and there a whole tree, than to lop and maim them all, that they may have room.

The planting of small trees in the midst of full grown ones does not answer so well for the small ones, as when the trees are all nearly of one size. A small tree among large ones has not an equal chance of exposure to the sun and air: Both of which are of great importance. So that it is of the less importance to replace a tree that dies in an orchard. And it is of no advantage to do it, when the nearest neighbouring trees appear to be rather too much crowded.

An orchard must be constantly well fenced, to keep out cattle. It should be enclosed by itself. Hungry sheep would peel the trees while they are young; and cattle will bite off all the limbs of young and old trees that are within their reach. But there is no danger in turning in a horse occasionally, when there is grass and no apples; and swine may be confined in an orchard that is grown up, so that the trees cannot be hurt by them, and when the fruit is not in their way.

Sheep sometimes get into an orchard that is well fenced, by means of high banks of snow, when they are stiff or crusted. I can think of no better way to prevent this, than to make the fence so open, with round poles, or pickets, that the snow will pass freely through it, and not rise in high banks. The latter kind of fence might be so constructed as to keep out such creatures as are apt to take fruit from the trees, without leave of the proprietor.

After an orchard is planted, it is best to keep the land continually in tillage, till the trees have nearly got their full growth; at least till they have begun to bear plentifully. The trees will grow faster, and be more fruitful. But great care must be taken that the roots be not disturbed by ploughing, nor the bark on the stems of the trees wounded. The ground near the trees, which the plough leaves, should be broken and made mellow with a spade, for two or three years, before the roots have far extended.

Severe prunings should generally be avoided. The limbs that interfere, and rub each other, must be cut out; but never shorten the shoots, nor cut off any of the bearing spurs. Take off all decayed and broken branches, close to the stems from whence they are produced; and cut away all suckers, as soon as they appear, whether from the roots, or any other parts.

Mr. Forsyth recommends the washing of trees annually, in the month of February or March, with the following mixture, which, he says, will destroy the eggs of insects, and prevent moss from growing on the trunks and branches: It will also help to nourish the tree, keeping the bark fine and healthy; and will have the same effect on it as a top-dressing has upon grass land.

Mix fresh cow dung with urine and soap-suds, and, with this mixture, wash over the stems and branches of the trees, as a white-washer would wash the ceiling or walls of a room; taking care to cut off the cankered parts, and to scrape off all the moss, before you lay the mixture on. In the course of the spring, or summer, you will see a firm new bark coming on. When the old bark is cankered, you must pare it off with a drawing knife, or such long knife as I have made for the purpose. Soft-soap, and lime and water have also been recommended for washing fruit-trees.——See the article *Fruit Trees*.

John Prince, Esq. of Jamaica Plains, Roxbury, gives an account of a "small white-ringed worm, about 3-4ths of an inch long, with a dark-coloured head," supposed to be the same, which attacks the peach trees at and just below the surface of the ground. Worms of this kind are got rid of "by digging round the tree, and clearing away the earth to the roots, and then with a sharp-pointed knife, a chisel or a gouge, (and a small wire to probe, if they are deep in the tree,) they may easily be destroyed." After taking out all that can be found, the wounds should be covered over with grafting clay, and a large proportion of dry wood ashes mixed, and the earth then returned to the tree.

The worms are discovered by the "*saw-dust* borings, which should be followed, and wholly extracted." Mr. Prince likewise observes, "I have also lost several mountain oaks and quince trees, I believe, by the same destroyer."—See *Mass. Agr. Repos.* vol. V. p. 360.

Orchards which are laid down to grass last longest; but it is necessary to keep the ground clear of weeds and grass for some little distance from the roots. They may be pastured with calves and swine, and sheep may be admitted, provided the trees are smeared with a coat of lime, or of Mr. Forsyth's composition, mentioned above; but large cattle which can reach up to the limbs of the trees should not be suffered to run in an orchard.

Judge Peters of Pennsylvania asserts that young apple trees, taken from a clay soil, and transplanted into one more suitable, throve better than other apple trees, apparently of better quality, taken from a nursery, where the soil was congenial to their growth.

It is said that a rope of straw, with one end tied round the body of a tree, and the other in a pail of water, will prevent the effects of a late frost in the spring.

Sir John Sinclair says, "It cannot be too strongly inculcated, that to permit young plants to bear fruit too early, is eventually to do essential injury to their future fruitfulness and duration."—*Code of Agriculture*, 316. *American Edition.*

"To bring an orchard as early as possible into profit, plant common wild trees, or what are commonly called crab apples, four or five years old; cut them down as soon as planted, and on their young shoots graft or inoculate such fruit as is desired."—*Massachusetts Agricultural Repository*, vol. III. p. 40.

In Tessier's Annales de l'Agriculture; tom. XLIII. page 185., the following method is mentioned of preserving the health of fruit trees in grass lands. In Germany they surround the roots of trees with hemp breakings, not only near the stock, but for some distance from the tree. The breakings of flax are made use of for the same purpose, and it is said they prove a defence against the canker worm. Spent tar applied to the roots of trees is likewise recommended.—See *Mass. Agricultural Repository*, vol. VI. 283.

ORE WEED, *sea weed, sea ware*, or *sea wreck*. These names are applied to all the vegetables which grow plentifully in the sea, and on the muddy and rocky parts of the shore below high water mark.

The sorts are chiefly three; the kali, or rock-weed, which strongly adheres to rocks, and which is allowed to be of the greatest value for manure. The alga, called eel grass, or grass wreck, is of the next rank as to its richness. But there is another sort, consisting of a broad leaf with a long shank or stem, of an inch diameter, by some ignorantly called kelp; this is said by Sir A. Purves to be of the least value of any of the sea weeds. However, none of them are unimportant for fertilizing the earth.

All vegetables when putrefied are a good pabulum for plants; for they consist wholly of it. But the value of marine vegetables is greater than that of any other; for, besides the virtues of the other, they contain a large quantity of salt, which is a great fertilizer. Mr. Dixon thinks those weeds which grow in the deepest water are the best. Perhaps they contain a greater proportion of salt than those which grow near the shore, as they are seldom or never wetted with fresh water.

A great advantage that these plants have above any other, is their speedy fermentation and putrefaction. The farmer has no need to wait long after he has got them, before he applies them to the soil. The rock weed may be ploughed into the soil, as soon as it is taken from the sea. This is practised in those parts of Scotland which lie nearest to the shore; by which they obtain excellent crops of barley, without impoverishing the soil. Neither have they any occasion for fallowing to recruit it. In hills of potatoes, it answers nearly as well as barn dung. I have known some spread it upon young flax newly come up, who say it increases their crops surprisingly. The flax may grow so fast, and get above this manure and shade it, so soon, as to prevent evaporation by the sun and wind; so that but a small part of it is lost; and flax is so hardy a plant that it does not suffer by the violence of salt, like many other young plants.

But I rather think it is best to putrefy sea weeds before they are applied to the soil. This may speedily be accomplished by laying them in heaps. But the heaps should not lie naked. Let them be covered with loose earth or turf; or else mixed in compost dunghills, or laid in barn yards with divers other substances. This substance will soon dissolve itself, and what is mixed with it, changing to a salt oily slime, very proper to fertilize light soils, and not improper for almost any other.

As to the eel grass, &c. the best way is to cart it in autumn into barn yards, filling the whole areas with it, two or three feet deep. It may be either alone, or have a layer of straw under, and another above it. When it has been trampled to pieces by the cattle, and mixed with their stale and dung, it will be the fittest to be applied to the soil. It being a light and bibulous substance, it will absorb the urine, which is totally lost by soaking into the earth, unless some such trash be laid under cattle to take it up, and retain it.

Farmers who are situated near to the sea shore have a vast advantage for manuring their lands. If they were once persuaded to make a spirited improvement, they might enrich their farms to almost any degree that they please. They should visit the shores after spring tides and violent storms, and with pitchforks take up the weeds, and lay them in heaps a little higher up upon the shore; which will at once prevent their growing weaker, and secure them from being carried away by the next spring tide.

Many are so situated that they can drive their carts on a sandy, hard beach, at low water, to the rocks; and fill them with weeds. Can they be so stupid as to neglect doing it? It is even worth while to go miles after this manure with boats, when it cannot be done more easily.

It has often been observed that manuring with sea weeds is an excellent antidote to insects. It is so, not only in the ground, but also upon trees. I have an orchard which has been for many years much annoyed by caterpillars. Last spring, about the last of May, I put a handful of rock weed into each tree, just where the limbs part from the trunk; after which I think there was not another nest formed in the whole orchard. April is a better time to furnish the trees with this antidote to insects. And the month of March is perhaps better still.

Putrefied sea weeds should, I think, be used for crops of cabbages, and turnips, and for any other crops which are much exposed to be injured by insects.

One disadvantage attending the business of farming in this country, is, that our cold winters put an entire stop to the fermentation, and putrefaction of manures. This may be in some measure obviated by the use of rock weed, which is so full of salt that it is not easily frozen: Or if frozen, it is soon thawed. I have been informed that some have laid it under their dunghills by the sides of barns; in which situation it has not frozen; but by its fermentation has dissolved itself, and much of the dung that lay upon it. There is undoubtedly a great advantage in such a practice.

Another advantage of this kind of manure, which must not be forgotten, is, that it does not encourage the growth of weeds so much as barn dung. It is certain it has none of the seeds of weeds to propagate, as barn dung almost always has. But some suppose that its salt is destructive to many of the seeds of the most tender kinds of plants; if it be so, it is only when it is applied fresh from the sea, at the time of sowing. But even this is doubtful.

This manure is represented in the Complete Farmer to be twice as valuable as dung, if cut from the rocks at low water mark; that a dressing of it will last three years; and that fruit trees which have been barren are rendered fruitful by laying this manure about their roots.

OSIER, *Salix*, *Sallow*, or *Willow Trees*. According to Mr. Miller there are fourteen species; the twigs of some of which are much used by basket makers in Europe.

A sort of grey or brown willow grows naturally in this country, in low moist places. But it is only a bushy shrub, of slow growth, and has not that toughness in its shoots for which some of the foreign willows are valued.

Two sorts are propagated in this country, which were brought from Europe. The young shoots of the yellow sort have a golden colour; but the trunks of the trees are almost black. The green sort bids fair to be more useful than the other. They will grow in almost any soil, and come to be large trees; but a moist soil suits them best. I have known the green sort to grow where the ground is some part of the year flowed with water, as in the borders of rivers and ponds.

It might be advisable for the people in some parts of the country to propagate them for the sake of the wood. I know of no other trees that increase nearly so fast as both these kinds do. A prodigious quantity of wood might be obtained from an acre planted with them. In less than twenty years they would be large trees. I have known sets, or cuttings of the smallest size, in ten years, grow to the size of thirty inches round, or ten inches diameter.

The trees are easily propagated by cuttings, or sets, either in spring or fall. If in spring, they should be planted early, as soon as the ground is thawed. Young sets should be three feet long, and two-thirds of their length in the ground.

Live hedges may be more cheaply and expeditiously made of osiers than of any other plants. Stakes or truncheons of seven or eight feet long may be set in a spungy or miry soil; they will take root and grow, and form a hedge at once. This saves the cost of securing a young hedge. It is with great pleasure that I observe some fences of this kind are begun in the country. It is a very cheap and easy method of fencing, which cannot be too much encouraged. The trimmings of the hedges will be of great value in towns where wood is become scarce, and may be had yearly. See *Willow*.

OVERFLOWING *of the* GALL, a disease in horned cattle, known by a copious discharge of water at their eyes. To cure it, take a hen's egg, open the end, and pour off the white, reserving the yolk; then fill up the cavity with equal quantities of soot, salt and black pepper; draw out the tongue of the animal, and with a slender stick push the egg down his throat. It should be repeated two or three mornings. It seldom fails to cure.

OUT HOUSES, slight buildings that belong to a mansion house, but stand at a little distance from it. When it can conveniently be so ordered, the out houses of a farmer ought to be so placed as to be all contiguous to the farm yard. Then all the dung, filth and rubbish they afford at any time, may be flung into the yard, without the trouble of carrying; where they will be mixed and mellowed by the trampling of beasts, and contribute to the increase of manure.

OX, a castrated bull. Till they are four years old, they are usually called steers, afterwards oxen. Oxen that are white, black and white, or a very pale red. are seldom hardy, or good in the draught. Red and white oxen are often good; but the darkest coloured oxen are generally best. Brown, dark red and brindled are good colours.

The signs of a good ox are these: Thick, soft, smooth and short hair; a short and thick head; glossy, smooth horns; large and shaggy ears; wide forehead; full, black eyes; wide nostrils; black lips; a thick fleshy neck, and large shoulders; broad reins; a large belly; thick rump and thighs; a straight back; a long tail, well covered with hair; short and broad hoofs.

Steers at the age of two years and a half, or earlier, may be yoked and trained for the draught. If it be longer delayed, they are apt to be restiff and ungovernable. They should not be worked by themselves, but in a team with other cattle which have been used to labour. Their work should be very easy at first, and only at short intervals, as they are apt to fret and worry themselves excessively. A gentle usage of them is best, and beating them should be avoided.

If oxen are worked in the yoke in wet and rainy weather, which sometimes unavoidably happens, their necks are apt to become sore. To prevent this, a little tallow should be rubbed on the parts of the yoke which lie upon their necks, and also upon the bows.

When steers come to be four years old, they have one circular ring at the root of their horns, at five two rings, and one ring is added each year; so that if you would know the age of an ox, count the rings on one of his horns, and add three, which amounts to the true number of his years. It is the same in a bull, and a cow. In very old cattle, these rings are sometimes rather indistinct.

When an ox has completed his eighth year he should be worked no longer, but be turned off to fatten. His flesh will not be so good, if he be kept longer. A little blood must be taken from him, that he may fatten the faster.

A valuable paper "On the Advantages of Oxen in preference to Horses," by the Hon. Timothy Pickering, was published in the Massachusetts Repository and Journal for July, 1820, vol. VI. No. 2., from which the following is extracted:

"I have been inclined to entertain the opinion (perhaps an erroneous one) that oxen might be trained (beginning with their first acquaintance with the yoke) to a greater quickness of movement than is common; and that this might be rendered habitual. I have seen a pair of oxen in a plough keep pace with another plough drawn by a pair of horses. And Sir John Sinclair, in his account of the Improved Scottish Husbandry, mentions two distinguished farmers of the name of Walker, who, contrary to the general practice of their neighbours, persisted in the use of ox-teams as profitable on their farms. Two oxen, harnessed like horses in a plough, performed the same labour, without losing a turn. After the experience of twenty years, these farmers pronounced oxen fit for every agricultural labour, travelling on *hard turnpike roads* excepted.

"It would seem that horse harness is generally, if not universally used for oxen in Scotland, the collars being reversed. Sir John Sinclair says, 'the principal objection to the use of oxen is the difficulty of shoeing them.' The facile mode of shoeing oxen in New-England would remove that objection."

In a letter, published on the next page of the same work, Mr. Pickering gives the testimony of an "experienced farmer of Pennsylvania, whose prejudices were opposed to the use of oxen for labour, and where horses, with very few exceptions, constitute the farmers' teams. This farmer, Mr. William Ashford, in a letter to John Vaughan, Esq. says:— "There is another thing in which I was wrong in not taking your advice, viz. not keeping oxen instead of horses: this spring all my horses became sick, and I was forced to buy a pair of oxen. I supposed I should be tired of them; but on the contrary I am tired of horses; as I find that with my two oxen, I can do more work than I could with four horses, and with half the expense. I have worked horses for forty years, and if I had used oxen in their place, they would have put five hundred pounds in my pocket."

To train oxen, put a broad strap round their necks: fasten one end of a cord to it, and the other end to a large log of wood: permit the ox to drag it about as he feeds in his pasture before he is put into harness, by which his docility is much forwarded.—*Transactions of the Society of Arts*, 1800.

OYSTER SHELLS are an excellent manure, but being large they should be burnt to lime before they are applied to the soil.

P.

PALE, a pointed stake, used in making enclosures, partitions, &c. Gardeners oftentimes have occasion to make pale fences, to secure choice apartments from the entrance of tame fowls, which will not often fly over a paled or picketted fence: as well as to prevent the intrusion of idle and mischievous people.

PAN, a stratum of compact earth under the soil. In some places it is so hard that it cannot be dug through without pickaxes or crows. If the pan be low, the soil is said to be deep and good; but if near the surface, the soil is thin and poor. The common depth in good land is from eighteen to twenty-four inches.

The deeper strata, or layers in the bowels of the earth, are supposed to have been formed, by the diurnal rotation of the earth, before it had become compact and solid. But this stratum being more constant and regular, the formation of it, if I mistake not, should be ascribed to other causes. If we suppose that this and the soil above were intermixed, and of one consistence after the creation, the pan must have been formed long before this time, by the subsiding of the more ponderous parts of the soil. For it has been often observed, that clay, chalk, and lime, which have been laid on as manures, after some years, disappear from the surface, and are found a foot or more beneath it. Rains, and fermentations in the soil, make way for the descent of the heaviest particles contained in the soil.

It is in favour of this hypothesis, that the pan under the soil most commonly bears an affinity to the soil itself. Under a gravelly soil, there is a large proportion of gravel in the pan; under a sandy one it usually is found to consist chiefly of sand; and under a stiff loam it is commonly clay: I think it is always found to be so.

But I suppose the operation of frost should be considered, as assisting in forming the pan. All the soil above it is usually hoven by the frost in winter. At least it is so in this latitude. We see rocks and stones below the surface when the ground is frozen, which before were on a level with it; and in a soft soil they do not rise quite up to their former situation, when the ground is thawed. The frost does more than tillage, and perhaps more than rains, or fermentation, towards causing the more ponderous parts of the soil (or ponderous bodies in the soil) to subside, or sink.

The frost may have another influence in increasing the compactness of the under stratum. As the frost expands the soil, the pressure of it downward is increased; by which pressure, the matter of which the pan consists, is made most close and hard, like earth that has been violently rammed. But this perhaps can take place, only when the frozen stratum is held down by strong objects, which reach far below the frost and pan; as the stumps of large trees deeply rooted, large rocks, &c.

But it will be objected, that some soils appear to have no pan under them. To answer this, it may be said, that perhaps some soils were originally made up of particles equally ponderous; so that one had no more tendency to subside than another. Or else the looseness and openness of the under earth in such places, was so great that it could not stop the ponderous parts of the soil in their descent; so that they have been dispersed among the loose earth, and part of them gone to a very great depth.

If I have given a just account of the formation of the pan, will it not follow, that this under stratum is less penetrable in cold than in warm latitudes, when made of like materials? So far as my observation has extended, this appears to be the case. It ought also to be lower in the earth, and the soil deeper; and future observations may convince us that this also is fact.

Another corollary may be, that deeper ploughing than is usually practised in this country would be proper. For it seems that nature designed all the strata above the pan to serve for pasture of plants. And it is well known that the more it is stirred and mixed, the fitter it is for this purpose; not only because it lies the more loose and open, but because the more of the food of plants will be contained in it.

Such a stratum, at a right distance from the surface, is a great benefit to the soil. For, as no manures can easily penetrate it, they must remain in a good situation to be taken up by the roots of vegetables. But where there is no compact under stratum, unless at a great depth, manures laid upon the soil are partly lost. Hence appears the great propriety of claying and marling such soils. In a long course of tillage, these dressings will subside, and do something towards forming the stratum that is wanted.

But to form a good under stratum at once, where it is wanted, let one hundred or more loads of clay be spread on an acre of sandy grass land. After it has lain, spread upon the surface one winter, let it be made perfectly fine and even by a bush harrow, and rolled. Afterwards turn it under with a very deep ploughing. This will greatly assist a weak dry soil to retain moisture, and to hold the manures that shall be given it. It will be a lasting benefit. But this ploughing should be done at a time, when the clay is so damp that it will turn over in whole flakes.

When a plot of ground intended for a garden wants an under stratum, it may be advisable to dig trenches four feet wide, and place a regular bed of clay in the bottom. The second trench may be contiguous to the first, and the first be filled up with the earth that is taken out of the second; and so on till the whole work is completed.

Some have put themselves to the expense of this operation, only with a view to get rid of all the seed of weeds in a garden which had long lain neglected, placing the upper part of the soil at the bottom.

PANAX, GINSENG, *or* NINSENG. As this plant is a native of our country, and is become a considerable article of commerce, I think it is necessary that every one should know how to distinguish it from all other plants when he meets with it. I desire therefore to entertain the reader with Mr. Miller's account of it.

"It hath male and hermaphrodite flower on distinct plants. The male have simple globular umbels, composed of several coloured rays, which are equal. The flower hath five narrow, oblong, blunt petals, which are reflexed, sitting on the empalement, and five oblong slender stamina inserted in the empalement, terminated by single summits. The hermaphrodite umbels are simple, equal, and clustered; the involucrum is small, permanent, and composed of several awl-shaped leaves. The flowers have five oblong, equal petals, which are recurved, and five short stamina terminated by single summits, which fall off, with a roundish germen under the empalement, supporting two small erect styles, crowned by simple stigmas. The germen afterwards becomes an umbilicated berry with two cells, each containing a single heart shaped, convex, plain seed.

"The species are, 1. *Panax quinquefolium, foliis ternis quinatis;* or panax with trifoliate cinquefoil leaves; called ninzin. 2. *Panax trifolium, foliis ternis ternatis;* or panax with three trifoliate leaves.

"Both these plants grow naturally in North America; the first is generally believed to be the same as the Tartarian Ginseng. It has a fleshy taper root, as large as a man's finger, which is jointed, and frequently divided into smaller fibres downward. The stalk rises above a foot high, naked to the top, where it generally divides into three small foot stalks, each sustaining a leaf composed of five spear shaped lobes, which are sawed on their edges; they are of a pale green, and a little hairy. The flowers arise on a slender foot stalk, just at the division of the foot stalks which sustain the leaves, and are formed into a small umbel at the top; they are of an herbaceous yellow colour, composed of small petals, which are recurved. These appear the beginning of June, and are succeeded by compressed heart shaped berries, which ripen the beginning of August. The Chinese affirm that it is a sovereign remedy for all weakness occasioned by excessive fatigues, either of body or mind; that it cures weakness of the lungs and the pleurisy; that it stops vomitings; that it strengthens the stomach, and helps the appetite; that it strengthens the vital spirits, and increases the lymph in the blood; in short, that it is good against dizziness of the head, and dimness of sight, and that it prolongs life in old age."

Mr. Miller found he could not propagate this plant by the seed, either raised in England, or brought from America. None of the seeds would grow. He believes the hermaphrodite plants should have some of the male plants standing near them, to render the seed prolific; for all the plants he saved seed from, had only hermaphrodite flowers.

PARING AND BURNING. Paring the surface of the soil, and burning the sods, is a process, highly recommended by writers on agriculture, and is useful in all cases where an excess of vegetable matter renders the soil too rank. It is scarcely possible to redeem fens and peat mosses from a state of barrenness without the assistance of fire. Old pastures, whose surface has become mossy, and unproductive, or covered with useless and coarse herbage, are best reclaimed in this way. A meadow overrun with rushes should first be drained, and then pared and burnt. But burning is prejudicial to sandy, dry and flinty soils, containing little animal or vegetable matter.

A new method of "burning without fire" has been lately discovered. This consists in substituting quick lime for fire. The lime in its most caustic state, fresh from the kiln, is laid upon the vegetable surface to be consumed; and before it is weakened by exposure to the air, a quantity of water, just sufficient to put it in powerful action, is applied. This process unites the advantages of burning and liming; and is probably the readiest and cheapest mode of fertilizing soils which super-abound with vegetable matter, and for which lime would prove a suitable manure. See *Code of Agriculture*. Likewise *Burning the Soil*.

PARSNEP, *Pastinaca*, an esculent root, of a sweet taste, and of a very nourishing quality.

Parsneps must have a mellow, rich and deep soil, not apt to be very dry. The best tasted roots are produced in a soil that is more sandy than loamy. When they are cultivated in kitchen gardens, the ground should be dug uncommonly deep; eighteen or twenty inches at least. No common ploughing will loosen the soil to a sufficient depth. The goodness of a crop of these roots depends much upon their length.

If they be set near together, they will not grow to a large size. I sow them in rows across the beds, 15 inches apart, and allow about 6 inches from plant to plant at the last thinning, which may be early, as they are not often hurt by insects. I have seldom known any to be destroyed by them.

The seeds should be sown as early as in March, if the ground be thawed, and not too wet. Some sow them in the fall; but that is not a good practice, because the ground will grow too close and stiff, for want of stirring in the spring; which cannot well be performed in gardens, without danger of injuring the roots. And weeds will be more apt to abound among them, if they be sown in autumn.

The manure that is used for parsneps should be very fine and rotten, and quite free from straws and lumps; otherwise it will cause the roots to be forked, which is a great damage to them. They require but little manure, as they draw much of their nourishment from a great depth. What manure is given them, should be spread before digging the ground, that some of it at least may go deep. They do not impoverish the soil. I have raised them near thirty years in the same spot, on a soil not naturally rich, and with a very slight yearly dressing. The crops are better than they were at first: And the earth is become very black to a great depth.

Parsneps will continue growing so late as till the tops are killed by the frost, if not longer. Some let them remain in the ground through the winter, expecting that they will grow larger in that season. But it is not possible they should grow at all, so long as they are enclosed with the frozen soil. They may possibly grow a little in the spring, before there is opportunity to take them up, if they escape rotting. But their growing will be chiefly sprouting at the top, which hurts them for eating. As soon as they begin to sprout, which will be as soon as the ground is thawed, they will begin to grow tough, and to have a bitterish taste.

The best way is to dig them up about the last of November, or in the beginning of December. Let them not be wounded, or so much as touched with the spade in doing it, if it can be avoided; neither should the tops be cut off very close to the roots, nor any of the lateral roots cut off. In either case the roots will rot, or become bitter.

Many lose their parsneps, or make them sprout, by putting them into a warm cellar. It is better to keep them in some out-house, or in a cellar that freezes; for no degree of frost ever hurts them. But to prevent their drying too much, it is best to cover them with dry sods, or else bury them in sand that has no moisture in it. Beach sand is improper, because the salt in it will make them vegetate.

It is said by European writers, that parsneps are an excellent food for swine, and useful for feeding and fattening all sorts of cattle.

If we would cultivate them for these purposes, the horse hoeing husbandry must be applied. The ground must be trench ploughed in October, and all the stones carefully taken out. The trench ploughing must be repeated before the end of November, the soil made fine by harrowing, laid in beds of from three to four feet wide, and sown by a line in drills on the middle of the beds. There may be either one or two rows on a bed. If there be two, they should be full twelve inches apart, and the intervals proportionably wider.

Autumnal sowing in the field culture is not amiss, as the ground is to be kept light by horse hoeing. In this operation the ground should be stirred very deep. The plough should go twice in a furrow. At the last ploughing, the furrows should be turned towards the rows.

PASTURE, according to the language of farmers in this country, means land in grass for the summer feeding of cattle.

To manage pasture land advantageously, it should be well fenced in small lots, of four, eight or twelve acres, according to the largeness of one's farm and stock. And these lots should be bordered at least with rows of trees. It is best that trees of some kind or other should be growing scattered in every point of a pasture, so that the cattle may never have far to go in a hot hour to obtain a comfortable shade. The grass will spring earlier in lots that are thus sheltered, and they will bear drought the better. But too great a proportion of shade should be avoided, as it will give a sourness to the grass.

Small lots thus sheltered, are not left bare of snow so early in the spring as larger ones lying bare, as fences and trees cause more of it to remain upon the ground. The cold winds in March and April hurt the grass much when the ground is bare. And the winds in winter will not suffer snow to lie deep on land that is too open to the rake of winds and storms.

It is hurtful to pastures to turn in cattle too early in the spring: and most hurtful to those pastures in which the grass springs earliest, as in very low and wet pastures. Potching such land in the spring, destroys the sward, so that it will produce the less quantity of grass. Neither should cattle be let into any pasture, until the grass is so much grown as to afford them a good bite, so that they may fill themselves without rambling over the whole lot. The 20th of May is early enough to turn cattle into almost any of our pastures. Out of some they should be kept later. The driest pastures should be used first, though in them the grass is shortest, that the potching of the ground in the wettest may be prevented.

The bushes and shrubs that rise in pastures, should be cut in the most likely times to destroy them. Thistles, and other bad weeds, should be cut down before their seeds have ripened; and ant hills should be destroyed. Much may be done towards subduing a bushy pasture, by keeping cattle hungry in it. A continual browsing keeps down the young shoots, and totally kills many of the bushes. Steers and heifers may mend such a pasture, and continue growing.

But as to cleared pastures, it is not right to turn in all sorts of cattle promiscuously. Milch kine, working oxen, and fatting beasts, should have the first feeding of an enclosure. Afterwards, sheep and horses. When the first lot is thus fed off, it should be shut up, and the dung that has been dropped should be beat to pieces, and well scattered. Afterwards, the second pasture should be treated in the same manner, and the rest in course, feeding the wettest pasture after the driest, that the soil may be less potched.

Something considerable is saved by letting all sorts of grazing animals take their turn in a pasture. By means of this, nearly all the herbage produced will be eaten; much of which would otherwise be lost. Horses will eat the leavings of horned cattle; and sheep will eat some things that both the one and the other leave.

But if in a course of pasturing, by means of a fruitful year, or a scanty stock of cattle, some grass of a good kind should run up to seed, and not be eaten, it need not be regretted; for a new supply of seed will fill the ground with new roots, which are better than old ones. And I know of no grass that never needs renewing from the seed.

A farmer needs not to be told, that if he turn swine into a pasture, they should have rings in their noses, unless brakes and other weeds need to be rooted out. Swine may do service in this way. They should never have the first of the feed; for they will foul the grass, and make it distasteful to horses and cattle.

Let the stock of a farmer be greater or less, he should have at least four enclosures of pasture land. One enclosure may be fed two weeks, and then shut up to grow. Then another. Each one will recruit well in six weeks; and each will have this space of time to recruit. But in the latter part of October, the cattle may range through all the lots, unless some one may become too wet and soft. In this case, it ought to be shut up, and kept so till feeding time the next year.

But that farmers may not be troubled with low miry pastures, they should drain them, if it be practicable, or can be done consistently with their other business. If they should produce a smaller quantity of grass afterwards, it will be sweeter, and of more value. It is well known, that cattle fatted in a dry pasture, have better tasted flesh than those which are fatted in a wet one. In the old countries it will fetch a higher price. This is particularly the case as to mutton.

Feeding pastures in rotation, is of greater advantage than some are apt to imagine. One acre, managed according to the above directions, will turn to better account, as some say who have practised it, than three acres in the common way. By the common way I would be understood to mean, having weak and tottering fences, that will drop of themselves in a few months, and never can resist the violence of disorderly cattle; suffering weeds and bushes to overrun the land; keeping all the pasture land in one enclosure; turning in all sorts of stock together; suffering the fence to drop down in autumn, so as to lay the pasture common to all the swine and cattle that please to enter; and not putting up the fence again till the first of May, or later. Such management is too common in all the parts of this country with which I am most acquainted. I would hope it is not universal.

Land which is constantly used as pasture, will be enriched. Therefore it is advisable to mow a pasture lot once in three or four years, if the surface be so level as to admit of it. In the mean time, to make amends for the loss of pasture, a mowing lot may be pastured. It will thus be improved: And if the grass do not grow so rank afterwards in the pasture lot, it will be more clear of weeds, and bear better grass. Alternate pasturing and mowing has the advantage of saving a good deal of expense and trouble, in manuring the mowing grounds.

Though pastures need manuring less than other lands, yet, when bushes, bad weeds, &c. are burnt upon them, the ashes should be spread thinly over the surface. The grass will thus be improved: And grass seeds should be sown upon the burnt spots, that no part may be vacant of grass.

Sheep, calves and horses, unless they are worked, it is said require no water in their pastures. The want of water induces them to feed in the night, when the dew is on and the grass the more nutritious. Cows however want pure water.

In pastures, which are on side hills water may generally be obtained by digging horizontally into the side of the hill, till it is found, and then carrying it out with a pipe.

PEACH TREE, *Amygdalus*, a well known kind of fruit tree. Mr. Miller reckons no less than 31 sorts, besides a number of less value. What we call the rareripe, is almost the only sort.

Peach trees should be cultivated near to or in the borders of gardens. When they are propagated by planting the stones, they should be taken from fruit that has thoroughly ripened on the tree, and be planted in October, three inches under the surface. The trees may also be propagated by inoculating upon plums and apricots. This will undoubtedly render them longer lived.

When the trees are transplanted, the downright shoot of the roots should be pruned very short, and the lateral ones be left at a good length; for if the trees draw much of their nourishment from a great depth, the sap will be crude, and the fruit not so good. As these trees are natives of a warmer climate they ought to have a southern exposure. They should also be screened from the direct influence of north, and north-easterly winds.

The soil that suits them best is a dry light loam; and the surface should be constantly tilled, and moderately manured with old rotten dung.

If too great a quantity of peaches appear on the trees, so as to crowd each other, they should be speedily thinned, by taking off the poorest: For if they be suffered all to remain on the tree, much of the fruit will drop off unripe: What remains will not be so perfect, and perhaps fewer in number.

As the fruit grows not on spurs,

but on the shoots made in the last preceding year, Mr. Miller directs, that the new shoots should be shortened, by cutting them yearly in October, leaving them from five to eight inches in length, according as they are weaker or stronger. I have practised this method of cutting in October for several years; which has caused trees, which were before barren, to bear some fruit. And I observe that the branches of the trees are not so often killed by the frost in winter. But the trees have now become sickly and barren.

Mr. Thomas Coulter of Bedford, Penn. gives some directions for cultivating peach trees, which he has practised with success in Pennsylvania and Delaware for forty-five years. See *Trans. Amer. Philad. Soc.* vol. 5.

Mr. Coulter advises to transplant peach trees, as young as possible when you mean them to stand; if in the kernel, so much the better. To plant them 16 feet apart both ways, except you would wish to take your waggon through the orchard to carry the peaches away; in that case give 24 feet distance to every fifth row, one way, after transplanting. In the month of March or April, in the third year after transplanting, *cut them all off by the ground*, and let all the sprouts and scions grow. From four to six will, generally, come to maturity; the rest will die, and should be cut away, taking care not to wound the stock. The sprouts growing all round the old stump, when loaded with fruit, will bend and rest on the ground in every direction, and will remain fruitful for many years. Three years after the trees are cut off by the ground, they will be sufficiently large and bushy to shade the ground so as to prevent grass of any kind from matting or binding the surface, so as to injure the trees; therefore ploughing is useless, as well as injurious; useless because nothing can be raised in the orchard by reason the trees will shade all the ground, or nearly so; injurious because either the roots, stock or branches will be wounded.

Mr. Coulter says that peach trees should have no manure, as it causes them to produce worse and smaller fruit.

A blue fly attacks peach trees from about the middle of July to the middle of September, and generally deposits its eggs in the bark at the surface of the ground. These eggs hatch into worms, which eat off the bark quite round the tree. To guard against this, raise a little hillock in the month of June round the tree, about a foot high, so as completely to cover that part of the bark, kept moist and tender at the surface of the ground. This hillock will not stand so long at one height as to make tender the bark above, as the rain will gradually wash it down level with the surface, and it must be raised again every summer.

To take out the worm, the roots must be uncovered, and the spot looked for where the gum oozes out; follow the cavity round with the point of a knife, until you come to the solid wood, and lay the whole open; the worm will be found, with a white body and black head; which must be destroyed, and the hole filled up with cow manure, rendered adhesive by lime sand and ashes, as directed by Forsyth. See *Orchard*.

Soap suds, heated after a family wash, and poured on the roots of the trees, about the middle of August, have been used with success in destroying the eggs, or the young worm.

Mr. John Ellis, of New-Jersey, directs taking away the dirt round the tree, so as to expose the root to the depth of about six inches,and surrounding the tree with straw about three feet long, and apply it lengthwise about an inch thick, the butt-ends of the straw resting on the ground at the bottom. This straw should be bound with three bands, one near the top, one at the middle, and the third at the surface of the earth. Fill up the hole at the root with earth, and press it closely round the straw. When the white frosts appear, the straw should be removed, and the tree remain uncovered until the blossoms put out in the spring. By this process the fly, it is said, is prevented from depositing its egg within three feet of the root, and although it may place its egg above that distance, the worm travels so slow, that it cannot reach the ground before frost, and therefore is killed before it is able to injure the tree.

Mr. Forsyth recommends thinning the peaches, when they are about the size of a small marble, which ought to be done according to the strength and size of the tree, and must be left to the judgment of the person, who performs the operation.

Mr. William Coxe, of Burlington, N. J. prefers a northern aspect for peach trees: "I have tried," he remarks, "with great success a northern aspect,and even the north side of a building for apricots, the tenderest and earliest of our fruits. The northern situation of the rising and setting sun, is sufficient to impart to them a large portion of heat in the long days of summer; while the blossoming is retarded by the greater shelter afforded by the position of the sun in the spring." *American Farmer*, vol. II. p. 71.

Peach trees are subject to a disease called the "Yellows," of which we have seen no particular description. The remedy however is wood ashes, scattered about the roots, which is said to be unfailing. *American Farmer*, vol. II. p. 187.

PEAR TREE, *Pyrus*. Pears have a nearer affinity to quinces than to apples: For a pear scion will grow upon a quince stock, but not so well upon an apple: And a quince scion will grow upon a pear stock.

The vast variety of pears, which are cultivated in the world, have been obtained from the seeds, which, like those of the apple, will produce fruit trees different from the parent tree. Seeds sometimes bring degenerate, and sometimes improved fruit trees. So that all the best grafted fruits have been, some time or other, produced by nature itself: And though the fruits vary, there is not a specifical difference.

Though the pear will grow upon the quince, or even upon the white thorn, it should not be grafted on the former, unless it be for dwarf trees, and even for these it is not good; and in no case upon the latter. The stock of the thorn will not grow to so large a size as the scion will: The trees will therefore be top heavy, and short lived, as I have found by experience. Therefore it is best in general, that pears should be grafted upon pears.

The propagation of pear trees from the seeds, and the culture of them in nurseries, do not differ from the propagation and culture of apple trees. See *Nursery*.

Pear trees bear fruit at the ends of the last year's shoots, as well as upon the spurs. Therefore, the new shoots should not be shortened, lest the fruit be diminished: And, for the same reason, these trees should never stand so near together as to crowd each other. But the distance at which the trees are to be set in an orchard, or in a grove, depends partly upon the nature of the trees, as some grow larger than others; and partly upon the fruitfulness of the soil. In general they may be allowed to stand nearer together than apple trees. These, as well as other fruit trees, should have the ground tilled about them, to promote their growth and fruitfulness, at least until they are become so large as to bear plentifully, and occasionally from time to time afterwards.

The Massachusetts Agricultural Repository, vol. IV. No. III. contains a description by Professor Peck, with drawings of the insects which destroy the young branches of the pear tree, and the leading shoots of the Weymouth pine. "The genus, to which this insect belongs is called Scolytus,of which it is an undescribed species. It is precisely $\frac{1}{10}$ or $\frac{10}{000}$ of an inch in length, $\frac{4}{100}$ in diameter; of a deep brown colour, the legs and antennæ pale, and of a rust colour; the thorax in front is rough with small tubercles which point upwards and is studded with erect bristles, as are also the elytra or wing-cases and other parts of the body. The elytra are striated with slightly impressed points, and between the series of points are rows of bristles. The plane of the anterior opening of the thorax, which receives the head, is nearly at right angles with that which joins the abdomen, so that the head is entirely underneath. The eyes are oblong, and the antennæ inserted at their lower and anterior edge.

"The mischievous effects of this minute insect are observed in June and July; the dead part of the branches of the pear tree should be immediately cut off and burnt without delay, as the insects have not then left them."

Mr. Forsyth gives a catalogue of more than seventy different kinds of this fruit.

A writer in the American Farmer, vol. II. p. 347, mentions a disease, which affects pear trees,which causes a withering of the interior bark, especially the limbs appearing in spring, in spots and bars, and more extended affections of the same kind, spreading to the destruction of the individual limb, though the superior parts of the limbs are often unaffected by the disease, and are only destroyed by being cut off from the main stock. This disease takes place, when a warm winter, especially a warm February is followed by a cold March. He thinks that manuring and high cultivation are the causes of the disorder, by causing the sap to flow too exuberantly when winter relaxes a little of its rigour. The remedy he proposes is to let the trees grow *naturally*, with shrubs and grass about their roots; but we doubt exceedingly both as to the supposed cause of the evil, and the proposed remedy.

PEASE, *Pisum*. The varieties are so numerous, that I shall not undertake to distinguish them. They are cultivated in gardens and in fields. The garden culture is thus: After the ground has been well dug, raked and levelled, mark it out in double rows one foot apart, and leave intervals of three feet between the double rows, so that when they are brushed, there may be a free passage through the intervals. Open the trenches three inches deep with the head of a rake, or with a hoe; scatter in the pease at the rate of about one to an inch, or nearer together if you have plenty of seed; and then cover them with a rake. Or small marks may be made for the rows, and the pease pricked in with a finger to the same depth, and the holes filled with a rake. The former method is best, as the mould about the pease is left lighter; and it is more expeditiously performed.

The ground should be hoed, and kept clear of weeds; and when the young plants are six inches high, the stems should be earthed up a little, and each double row filled with brush wood, so that each plant may climb, and none of them trail upon the ground. The brush should be set strongly in the earth, or they will not bear the weight of the plants in windy weather. I set the larger bushes strongly between the rows, making the holes with a crow bar; and then the smaller bushes in the rows as leaders. The latter may be sharpened a little at the points, and pushed in by hand. They will be the more fruitful for brushing or sticking, as well as more sightly, and more conveniently gathered. But the low dwarf kinds seldom need any supporting.

Whatever be the sort, no weeds should be suffered to increase among them; and the alleys should be hoed deep once or twice after brushing. But the soil should not be very rich, lest the plants run too much to haulm. The most hungry part of a garden answers well for pease.

The earliest sorts of pease will somtimes be ripe in June: So that a crop of potatoes, turnips, or cabbages, may be had after them.

For field pease, land that is newly ploughed out of sward is generally accounted best; and land which is high and dry, and has not been much dunged. A light loamy soil is most suitable for them; and if it abound with slaty stones it is the better. But they will do in any dry soil. The sorts that grow large should have a weaker soil; in a stronger soil the smaller sorts answer best. The manures that suit pease best are marle and lime.

Horse-hoeing husbandry applied, if it were practicable, would greatly assist the growth of pease. They soon begin to trail upon the ground, that the season in which this culture can be applied, is extremely short. But some have obtained very good crops in this way. Much of the seed at least might be saved.

Our farmers do not commonly allow a sufficient quantity of seed for pease, in broad-cast sowing. When pease are sowed thin, the plants will lie upon the ground, and perhaps rot: When they are thick, the plants will hold each other up, with their tendrils, forming a continued web; and will have more benefit of the air.

At Fryburgh and Conway, as I am informed, the farmers sow three bushels on an acre, according to the practice in England; and their crop, one time with another, is upwards of twenty bushels. This is certainly better for them, than to sow one bushel, and reap fifteen: But he that sows one bushel only on an acre, must not expect, one time with another, to reap twelve.

The only insect that commonly injures our pease, is a small brown bug, or fly, the egg of which is deposited in them when they are young, and the pods easily perforated. The insect does not come out of his nest, till he is furnished with short wings. They diminish the pease in which they lodge to nearly one half, and their leavings are fit only for the food of swine. The bugs, however, will be all gone out, if you keep them to the following autumn. But they who eat buggy pease, the winter after they are raised, must run the venture of eating the insects.

If sown in the new plantations, to which this bug has never been carried, pease are free from bugs: For the insects do not travel far from their native place. Therefore, care should be taken not to carry them, as some are apt to do, in seed, from older settlements. Even in the part of an old farm, near to which pease have not for a long time, if ever, been sown, a crop of pease are not buggy, if clean seed be sown. Therefore, in such places, one may guard against this insect, by sowing pease which are certainly known to be clear of them. But if the contrary be known, or even suspected, let the pease be scalded a quarter of a minute, in boiling water; then spread about, cooled, and sown without delay. If any of the bugs should be in the pease, this scalding will destroy them: And the pease, instead of being hurt, will come up the sooner, and grow the faster.

All pease that are sown late, should be steeped, or scalded, before sowing. They will be forwarder. But pease should always be sown as early as the ground can be got into a good tilth, without any silly regard to the time of the moon; by which I have known some miss the right time of sowing, and suffer much in their crop. The real causes of a crop not ripening equally, are bad seed, poor culture, and sowing too thin. If the ground be ploughed but once, it should be harrowed abundantly. But on green sward ground, I think it should be ploughed early in autumn, and cross ploughed and harrowed in the spring. In old ground, as it is called, it is no bad way to plough in the seed with a shoal furrow: It will be more equally covered, and bear drought better; and I should think the crop would ripen more equally. There is no danger of their being buried too deep, in our common method of ploughing. The European farmers think six inches is not too great a depth for pease to be covered in moist soils, and four inches not too deep in clay.

Changing the seed is a matter of very great importance; for pease are apt to degenerate more rapidly than almost any other plants. Seeds should be brought from a more northern clime; for those which ripen earliest are best. I would change them yearly, if it could be done without much trouble or cost. Once in two or three years is necessary.

If weeds come up among field pease, while they are young, they should be weeded. But when they are grown up, they will hinder the growth of weeds by their shade, unless they are sown too thin. Pease sown thick form so close a cover for the soil, that they cause it to putrefy; they are therefore called an improving crop: But they also draw a greater proportion of their nourishment from the air, than most other plants; for it is observable that they continue their greenness long after the lower parts of the stems are dead to appearance.

Garden pease are harvested by picking them off as they ripen; but field pease must unavoidably be harvested all at once. They should be carefully watched, and harvested, before any of them are so ripe as to begin to shell out. Those among them which are unripe, will ripen, or at least become dry, after they are cut or pulled up; and such pease, well dried, are not commonly bad for eating, though ill coloured. To dry them, they should be laid on the ground in small heaps, as light and open as possible, the greenest of the straw and pods uppermost. The heaps should never be turned upside down, though rain should fall, but they may be gently lightened up, if they settle close to the ground. This will be sufficient. When thoroughly dried, they should be carefully removed to the barn, at a time when the air is not dry, and thrashed without delay. But if the thrashing must be delayed, it is better to keep them in a stack than in a barn.

After winnowing, pease should lie on a floor, three or four inches thick, and air should be let into the apartment, that they may be dried; which they will be in two or three weeks, the weather being generally dry. After this they may be put into casks to keep.

Our common method of pulling up pease by hand, is too laborious.

They should be cut or pulled up with a sharp hook in the form of a sickle; fastened to a long handle. Some perform it expeditiously with a common sickle. But this is little, if at all, less laborious than doing it with the hand.

When land is in suitable order, field pease may be cultivated according to the new husbandry, with advantage. M. Eyma found his crops were half as large again in this way, as in the old husbandry, besides saving half the seed. The intervals between the double rows should be near four feet wide, or there will not be sufficient room for horse-hoeing. And this should be done with, before the plants begin to trail on the ground.

Pease are said to be much improved for the purpose of feeding hogs by steam boiling. The Domestic Encyclopedia recommends grinding them for that use.

" It is a great error in those persons, who sow the rows of tall growing pease close together. It is much better in all those sorts which grow six or eight feet high to have only one row, and then leave a bed ten or twelve feet wide for onions, carrots, or any crops, which do not grow tall.

" The advantages which will be derived are, that the pease will not be drawn up so much; be stronger; will flower much nearer the ground, and in wet weather can be more easily gathered without wetting yourself."—*Domestic Encyclopedia.*

The same work recommends sowing pease in rows of circles, three feet in diameter, with a space of two feet between each circle; and if you want more than one row of circles, leave a bed of ten or twelve feet before you begin another.

PEAT, a kind of earth, or rather a fossil, used in some countries for fuel.

It is often found in low, miry, and boggy places, that lie between hills. That which is the most solid is the most valuable. It lies at different depths; sometimes, very near the surface; sometimes eight or ten feet below it. The best way to find it is by boring. The stratum above it is most commonly mud, or moory earth.

I suppose many places where it is found to have been originally ponds; and that they have been, either suddenly, at the time of Noah's flood, or gradually since, filled up with wood, and other vegetable substances, which, by a slow putrefaction, have been changed into the substance we call peat. For some undissolved trunks of trees, bark, &c. are found among it.

It is sometimes found in interval lands, and near to the banks of rivers. In these places, the shifting of the beds of rivers, caused by the choking of the old currents, will afford a probable account of its formation.

Peat is distinguishable by its cutting very smooth, like butter or lard, by its being free from grit, and its burning freely, when thoroughly dried. It will not dissolve when exposed to the air for a long time, but become hard like cinder.

A dry season is the best opportunity for digging it, as the labourers are but little incommoded by water. They who dig peat for fuel, should have long angular spades, the blades of which should be spiked like a carpenter's bur, with which it may be easily cut out of the pits, in pieces four inches square, and twenty inches in length. These should be laid singly on the surface to harden. When they are partly dry they are piled open, athwart each other: And in a few days of drying weather, they will be fit to cart, and store for fuel. This fuel must be constantly kept in a dry place.

It has been found by trials that the ashes of peat is a very important manure, of three times the value of wood ashes. Fifteen bushels are allowed to be a sufficient top-dressing for an acre. It is an excellent manure for cold grass lands; and for all such crops in any soil as require much heat. They should be sowed by hand, as they can thus be more evenly spread. It may be done in winter with the least danger of hurting plants by its heat. If sown in summer, it should be just before rain, when it should be immediately deprived of its burning quality.

The method of burning peat to ashes, I will give from the *Museum Rusticum*, as I have had no experience in it myself.

"As soon as it is dug, some of it is mixed in a heap regularly disposed with faggot wood, or other ready burning fuel: After a layer or two of it is mixed in this manner, peat alone is piled up to complete the heap. A heap will consist of from one hundred to a thousand loads.

"After setting fire to it at a proper place, before on purpose prepared, it is watched in the burning: And the great art is to keep in as much of the smoke as possible, provided that as much vent is left as will nourish and feed the fire.

"Whenever a crack appears, out of which the smoke escapes, the labourer in that place lays on more peat; and if the fire slackens too much within, which may be known by the heat of the outside, the workmen must run a strong pole into the heap, in as many places as is needful, to supply it with a quantity of fresh air. When managed in this manner, the work goes on as it should do. It is noticed, that when once the fire is well kindled, the heaviest rain does it no harm whilst it is burning." To preserve the ashes for use, this writer proceeds thus:

"It is necessary to defend the ashes from the too powerful influence of the sun, air, dews, rain, &c. or great part of their virtue would be exhaled and exhausted. If the quantity of ashes procured is not very great, they may be easily put under cover in a barn, cart lodge, or hovel; but large quantities must necessarily, to avoid expense, be kept abroad; and when this is the case, they should be ordered as follows:

"A dry spot of ground must be chosen; and on this the ashes are to be laid in a large heap, as near as possible in the form of a cone standing on its base, the top as sharp-pointed as possible: When this is done, let the whole be covered thinly over with a coat of soil, to defend the heap from the weather: The circumjacent earth, provided it is not too light and crumbly.

"When thus guarded, the heap may very safely be left till January or February, when it is in general the season for spreading it. But before it is used, it is always best to sift the ashes, &c."

Mr. Eliot supposed it was necessary to dry the peat before burning: But perhaps he never tried the above method. He says, if it be stifled in burning, it will be coal instead of ashes; and that the red sort makes better charcoal than that made of wood.

It is happy for mankind, that bountiful Providence has prepared and preserved this precious treasure, containing the essence of vegetables, by which they may be supplied with fuel in their houses, manure for their lands, and coal for smiths' forges. But in vain it is provided, unless men will search for it, and make use of it. There is no reason to doubt of its being as plenty in this country, as in any other. When Mr. Eliot searched for it, he tells us he soon found it in seven different places. Peat is now found in great abundance and is much used.

The ashes are said to have a better effect upon winter, than upon summer grain; and to be not good for leguminous plants, as it makes their haulm too luxuriant. The good effects of a dressing are visible for three years; and they will not leave land in an impoverished state.

PERRY, a liquor made from pears, in the same manner as cyder is from apples. The pears should, in general, be ripe before they are ground. They will not bear so much sweating as apples. The most crabbed and worst eating fruit, is said to make the best perry. After perry is made it should be managed in all respects like cyder. Boiling has a good effect on perry, changing it from a white to a flame-coloured and fine flavoured liquor, which grows better by long keeping and bottling.

PERSPIRATION *of* PLANTS, the passing off of the juices that are superfluous, through pores prepared by nature on their superficies for that purpose.

The analogy which plants bear to animals, is in no instance more remarkable than in this evacuation. The parts of a plant which contain the excretory ducts, are chiefly the leaves. For we find, that if a tree be continually deprived of its leaves for two or three years, it will sicken and die, as an animal does when its perspiration is stopped. But smear the bark on the stems with any glutinous substance sufficient to stop any pores, and no great alteration will be observed in the health of the tree, as has been proved by experiment. And as M. Bonnet has proved that leaves generally imbibe the moisture of the atmosphere on their under surface, it is not reasonable to suppose that the pores for transpiration are placed on their upper surface? But that the stems of plants contain some bibulous pores, seems evident from this, that when placed in the earth, they will send out roots. But these pores in the stems are so few, that the stopping of them does not materially injure a plant.

As animals have other ways of throwing off those parts of their food which are not fit to nourish them, it is no wonder that plants have been found to perspire insensibly a far greater quantity than animals. Plants cannot choose their food as animals do, but must take in that which is presented by the earth and atmosphere, which food in general is more watery, and less nourishing, than that of animals; and for these reasons, also, it might be justly expected, that the matter perspired by a plant should be vastly more than that perspired by an animal of the same bulk; and this has been found to be the case. See the article *Leaves*.

A practical inference or two from the copious perspiration of plants may be, that the plants we cultivate should not be set too close, that they may not be incommoded, or rendered sickly, by the unwholesome steams of each other. They are as liable to be injured this way, for ought that appears to the contrary, as animals are. And the water that drips from trees upon smaller vegetables is known to be not healthy for them; the reason is, because this water contains some of the matter which is perspired from the trees. But if the perspirable matter of plants be injurious to plants, it does not follow that it is so to animals. It is thought to be not so in general, but the reverse. The effluvium of poisonous plants is an exception.

PLANTS. Sir Humphrey Davy says, that every plant examined as to external structure, displays at least four systems of organs, or some analogous parts. First, the *Root*; secondly, the *Trunk* and *Branches*, or stem; thirdly, the *Leaves*; and fourthly, the *Flowers* or *Seeds*. The roots of plants in their anatomical division are very similar to the trunk and branches. The root may indeed be said to be a continuation of the trunk terminating in minute ramifications and plamenta and not of leaves.

The bark is covered by a thin cuticle or *epidermis*, composed of a number of laminæ, or scales. Immediately beneath the epidermis, is the *parenchyma*, a soft substance, consisting of cells filled with fluid, having almost always a greenish tint. The innermost part of the bark is constituted of *cortical layers*, and their numbers vary with the age of the tree.

It has been shewn by the experiments of Mr. Knight, and those made by other physiologists, that the sap descending through the bark, after being modified in the leaves, is the principal cause of the growth of the tree; thus, if the bark is wounded, the principal formation of new bark is on the upper edge of the wound; and when the wood has been removed, the formation of new wood takes place immediately beneath the bark.

The wood of trees is composed of an external or living part, called *alburnum*, or *sap-wood*, and of an internal or dead part, the *heart-wood*. The alburnum is white and full of moisture, and in young trees, and annual shoots, it reaches even to the pith. The alburnum is the great vascular system of the vegetable through which the sap rises, and the vessels in it extend from the leaves to the minutest filaments in the roots.

The *pith* occupies the centre of the plant; its texture is membranous; it is composed of cells, which are circular towards the extremity, and hexagonal in the centre of the substance. In the first infancy of the vegetable, the pith occupies but a small space. It gradually, and in annnal shoots and young trees, offers a considerable diameter. In the more advanced age of the tree, acted on by the heart-wood, pressed by the layers of the alburnum, it begins to diminish, and in very old forest-trees disappears altogether.

The *leaves*, though infinitely diversified in their forms, are in all cases similar in interior organization, and perform the same functions.

The alburnum spreads itself from the foot-stalks into the very extremity of the leaf; it retains a vascular system and its living powers; and its peculiar tubes, particularly the tracheæ, may be distinctly seen in the leaf.

PLASTER OF PARIS. Great differences exist among agriculturists respecting the uses of gypsum, and the manner of its operation. Some have supposed that its efficiency as a manure is altogether owing to its power of attracting moisture from the air. But Sir Humphrey Davy is of opinion, that but little effect can be produced by such attraction. "When *combined* with water," he observes, "it retains that fluid too powerfully to yield it to the roots of the plant, and its adhesive attraction for moisture is inconsiderable; the small quantity in which it is used likewise is a circumstance unfavourable to this idea." Some have supposed that gypsum assists in promoting the putrefaction of animal substances, and the decomposition of the manure in the soil. Sir Humphrey Davy, however, has proved by repeated experiments, that it rather retards than accelerates petrefaction. This philosopher likewise says, "In examining the ashes of sainfoin, clover and rye-grass, I found that they afforded considerable quantities of gypsum; and this substance, probably is intimately combined as a necessary part of their woody fibre. If this be allowed, it is easy to explain the reason why it operates in such small quantities; for the whole of a clover crop, or sainfoin crop, on an acre, according to my estimation, would afford by incineration only three or four bushels of gypsum. The reason why gypsum is not generally éfficacious, is probably because that most cultivated soils contain it in sufficient quantities for the use of the grasses. In the common course of cultivation gypsum is furnished in the manure; for it is contained in stable dung, and in the dung of cattle fed on grass; and it is not taken up in corn crops, or crops of pease and beans, and in very small quantities in turnip crops; but where lands are exclusively devoted to pasturage or hay, it will be continually consumed. Lord Dundas informs me, that having tried gypsum without any benefit on two of his estates in Yorkshire, he was induced to have the soil examined for gypsum, and this substance was found in both the soils."—*Elements of Agricultural Chemistry*, p. 224, 225. *Philadelphia Edition.*

It has been made a question, whether burning and calcining gypsum make any difference with regard to its fertilizing properties. This is said to be the practice of the French cultivators, and was likewise recommended by Mr. Deane. But an English writer on agriculture says, that "calcining is not likely to make any difference, because the sulphuric acid in gypsum cannot be expelled by the most violent heat of the furnace; and an experiment of Arthur Young countenances the assertion, that the effects of gypsum are the same, whether calcined or rough." Others have supposed that the sulphuric acid, which forms one ingredient of gypsum, "diluted with water by the chemistry of nature," may be instrumental in converting the starch of plants into sugar. "As starch boiled in water with sulphuric acid, and thereby changed into sugar, increases in weight without uniting with any sulphuric acid or gas, or without forming any gas, we are under the necessity of ascribing the change solely to the fixation of water. Hence we must conclude that starch sugar is nothing else than a combination of starch with water in a solid state. The sulphuric acid is neither decomposed, nor united to the starch as a constituent; nevertheless it is found that long boiling in pure water does not convert the starch into sugar.* This fact opens a large field for rational speculation on the physiology of vegetables; as it renders it possible that some of the mineral acids in the sap of plants, after acting chemically on the juices concocted into pulp, may be thrown out unchanged; they may alter the flavour without entering into the essence.

"Another step in the process of conversion brings us to pure *sulphur*. Some plants yield this on analysis. Seeds sown by way of experiment on nothing but this mineral, have produced healthy plants; and many soils, which nature has impregnated with sulphur, are highly fertile."†

It is possible that the sulphuric acid contained in gypsum may give that substance its principal value as a manure. And it would be well to ascertain, by experiment the efficacy of sulphuric, and other mineral acids, by applying them, in a very diluted state, as a liquid manure to plants, and likewise to

* See a Translation of the original Paper, in *Annals of Philosophy*, for December 1815. (No. XXXVI. pp. 425, 426.)

† See a Treatise on Soils and Manures, appended to the Philadelphia Edition of Elements of Agricultural Chemistry, Article *Gypsum*.

mix them with composts, in such proportions as may be most likely to ascertain their fertilizing qualities, if any such exist.

It has been affirmed that gypsum is not useful as a manure in the vicinity of the sea. And the writer last above quoted observes, that "on the hypothesis that sea-air destroys the fertilizing principle in gypsum, Mr. R. Bakewell, a correspondent of the Monthly Magazine, proceeds to account for its failure as a manure in so many parts of the country. It is enough to dispel this opinion, to name the county of Kent, in England, as the place where it has most fully succeeded." It has likewise been used on Long Island, in America, in the vicinity of the sea, with good effect, and in the state of Maine on the margin of the sea.

Dr. Jos. E. Muse, of Maryland, in an essay on the subject of gypsum, and its mode of operation, published in the American Farmer, vol. I. p. 338, gives it as his opinion "That the chief, if not the whole cause of the efficacy of gypsum in promoting vegetation, is to be found in its tendency to become phosphoric. This gentleman produces many chemical facts, and deductions therefrom, to shew that gypsum by exposure to the atmosphere, becomes phosphoric; and that phosphorus exists in vegetables.

Dr. Gorham, in a paper read before a Society in Boston, July 16th, 1813, and published in the Massachusetts Agricultural Journal, vol. III. No. 2., gives it as his opinion, that "when the plaster of Paris is applied to the seed, it stimulates the little rod, the action of the vessels is thus increased, absorption goes on more rapidly, and it acquires more nourishment in a given time than in ordinary circumstances; the consequences are a quick growth and enlargement of the organs," &c.

Col. Taylor, of Virginia, observes, in substance, that he sows of plaster from three pecks to one bushel to the acre. Sown on clover in the spring, it benefits it considerably. And in any other mode he ploughs it in. The best way of using it is in the spring upon the long manure of the preceding winter, to be ploughed in with it. He thinks it a valuable ally, but by no means a substitute for manure. That there should be intervals of two, three, or four years between applying it to the same land. That its effect is graduated by the quantity of vegetable matter on which it is sown. That upon close grazed land it does little good at first, and repeated would become pernicious; and that it must be united either with long manure of the winter, or the ungrazed vegetable cover produced in the summer. That all crops are ultimately improved by its improving the soil, even when its effects are not immediately visible, but he does not recommend it as a top-dressing, except for clover.

M. Canolle, a French writer, observes, that plaster acting, or operating chiefly on the absorbent system of plants, its effects are not like those of manures buried in the soil, which act principally on the roots. The latter according to their *particular nature* divide, soften, enrich, warm, or stiffen the soils with which they are mixed. The quantity of plaster spread upon the land is so trifling, that it can have little effect on the soil. ' I speak from experience. Plaster buried in the earth where sainfoin has been sown, has produced no visible alteration; whilst the same quantity of plaster spread over the same surface of sainfoin, has produced the most beautiful vegetation.

"From this experience, so uniform in the application of plaster, I am led to believe, that one must consult as well the nature of the soil, as the kinds of plants to which we apply plaster. Thus, whatever may be the soil on which clover, lucern, and sainfoin naturally flourish vigorously, or with that vigour which encourages us to apply manure, there is no risk in trying plaster.

" It is to be remarked, that plaster operates on plants in a direct ratio to the size and number of their leaves. I have spread plaster on land, where sainfoin was mixed with the common grasses which compose our meadows. The growth of the sainfoin and wild honey-suckle has been, beyond comparison, greater than that of the common grasses. It is to this cause I attribute the failure of success on grass-ground, chiefly filled with common grasses. I have a field of lucern separated from a natural meadow only by a brook. I have greatly increased the lucern by the plaster, whilst the effect of a like quantity on the adjoining grass-land, was scarcely, if at all, perceptible."

PLOUGH. It would not be possible to give any accurate idea of modern improvements in the construction of this very useful implement, without long descriptions, and expensive drawings, incompatible with the general design of this work, we can, therefore, only give some general rules for the construction of ploughs, and refer our readers to more voluminous and elaborate works for further and more minute information on this subject.

" The great points to be attended to in ploughing, are, 1., to open a fair regular furrow; and 2., to do this with as little resistance as possible. It is believed that these advantages are to be obtained, by the use of a plough, to which the mould-board, invented by THOMAS JEFFERSON, is affixed, and of which the annexed views will give a clear idea.

" The following account of this mould-board, and of the principles upon which it is constructed, are taken from a communication, addressed to Sir JOHN SINCLAIR, in 1798, then President of the British Board of Agriculture, and inserted in the 4th volume of the *Transactions of the American Philosophical Society*, vol. IV. p. 314.

" The mould-board should be a continuation of the wing of the ploughshare, beginning at its hinder edge, and in the same plane. Its first office is to receive the sod horizontally from the wing; to raise it to a proper height for being turned over; and to make, in its progress, *the least resistance possible*, and consequently to require a minimum in the moving power. Were this its only office the wedge would offer itself as the most eligible form in practice. But the sod is to be turned over also. To do this the one edge of it is not to be raised at all; for to raise this would be a waste of labour. The other edge is to be raised till it passes the perpendicular, that it might fall over of its own weight. And, that this may be done, so as to give also the least resistance, it must be made to rise gradually from the moment the sod is received. The mould-board then, in this second office, operates as a transverse, or rising wedge, the point of which sliding back horizontally on the ground, the other ends continues rising till it passes the perpendicular. Or to vary the point of view, place on the ground a wedge of the breadth of the ploughshare, of its length from the wing backwards, and as high at the heel as it is wide: draw a diagonal on its upper face, from the left angle, at the point, to the right upper angle of the heel: bevil the face from the diagonal to the right bottom edge, which lies on the ground. That half is then evidently in the best form for performing the two offices of raising and turning the sod gradually, and with the least effort; and if you will suppose the same bevil continued across the left side of the diagonal; that is, if you will suppose a straight line, whose length is at least equal to the breadth of the wedge, applied on the face of the first bevil, and moved backwards on it, parallel with itself and with the ends of the wedge, the lower end of the line moving along the right-bottom-edge, a curved plane will be generated, whose characteristic will be a combination of the principle of the wedge in cross directions, and will give what we seek, the *mould-board of least resistance*. It offers too this great advantage, that it may be made by the coarsest workman, by a process so exact that its form shall never be varied a single hair's breadth. One fault of all other mould-boards is that, being copied by the eye, no two will be alike. In truth, it is easier to form the mould-board I speak of with precision, than to describe that method either by words or figures."

To this is appended a more minute description, applicable to drawings, with letters of reference, &c.—See *Domestic Encyclopedia*, vol. III. p. 113., and American Farmer, vol. II. p. 185.

Robert Smith, of the township of Buckingham, Buck's county, Pennsylvania, gives the following rules for constructing ploughs.

"In constructing ploughs, the beam ought to be placed directly over the land side of the plough, so that the cut of the coulter may be square with the cut of the share; and the land should be given to the plough, between the coulter mortice, and the fore end of the beam: for if the cutting of the share and coulter makes an acute angle with the land, then the plough will incline to fall to the right; but if it makes an obtuse angle, then it will incline to fall to the left. A plough for two horses ought not to be less than nine inches, nor more than ten inches wide in the bottom, and for three horses from eleven to twelve inches wide. The share should never differ much in width from the plough. The cut of the share and bottom of the plough, should be exactly in one plane. A three horse plough requires no land in its construction. A crook of three inches and a half in the beam, before the coulter mortice to the right, will suffice for the land of a two-horse plough. A plough with a long beam runs the steadiest, and it being long prevents the plough from kicking; and long shafts give the ploughman a greater command of its direction. The cast-iron plate ought to be scoured with a grit stone before it is used."

Freeborn's patent plough is highly recommended, but we are not able to give a description of it; and perhaps it is so well known among practical farmers, that a description is not necessary. The Hon. Josiah Quincy gives this plough the following recommendation.

"Concerning its superiority, I have had the opinion of every practical farmer, who has witnessed its operation, I believe without exception. The effect upon my farm is this; that I now break up, with care, the same quantity and qualities of land, say one acre in a day, with one yoke of oxen and one man, who both holds and drives, which was never before to my knowledge, broke up with less than two yoke of oxen and two men. My ploughmen agree, that it takes one-third less power to do the same work than common ploughs require."—*Massachusetts Agricultural Repository*, vol. V. p. 262.

PLOUGHING, the operation of turning, breaking and loosening the earth with a plough.

One rule to be regarded in ploughing is, that no land, excepting green sward, should be ploughed when it is so wet that it will not easily crumble. For the principal design of ploughing is to break the cohesion of the soil, and set the particles of it at such a distance from each other, that even the smallest and tenderest roots of plants may find their way between them in quest of their nourishment.

When, in ploughing, of land in tillage, the furrow turns over like a dead mass of mortar, ploughing can be of no advantage at all. The soil becomes no lighter or looser by it, but rather heavier, and more compact. On the contrary, land should not be ploughed when it is too dry; because it requires the more strength of team to perform it, nor can the furrows be so well turned over.

The plough should be used much more than it is in this country. When a crop of barley or wheat is designed, the ground should, at least, be thrice ploughed; for a crop of Indian corn, twice is not too much. The extra expense will be repaid by the increased crops. The advantages of frequent ploughing have not been duly considered. By often repeated ploughings, land may be brought to any degree of richness almost that is desired. Frequent ploughings are destructive to weeds, and save much labour in hand hoeing and weeding; besides making a greater quantity of pasture for plants, preparing the vegetable food the better to enter the roots of plants, and disposing the soil to imbibe the rich and fertilizing particles of the atmosphere.

As it is known that repeated ploughings supply the place of manure; where manure is scarce, farmers have need to plough the more frequently. Mr. Tull was of opinion that it was a cheaper method to enrich land by ploughing than by manuring. In some situations it undoubtedly is so. But it is best that land should have enough of both, when it is practicable.

And the more to promote the fertilization of the soil by ploughing, let the farmer plough as much of his ground as possible while the dew is on it, because dew contains much nourishment for plants. The early riser has the advantage of his sluggish neighbour; not only in ploughing, but also in harrowing and hoeing, to greater advantage.

When land is to be ploughed that is full of stumps of trees and other obstacles, as land that is newly cleared of wood, or that is rocky, the strong plough should be used; and the strength of the team must be proportioned to the strength of the plough; and the plough to the condition of the soil.

It is sometimes advisable, to cut off close to the bodies of stumps, before ploughing, the horizontal roots which lie near the surface; especially if there be no stones, nor gravel in the way, to hurt the edge of an axe. When this is done, the strong plough will be apt to take out the most of the roots so parted. And the frosts of a few winters will be the more likely to heave out the stumps, or so to loosen them that they may be easily removed. I have conquered the stumps of white pine in this manner; but stumps which rot very soon it is not of such importance to manage in this way.

The plough must go deeper in breaking up new ground, than old. Otherwise the little hollows will go unploughed; and there will not be mould enough raised in the hillocks to level the surface, and leave sufficient depth for the roots of plants to extend themselves.

The last of summer, or the beginning of autumn, is the right season for ploughing new ground. For it will be best to harrow, and cross plough it, before it is seeded, that the soil may be thoroughly mixed and pulverized. Therefore, the first ploughing should be performed so long beforehand, that before the second, the turfs may ferment and become partly rotten. But this is not to be expected, if the ground be ploughed late in the fall; because the sun, at that late season, will not warm the ground enough to bring on any fermentation before the following spring, when the ground is to be sowed. For fall sowing, the ground should be broken up still earlier;

either in spring or summer will do very well, if time can be spared for it.

But it is best that the most or all of our tillage land should be ploughed in autumn, both in new and old ground. It saves time and labour in the following spring, the hurrying season, when more work is to be done than we can well get time for; and when our teams are usually much weaker than they are in the fall. But land ploughed in the fall must be again ploughed in the spring; and a weaker team will perform it for its having been ploughed in autumn. In very light old ground a single horse may perform it; and two ploughings are better than one in most cases.

Green sward land may be ploughed at any season of the year, if it be not too dry, nor too much frozen. In the former case the plough will go very hard; in the latter ploughing is impracticable, which is the case for four months together, commonly, from the first of December to the last of March.

Farmers generally choose to plough green sward ground when it is pretty wet, if it be not miry; because the labour is more easy for man and beast.

The English farmers practise ploughing green sward in January, not only because they have leisure, but because it is so wet as to plough easily. They call it *ploughing in lays;* and it is said to be well performed, when the sward is all completely turned over, without lapping one furrow on another. The depth that the plough should go is a matter that ought to be attended to. The depth should be governed in some measure by the staple of the soil. Where the soil is deep, deep ploughing is best.

But where the soil is very thin, shoal ploughing is necessary; for if the plough turn up much of the under stratum, and mix it with the soil, it will be rather hurtful, at least for some years after.

Land should always be ploughed out of sward with a deeper furrow than will be necessary afterwards, through the whole course of tillage. All the after ploughings will be the more easily performed.

Mr. Young, by attending particularly to the depth of ploughing in various towns in England, found that the average depth in sandy soils was four inches, in loamy soils four and three quarters, and in clayey soils three and an half. But in Ireland they plough much deeper; sometimes not less than nine or ten inches.

Our farmers are sometimes led to plough too shallow, to save a little labour. And some are too much afraid to turn up what they call dead earth. But they should know that all the soil above the hard pan may be well employed in tillage, for some crops or other; and that if they turn up a red soil, it will in a year or two become dark, and fit to nourish plants, by being exposed to the sun and the weather, and imbibing rich particles from the atmosphere.

Trench ploughing is sometimes practised to advantage; and the culture of some plants with tap roots requires it. This is done by passing a plough twice in a furrow. Ground may be thus ploughed to the depth of twelve or fifteen inches. But, instead of this double labour of the plough, where labourers are plenty, the furrows may be deepened with shovels, by a number of hands following the plough.

In old countries, where lands have been tilled for a thousand years, and have been frequently manured, the rich black soil has been growing deeper and deeper. So that trench ploughing by this time may be very proper in many of their fields; and even necessary to bring up the strength of manures, which has subsided to a greater depth than common ploughing reaches.

But there is only a small proportion of our land in this country, to which trench ploughing is suitable, or which will well pay the cost of it. In most of our soils, even where the hard under stratum, or pan, lies deep, trench ploughing would throw up so much cold hungry earth, and bury the upper mould so deep, as to render the land very barren at first. The places where it would answer best, are hollows, into which much vegetable mould has been washed down from the neighbouring heights, which has a black moory soil to a great depth; and such spots as have been used as gardens, and have been often dug with the spade.

If labour of men and teams were as cheap as it is in some countries, it would be advisable, to give more of our deepest soils this culture than we do at present. But wherever it is once begun, it should be continued, at least through a course of tillage; or else the first ploughings will be worse than lost. The best of the soil would be buried at such a depth as to become almost useless, unless it were alternately brought near the surface, by after ploughings equally deep.

Regard should be had to the shape of the land in ploughing. They who plough a steep hill up and down injure their cattle, and miss of ploughing their land to advantage. The furrow that is drawn up hill must be excessively shoal; or the team much stronger than common. For this reason a hill should be ploughed horizontally; with furrows as nearly parallel to the base as possible. This may be easily done when all the sides of a hill are to be ploughed at once. The rains will carry much of the finest of the soil to the bottom of the hill if the furrows are made up and down. But ploughed the other way, the hentings, or parting furrows, will be sufficient drains; and the water will move so slowly in them, that none of the soil will be washed away. But when a hill is very steep, no turning of a furrow upwards should be attempted. And if only one side of a steep hill is to be ploughed, the furrows should be all cut the same way, the team returning light after each furrow.

The reader will perceive, that what is commonly called cross ploughing on hills' sides is not approved. But cross-ploughing of land that is level, or gently sloping, is oftentimes very proper. Land in general should be ploughed one way and the other alternately, that it may be the more thoroughly pulverized and mixed; that is, when the shape of the ground and the dimensions of a lot admit of it.

Green sward ground, that is broken up in the fall is usually cross-ploughed in the spring following. But this should not be done without caution. For if the turf be not considerably rotted, cross-ploughing will only drive it into heaps, instead of cutting it to pieces: Neither will the harrow reduce the turf to powder. In this case it will be best to omit the cross-ploughing: And after a heavy harrowing lengthwise of the furrows, seed the land with pease, potatoes, maize, or any thing that will do well with such culture.

Some plough green sward in the spring and seed it without delay. It sometimes does well for maize, oats, and flax, if well dunged; or for pease and potatoes without much dunging. Potatoes seem to do better than any thing else. But the holes must be made quite through the furrows, whether dunged or not. As this crop requires the greatest part of its nourishment in the latter part of summer, about that time the turf comes to be in its best state for yielding nourishment to plants.

For a crop of winter wheat the tillage ground should be ploughed in the spring, again in June, and lastly just before sowing. Whatever manure be put on, it should be just before the last ploughing, and ploughed in immediately. If the grain be ploughed in with a shoal furrow, it will not be so apt to be killed by the winter. The roots will lie deeper than those of harrowed grain; and it will the better bear drought in the following summer, if that should happen.

For other seeding in general, or for whatever is planted or sown in the spring, on what we call old ground, it should be ploughed near the time of seeding, although it were ploughed in the fall; and the nearer to the time of seeding the better. The seeds will be the better supplied with moisture to make them vegetate; and the crop will have the better chance of being able to outgrow and stifle the weeds, and have the benefit of a looser soil, during the whole of its growth. These autumnal ploughings, I have found to be greatly advantageous, especially in clays, and in stiff loams.

Many, to save labour, plough their land so shallow for sowing, as scarcely to take up the roots of the weeds. Men of common understanding, I should suppose, need not be told that this is bad husbandry: For it may rationally be expected that there will be a larger crop of weeds, than if it had not been ploughed at all; and that the roots of the plants will not have sufficient room to extend themselves. Ploughing the ground in autumn will have a tendency to prevent this most absurd conduct in the spring, which many go into that they may favour their teams in a faint season.

That seed may be sown as early as possible, many are led to make the seed furrow before the ground is sufficiently dry. If the crop should be a little earlier, it will be the poorer. It will be slower in coming up; more of the seeds will fail; the blade will be more slender; nor will it grow so fast as if it were sowed later, when the ground is warmer. Sometimes it will not grow at all for a long time, but become so stinted, that a crop must be despaired of. No practice can be worse than to give the seed furrow in stiff soils, before the ground is sufficiently dried.

Land that is low and flat, and therefore apt to be too wet and heavy, ought to be ploughed in ridges. The ridges may have two, three or four furrows on each side, according as the ground is wetter or drier. The wettest ground should have the narrowest ridges; but they should never be narrower than four furrows in a ridge. The rows will be between four and five feet apart, if one row of plants be set on each ridge. But if there be six or eight furrows in a ridge, it may admit of two rows, one on each side of the veering.

After lying in ridges through the winter, the ridges should be thrown into the hollows by another ploughing in the spring; which will bring it into good order for seeding.

Or if it should be too miry to be ploughed in the spring, either Indian corn or potatoes may be planted on the ridges; and what is wanting of the proper tillage, may be made up after the ground is become drier, by frequent and deep horse-hoeings. Good crops of corn have been obtained in this method, on land, which, with plain ploughing, would have produced next to nothing.

Most of our clay soils, which lie level, require this sort of culture; for this more than any other soil is liable to be injured by overmuch wetness. And the drier it lies the weaker will be the cohesion of its parts.

Some soils which lie gently sloping are so wet as to need ridging. It is not best to make the ridges directly up and down the slope, nor horizontally, but on a medium between both. But where the land will admit of it, the ridges should lie north and south. It is no bad practice to lay lands to grass in ridges or beds. For too much wetness is apt to hurt grass lands, as well as lands for tillage, whether they are used for mowing or pasturage. In the former, the grass will be too sour to make a good hay; in the latter, not only the grass will be bad, but the soil so soft as not well to bear the tread of cattle. I have found that not only better grass, but a greater quantity, will be produced in this method. Nor will the soil so soon become hard and bound.

Nor is it a bad practice to split the hills with the light plough in autumn, after a crop of corn; even though the ground be not seeded till the following spring. One side of a row of hills is ploughed off with one furrow, and the other side ploughed off the contrary way by another furrow, so as to form veerings, or ridges, in the intervals. It is performed with less than half the expense of a plain ploughing; and nearly the whole of the surface is either taken up or covered. European writers think land should be ploughed immediately after a crop of corn, to prevent the stubs from robbing the soil of its juices. Be this as it may the ploughing is at least as useful as other autumnal ploughing; and where dung has been put in holes, it mixes it with the soil; not to mention the burying of some of the stubs and leaves of the corn, which is of some advantage towards enriching the soil.

There is another way of ploughing, called ribbing; which is making furrows unconnected with each other, three feet or more asunder. It is but a fourth part so much work as ploughing plain. One very considerable advantage of it is, increasing the superficies of the soil, by which it is more exposed to the action of frost, air, and dews, and absorbs the largest quantity of nutritive particles.

In tillage land that is steep, ribbing is a further benefit to the soil, as it prevents the washing down of the vegetable mould, and the strength of manures. With this view the operation should be performed in autumn. And the plough must pass horizontally, or nearly so, not up and down the steep.

In pastures or grazing land, declivities would produce the more grass, if they were ribbed; as the benefit of sudden rains would not so soon be over, by means of their quickly running down into the vallies. At the same time, the vallies would not so often be overcharged with water. Furrows eight or ten feet apart would answer, and the ribbing would not want to be repeated for a long time. The furrows should be as nearly horizontal as possible, as well as in tillage land.

The following maxims respecting the proper depth of ploughing are chiefly derived from the *Code of Agriculture*.

1. The depth to which land ought to be ploughed, must first be regulated by the depth of the soil. On these soils, more especially on a rocky substratum, the ploughing must necessarily be shallow. 2. The depth ought likewise to depend on the means of improving the soil;—for where the land is poor, and the means of enriching it is scanty, the depth of ploughing ought to be in proportion to the quantity of manure which can be obtained. 3. Deep ploughing is highly advantageous upon every soil excepting those where the substratum is of an ochry sand. In fact such are scarcely worthy of being cultivated, unless in situations where much alluvial compost, or short town manure can be procured. 4. It is a general rule, never to plough so deep as to penetrate below the soil that was formerly manured and cultivated, excepting upon fallow, and then only when you have plenty of lime or dung to add to, and improve the new soil. 5. Many farmers recommend, when fallowing land, to go as deep as possible with the first furrow; by which the subsequent

furrows will be more easily done, and to expose the soil to the winter's frost, and to the summer's heat: but when the land is ploughed in spring, for a crop of oats, *a strong soil* cannot be ploughed with safety, above five or six inches. 6. Deep ploughing is advisable on moorish cold soils, as it affords a greater scope for the roots of plants to procure nourishment, admits the superabundant moisture to subside from them, and prevents the summer drought from making any injurious impression on the growing crops; for, on such land, shallow ploughing exposes vegetation to be starved or drowned in wet weather, and to be scorched or withered in dry. 7. It is unnecessary to plough deeper, when the seed is sown, than where there is a fair probability of the different kinds of plants sending their roots; and as beans, clover, and turnips, the only tap-rooted kinds usually cultivated in this country, seldom send their shoots above seven or eight inches down into the soil, and the culmiferous species not so far, it is probable from these circumstances that from seven to eight inches may be deep enough for all the purposes of ordinary culture. Occasionally, however, ploughing even ten inches in the course of a rotation, during the fallow process is advisable.

"Deep ploughing is not to be recommended: 1. When lime or marle has been recently applied, as they have such a tendency to sink from their weight, and the moisture they imbibe. 2. Where turnips have been eaten off by sheep on the land where they were grown. 3. When grass only two or three years old, more especially where it has been pastured with sheep is broken up; because, owing to the extreme condensation of the soil, by the trampling of the sheep, a furrow, even of a moderate depth, *to appearance*, will make the plough penetrate below the staple that had been cleared, by the culture given during the previous fallow."

The advantages of deep ploughing, according to the same writer are, "1. Bringing up new mould, which is peculiarly favourable to clover, turnips, beans, and potatoes; and indeed without that advantage, these crops usually diminish in quantity, quality and value. 2. Deep ploughing is likewise of great consequence to every species of plant, furnishing not only more means of nourishment to their roots, but, above all, by counteracting the injurious consequences of either too wet, or too dry a season. This is a most important consideration, for if the season be wet, there is a greater depth of soil to absorb the moisture, so that the plants are not likely to have their roots immersed in water; and in a dry season it is still more useful, for in the lower part of the cultivated soil, *there is thus a reservoir of moisture*, which is brought up to the roots of the plants, by the evaporation which the heat of the sun occasions. 3. By deep ploughing animal and vegetable manures, which have such a tendency to rise to the surface are properly covered. This cannot be done by shallow ploughing, in consequence of which much of the value of such manures is lost. And 5. By deep ploughing, a heavier crop is raised than can be got from a shallow furrow. An intelligent farmer, indeed, after pointing out that deep ploughing increases the staple of the soil, keeps the roots of the corn from being injured by wetness, and also enables the crop longer to resist drought, adds, "*I have ever found deep ploughing attended with great crops, when ridges, shallow ploughed, in the same field were but indifferent*:" Which seems a decisive proof in favour of deep ploughing."

It is stated in a communication for the American Farmer, by George W. Jeffreys, Esq. of North Carolina, that "Deep ploughing is becoming much more general every day, and this is greatly facilitated by the use of cast-iron mould-boards, which are now generally used here. By a little use they become bright and smooth, the obstruction is consequently less, and deep ploughing is more easy. The advantages of deep ploughing in corn crops are very great. The deeper the soil is ploughed the greater may be the quantity of corn planted on an acre, or any given quantity, and the crop thereby greatly increased. At the commencement of my system of farming, my corn was planted about six feet by three. I now plant it on the same land, five feet by two, in many places nearer, with two stalks in a hill, being near doubly as close as formerly, and the crops are nearly in the same proportion." *American Farmer*, vol. II. p. 15.

The Hon. Timothy Pickering, in an address to the Essex Agricultural Society observes: I entertain no doubt of the utility of deep ploughing; not at once, on our lands in general, but by an increase of two or three inches at every annual ploughing, till the earth be stirred and pulverised to the depth of ten or twelve inches.

A loose, sandy soil may be ploughed too much; see the article *Sandy Soil*. Stiff clayey soils can hardly be ploughed too often, and they will likewise require to be rolled, and harrowed often.

PLOUGHING HORIZONTALLY. In a letter from Mr. Jefferson to T. Dalton, Esq. published in the Massachusetts Agricultural Repository, vol. IV. p. 320. Mr. Jefferson observes that "A method of ploughing our hill sides horizontally, introduced into this most hilly part of our country, by Col. J. M. Randolph, my son-in-law, may be worth mentioning to you; he has practised it a dozen or fifteen years, and its advantages were so immediately observed, that it has already become very general, and has entirely changed and renovated the face of our country. Every rain before that, while it gave temporary refreshment did permanent evil, by carrying off our soil, and fields were no sooner cleared than wasted. At present we may say, that we lose none of our soil, the rain, not absorbed in the moment of the fall, being retained in the hollows between the beds, until it can be absorbed.

"Our practice is, when we first enter on this process with a rafter level of ten feet span, to lay off guide lines conducted horizontally around the hill, or valley, from one end to the other of the field and about thirty yards apart. The steps of the level on the ground, are marked by a stroke of the hoe, and immediately followed by a plough to preserve the trace. A man or a lad with the level, and two small boys will do an acre of this in an hour, and, when done it remains forever. We generally level a field the year that it is put into Indian corn, laying it in beds of six feet wide, with a large water furrow between the beds, until all the fields have been once levelled. The intermediate furrows are run by the eye of the ploughman. Governed by these guide lines the inequalities of the declivity in the hill will vary in places the distance of the guide lines, and occasion gores, which are thrown into short beds. As in ploughing very steep hills horizontally, the common plough can scarcely throw the furrows up hill, Col. Randolph has contrived a very simple alteration of the share, which throws the furrow down hill, both going and coming. It is as if two *shares* were welded together, and at a right angle to each other. This turns on its bar as on a pivot, so as to lay either share horizontally; then the other, becoming vertical acts as a mould-board. This is done by the ploughman, in an instant, by a single motion of the hand at the end of every furrow."

A drawing and description of the rafter level referred to in the preceding extract may be seen in the "*American Farmer*," vol. I. page 358.

PLUM TREE, *Prunus*. The Plum is generally supposed to be a native of Asia, and the Damascene to take its name from Damascus, a city in Syria.

Mr. Forsyth reckons more than thirty varieties of this fruit, for which see his treatise on Fruit Trees. The soil in which they are planted should be made two feet and an half, or three feet deep, of good light fresh loam. The trees should not be headed down till they begin to throw out fresh shoots. Strong trees should be cut a foot from the ground; and those that are weak, about half that length. They should not be headed down till they begin to bud, and then cut as near the eye as possible, that the young shoot may cover the cut. They should not be headed at the time of planting, as that practice often proves fatal to them. Plum trees, as well as all other fruit trees, should be pruned in the spring instead of autumn or winter.

Mr. Yates, a respectable farmer of Petersham, Mass. informs that the insects which occasion black bunches on plum trees, are prevented by digging around the roots in the spring, and putting in half a bushel of ashes, and covering them with earth. Also, that the slug worm is destroyed by putting tanners' bark round pear trees; the bark used was mostly hemlock, with a little oak bark mixed with it.

POLL EVIL, "an imposthume on the poll of a horse. At first it requires no other method of cure than what is common to other boils, and inflamed tumours. But sometimes it degenerates to a sinuous ulcer, through ill management or neglect.

"There is a small sinus under the noll bone, where the matter is apt to lodge, unless care be taken to keep the part firm with a bandage: But instead of that the farriers generally use to thrust in a long teat, which raises the flesh, and opens a way into the sinus. And thus an ulcer is created where there needs be none. All therefore that is further necessary on this head is, to caution the practitioner against such ill methods. And if the tumour has a very large cavity, it is better to lay it open, than to thrust foreign substances into it. And if it acquires an ulcerous disposition, it must be treated as such." *Gibson's Farriery.*

The following is extracted from a valuable work entitled "The Gentleman's New Pocket Farrier;" by Richard Mason, M. D. published in Richmond, Virginia, in 1820.

"The poll evil, like the fistula, proceeds from some blow, bruise, or external injury, and its consequences are much to be dreaded. A horse thus diseased would be well sold almost at any price, though the cure is tolerably certain, yet extremely slow. The poll evil is an abscess or swelling found in the sinews, between the noll bone and the uppermost vertebræ of the neck. When this swelling first makes its appearance, bathe it frequently with hot vinegar; and if the hair be fretted off, with an oozing through the skin, make use of equal parts of vinegar and spirits of wine; but if there be an itching, with heat and inflammation, he safest way will be to bleed plentifully and apply a red oak poultice, which will sometimes disperse the swelling, and put an end to the disease. But whenever the tumour is critical, having all the signs of matter, and appears not benefitted by the applications already recommended, it will be advisable to bring it to a head as speedily as possible, with the following poultice; corn-meal, marsh-mallows, oil turpentine and hog's lard. When the tumour becomes ripe, or full of matter, it may be either opened or permitted to break of itself—if opened with a knife, great care should be used to prevent wounding the tendinous ligament that runs along the neck under the mane. When the matter appears to be on both sides, the tumours must be opened on both sides, and the ligament between remain undivided; if the matter flows in great quantities, resembling melted glue, and is of an oily consistence, it will require a second incision, especially if any cavities are discovered by the finger or probe; these should be opened by the knife, and the wound should be dressed with spirits of turpentine, honey, and tincture of myrrh, until light and thick coloured matter is found. Cleanse the sore well with strong soap-suds and a sponge; then take of verdigris half an ounce, oil of turpentine four ounces, of blue stone two ounces, of green copperas half an ounce; mix them well together, and hold them over a fire until they are as hot as the horse can bear them; then pour them into the abscess, and close the lips by one or two stitches; this is to remain several days without any other dressing except bathing with spirits of wine. Should matter flow in great abundance, and of thin consistency, the above application must be again repeated until the matter decreases in quantity and becomes of a whitish colour and healthy appearance.

POND, a collection of still water. A mill pool is so called, though it gradually receives water in one part, and discharges it in another: So that it is not perfectly still water. The water is so often shifted, that it is not apt to putrefy.

Pastures that are destitute of water, should have artificial ponds made in them, for watering places. "Observe where rushes, reeds, flags, and other aquatic plants grow spontaneously; or where frogs are observed to lie squatted down close to the ground in order to receive its moisture. Or observe where a vapour is frequently seen to rise from the same spot. Some say, wherever little swarms of flies are seen constantly flying in the same place, and near the ground, in the morning after sun-rise, there is water underneath." "If a well is made in a sloping ground, and the declivity is sufficient to give it a horizontal vent, it will be worth the husbandman's while to dig such a passage, and by means of pipes, or any other conveyance, to carry the water across the light soil, through which it might otherwise sink. The greatest quantity of water will be obtained in this manner, because there will be a continual stream." There is no difficulty in making a durable pond in a clayey soil. Let a large hollow basin be made in such earth, and it will preserve the water that falls in rain. But it is apt to be thick and dirty, if some pains be not taken to prevent it. The declivity, by which the cattle enter, should be paved, and gravel should be spread on the bottom. Or it might be better if the whole were paved.

There are many large natural ponds, which have outlets in one part, and are supplied by brooks or rivers in other parts; but a greater number of smaller ponds which are perfectly stagnant, unless when they are agitated by winds. Such ponds as the latter, in hot seasons, are apt to become putrid, and contaminate the air about them. For this reason they should, if possible, be drained. And when the water is not deep, and an outlet can be made without too much cost, they should be drained for the sake of reclaiming the soil. This will be of great value, as it commonly is found to be extremely rich, being made up of the finest particles of soil, wafted into them by winds, and of decayed vegetable substances, besides the fine mould washed into them by rains.

Many farms contain little sunken spots, which are most of the year covered with water, and produce some aquatic bushes and weeds. These are notorious harbours for frogs; and are therefore called frog ponds. They should be drained, if it be practicable. It is commonly the case, however, that draining them in the common way, by making an outlet, would cost more than they would be worth when drained, because of the height of the land on every side. But in this case, if the banks be not clay, they may be drained in the following manner.

Take notice on which side land that is lower than the pond is nearest. On that side, in the bank near the pond, dig a kind of cellar, two or three feet deeper than the surface of the pond; do it in a dry season. If a hard stratum appear, dig through it; and leave digging where the bottom is loose gravel, or sand. Then make an open or a covered drain from the pond to the cellar. The water will be discharged from the pond, and soak into the earth through the bottom of the cellar, till a scurf is formed on the bottom that will stop the water from soaking into the earth. This scurf should be broken from time to time, and taken away with a long handled hoe. Or, the cellar may be filled up with refuse stones, which I think is preferable to the other method.

If the pond should not then become sufficiently dry, a small ditch should be drawn round it, and discharge itself into the cellar. The land that is thus gained will be rich muck, much of which may be carted away for manure; and common earth, or sand, may replace it, without detriment to the soil.

POTATOE, *Solanum*, a well known vegetable. It was reported to have been first brought to Europe by Sir Walter Raleigh, but the better opinion is, that it is a native of the high table lands of South America.

No plant is so prolific in varieties as the potatoe, and no one varies so much as this does, in the several requisites of productiveness, and of excellent esculent qualities; some being very dry and farinaceous, and others watery. The attentive and scientific observers of Europe have maintained that the varieties of the potatoe soon run out; and they have gone so far as to assert that their duration extends only to fourteen or fifteen years, after which period they degenerate in quality and productiveness. This, certainly, is very much corroborated by the fact, that not one of the varieties, recommended by Dr. Deane in his last edition is now even known in our country. Since his time we have had many kinds, which have had their day of success, and have entirely disappeared. Perhaps there is no one fact better proved than this. We shall quote one example only. The red, or cranberry potatoe, which 20 years since was the favorite, and almost the only one in the Boston market, has entirely disappeared. We speak from experience, when we say that it gradually diminished in its productiveness till it entirely disappeared.

The foregoing facts would lead us either to seek new and improved varieties from other countries, or to endeavour, as the Europeans do, to raise them from the seed of the berry.

I shall here give the method of doing it. Take the apples in the beginning of October, before the frost has hurt them: Hang them up by the foot stalks in a dry closet, where they will not freeze: Let them hang till March or April: Then mash the apples, wash the seeds from the pulp, and dry them in a sunny window. Sow the seeds in a bed, about the first of May. When the plants are four or five inches high, transplant them into ground well prepared, one or two plants in a hill. They will produce full grown apples, and some of the roots will be as big as hens' eggs. But if the seeds were sown in autumn, some of them would come up in the following spring. Nothing is more common than their appearing in fields, where potatoes have been raised the preceding year.

As potatoes are come to be of more importance in this country, than any other esculent root, and are even an article of exportation, I shall be the more particular in pointing out the best methods of cultivating them.

The land should be ploughed deep for this crop; because roots will commonly grow as low as the soil is stirred, and no deeper. And the more the ground is pulverised before planting, the better will be the crop.

Perhaps green sward ground ought to be mentioned here as an exception. I have had the largest crops on such land, even with one ploughing, and that just before planting. I account for it thus: Potatoes want air; such land affords it from the hollows under the furrows, in no small quantity, both fixed and putrid, and in the greatest abundance towards the end of summer, when they require the greatest quantity of nourishment.

No dung is found to be more

suitable for potatoes than hogs' dung, mixed with a great deal of straw, or other rubbish. This dung is late in fermenting. and therefore affords the roots plenty of nourishment, when they most need it. And as they want air and room, rubbish, and even sticks and chips, or any thing that makes the ground lie light and hollow, encourages their growth.

But those roots are accounted best for eating, which are raised without dung. I once had a middling crop, by putting a handful of old weather beaten salt hay in each hill. New land, burnt, produces excellent roots, and a large crop, without any manure but what is made by the burning; sometimes not less than a peck in a hill.

The potatoe is so hardy a plant, that it will grow in any kind of soil, and even with the poorest culture. It is a great improver of land; not only by the rotting of its succulent stalks, which should be buried in the soil at, or immediately after digging; but the digging itself is a further improvement. A crop of potatoes is good to prepare land for other crops. It is not uncommon, on poor land, with very little cultivation and without manure, to obtain one hundred bushels per acre. But in Ireland, with deep ploughing, or digging, with manure, four times that quantity is common: And Mr. Young mentions one instance of an acre in England, producing a thousand bushels. As they will grow almost any how, we are tempted to neglect them; but no crop that I know of will better pay for good cultivation.

The first of May is perhaps the right season for planting potatoes, in a dry warm soil: But they will sometimes produce well, though planted at the last of June. An early crop will be better ripened, and more dry and mealy. A late one is unsolid and watery, as the roots do not arrive to their full maturity.

From experiments which have been made since the time in which Mr. Deane wrote, it has been found that whole potatoes are best for planting. The Hon. Josiah Quincy in a letter to the Corresponding Secretary of the Massachusetts Agricultural Society, observes; "I had directed my farmer to plant a field of about six acres, with the large red potatoe, called the River Plate potatoe. Being not present when he began the labour, he had planted a part of two rows, with the *potatoe whole*. Coming upon the field, I objected to the practice as wasteful, and directed him to cut the residue of the potatoes, and to put a mark so as the place where the whole potatoes were planted, might be known. In the whole course of vegetation, the whole potatoes had a decided superiority over every other part of the field, in the vigour and size of the tops; and, at harvest, in comparing these rows with the adjoining rows, the product of the rows planted whole exceeded an equal extent of the adjoining rows, planted with cut potatoes, *more than one third*. There was nothing in the cultivation, or state of the land, which could produce this difference, except the circumstance of the one having been planted whole, and the other cut." *Mass. Agr. Rep. vol. V. p.* 64.

The shooting parts exist in a potatoe, in the form of a tree, of which the stock is at the but, or root end. I therefore take care to cut athwart these parts as little as possible: For though they will grow any way, the greater length of shooting stem there is in a set, the more strong and vigorous will be its growth at first.

If dung be used, it may be spread before the second ploughing, or else laid under the sets. The latter method will give a larger crop. Dung laid under the sets, will produce more than if laid above them; as Mr. Wynn Baker proved by accurate experiments. The feeding roots should go into the dung, not directly into hungry earth below; and these roots strike downwards; and therefore need some loose earth under the dung to extend themselves into.

The fashionable way of planting potatoes in hills, may be as good as any in rough ground, or that which is not well subdued. But in a rich, mellow soil, well pulverised, the drill method is to be preferred. The setts may be either in single rows, three feet, or double, one foot apart, and from seven to nine inches asunder in the rows. One of my neighbours planted in his garden, drills and rows of hills alternately of equal length, and equally manured; when he dug them he found the drill rows produced twice as much as the other. It is not more labour to lay the dung in drills, than in hills; and the labour of hoeing is not increased. My trials in the drill way, have produced only half as much again. But I did not put dung in the furrows, but always put dung in the hills. My method has been, in dry ground, first to plough in the dung; then harrow; raise the ridges, and dibble the sets into the ridges.

The lazy bed method, or trenching, is most practised in Ireland. I have tried it several times, and am convinced, that a greater quantity on the same ground may be raised in this way, than in almost any other. But the labour is so great, as it must be performed with the spade, that I dare not recommend it, unless in particular cases, or to those who have but little land.

It is a good, and very effectual method, to subdue bad weeds in the border of a field, which cannot well be ploughed. But the soil should be deep, that the trenches may not go into the under stratum of hard earth, nor too near to it.

And in this way good crops may be got in springy and miry places, which are too wet for other tillage. But the work must be begun in autumn. In October, mark out the beds, five feet wide, leaving two feet between each bed for the trench: Spread the dung upon the beds: Dig the trenches, and with their contents cover the beds to the depth of about five inches. In May following, dibble the sets into the beds, quite down to the dung, and fill the holes with earth. Besides getting a good crop, the soil will be thus drained and subdued, and fitted for ploughing, and tillage crops.

An expeditious way of planting potatoes is as follows. After the ground is prepared, by ploughing and harrowing, cut furrows with the horse plough, forty inches apart; drop the sets in the furrows; then pass the plough along the back of each furrow, which will throw the earth of both furrows upon the sets; and afterwards level the ground with the back of a harrow, or with a harrow that has short tines if you will; but it is of no great consequence whether it be levelled at all. Another method of planting is, to plough the ground plain, keeping the furrows straight and regular, and drop sets in every third or fourth furrow. But before this is done, the ground should be ploughed and made level and fine with the harrow.

But the method last mentioned is fit only for a dry soil, where the seed needs to be laid deep. Where the soil is moist, a better way would be to furrow the ground, and lay the sets on the surface, close to the backs of the furrows, and cover them by turning another furrow towards each. If this should bury the sets too deep, the ridges may be easily lowered, with a hoe or a rake; but I do not apprehend it would be necessary. The ridges may remain as the plough leaves them.

As soon as rows of potatoe plants are grown to the height of four or five inches above the surface, or earlier if the ground be weedy, the cultivator, with two mouldboards, should be passed between them, as deep as one horse can draw it. For want of a cultivator, a common light plough should go and return in an interval, turning the earth at the first ploughing from, and then at the second towards, the rows. After each ploughing the plants should be weeded, and a little of the fresh earth drawn close to their stems, uncovering those which chance to be covered by the cultivator, or plough. This operation should be repeated three times, taking care not to earth the plants too much, as some are apt to do where the ground is light and mellow: For potatoes will not grow well more than about five inches under the surface, being too far removed from the influence of the sun. The ridges, or hills, should be rather broad than steep; flat on the top, that the water, which falls in rain may not be too much diverted from the roots.

The last hoeing should be finished before the plants are in blossom; and before the branches begin to trail upon the ground. Otherwise a new set of roots will be formed, too late to get their full growth, and which will rob the former sets of their nourishment. But if killing weeds be necessary after blossoming, it may be done with the hand hoe, observing not to earth up the plants at all.

Cattle should be kept from a field of potatoes, till the roots have got their full growth, as carefully as from a field of corn. For potatoes will not grow after the tops are browsed. They doubtless receive as much of their nourishment through the tops, as almost any plants.

As soon as the tops are dead, either by ripeness or by frost, the roots may be taken up. If they lie in the ground till they are soaked by the heavy autumnal rains, they will be the worse; and the labour of digging will be increased. Those that do not much adhere to the tops, may be thrown up by the cultivator, or by the horse plough, which will facilitate the digging. But the tops should be pulled out, and the fruit that comes out with them gathered, before the plough is passed under the rows. Some recommend a four or five pronged fork, as the best instrument to dig them with.

There is no difficulty in keeping them through the winter, in a cellar that is free from frost. Caves, dug in a dry soil, preserve them very well. They should be covered with two feet of earth over them. If they are in danger of frost in a cool cellar, they should be covered with a little salt hay. This any farmer may easily do, who has a maritime situation.

In cellars, they are more forward to sprout in the spring, than in caves. Those which are for summer eating, should be attended to in May, the sprouts rubbed off and put into a dry and dark place above ground. A barn floor is well adapted for the purpose. They do better than in a cellar. They will thus keep well till new potatoes are grown. But if any light come to them, they will send out long shoots towards the place where it enters.

Raw potatoes will keep swine alive through the winter: But they will not grow much with this food alone. Parboiled, they are an excellent food for swine, and will almost fatten them. The English farmers parboil them, not only for swine, but for horned cattle. I know of no food that will more increase the quantity of milk in cows; and they give milk no ill taste, whether boiled or raw. In either way cows are very fond of them. For horses they should be boiled. The largest and smallest, the irregular shaped and the cut ones, should be put by for the cattle: For middling roots are best both foreating and planting. Overgrown ones are apt to be hollow and watery; and wounded ones rot, oftener than sound ones.

As a further recommendation of this useful root, I may add, the farinaceous part of it makes an excellent starch, much superior, as some say, and not half so costly, as that made of wheat. The method of making potatoe starch, according to Mr. Weston, is as follows: "Wash and pare them, grate them upon large tin graters, and fill tubs about half full with the pulp: Then fill them up with water: Stir it well once a day, for three or four days, and take off all the scum. About the 5th day take out the pulp, and put it into shallow earthen pans, such as are used for milk, as much as will cover the bottom an inch think, and put water upon it. Every morning pour off the water, break up the starch, and add fresh water. When it is thus become very white, leave it in the pans till it is quite dry, then put it into paper bags, and put it in a dry place to keep."

This sort of starch has been made and used in my house, for twenty or thirty years past. The making of a quantity that will serve for a year is always begun and finished in a day or two. As soon as the starch is settled to the bottom, which it does in twenty minutes, the water is renewed; and instead of its standing in tubs, and being skimmed, we strain it through a cloth. Which of these methods is to be preferred, I do not determine.

Some suppose this starch is apt to rot the things which are stiffened with it; but this is a great mistake.

In an abstract of the *Memoirs of the Swedish Academy*, the above writer found the following account of one of their methods of using potatoes. "Mr. Charles Skytse has proposed to distil brandy from potatoes, in order to save the corn, which is so dear in Sweden; and finds by experience, that an acre of land set with potatoes, will yield a much greater quantity of brandy, than when sown with barley." It is asserted that a gallon of good strong spirit may be taken from six pecks of boiled potatoes, by distillation.

The account given by Dr. Anderson of his success in extracting potatoe spirits is this: He boiled 72 pounds of potatoes, they were then bruised, and passed through a riddle along with some fresh water. The pulp was then mixed with cold water, till the whole amounted to about 20 gallons. This was allowed to cool, till it attained to such a temperature, as would be proper for mixing yeast with wort, when some yeast was put to it. In ten or twelve hours the fermentation began, which continued very briskly for as many hours. After waiting some time, and in vain warming it a little, with a view to renew the fermentation, he stirred it briskly, which renewed the fermentation. Stirring it daily, the fermentation went on for a fortnight, and then abated, and could not be renewed by agitation or otherwise. It was then distilled with due caution, care having been taken to stir it in the still, until it began to boil, before the head of the still was applied; and the fire was afterwards so kept up as to keep it boiling briskly, till the whole was run over. In consequence of these precautions and due rectification, he obtained an English gallon of pure spirit, considerably above proof, and about a quart more of a weaker kind, a good deal below proof. The Dr. says, it was in every respect the most agreeable vinous spirit he ever saw; and that in taste it somewhat resembled fine brandy. According to this account, one acre of potatoes might yield 300 gallons of good strong spirit, worth at least 90*l*.

My new method of planting potatoes is this. After the dung is spread and ploughed in, and the ground levelled with the harrow, I raise the ridges about three feet and a half apart, with the cultivator; and then dibble in the sets along the tops of the ridges, about seven or eight inches apart, laying each set about as low as the surface was before the ridges were made. I have had as good crops in this way, as in any other.

The method of raising potatoes under straw, is very simple and easy. Lay the sets about eight inches apart each way, on any kind of soil that is not too rich: Cover them with straw, or refuse hay, to the depth of about twelve inches. Nothing more is to be done to them till they are taken up. They will be very clean, and the crop considerable.

Potatoes may be spoiled by bad management in harvesting. They should be dug in cool overcast weather, gathered as soon as possible, and kept moist with a quantity of dirt about them. If they are dug in warm and clear weather and exposed to the rays of the sun, they will sweat, and become strong, and in time will turn green and become poisonous.

In feeding stock with potatoes, it is best to steam-boil or bake them. Sir John Sinclair asserts that "there is something injurious in the juices of the potatoe in a raw state, which cooking eradicates or greatly dispels."

The following method of raising this crop is pursued in England, and probably will answer well in this country.

Break up sward ground in autumn, harrow it in the spring, and spread barn dung, or other suitable manure at the rate of about twenty loads to an acre. In planting run a furrow of proper depth, and draw in the dung from a distance of about twenty-seven inches next the furrow, and distribute it evenly along its bottom. On this the potatoes are placed, about eighteen inches apart if whole, but proportionally nearer if cut in pieces, and then covered by the next furrow. Then two more furrows are run, the second of which is to be the bed for the next row, and the manure is again drawn into this, the seed placed as before, and covered by the next furrow. The rows in this way are every third furrow,or about twenty-seven inches apt.

When the plants are about six inches above the surface of the ground, it should be harrowed across the rows, and the hoe should follow to set the plants right where covered, and drawing some earth around them. In due season, a furrow, with a one-horse plough, should be run on each side of the rows, with the earth thrown up to the plants, which is to be followed with the hoe, which completes the tillage of the plant. If any weeds afterwards rise, they should be cut up with a hoe,so as not to go to seed. It has been affirmed that plucking the blossoms from potatoe-vines increases the crop. But Mr. Justin Ely, of West-Springfield, asserts that no benefit is derived from that process. See *Massachusetts Agricultural Repository*, vol. IV. p.324.

In gathering the crop, run furrows on each side of the rows and then a pretty deep one through the middle, which turns up most of the roots to the surface, for the purpose of picking up by hand. A fork, with four prongs, with the addition of what may be called a fulcrum, fastened by a pivot to the back part of the handle may be used for raising the potatoes not turned out by the plough. See *Farmer's Assistant*, p. 301—2.

"In the report on the agriculture of the county of Hereford drawn up for the British Board of Agriculture, is a description of an excellent implement, invented by Mr. Yeldall, for taking up potatoes, having four prongs, on barbs of iron, with a fang, in the form of a double mould-board, drawn by three horses or four oxen. It enters the ground, under the bed of potatoes, and throws the whole to the surface." *Domestic Encyclopedia.*

Fowls of every sort may be profitably fed on boiled potatoes and meal mixed. A little alum, Armenian bole, or other astringent substance, has been recommended to mix with potatoes given to hogs to stop their tendency to purge.

Potatoes may be preserved almost any length of time by slicing and drying them in an oven or kiln.

" *To boil Potatoes*—Put them in cold water with the top of the pot loosely on. Let them boil gradually in water that will just keep them covered; when the water has boiled, put on the cover close, and continue boiling them for some time longer, till they are soft. Pour off the water and let the potatoes drain in a cullender. Then put them in the pot again without water, and drive off the moisture till they split. *Domestic Encyclopedia.*

Payson Williams, Esq. of Fitchburgh, Mass. received a premium from the Mass. Agr. Society for the best crop of potatoes raised in 1820, being six hundred and fourteen bushels on an acre.

POTATOE, SWEET, *Convolvulus Batata*. The following is the method used by Mr. Joseph Cooper, of New-Jersey, for cultivating this root.

He plants his seed-potatoes in a hot-bed, and cuts off the sprouts, and plants them in their destined spots in the field, three or four sprouts in a hill. In this way potatoes are raised earlier, much seed is saved, and much trouble of weeding is avoided; for the sprouts grow so fast, that they suffocate weeds, &c.

The sweet potatoe grows best in a mellow sandy soil, and the richer the soil the better the crop.

"Sweet potatoes, especially those with red skins, yield a considerable quantity of farinaceous matter, which forms a fine nutritious jelly, when mixed with water. *Bowen's* patent sago is this powder, which was for a long time prepared by the above person, at Savannah, in Georgia, chiefly for the British army. The process was similar to that which is known in Europe, for procuring the flour of the common potatoe, viz. grating the clean roots, washing the mass through brass sieves of different sizes, and collecting the flour at the bottom of the vessel which receives the fluid; finally drying it in pans either in the sun or by a fire." *Domestic Encyclopedia.*

J. Lowell, Esq. and some others have succeeded in raising this root for several years, in Massachusetts. See Agr. Rep. vol. VI. p. 265.

POULTRY, all kinds of tame birds, as hens, geese, ducks, turkeys, &c.

These may be considered as part of a husbandman's stock: But the keeping of great numbers of dunghill fowls will not turn to his advantage; as it is certain they will never indemnify him for the corn and grain that are requisite for their support. Yet on a farm a few of them may be useful, to pick up what would otherwise be lost. And in this view they seem to be profitable only part of the year. If confined they will not prosper, though they have a yard of some extent; if not confined they will be mischievous to the garden and field.

Poultry are most easily fattened when kept in a dark place. They should be furnished with gravel, but no water. Barley meal is said to be their best food, which should be mixed so thin with water as to serve them for drink. Their thirst induces them to eat more than they otherwise would, in order to extract the water, which is in their food. This should not be put in troughs, but laid on a board, which should be washed every time fresh food is put upon it. Foul and heated water, says the Domestic Encyclopedia, is the sole cause of the pip.

Hens, which do not lay in the winter, should have access to slacked lime, pounded bones, or oyster shells, as something of the kind is necessary to form their shells, which are composed of the phosphate of lime. When corn is given to fowls it should be crushed, or soaked in water.

A correspondent of the American Farmer is of opinion, from actual experiment, that pot-liquor is a cure for the *pip* in poultry. vol. I. p. 412.

PRONG HOE, a hoe with prongs instead of a blade. Four or five prongs are found best. It is easily struck into the ground; and as the tines are six or seven inches long, it will stir the ground to the same depth that a plough does. It is useful in taking up all roots.

PUMPKINS. Pumpkins will grow on any kind of soil which is proper for hoed crops, but the land cannot easily be made too rich for them. The "*Farmer's Assistant*" thinks they will grow better, when planted by themselves, than when raised, as usual together with Indian corn. The hills in such case should stand about seven feet apart each way, and a number of seeds should be planted in each hill, to make an allowance for what may be destroyed by insects.

A new and very large species of pumpkin has lately been introduced into this country, of which it is said more than five hundred pounds can be raised from a single seed. But these large sorts are not so profitable as the common ones. They are more tender, and will not keep so well.

"Pumpkins are excellent for fatting horses. They, however, do not relish them at first; and therefore must be kept from feeding, till they are hungry, before the pumpkins are offered to them; and let a little salt be first sprinkled on this food; when they will soon grow fond of it, and eat it readily without salt." *Farmer's Assistant.*

Q.

QUAKING MEADOW, or MARISH, low boggy land, that shakes and settles under any one in passing over it.

It has a sward that is tough, being a web of the roots of strong grasses; but the mud under the sward is very soft and yielding.

Such places should be drained when it can be done without too much expense. For its natural produce consists of the worst water grasses, cranberries, &c. but the soil is always deep, and rich. See the article *Draining*.

Mr. Eliot drained such a piece of ground, and soon made it fit for tillage.

QUINCE TREE, *Pyrus Cydonia*, a small fruit tree, bearing a large yellow fruit, useful in cookery and medicine, but not fit to eat raw.

It is easy to propagate the trees by suckers, layers or cuttings, but they require a moist soil. The cuttings should be planted early in autumn. The trees require very little pruning; the principal thing is, to keep the stems clear of suckers, and thin the branches where they cross each other. Upright luxuriant shoots in the top should also be taken out, that the trees may not have too much wood, which is bad for all sorts of fruit trees.

Quince trees are highly esteemed as stocks on which pears may with great advantage be grafted or budded. This operation greatly improves the taste and flavour of these pears, which arrive at maturity in the summer and autumn; but it is by no means proper for winter fruit, which is thus rendered hard and stony. This fruit is generally boiled and eaten with sugar, in which form it may also be usefully employed in cases of dysentery. Five gallons of the juice of quinces, mixed with twenty-five pounds of sugar, and fermented, make a delicious wine. *Domestic Encyclopedia.*

QUINCUNX ORDER, according to Mr. Miller, is applied to a plantation of trees, disposed originally in a square, consisting of four trees, one at each corner, and a fifth in the middle; which disposition, repeated again and again, forms a regular grove, wood, or wilderness; and, when viewed obliquely, presents straight rows of trees, and parallel alleys between them.

QUITCH GRASS, otherwise called Couch Grass. The only effectual way of extirpating this grass is by horse and hand rakes, and burning it.

R.

RACK, a frame made to hold fodder for cattle, to prevent their trampling it under foot, and wasting it.

Those racks which are under cover, as in sheep houses, horse stables, &c. may be constructed of almost any kind of wood; but those which stand abroad should be of such timber as lasts long in the weather. The rails may be larch, or white cedar, and the cross sticks white oak. Such a one will endure the weather many years.

RADISH, *Raphanus*, a pleasant root, which has an attenuating virtue, and is a good antiscorbutic.

I have had better success with those sown as late as June or July, than with those sown in the spring. The earliest are apt to be destroyed, or greatly injured, by the white maggots; to which sea water is an antidote; but with respect to this root not quite effectual.

To have a constant succession of radishes at table, the seeds should be sown once a fortnight, from April to August. But in midsummer they sooner grow sticky and strong, than in spring or fall. They must therefore be eaten while they are young. I have had better success with those sown in August, than in any other month. In hot houses they may be raised any month in the year. Or those raised in autumn may be kept in dry sand, fit for eating in the winter.

As radishes are uncertain in their growth, the best method is to put in the seeds between rows of other plants; and they are so soon pulled up, that they will not incommode the plants among which they grow.

Radishes that are for seed require much room, as they grow to a large size. For this purpose some of the most thrifty ones should be left standing; or else be transplanted to a place where each shall have as much room as near a yard square. The ripeness of the seed is known by the pods turning brown. For this purpose the seeds must be sown early in the spring, because they ripen slowly.

"This root being liable to be eaten by worms, the following method is recommended for raising them: Take equal quantities of buck-wheat bran, and fresh horse dung, and mix them well and plentifully in the ground by digging. Suddenly after this a great fermentation will be produced, and numbers of toad-stools will start up in forty-eight hours. Dig the ground over again and sow the seed, and the radishes will grow with great rapidity, and be free from the attacks of insects. They will grow uncommonly large. Buck-wheat bran is an excellent manure of itself." *Farmer's Assistant.*

RAIN, condensed vapour, which falls in drops, and waters the earth. This is of more advantage to the husbandman than all his labour and care. No kind or degree of culture will secure a crop,if the ground do not receive a considerable quantity of moisture from the clouds; for if the earth be not frequently moistened, the food of plants in it will become fixed; and there will be no fermentation in the soil; so that the roots of plants cannot receive any nourishment. Was it not for the falling of dews the want of rain would be much oftener destructive to plants than it is. Dews are often great in a dry season; and from dews plants receive a considerable part of their nourishment.

The due quantity of moisture might indeed be supplied by watering by hand, as long as wells,springs and rivers were not dried up. But the labour of doing it would be worth more than all the crop. Neither would artificial watering have so good an effect as rain,on account of the inferior quality of the water for this use, and the mode of applying it. The gentlest rains are generally most conducive to the growth of plants, and the fruitfulness of the soil, as all parts are more equally soaked; and cloudy weather,which most commonly happens before rain, helps to predispose the earth, and its vegetables, to receive the greater advantage from the water that falls. It is also believed the electric fluid, which is conducted to the earth by rains, conduces much to the invigoration of plants.

Rain not only gives fluidity and motion to the food of plants contained in the soil, but contains in itself more or less of the ingredients of it. The atmosphere contains abundance of saline, earthy and oleaginous particles; so that rain water cannot fail of being impregnated with them.

It has been proved by a variety of experiments, that a much greater quantity of rain falls at the surface of the ground, than at the top of a house, or other building; which may be partly owing to the vapour contained in the lower part of the atmosphere, which is joined to the drops in their descent.

Perhaps the action of the sun's heat is proportionably greater in vallies than on summits of hills; if so,there is a happy balance between heat and rain on all parts of the surface of the earth. Though it is often regretted that low hollows are overcharged with water, it is commonly soon exhausted by the heat of the sun in summer, which is much greater in vallies than on hills.

It may be asked, would it not have been better, if a greater proportion of rain had fallen on hills than on vallies? But they need it not so much, because of the greater coolness of the air on hills. More of the fine mould would have been washed down into the hollows, and deeper channels would have been made in the soil by the running of water, which would have been considerable inconveniences.

The quantity of water that falls in a year may be from twenty-five to thirty inches. If the whole were to fall at once, destructive deluges would be experienced,and droughts equally destructive. It is the frequency of rains that renders the earth fruitful. To some soils, as stiff clays and loose sands, frequent rains are more needful than to others. The former imbibes the water too slowly ; the latter parts with it too speedily. These two kinds of soil, therefore, need the most frequent showers.

In some years the rains are so ordered, as to make the seasons most fruitful. A moderate quantity in each week through the summer will be apt to supply so much moisture, and keep up such a degree of fermentation in the soil, as is most conducive to the progress of vegetation.

Farmers in this climate generally wish for but little rain in April, and for much in May and part of June ; then less in hay time, and English harvest. But as it is not left to us to order this matter, we should endeavour to accommodate ourselves to the seasons ; and to assist nature whenever we have opportunity for doing it, draining land which is too wet, watering that which is too dry, and applying more manures to dry soils, which will make them more retentive of water.

RATS, a mischievous kind of vermin too well known to the farmer. No walls that I know of have been found to be sufficient barriers against them.

The same poison which I prescribed for mice, will well serve to destroy these animals. But the best way is to catch them in a cage made of wire, in a cubical form, enclosed in a wooden box. Each side of the cage should be a plane of about fifteen inches square.

Take one quart of oat meal, four drops of oil of rhodium, one grain of musk, two nuts of nux vomica powdered ; mix the whole together, and place it where the rats frequent ; continue to do so while they eat it,and it will soon destroy them : Or, take equal quantities of unslacked lime, and powdered oatmeal ; mix them by stirring, without adding any liquid, and place a small quantity in any place frequented by rats. They will eagerly swallow the preparation, become thirsty, and the water which they drink will cause the lime to swell, and thus destroy them.

RICE, or Oryza, is a genus of plants consisting of several species, which, however, may be divided into two varieties, viz. Mountain-rice, which grows on dry elevated soils, and Marsh-rice sown in low swampy districts. According to the Domestic Encyclopedia, Mr. Bordley raised the former kind of rice on dry sandy land at Annapolis, in Maryland. It is not impossible but this plant may be gradually introduced into the northern States, and made to grow on dry and elevated ground. This is a great desideratum, as rice is the cheapest nutriment known ; containing, it is said, 96 parts in 100 of nutritive matter.

"The following directions respecting the *water culture of rice,* are furnished by one of the most successful cultivators of that article in the State of South-Carolina.

"Begin to plant about 25th of March, and trench shallow and wide, and scatter the seed in the row; make 72 or 75 rows in a task, and sow two bushels to an acre.

"1st. Hoe about the end of April, or beginning of May, when the rice is in its fourth leaf; then flood, and clear the field of trash. If the planting be late, and you are likely to be in grass, flood before hoeing; but hoeing first is preferable. The best depth to flood is three or four inches. It is a good mark to see the tops of the rice just out of the water; the deep places are not to be regarded; the rice will grow through in three or four days. Observe to make a notch on the frame of the trunk, when the water is of a proper depth; if the rains raise the water above the notch, or it leaks out, add or let off accordingly. This is done by putting a small stick in the door of the trunk, about an inch in diameter; if scum or froth appear in eight or ten days, freshen the water, take off the trunk doors, run off the water with *one ebb*, and take in the next *flood;* then regulate as before. Keep the water on about fifteen or seventeen days, according to the state of the weather; that is, if a hot sun, fifteen days, if cool and cloudy, seventeen days, counting from the day the field is flooded; then leak off with a small stick for two days, then run off the whole, and keep the field dry. In four or five days after, hoe the second time, stir the ground, whether clean or not, and comb up the fallen rice with the fingers. Keep dry, and hoe through the field. Hoe the third time and pick clean. This will be about the beginning of July. Then flood as you hoe. Let the water be the same depth as before. If any grass has escaped, it must be picked in the water after it shoots out This is called the fourth hoeing, but the hoe is never used except for some high places, or to clean the dams. If the rice is flaggy and likely to lodge, flood deep to support it, and keep it on until fit to harvest." *Domestic Encyclopedia*, vol. — p. 175. 3d edition.

REED, *Arundo*, "the name of an aquatic plant, infesting low grounds. The best method of destroying them, is by draining the land. Ashes and soot will kill them. So will ploughing the land, and laying it in high ridges. They always indicate a good soil." *Complete Farmer.*

RIDGLING, a male animal, half castrated. A horse of this kind is as troublesome as a stallion, or more so; but is not fit to be depended on as one. A ridgling hog will never be fat, nor grow so large as a barrow, till his castration be completed; as it may be by making an opening in the belly, when the case is the most difficult. They should be either killed young, or completely castrated. The flesh of a young ridgling pig is good; but that of an old one disagreeable.

ROLLER, a cylindrical instrument to pass over lands, to answer several good purposes in husbandry.

Those rollers which are cut out of free stone, being heavier than wooden ones, are best to smooth, and harden, the alleys in gardens, walks, &c. But wooden ones answer better in tillage, when they are sufficiently large. A roller for field husbandry should be five or six feet long; so that it may perform much in a short time, being drawn by a horse or a yoke of oxen, for either of which it may be easily harnessed. It should be made perfectly round and smooth, that it may be drawn the more easily, and press the ground the more equally in all parts. And it should be from eighteen to twenty-four inches diameter. Being large, the pressure will be greater; and the surface will be left the more level.

A spiky roller, or a roller filled with spikes, six or seven inches long, sharp pointed at the outer ends, is sometimes used in the old countries, to pulverise cloddy land in tillage, or to brake and open the sward of grass land when it is bound, and too compact. After grass land is so broken, a top dressing will have the better effect. A roller is sometimes armed with circular knives, four or five inches broad, put on in the manner of hoops, the edges at right angles with the axis of the roller, twenty inches from each other, They use these instruments to cut the sward into strips, in order to cut up the turfs with a sharp ironed plough for burn beating. This manner of doing the work, is far less expensive than cutting up the turfs with the beating axe. But the sward of land to which this instrument is applied, ought to be extremely level, and free from stones and strong roots.

Sir John Sinclair has the following remarks on this subject. "The roller is the most useful implement, for breaking hard clods expeditiously, and smoothing the surface of land when in tillage, ever yet invented. It is likewise of use to grass lands laid down for hay; and heavy rollers would prevent those ant hills, by which so many pastures are deformed. Rollers are made of various substances; as wood, freestone, granite, or cast iron; but on the whole the two latter are to be preferred. It is of importance that the weight of the roller should be in proportion to the extent of the surface on which it rests, and the nature of the land on which it is to be employed. The best plan is that of having two rollers, each about two feet and a half in length, and both placed in one frame, so as to roll clear of one another. This is the most suitable both for corn crops and sown grass, as it neither tears up the tender soil, nor injures the young plants. Besides the labour in turning is much less severe on the frame, and on the cattle. Every farm ought to be provided with rollers of different diameters and weights, so as to suit the several purposes to which they are destined; those of a small diameter are generally applied to land in tillage; and those of a large diameter, with double shafts to grass lands. Heavy rollers are of great use for destroying worms, slugs, and other vermin in the soil." *Code of Agriculture.*

ROLLING, smoothing and moderately hardening the surface of land, by drawing a roller over it.

The rolling of land in tillage

should be done only in dry weather; never, when the soil is so wet as to stick to the roller.

No soil will admit of rolling that is very uneven, or much rocky or stony. But small round pebbles in a mellow soil, well pulverised,need not prevent rolling: For the roller will press them all into the soil. Land that is apt to have a stiff crust formed upon it, by lying only a few weeks, I think should not be rolled; because it will cause the crust to be the more hard and stiff. But the advantages of rolling in a light and rich soil are so great, that it is pity that the practice of it is so generally laid aside in this country.

Rolling, after sowing and harrowing, will cause the mould to enclose the seeds; much of which, otherwise, lying in cavities that soon become dry, is apt to fail of vegetating.

Rolling also fills up ten thousand little cells, which, when left open, are haunts and harbours for flies and other noxious insects; besides, it has the advantage of destroying some kinds of insects in the operation. It is peculiarly beneficial on this account to a crop of turnips. And some recommend passing the roller over turnip ground, not only when the seed is newly sown, but after the plants are up.

When a clay soil is sown, rolling breaks many lumps, or hard clods, which have escaped the plough and the harrow. But an over light soil, which is apt to dry too fast, needs rolling more than any other. It serves to prevent the evaporation of moisture, by making it less porous.

Some of the European farmers prefer rolling after the grain has risen to the height of four or five inches. But of the utility of this we are not yet convinced by a sufficient number of trials.

In all kinds of soil that are laid down to grass, rolling is necessary, to lay the surface so smooth and even as to facilitate mowing and raking. And those kinds of sowed corn which are to be cut with the scythe, and raked, should be rolled, that loss may be prevented in harvesting. Without it, a crop of barley cannot be well taken up clean with a rake, especially when the corn is short and small, as I have often found to my loss. Some writers on husbandry think a crop of barley, in particular, will be considerably larger for rolling it, as it is a dry seed, that needs to be well enclosed with mould, in order to its vegetating. Lands that are in grass, may be kept even by a yearly rolling, which will press down mole hills and other unevennesses, and cause the grass to grow thicker. It will also be an advantage to be able to mow it the more closely.

"An intelligent farmer maintains that if draining is the first,manuring the second,and cultivation the third, rolling ought to be considered as the fourth principal operation in the process of agriculture. Its importance indeed, is every day becoming more apparent, and new advantages are derived from its use both on arable and on grass lands.

"Wheat should always be rolled in the spring, after frosts,as it makes the soil adhere more closely to the roots of the plants, encourages vegetation, and strengthens the stems, and renders the grain more perfect. When any crop of grain is sown with artificial grasses, rolling is particularly necessary, to make an even surface, bruising all clods,and pressing down any stones, it may not be thought necessary to carry off, to facilitate the future operation of the scythe. Oats in a light soil, may be rolled to advantage, immediately after the seed is sown, unless the soil is so wet as to cling to the roller. After turnips are sown in drills, they ought immediately to be rolled; to make the soil compact, and to promote their speedy germination. Not only for turnips, but for all other crops, rolling during the night is found to be an efficient means of destroying slugs, snails, the wire-worm, and other vermin, so destructive to young plants. Flax ought to be rolled immediately after sowing; it makes the seed vegetate equally, and prevents after growth; the bad effects of which are visible in every step of the process of dressing flax.

"The other advantages in rolling arable land are, that it renders a loose soil more compact and solid. This encourages the growth of plants, by pressing the soil to their roots. It likewise keeps in moisture and prevents drought from penetrating. When the soil is worked up lightly, moisture either filters through it too quickly, or is too easily evaporated. In a dry season this may occasion a very material difference in the crop, more especially in a light soil. Rolling is executed to most advantage, across the direction of the ridges, because more adapted to ensure full benefit to the furrows,which otherwise may not be properly gone over."

ROOTS, the parts of plants that are under the surface of the earth, which imbibe the nutritious juice of the earth, which feeds and increases the plants.

Botanists distinguish roots into divers sorts, according to their different forms. But the only distinction to which the husbandman needs to give his attention, is, to consider roots as of the tap, bulbous, or fibrous kind. Of the first kind are the carrot, parsnep, beet, &c. of the second, the potatoe, onion, turnip, and several others; of the last, wheat and other kinds of grain, and many grasses.

But still there are, perhaps, but few plants which have only one of these kinds of roots, though the form that is most obvious is the one which gives the name to the root. Carrots, and other tap rooted plants, send out horizontal fibres to a considerable distance. Trees in general have both tap and fibrous roots. A turnip has the three kinds of roots, having a bulb, a tap, and many lateral fibres from the tap.

Mr. Mills, on this subject says, the roots that proceed immediately from the seed, are always of the carrot or tap kind. Tap roots strike down perpendicularly into the earth, till it becomes too hard to admit of their farther passage; but when the soil is deep,and easily pierced,they penetrate sometimes to the depth of several yards, unless they are cut or broken; in which case they alter their direction. This is frequently observed; particularly in plants raised in water only. The tap roots shoot out branches which extend horizontally; and these branches are stronger, in proportion as they are nearer to the surface of that layer of earth, which is stirred by the plough or spade.

These are the roots which we calling creeping or fibrous. They extend sometimes to a considerable distance from the plant that produced them; but then they become so minute, that the naked eye can no longer trace them; especially when they have taken the tincture of the earth that surrounds them, as they generally do.

A carrot, for an example, which seems to have only one great root, furnished with some fibres, pushes its roots, according to Mr. Tull, to a considerable distance; but they grow so very slender, that they cannot be distinguished from the earth that covers them, without great attention. The case is the same with almost all plants.

To convince the reader of this, and at the same time to shew how far the roots of plants can extend in ground that is well loosened, he recommends the experiment which I have mentioned under the article, *Pasture of Plants*, which see.

The following instances, says M. Duhamel, shew what effort trees will make, to find a proper soil for the extension of their roots. On examining those of a hedge, at the side of which a ditch had been dug, it appeared, that after passing underneath the ditch, they reascended, and spread themselves in the ploughed earth on the other side.

He made the same observation on a row of elms, which were very near being killed by the digging of a deep ditch pretty near them, in order to prevent their roots from damaging an adjacent piece of ground. The elms shot out fresh roots in the loose mould that dropped into the ditch; these roots reascended on the other side of the ditch, and spread in the ploughed ground, and the elms soon recovered their former vigour.

He likewise observed, that on digging a trench at a small distance from a young elm, and filling it with good mould, the roots of that elm took their direction towards the trench, and grew to a great length in it.

These observations prove that horizontal roots extend far, especially in loose mould: And as a plant thrives in proportion to the length of its roots, Mr. Tull justly infers the necessity of keeping the earth in tillage in a light state, that the roots may easily penetrate it.

A root that has been cut or broken, never grows longer, but soon produces several new roots, all of which gather the proper food of the plant. Its means of subsistence are therefore increased, by the breaking of its roots, in digging or ploughing, rather than otherwise. In the horse-hoeing husbandry many of the fibrous roots of the growing plants are undoubtedly cut off by the plough. But it occasions the multiplying of the roots, and consequently the greater nourishment of the plants.

A writer in the Massachusetts Agricultural Repository, Vol. V., page 326., recommends the field culture of roots for the consumption of stock, and as the principal dependence of the farmer for its winter's support. He remarks, that 'the connexion of root cultivation with a prosperous agriculture is not a recent discovery, although it is the great boast of the farming interest of England and Scotland. More than fifty years ago, a Dutch farmer asserted to Joseph Marshall, an intelligent Englishman, at that time travelling through the northern kingdoms of Europe for agricultural information, "that the foundation of all good agriculture was the raising roots as the winter food for cattle." The reason is obvious. By this system the greatest quantity of food is raised upon the smallest extent of land with the least labour. What other cultivation can enable a farmer to raise fifty, sixty, and even one hundred tons of food of an excellent quality upon a single acre?

ROT, a disease in sheep, similar to a pulmonary consumption in men. A writer in the Scots Farmer thinks that if the disease have not proceeded far, the animals may be cured by feeding on turnips. But this is rather to be doubted. It is said to be caused by keeping them in a pasture that is too moist, producing rank and watery grasses. The raging of this distemper in a flock, is stopped by removing them to a dry situation: But the individuals which are deeply seized with it, are seldom cured. Cough is a constant symptom. The lungs decay, and the whole body droops and languishes, in the same manner as persons in a hectic. The sick of the flock should be removed from the sound sheep, that the infection may spread no further among the flock. See *Sheep*.

ROTATION *of* CROPS, a course of different crops in succession, on the same piece of ground.

The matter has not yet been sufficiently attended to by New England Farmers. This appears by their often being necessitated to lay their tillage lands waste for a considerable number of years, that they may get recruited. The expense of recruiting worn out land is so great, that such a course of crops ought to be preferred as the soil will bear without material injury, or without being too much exhausted. And, when other things are equal, such a course should be adopted as requires the least labour, or cost of manures and cultivation. When a course is well chosen, it may be repeated on the same spot perpetually, without damage to the soil.

It is not to be expected, that the best rules concerning this matter can be established, but from the experience of many years. For though it may be easy to compare the respective advantages of different courses, in a few years, so as to find which is more productive; it will take a much longer time to determine which course will be best on the whole. For the state of the soil, at the end of a long course, is to be taken into the account. And it is to be remembered that a course that is suitable for one soil, may not be so for another.

In countries where a spirited attention to agriculture has for a long time subsisted, one would expect, that people have most probably adopted the best courses. It is not amiss, therefore, to observe what courses they generally prefer in Britain and Ireland, taking care not to go into a rash and inconsiderate imitation of them, without making allowance for local differences, &c.

A common course in Ireland is, turnips, barley, clover, wheat: or, potatoes, barley, clover, wheat.

From the account that Mr. Young gives of the courses in different places, which he passed through in his northern tour, the following things are observable. That where they do not fallow, green and white crops follow each other alternately; and that wheat follows clover oftener than any other crop: That where fallowing is practised, wheat is next, and after it sometimes another white crop; but not generally. It ought to be never.

The courses of crops in Ireland, will furnish nearly the same observations.

The judicious farmer knows, that some regard must be had to the nature of the soil in a course of crops. Those crops which require a light soil, should make no part of the course in a stiff one, and *vice versa*.

But supposing the European courses to be the best that can be, some variation is surely to be made in this country; what that variation is, experience must discover. Not only our climates, but also our crops are different. We raise some crops that they do not, and not all that they do raise. But a rule that is fit to be extended to all countries, is, that two impoverishing crops should seldom, or never, succeed each other in a course. And it is certain, that white or grain crops in general, are apt to impoverish the soil, as they continue to draw nourishment from the earth, for some time after the leaves are dead, and cease to receive nourishment from the air. And all plants that bear an oily seed, rob the soil of much of its vegetable food: Such are flax and hemp, supposing them to continue in the soil till the seed is ripe.

Reasoning from experience and observation, I am led to believe, that the following are as good courses, as may be expected to be introduced in this country. On light warm soils, the first year, corn dunged, pease, or potatoes; 2d year, rye, barley, or buck wheat: the 3d, and 4th, clover: the 5th, wheat: the 6th, and 7th, clover. On cold and stiff soils, first, oats or potatoes: 2d, potatoes well dunged: 3d, flax, or wheat: 4th, grass, and so on till it needs to be broken up again. Though these may serve for general rules, yet as there is a great variety in soils, and some farmers can obtain manure in greater plenty than others, each farmer must endeavour to accommodate his courses to his soil and other circumstances.

The writer of a "Treatise on

Agriculture," originally published in the Albany Argus, recommends the following rotations of crops as best adapted to the northern parts of the United States.

"Medium course in sandy soils: 1st year, potatoes dunged; 2d. rye. with turnips after harvest, consumed on the fields; 3d, oats and clover, or barley, and clover; 4th, clover; 5th, wheat, with turnips after harvest consumed on the field; and 6th, pease, or lupins. or lentils. We have by this course *eight* crops in *six* years, and five of these ameliorating crops.

"Medium course in loamy soils: 1st year, potatoes dunged; 2d year wheat, with turnips as in the preceding course; 3d year, Indian corn and pumpkins; 4th year, wheat and turnips as before. In this course we have *nine* crops in six years—five of which are ameliorating crops. And

"Medium course in clay soils; first year, oats with clover; 2d. clover; 3d, wheat; 4th, beans dunged; 5th, wheat; 6th, the yellow vetchling."

An able writer in the Massachusetts Agricultural Repository, Vol. V. p. 334, recommends the following rotation of crops, as best adapted to the general state of agriculture in Massachusetts; viz.

1st year. After breaking up the sward, oats, sown thick, to be cut for fodder.

2d year Potatoes or Indian corn or both.

3d year. Carrots or turnips or both.

4th year. Barley or wheat, sown with clover and herds'-grass or red top.

5th year. Clover mowed.

6th year. Herds'-grass and clover.

In the autumn of the sixth year land to be broken up. and on the seventh year the same rotation recommences. See *Indian Corn* and *Flax*.

ROWEL, a kind of issue, or artificial wound, made in the skin of a horse, by drawing a skein of silk, thread or hair, through the nape of the neck, or some other part, answering to what surgeons call a seton.

Horses are rowelled for inward strains, especially about the shoulders or hips, or for hard swellings that are not easily dissolved. The rowel may be made in almost any part, and should always be not far from the diseased part, and about a hand breadth beneath it. The two ends of the rowel should be tied together, that it may not come out, and be smeared with lard, or fresh butter, before it is put in. Afterwards, it should be daily smeared again, and drawn backwards and forwards, that the putrid matter may discharge itself.

What are called rowels by the English Farriers are made as follows: An incision is made through the skin, about three eighths of an inch long. Then the skin is separated from the flesh with the finger, or with the end of a blunt horn, as far as the finger will easily reach. Into this a piece of leather made very thin, and round shaped, is introduced, about the size of a crown piece, having a large round hole in the middle of it. Previous to introducing the leather, it is covered with lint or tow, and dipped in some digestive ointment. Also a pledgit of tow, dipped in the same ointment, is put in the orifice, to keep out the cold air. See *Clark's Farriery*.

RUNNET, *or* RENNET, an acid juice, contained in the maw of a calf that has fed on nothing but milk. When the rennet is to be preserved for use, the calf should be killed soon after he has sucked; for then the curd is entire and undigested.

Dairy women usually preserve the maw, and the curd contained in it, after salting them; and then by steeping this bag and curd, make a rennet to turn their milk for making cheese. But a method which seems to be more simple, and is equally good in every respect, is, to throw away the curd, and after steeping it in very strong pickle, stretch out the maw upon a slender bow inserted into it, which will soon be very dry, and keep well for a long time. Take an inch or two of the maw thus dried, and steep it over night in a few spoonfuls of warm water; which water serves full as well as if the curd had been preserved, for turning the milk. It is said that one inch will serve for the milk of five cows.

In the Bath papers, Mr. Hazard gives the following receipt for making rennet: "When the raw skin is well prepared and fit for the purpose, three pints of soft water, clean and sweet, should be mixed with salt, wherein should be put sweet brier, rose leaves and flowers, cinnamon, mace, cloves, and almost every sort of spice; and if these are put into two quarts of water, they must boil gently, till the liquor is reduced to three pints, and care should be taken that this liquor is not smoked. It should be strained clear from the spices, &c. and when found to be not warmer, than milk from the cow, it should be poured upon the cell or maw; a lemon may be sliced into it, when it may remain a day or two; after which it should be strained again, and put into a bottle, where, if well corked, it will keep good for twelve months. It will smell like a perfume; and a small quantity of it will turn the milk, and give the cheese a pleasing flavour." He adds, "If the maw be salted and dried for a week or two near the fire, it will do for the purpose again almost as well as before." Another receipt is as follows: After the maw has been well cleaned and salted, and dried upon sticks or splints, take boiled water two quarts, made into brine that will bear an egg, let it be blood warm, put in the maw, either cut or whole; let it steep twenty-four hours, and it will be fit for use. About a tea cup full will turn the milk of ten cows. It should be kept in glass bottles, well corked.

An ingenious correspondent, who has made strict inquiry into this subject, recommends the following method of preparing a rennet, which he has found to be better than any other. "Throw away the natural curd, which is apt to taint, and give the bag a bad smell: Then make an artificial curd, or rather butter, of new cream, of sufficient quantity to fill the bag. Add three new laid eggs well beaten, one nutmeg grated fine, or any other good spice: Mix them well together, with three tea cup fulls of fine salt: Fill the rennet bag with this substance: Tie up the mouth: Lay it under a strong brine for three days, turning it over daily: Then hang it up in a cool and dry place for six weeks, and it will be fit for use. When it is used, take with a spoon out of the bag, a sufficient quantity of this artificial butyrous curd for the cheese you purpose to make: Dissolve it in a small quantity of warm water, and then use it in the same manner, as other rennet is mixed with the milk for its coagulation."

Whatever kind of rennet the dairy woman chooses to prepare, she should keep it in mind, that this animal acid is extremely apt to turn rancid and putrefy, and take care to apply a sufficient quantity of salt to preserve it in its best state. It should be as much salted as possible. The strongest kind of salt should be used. For it is probable that the rank and putrid taste, which is so often in cheeses made in this country, is owing to a putridity in the rennet.

RUSH, *Juncus*, a troublesome sort of plant, commonly found growing in wet and miry land.

"Rushes always indicate a good soil. They may be destroyed by lime, even after it has been slaked, by sea coal ashes, or by draining the land. Rushes thrive most in land that is too cold and moist for most other plants. Ashes, and other warm manures of various kinds, laid on plentifully, will keep down the rushes for a time: But to eradicate them perfectly, it is necessary to drain the land."—*Complete Farmer*.

RUST, dark spots, of the colour of the rust on iron, that appear on the stems and leaves of blighted grain. See the article *Mildew*.

Some sorts of grass are also subject to the same distemper.

RYE, *or* RIE, *Secale*, a well known grain, that is much cultivated in this country.

Though rye by itself makes a dark-coloured, clammy, and unsavoury kind of bread, it is better to mix with Indian meal in bread than any other kind of English grain; and for this reason, our farmers are the more fond of cultivating it.

Rye is as liable to suffer by rust, as wheat; but it is seldom known to be smutty. It is, however, sometimes hurt by a distemper called the *Spur*. See that article.

Mr. Miller thinks there is but one sort of rye, though distinguished by farmers, into winter and spring rye. The winter rye is larger and heavier than the other, and is commonly more profitable to the farmer. This is sown in autumn, at the same time as wheat. The spring rye should be sowed as early in the spring, as the ground will admit of it.

Some sow their winter rye at the last hoeing of Indian corn, and hoe it in. This is a good practice, when it is sown on flat land, or on a rich or heavy soil, where grain is apt to suffer by the frost of winter. For the plants of rye will be mostly on the corn hills, and so escape injury from frost: At least they will most commonly escape, or so many of them as are necessary to give a good crop. The plants that are killed will be those in the low spaces between the hills.

Sandy and gravelly soils are most suitable for rye. It commonly prospers much better on such, than on richer soils: The principal reason of which may be its ripening earlier, and so escaping the blight. Weak land has strength enough to produce rye, and it does not exhaust the soil so much as other corn.

I have known the same spot produce twenty crops of this grain in succession, (excepting that it was planted with Indian corn once or twice, to subdue the weeds) the crops yearly increasing, instead of diminishing. The right method is, to plough in the stubble as soon as the crop is off; and in a fortnight or three weeks, according as weather and circumstances favour, cross plough the ground, and sow the seed. The stubble, so early buried in the soil, serves as a manure. It will need no dung.

It is said by some writers, that sowing rye two or three years on a warm dry soil, it will be forwarded, so as to ripen a month earlier than that which has been long cultivated in other soils. This ought to be attended to by farmers in this country, where grain that ripens late, is so apt to be blasted. But this observation, possibly, may not be founded in truth.

The quantity of seed to be sowed, is recommended by some, to be two bushels per acre. But when the grain is small, five or six pecks may be a sufficient quantity. For the smaller the grain the greater the number of seeds.

The signs of ripeness are, the yellow colour of the straw, the hanging of the ears, and the hardness of the grain. But some choose to cut it when in the milk, because the flour will be whiter. The quantity, however, will be less, unless it lie a good while on the ground to ripen, which it may safely do in good weather, if care be taken to keep the top ends from the ground. Winter rye is some times fit to harvest by the middle of July even in the northern parts of New-England: Spring rye is always later.

Some recommend sowing winter rye for grazing and fodder. It affords very early feed for cattle in the spring. Or it may be mowed for hay two or three times in a summer. In countries that are dry, and do not naturally produce much grass, this may be considered as a good piece of husbandry.

Spring rye may be sowed very late in the fall at first, and a little earlier each succeeding year, until it may even be sown the May preceding the year in which it is harvested, and used the first season for pasturing or mowing. And winter rye may be sowed later and later each fall, and at length sowed in the spring, when it will become Spring rye.

The *Farmer's Assistant* recommends sowing winter rye and spring rye alternately, in order that the ground might every other year be benefitted by the application of gypsum. "The growing crop of rye," says the author of that work, "receives no benefit from the application of this manure; but it quickly covers the ground with a fine sward of white clover; and as soon as ground is thus swarded, it is in good condition for bearing any crop. Let the gypsum, therefore, be sown in the spring, on the growing crop of winter rye; and by the middle of October following, the ground will be covered with white clover; turn this sward over in the latter end of the fall and in the spring sow a crop of spring-rye; and, as soon as that is taken off, turn the ground over again for a crop of winter-rye; and in the spring repeat the process of manuring with gypsum as before, for a crop of spring-rye, and thus proceed with these crops alternately.

"In such a mode of culture, however, particular attention should be paid to turning under the stubble of the crop of spring-rye as quick as possible, in order that the seeds of the growing weeds be prevented from ripening, as well as for enabling other seeds, which may be buried in the soil to vegetate, and thus be destroyed by the after-ploughing and harrowing for putting in the next crop.

"Where winter rye is early sown, a bushel to the acre is probably sufficient; but for spring-rye, a bushel and a half to the acre, or perhaps more should be sown."

It has been said that a peck of oats, sowed with a bushel of spring-rye, will prevent its blasting. It is easily separated by winnowing.

Mr. Gardner Whiting, of Charlestown, Massachusetts, raised in the year 1820, on one acre of land, on Bunker Hill, forty-eight bushels and four quarts of rye.

RYE GRASS, *Lolium*, a sort of grass propagated in England for hay, sometimes called *Ray grass.*

This grass is in good esteem in England, and has been but little cultivated in this country, as we believe chiefly from the circumstance of its seeds not being for sale with us. We know one, and it is the only instance in which the experiment has been extensively tried. In this the ray-grass was sowed alone, and the field has continued for fifteen years to yield an ample crop of excellent grass. See *Grasses.*

S.

SAINFOIN, *Hedysarum Onobrychis.* This is a species of grass which will grow on dry and strong soils. which are unfit for other cultivation, and will produce a ton of hay to an acre, besides after growth, on poor and shallow soils. It is an early grass, and therefore useful for soiling, for which purpose it may be used for the fore part of the season and mowed the latter part. It will yield good crops for ten or fifteen years, at the expiration of which time it affords an excellent pasture for sheep. The hay, it is said, will fatten horses without oats. It increases the quantity of milk in cows, and improves the quality of the butter. It should be sown early in the spring, either by drilling it in rows two feet asunder, or broadcast. By the former method half a bushel of seed is sufficient to stock an acre; and at least four bushels are required for the latter. During the first season of its growth, no cattle should be allowed to feed upon it, nor should sheep be permitted to graze upon it the second season. At the end of five or six years it will require dung, gypsum, or composts of any kind, used for other grasses.*

SALT. Salt is one of the essential ingredients of the nourishment of plants; and some kind of salt is contained in every plant.

Salt is of essential importance to the farmer as a manure. It may be applied to the soil, either by itself, or mixed and dissolved in compost. In the latter method, I have found it to be a great fertilizer of land.

But if salt be applied unmixed and undissolved, it will endanger the existence of tender plants.

In June, 1786, I salted one bed of my onions, one bed of my carrots, and one bed of my early turnips; laying the salt under the surface, in the centres of the intervals, between the rows; at some distance, perhaps six inches, from the plants, that the salt might have time to be dissolved, and altered, before the fibrous roots should reach it. The carrots of the salted bed, evidently grew much larger and better than the rest of the carrots; but I could not perceive that the salt was at all beneficial to the onions, or to the turnips.

According to Mr. Ford's experiment in salting flax ground, salt seems to be highly beneficial to that crop. He spreads the salt over the ground, at the time of sowing the seed; and thinks that the quantity of salt should be double that of the seed. From three acres in flax salted, he had fifty bushels of seed, and an excellent crop of flax. It was thought that the advantage of salting appeared more in the seed than in the harle.

Mr. Eliot speaks of five bushels of salt being applied to one acre of flax, which is a much larger proportion, and that it had an extraordinary effect: And also of a crop of wheat being increased by salt.

Salt is esteemed more efficacious when mixed in composts than when applied to the soil in its crude state. All the salts will operate as manures, but if too much common salt be applied to lands at once it will prove injurious, for awhile at least. "Salt," says Sir John Sinclair, "is particularly useful when mixed with a dunghill, or strewed over farm-yard manures, at the time they are carried to the field." A mixture of soot and salt is recommended by the same writer, as preferable to any other manure. Salt is likewise of use in preventing mildew. See *Mildew*.

It is said to be an excellent practice to keep salt under cover in such a situation that cattle or sheep may have recourse to it at plea-

* Sainfoin will not answer in Massachusetts, nor, we suspect, in any of the New-England States. It is too apt to be winter killed.

sure. Those cattle, however, which have not been accustomed to so free an use of salt should be brought to it by degrees.

Upon the occasion of an application by the landed interest to the British parliament in 1817 for a reduction of the excise duties on salt, on the ground of its great value as a manure, a committee of parliament was appointed to take evidence in the case; and the result of that evidence seems to establish, beyond all doubt, the great benefit of salt as a manure It is still a question, whether the price in this country will permit its use to a great extent. We shall, however, state some of the most interesting parts of the evidence. Of course, in a work like this, we can give only a brief abstract.

The late Mr. Hitt, well known for his treatise on fruit trees, declares it to be of great use in destroying insects which affect them. He found also that water impregnated with salt, in the proportion of one ounce to a gallon of water, produced a visible effect on grass land. materially improving it.

Dr. Holland, in his survey of the county of Cheshire, states, its advantage to rushy meadow land. applied at the rate of eight bushels to the acre. It killed the rushy growth in April; but the latter end of May, a flourishing crop of rich grass made its appearance.

Lord Erskine, the present distinguished orator, also speaks of salt, as a manure, in strong terms.

There is an abundance of evidence furnished, as to the utility of salt for cattle; but this is so universally known and practised in this country, that it cannot be necessary to say any thing more, than that, we fear, a misplaced economy prevents as free an use of it, as would be desirable. The earl of Dundonald, in his evidence before a committee of the house of commons in 1801, speaking of its use as a manure, said, that when the benefit of salt and saline substances as manures should be fully understood, the consumption of it for these purposes would amount to five times the quantity consumed in Britain for all other purposes.

Its use was known to the ancient Romans. Pliny speaks of it as favourable to cattle. Lord Bacon mentions sea sand as an excellent manure on account of its salt.

In more modern times, the testimony has multiplied to an extent, which could not be given in a work like this. In proportion as the science of agriculture has improved, the sentiment of the value of salt as a manure has gained ground. Mr. Hollinshead, in a work printed in 1802, declares salt to be the cheapest best, and most durable manure ever used. It is admitted, however, by all, that, used in excess, it is fatal to all vegetation. The corrector of this article used it on grass land in Roxbury at the rate of a bushel per acre, and it did not affect the vegetation injuriously. He intends to extend its use, being satisfied of its salutary effects. On asparagus, its effects he proved this year to be very useful. That plant, however, grows in and near salt marshes.

It is affirmed to be an admirable

manure for onions. For meadow lands, six bushels per acre is recommended to be sown as soon as the grass is off. By meadow lands, the English writers mean all grass lands, whether upland or lowland.

This is not one tenth part of the whole mass of evidence on the subject, yet we are constrained to say, that there is a defect, to wit, that of a well conducted course of experiments on the principles, on which all modern philosophy reposes. Facts, not insulated ones, but a mass of them in continuity, should be collected. Trials on all soils, and for several years, should be made. And we hope our country will have the honour, as with respect to gypsum or plaster of Paris, of leading the way to an extensive use of this natural product. Though costly, the quantity required is so small, as to amount to less than many other manures in actual use.

SALTING *of* MEAT, the method of preventing its corruption for a long time, by the application of common salt, &c.

As farmers are most commonly too far distant from market places, to be supplied from them with fresh meat, and as it is most convenient for them to kill only at certain seasons, they ought to be well acquainted with the best methods of keeping meat in good order, by salting.

The common method of preserving pork, reserving the lean parts for use in the cold season, and applying a large quantity of salt to the fat, is perhaps as good as any can be. But beef is greatly injured, and rendered unwholesome by a severe salting.

A good method of preserving beef, which I have known to be practised for several years past, is as follows: For a barrel of beef of the common size, reduce to powder in a mortar four quarts of common salt; then eight ounces of salt petre, and five pounds of brown sugar. Let the salt be well rubbed into the pieces, pack them close in the barrel, and sprinkle the saltpetre and sugar evenly over each layer. No water at all is to be applied. The juices of the meat, if well packed, will form a sufficient quantity of brine; and the beef will keep sweet and good through the following summer, supposing it killed and packed in the beginning of winter, or late in autumn; and will not be too salt to be palatable. Draining off the brine and purifying it by boiling and scumming, with the addition of a little salt in the beginning of summer, and returning the brine upon the meat, will be a real improvement.

Dr. Anderson recommends a similar method for preserving butter. Take of sugar one part, of nitre one part, and of the best Spanish great salt two parts. Beat the whole into a fine powder, mix them well together, and put them by for use. One ounce of this is to be thoroughly mixed with a pound of butter, as soon as it is freed from the milk, and then immediately put into the vessel designed to hold it. After which it must be pressed so close as to have no air holes; and then so closely covered that no air can come to it. If all this is done, he thinks the butter may be kept perfectly sound and good for many years. For he has seen it at two years old, in every respect as sweet and sound as when only a month old.

The following receipt for pickling beef has been highly recommended.

For an hundred pounds of beef, take sixteen pounds of bay or fine salt, two pounds brown sugar, four and an half gallons water, and six ounces salt petre.

The salt, sugar, and water should be put into a brass or copper kettle over a fire. Stir the salt often, and continue stirring it until the salt has all dissolved, and the scum has done rising. Take off the scum as it rises. After this add the salt-petre. Let the pickle stand till it is about cold, or blood warm. Have the beef cut in smallish pieces, and packed pretty close, free from any bloody pieces. Add the pickle and cover it tight from the air. Should there be any appearance of mould on the surface of the pickle, at any time, add a handful or two of fresh salt.

SAND, is described as a genus of fossils, found in minute concretions, forming together a kind of powder, the genuine particles of which are all of a tendency to one particular shape, and appear regular, though more or less complete, concretions; not to be dissolved, or disunited by water, or forming into a coherent mass by it, but retaining their figure in it: transparent, vitrifiable by extreme heat, and not dissoluble in, or effervescing with acids.

"These are subject to be variously blended and intermixed, either with homogene or heterogene particles, particularly with flakes of talc; and, according to these, and their different colours, are to be subdivided into different kinds, as red, white, &c.

"As to sand, its use is to make the clayey earth fertile, and fit to feed vegetables: For such earth alone, we find, is liable to coalesce, and gather into a hard coherent mass, as is apparent in mere clay. The earth thus embodied, and as it were, glued together, is no ways disposed to nourish vegetables. But if with such earth, a sufficient quantity of sand be intermixed, it will keep the pores of the earth open, and the earth itself loose and incompact; and by that means give room for the juices to ascend, and for plants to be nourished thereby.

"Thus a vegetable, planted, either in sand alone, or in a fat glebe, or earth alone, receives no growth or increment at all, but is either starved or suffocated: But mix the two, and the mass becomes fertile. In effect, by means of sand, the earth is rendered, in some manner, organical: Pores and interstices being hereby maintained, something analogous to vessels, by which the juices may be conveyed, prepared, digested, circulated, and at length excerned, and thrown off into the roots of plants.

"Grounds that are sandy and gravelly, easily admit both heat and moisture: But then they are liable to these inconveniences, that they let them pass too soon, and so contract no ligature, or else retain it too long, especially where there is a clay bottom: And by that means it either parches or chills too much, and produces nothing but moss and cankerous infirmities. But if the sand happens to have a surface of good mould, and a bottom of gravel, or loose stone, though it do not hold water, it may produce a forward sweet grass; and though it may be subject to burn, yet it quickly recovers with the least rain.

"Sea sand is accounted a very good compost for stiff ground: For it effects these two things; it makes way for the tree or seed to root in stiff grounds, and makes a fume to feed it.

"Sand indeed is apt to push the plants that grow upon it, early in the spring, and make them germinate near a month sooner than those that grow upon clay, because the salts in the sand are at full liberty to be raised, and put into motion, upon the least approach of the warmth of the sun. But then, as they are hasty, they are soon exhaled and lost.

"The best sand, for the farmer's use, is that which is washed by rain from roads, or hills, or that taken from the beds of rivers. The common sand, that is dug in pits, never answers nearly so well. Sand mixed with dung, is much better than laid on alone: And a very fine manure is made, by covering the bottom of sheep folds with several loads of sand every week, which are to be taken away, and laid on cold stiff lands, impregnated as they are, with the dung and urine of sheep.

"Besides clay land, there is another sort of ground very improveable by sand. This is that sort of black foggy land, on which bushes and sedge grow naturally, and which they cut into turf in some places. Six hundred loads of sand, being laid on an acre of this land, meliorate it so much, that it will yield good crops of oats, &c. though before, it would have produced scarce any thing. If after this crop is taken off, the land be well dunged, and laid down for grass, it will yield a large crop of sweet hay.

"Sea sand, which is thrown up in creeks and other places, is by much the richest of all sand for manuring the earth: Partly its saltness, and partly the fat and unctuous filth that is mixed among it, give it this great virtue. In the western parts of England, that lie upon the sea coast, they make great advantage of it. The fragments of sea shells also, which always abound in this sand, add to its virtues: And it is always the more esteemed by the farmers, the more of these fragments are among it.

"Sea sand is best, which is taken up from under the water, or from sand banks which are covered by every tide. The smallest grained sand, is the most sudden in its operation, and is therefore best for the tenant, who is only to take three or four crops: But the coarse, or large grained sand, is much better for the landlord, as the good it does lasts many years."—*Complete Farmer*.

Sand entirely changes the nature of a clayey soil; so that it will scarcely ever become so compact, as it was before sanding. Nor is any other manure so good as sand, to loosen and soften it. No other will have so lasting an effect. From being the least productive, a soil of clay, by sanding, comes to be the most fruitful of any, when it is sufficiently sanded; for it has more of the food of plants in it than any other soil, wanting only to have its cohesion sufficiently broken, to give a free passage to the roots of vegetables. For this purpose, a very small dressing of sand will not seem to produce any effect. A layer of two and a half or three inches will not be too much for land in tillage, if it be a stiff clay.

The benefit of sanding does not appear so much the first year or two as afterwards: For the oftener the land is tilled, the more thoroughly is the sand mixed with the clay; by which the vegetable pasture is more and more increased.

But sand laid upon clay land in grass, will have a great effect, without mixing it with the soil. I have known half an acre of clay land laid to grass, which became so bound and stiff, as to produce only two or three cocks at a mowing, with a mixture of low moss and other trash. The owner, in October, 1783, with one yoke of oxen, carted on eighty loads of yellow sand from the road, which was about equal to forty carts full; levelled it with a harrow, and threw in some hay seed. The following year it produced ten hundred weight of good hay: Last year it produced twenty hundred; and it is expected, that about thirty hundred will be the weight of the crop in the present year, 1786. The sand not only added warmth to the soil, but prevented the clay from becoming so dry and hard as to prevent the roots of the grass from extending themselves in it.

SANDY SOIL, a soil in which sand is the predominant ingredient.

It is seldom found unmixed with other ingredients. Wherever it is so, it is extremely barren, and of little or no value. It will scarcely produce weeds.

Some barren sands consist of very fine particles, and have no sward over them. The wind drives them before it, and makes what are called *sand floods*, which bury the neighbouring lands and fences. The fences near them should be tall hedges to abate the force of winds: And trees which require but little nourishment from the earth, should be planted in these sands, that a sward may be obtained upon them. See *Locust Tree*.

When a sandy soil is used in tillage, it should be for those crops which require the most heat, and are least apt to suffer by drought; as maize, tobacco, rye, pease, &c.

The best manures for a sandy soil, are marle, cow dung, and swines' dung; mud from flats, swamps, ponds, rivers, &c.

Clay is as beneficial to a sandy, as sand is to a clayey soil. A dressing of clay two or three inches thick, laid on a sandy soil, and well mixed, will make it fruitful for many years after, as I have found by experience. It brings the soil to the right consistence, renders it less porous, and causes it to retain its moisture. At the same time it is more retentive of manures applied to it: Perhaps the benefit received from the clay will never be wholly lost. Though the clay is continually sinking further into the earth, by means of every rain, deep ploughing will return it to the surface; so much of it at least as is necessary. And repeated dressings of clay may be needed.

"Rich sandy soils, under a regular course of husbandry, are invaluable. They are cultivated at a moderate expense; and at all seasons, have a dry soundness, accompanied by moisture, which secures excellent crops, even in the driest summers.

"The management of sandy land, according to the system adopted by the celebrated Ducket of Petersham and Esher, in Surrey, has been strongly recommended by an eminent author. It was founded on three principles:—1. Ploughing very deep: a due degree of moisture was thus preserved in his light land, by means of which his crops were flourishing in seasons of drought, which destroyed those of his neighbours. 2. Ploughing seldom, but effectually, by a trench plough, or what he called a skim-coulter plough, with which he buried the weeds that grew on the surface; he has been known to put in seven crops with only four ploughings: and 3. Occasionally raising a crop of turnips the same season, after a crop of wheat or of pulse."—*Code of Agriculture.*

SAP, the fluid contained in plants, which is drawn from the earth and atmosphere, by which plants are nourished, augmented, and rendered fruitful. It answers the same purposes as the blood and other circulating juices in animals. It conveys nourishment to all the parts.

In all plants there exists a system of tubes or vessels, which in one extremity terminate in roots, and at the other in leaves. It is by the capillary action of the roots that fluid matter is taken up from the soil. The sap, in passing upwards becomes denser, and more fitted to deposit solid matter; it is modified by exposure to heat, light, and air in the leaves; descends through the bark; in its progress produces new organized matter; and is thus in its vernal and autumnal flow the cause of the formation of new parts, and of the more perfect evolution of parts already formed."—*Agricultural Chemistry.*

SCRATCHES, or *Selenders*, a disorder between the hinder pastern joints and hoofs of horses, consisting of cracks and soreness, with suppuration. It is troublesome commonly in the winter season only.

"Nothing tends so much to prevent grease and swelling of the legs, as frequent hand-rubbing, and washing the heels carefully with soap suds, as soon as a horse comes in from exercise. In inveterate cases, where the disease appears to have become habitual in some degree, a run at grass is the only remedy; if a dry pasture be procured where a horse can be sheltered in bad weather, and fed with hay and oats, it will be found extremely convenient, as in such circumstances he may perform his usual labour, and at the same time be kept free from the complaint." —*The Complete Farrier.*

SEA WATER. this fluid, besides water and particles of common salt, contains, according to Dr. Russel's account, sulphur, nitre and oil.

As it undoubtedly contains much of the essence of animal and vegetable substances, by means of the perishing and consuming of both in it, it is fitter than mere salt to be used as a manure, whether by it self, or in compost.

In the year 1786, one hundred hills of potatoes near the shore were watered with sea water, about two quarts on a hill, being one hour's work of a man. The crop was half as much again, as in the same number of hills adjoining. The water was applied to the soil just after planting the sets, which I suppose to be the best time for doing it, as there can be no danger of burning the young shoots, and as the salt will be mixed with rain and the moisture of the earth, before shoots are produced.

In the year 1787, alternate rows were watered in the same manner with sea water. The result of this experiment was uncertain; because by ploughing off and on alternately between the rows, the earth of the watered and unwatered rows was blended together. But altogether, a good crop was obtained.

The same year a piece of flax was, in the month of June, very short and yellow on one side of the piece; but of a good colour on the other, and much taller: This induced the owner to water the poor side from the sea. In ten days it was equal in length and colour with that on the other side, though very little rain fell in the time. At pulling, the watered side was evidently better grown than the other. This was a sufficient demonstration of the advantage of sea water, when the land lies adjoining to the sea shore; so that the labour of applying it is inconsiderable.

The above experiments were made in a clayey soil.

In a sandy soil, the same year, watering the ground where French turnips were just sown, had an excellent effect. Though it was a spot where the turnips had been destroyed by insects, several years successively, they generally escaped this year. Not more than one pailful was applied to a drill row two rods in length, wetting the ground over the seeds, soon after sowing.

Salt water applied to tender plants, most commonly proves too strong for them, if applied when the ground is dry. But if it be wet, the strength of the water is abated by mixing with the juices in the soil, before it is taken up by the roots, and thus it is rendered innocent and safe, as I have found by experience. The seeds bear the application of the sea-water, better than the young plants do.

"Sea water might be carried from the sea some distance on the land, to advantage, in the following manner:—Take a one horse cart, and suspend a tight box, rightly shaped, under the axletree; the box having a valve in the underside; drive the cart into the water, and the valve opens, and lets that fluid into the box; and, when the cart is driven out, the valve closes and holds the water.

"When the cart is driven out to the ground on which the water is to be spread, this operation may be performed in the manner we shall next describe: A tube is to be provided, say, twelve feet in length, with small holes bored into it at the distance of six inches apart, and the ends of the tube closed; attach this to the under side of the box, crossways, at either end, so as to be out of the way of the wheels of the cart.

"When you come to where the water is to be spread, it is to be let out of the box into the tube, by an aperture for the purpose; and as the cart moves along, the water runs out of each of the small holes in the tube, and thus sprinkles over a piece of ground of twelve feet wide, till the whole is exhausted.

"With the next load; begin where the water ceased running before, and thus continue the watered strip across the field. Then take another strip of twelve feet wide, adjoining that already watered, and thus proceed till the whole has been gone over.

"In this way a man would carry out, say, forty cart loads a day, at the distance of half a mile, or half that number if a mile; as but little time need be spent, either in loading or unloading. About ten loads of a hundred gallons each would probably be sufficient for an acre at any one time."—*Farmer's Assistant.*

SEEDS *of Vegetables*, "their last product, by which their species are propagated; being frequently all the fruit of a plant, but sometimes only a part included in the fruit.

"Every seed contains a plant in embryo. The embryo, which is the whole future plant in miniature, is called the germ or bud; and is rooted in the cotyledon, or placenta, which makes its involucrum, or cover. The cotyledon is always double; and in the middle, or common centre of the two, is a point or speck, viz. the embryo plantule, which being acted on by the warmth of the sun and of the earth, begins to protrude its radicle, or root, downwards, and soon after, its plumula, or bud, upwards; and as the requisite heat continues, it draws nourishment by the root, and so continues to unfold itself and grow.

"The two cotyledons of a seed, are a case to the little embryo plant; covering it up, and sheltering it from injuries, and feeding it from its own proper substance; which the plantule receives and draws to itself by an infinite number of little filaments, which it sends into the body of the placenta.

"The cotyledons for the most part abound with a balsam disposed in proper cells; and this seems to be oil brought to its greatest perfection, while it remains tumid, and lodged in these repositories. One part of the composition of this balsam is oily and tenacious, and serves to defend the embryo from any extraneous moisture; and, by its viscidity, to entangle and retain the fine, pure, volatile spirit, which is the ultimate production of the plant. This oil is never observed to enter into the vessels of the embryo, which are too fine to admit so thick a fluid. The spirit, however, being quickened by an active power, may possibly breathe a vital principle into the juices that nourish the embryo, and stamp upon it the character that distinguishes the family; after which, every thing is changed into the proper nature of that particular plant.

"Now when the seed is committed to the earth, the placenta still adheres to the embryo for some time, and guards it from the access of noxious colds, &c. and even prepares and purifies the cruder juice which the young plant is to receive from the earth, by straining it through its own body. This it continues to do, till the embryo plant being a little enured to its new element, and its root tolerably fixed in the ground, and fit to absorb the juice thereof, it then perishes, and the plant may be said to be delivered; so that nature observes the same method in plants as in animals in the mother's womb.

"Many sorts of seeds will continue good for several years, and retain their vegetative faculty; whereas others will not grow after they are one year old: This difference is in a great measure owing to their abounding more or less with oil; as also to the nature of the oil, and the texture of their outward covering. All seeds require some share of fresh air, to keep the germen in a healthy state; and where the air is absolutely excluded, the vegetative quality of the seeds will be soon lost. But seeds will be longest of all preserved in the earth, provided they are buried so deep as to be beyond the influence of the sun and showers; since they have been found to lie thus buried twenty or thirty years, and yet vegetate as well as new seeds. How the vegetative life is so long preserved, by burying them so deep, is very difficult to explain; but as the fact is very well known, it accounts for the production of plants out of earth taken from the bottom of vaults, houses, &c.

"In the common method of sowing seeds, there are many kinds which require to be sown soon after they are ripe; and there are many others which lie in the ground a year, sometimes two or three years, before the plant comes up: Hence, when seeds brought from distant countries are sown, the ground should not be disturbed, at least for two years, for fear of destroying the young plants.

"As to the method of preserving seeds, the dry kinds are best kept in their pods or outer coverings; but the seeds of all soft fruits, as cucumbers, melons, &c. must be cleansed from the pulp and mucilage which surround them; otherwise the rotting of these parts will corrupt the seeds.

"When seeds are gathered, it should always be done in dry weather; and then they should be hung up in bags in a dry room, so as not to deprive them of air."—*Dictionary of Arts.*

"The seeds of plants exalted by cultivation always furnish large and improved varieties; but the flavour, and even the colour of the fruit, seems to be a matter of accident. Thus a hundred seeds of the golden pippin will all produce fine large-leaved apple trees, bearing fruit of considerable size; but the tastes and colours of the apples from each will be different, and none will be the same in kind as those of the pippin itself. Some will be sweet, some sour, some bitter, some mawkish, some aromatic, some yellow, some green, some red, some streaked. All the apples will, however, be more perfect than those from the seeds of the crab, which produce trees all of the same kind, and all bearing sour and diminutive fruit."—*Agricultural Chemistry.*

It has been recommended when seeds are intended to be sent a great distance, or it is wished to preserve them a long time, to wrap them in absorbent paper, and surround them by moist brown sugar.

Mr. Humboldt has found, that seeds, which do not commonly germinate, become capable of germinating when immersed in *oxygenated muriatic acid gas* mixed with water. If the liquid be a little warmed, it will quicken the vegetation of seeds surprisingly. Cresses thus treated exhibited germs in three hours. Seeds which were more than an hundred years old, were also made to vegetate by those means.

Old seeds may likewise be made to germinate by immersing them in water nearly boiling hot for about half a minute, and cooling them suddenly by exposure to air. But if such seeds are sown when the earth is cold, they will rot in the ground.

SEEDING, the same as sowing of seed. See the article *Sowing*.

SEEDLING, a root that springs from seed sown. The name is applied also to the tender tops of plants that have newly come from seed. The little plants are thus distinguished from cuttings, layers, and slips.

SEMINATION, the manner in which plants shed and disperse their seeds.

Some seeds are so heavy, that they fall directly to the ground; others are furnished with a pappus, or down, that they may, by means thereof, be dispersed by the wind; and others again are contained in elastic capsules, which, bursting open with considerable force, dart or throw out the seeds to different distances. Some of the second sort are wafted over vast tracts of land, or even carried to remote countries. The weed that is peculiar to burnt land, and is called fire weed, has such a kind of seed: It is not strange, therefore, that we see it grow in burnt places, many miles from where it has grown before.

SHADE, a shelter or defence against the heat of the sun. Cattle need not only to be sheltered against cold and wet weather in other seasons, but against heat in summer. Therefore the pastures in which they feed, should have trees in them, that they may repair to their shadow in the hottest hours. Clumps are preferable to single trees, as they not only afford a cool shade, but may screen the cattle from the violence of rain and storms, some of which happen in the time of grazing.

SHED, a slight roof or covering, of boards or other materials, for temporary purposes. Where boards are not easily obtained, sheds may be covered with straw, which will last a few years; or with the bark of trees, which will be far more durable.

SHEEP, a well known tame animal.

They multiply fast; they are subject to but few diseases in this country; their flesh is excellent food, and their wool of the greatest importance to this nation; in which the woollen manufactory ought to be encouraged, and may be carried on to great advantage.

Mortimer says, "The farmer should always buy his sheep from a worse land than his own, and they should be big boned, and have a long greasy wool.

"For the choice of sheep to breed, the ram must be young, and his skin of the same colour with his wool; for the lambs will be of the same colour with his skin. Those ewes which have no horns, are found to be the best breeders."

The farmers of Europe know how to distinguish the age of sheep by their teeth. When a sheep is one sheer, as they express it, that is, has been sheared but once, or is in its second year, it has two broad teeth before: When it is two shear, it will have four: When three. six: When four shear, or in its fifth year, it will have eight teeth before. After this, their mouths begin to break.

"The fat pastures breed straight tall sheep, and the barren hills square and short ones. But the best sheep of all, are those bred upon new ploughed land, the reason of which may be easily guessed, as such land is commonly the most free from bad grasses.

"All wet and moist lands are bad for sheep, especially such as are subject to be overflowed, and to have sand and dirt left on them. The salt marshes are an exception from this general rule: For their saltness makes amends for their moisture; any thing salt, by reason of its drying quality, being of great advantage to sheep. The best time for sheep to yean, which go twenty weeks with lamb, is in April, unless the owner has any forward grass, or turnips. Ewes that are big, should be kept but bare; for it is dangerous for them to be fat at the time of their bringing forth their young. They may be well fed, indeed, like cows, a fortnight beforehand, to put them in heart."

M. Buffon says, "One ram will be sufficient for twenty-five or thirty ewes; but that he should be remarkable for strength and comeliness: That those which have no horns are very indifferent: That the head of a ram should be large and thick, the forehead broad, the eyes large and black, the nose short, the neck thick, the body long, the back and rump broad, the testicles large, and the tail long: That the best are white, with a large quantity of wool on the belly, tail, head and ears, down to the eyes: That the best sheep for propagation, are those which have most wool, and that close, long, silky and white; especially if, at the same time, they have a large body, a thick neck, and are light footed."

He says, "that ewes fatten very fast during their pregnancy; that as they often hurt themselves, and frequently miscarry, so they sometimes become barren; and that it is not very extraordinary for them to bring forth monstrous productions But when properly tended, they are capable of yeaning during the whole of their life, or to the age of ten or twelve years. But most commonly when they come to be seven or eight years old, they begin to break, and become sickly; and that a ram is no longer fit for propagation after eight years, at which time he should be knit, and fattened with the old sheep."

According to the same writer, "sheep should in the summer be turned out early in the morning to feed; and in four or five hours, after watering, be brought back to the fold, or to some shady place. At four o'clock, P.M. they should be turned to their pasture again, and continue there till evening; and were it not for the danger of wolves, they should pass the night in the open air, which would render them more vigorous, clean, and healthy. As the too great heat of the sun is hurtful to them, shady pastures are best for them; or else to drive them to a place with a western descent in the morning, and the contrary towards evening." That their wool may be saved, they should not be pastured in bushy places, or where there are briars. Sheep are often thus deprived of most of their fleeces; which besides the loss of the wool, is very hurtful to the animals, when the weather is not warm.

The above writer directs, "that every year a flock of sheep should be examined, in order to find out such as begin to grow old, and ought to be turned off for fattening. As they require a particular management, so they should be put in a flock by themselves. They should feed while the grass is moistened with dew in the morning. Salt should be given them to excite thirst, as the more they drink, the faster they will grow fat. But to complete their fattening, and make their flesh firm and solid, they should have some corn or grain given them." They may be fattened in the winter; but it is commonly too expensive, as they will require a good deal of richer food than hay. When sheep are once become fat, they should be killed; for it is said they cannot be made fat a second time. The teeth of ewes begin to decay at five, those of wethers at seven, and those of rams not until eight.

We shear our sheep in general too early in this country. In England, where the spring is more forward than in this country, the approved time of shearing is from the middle to the latter end of June. They should be washed in a warm time. After this they should run three or four days in a clean pasture, before they are shorn. It is good for them to have time to sweat a little in their wool, after washing.

In shearing, great care should be taken not to wound, prick, or cut their skins with the shears. In England, after shearing, the farmers smear their sheep with a mixture of tar and fresh butter. This not only cures any little wounds that may chance to get in shearing, but is supposed to fortify their bodies against cold, and cause their wool to grow again the sooner.

If any cold rain happens soon after shearing, the sheep should be put up in a warm house. For if they be left abroad, it is apt to be fatal to them.

But Mr. Young thinks they are so apt to be hurt by being kept very warm, that they should never be confined to a house, but always have the door open, that they may be in the house or the yard as they choose. They will undoubtedly prefer the warmer place when they are newly shorn, if the air be colder than common. Small flocks commonly prosper better than large ones, as they are not often so overheated by crowding each other.

In France, fifteen pounds of salt per annum are allowed to a sheep, and fifty for each head of cattle. The truth is, that in the inland parts of this country, both sorts should have salt often, and be allowed to eat as much as they please, their health requires it, and they will pay well for it to the owner.

Some are fond of having black sheep in their flock. But their wool is seldom so fine, or so strong, as that of white ones. Nor is the wool ever a perfectly good black, and it is found difficult to give it any good durable colour by dying.

There are a great many varieties of sheep with differences more or less marked. To give even an abridged account of all the kinds described in foreign publications would require a volume. In England, the principal division of sheep is into the LONG WOOL, and the SHORT WOOL kinds. Among those bearing long wool, are the *Teeswater*, the *Lincolnshire*, the *South Downs*, the *Bakewell* or *Dishley* breeds, &c. &c. The origin of the last mentioned breed of sheep is thus described by an English writer.

" Mr. Bakewell selected from his own flock, and from the flocks of others those sheep to breed from, which possessed in the greatest degree that perfection of form he was desirous to retain and perpetuate. By judiciously crossing them, and selecting the most perfect of their progeny, he at length succeeded in forming the breed, which has been distinguished by the name of the New Leicester, or Dishley breed; and having attained his object, he carefully guarded against any future intermixtures with other breeds. This breed exceeds all others in its propensity to fatten; and by crossing by rams with this breed, a very considerable portion of the long-wooled sheep in England have been greatly improved in this respect.

"The peculiar characters of these sheep have been well described by Mr. Culley, an eminent grazier in Northumberland, who introduced the breed into that part of England. The Dishley breed are distinguished from other long-wooled breeds by their fine lively eyes, clean heads, straight, broad, flat backs, round (barrel like) bodies, very fine small bones, thin pelts, and inclination to fat at an early age. The last property is probably owing to the before specified qualities, which, from observation and experience, there is reason to believe extend generally through every species of domestic quadrupeds. The Dishley breed is not only peculiar for the mutton being fat, but also for the fineness of the grain: the flavour is superior to the mutton of most other long-wooled breeds. The weight of the carcase may be stated in general; ewes three or four years old from eighteen to twenty-six pounds per quarter; wethers, two years old from twenty to thirty pounds."

Among the *short-wooled* sheep, the English possess, besides the Merino breed, the *Ryeland*, the *Shropshire*, the *Shetland* the *Dorset*, *Wilts*, &c. &c. Merino sheep were first introduced into Great Britain in the year 1787; and although it was formerly a general opinion, that the excellence of their fleece depended in a great degree upon the temperature of the Spanish climate, it has been ascertained that the fineness of the Spanish wool is not impaired by breeding the sheep in England, France, Saxony, Hungary, &c.

Mr. Custis, of Virginia, is rearing a new breed of sheep, which he calls the Arlington sheep, with long-wooled fleeces for worsted manufactures. They are a mixture of the Bakewell breed with a ram from Persia, imported by General Washington.

The common practice of smearing the bodies of sheep with a mixture of tar and fresh butter or other oily substance is not approved of by Mr. Luccock, an English writer, who says, "The oil which the mixture contains is most certainly useful, but the tar, a dirty and tenacious substance, adheres to the wool so closely as frequently to corrode the hair, rendering the part to which it was immediately applied thin, rough and weak. When affected by the filthy custom of smearing, the pile is less capable of acquiring the softer and more delicate tints, which it is so often desirable to communicate to the different articles of the woollen manufacture. A portion of that dirt, which it obstinately retains through every previous process, is dissolved among the ingredients of the dying vat, and disqualifies them from communicating that vivid lustre, which they would have afforded to a purer wool, even though the artist supply his pans with a much larger proportion of the colouring materials. In the subsequent processes of the manufacture, this filthy staple produces much greater inconvenience, and is subject to more considerable waste than the purer pile, even though we make every reasonable allowance for the weight of dirt which it obviously contains; in the jenney and the loom, the machines employed in spinning and weaving it, more dexterity and patience are required of the work people, and the cloth which it produces is inferior in its quality, and smaller in its quantity than might have been obtained from the same pile in a pure state. These objections to tar, when it is applied to wool as a substitute for the yolk of the sheep are collected chiefly from the clothiers' account of it, and appear abundantly sufficient to prompt him to require a less pernicious mixture. The only circumstance, which can be mentioned as a counterbalance to these objections, is the consistency which it gives to oil or other greasy substances, with which it is mingled, whereby they are retained among the pile, although exposed to the heat of the animal and the detersive influence of the rain. But if it be desirable in all substitutes of this kind to imitate as nearly as possible the combinations of nature, we should apply to the growing pile a thick coating of soap in all cases where the sheep is incapable from the peculiarity of its constitution of yielding a sufficient quantity of yolk to secure a valuable fleece." *Luccock's Essay on Wool*, 131 2.

Large numbers of sheep should not be penned together during shearing, nor should their legs be tied together in that process.

The wool of yearling sheep should be kept by itself; because it is said to make the cloth shrink unequally, if mixed with the wool of older sheep. The fleeces may be sorted at shearing time, making separate parcels of the thighs the belly, the back, and sides. Wool should not be kept long without washing, as in that case it is liable to ferment and spoil in hot weather. After shearing, the horned sheep should be examined to see that the horns do not press on the scull, or endanger the eyes. In such case they should be sawed off, tar applied to the stums, and a double linen cloth tied over them to keep off flies. At this time lambs should be docked, castrated and marked.

When ewes are intended to be fatted, the lambs must be weaned early, and then if the ewes take the ram again, they will fatten the better. Lambs thus weaned should be put in a pasture of young and tender grass out of the hearing of their dams, and an old wether or ewe should be put with them. The ewes should be milked every day or two for the first week, till their milk dries up.

Lambs should be shorn the first year. Lambs, which do not come until the snow be gone, are most easily reared. The rams should, therefore, be kept from the ewes till about five months previous to the time the ground is generally bare in the spring.

In Spain twenty-five ewes are allowed to a ram. Mr. Livingston, of New-York, (who has written an excellent treatise on sheep, from which many of these directions are abridged) thinks forty ewes are not too many for one ram: and says that instances are not unfrequent, where one ram has served double and even treble that number. If the ram, however, be not well kept, where so many ewes are allowed, he will be in danger of being injured.

Sheep need no water in the summer, as they feed when the dew is on. In winter they should have access to it. When sheep have colds and discharge mucus from the nose, good feeding, together with pine boughs, given them occasionally, will cure them; or tar spread over a board, over which a little fine salt is strewed, will induce sheep to lick up the tar, and this will cure a cold.

If sheep are fed with a little Indian corn, about half a gill a day to each, it will keep them in good heart, prevent the wool from falling off, and enable the ewes to rear their young much better. They should have the best of hay through the winter, and red clover is esteemed the best. About a peck of salt applied to every ton of hay is recommended. The rack in which the hay is put should be upright, so that in feeding the seeds, chaff, &c. may not fall into the wool about their necks. Under the rack should be a trough for catching the seeds of the hay, and feeding the sheep.

Sheep should have a yard by themselves, which should be spacious in proportion to the number of the flock. They should have a shelter, close over head, but not so close at the sides as to preclude the free circulation of air. It should be so large as not to cause the sheep to be crowded. It should contain some small apartments in which to keep the ewes a few days previous to, and after they have lambs. These should be fed during this time with succulent food, and their apartments kept well-littered. The fence round the sheep yard should be such as to keep out dogs. If the flock be large, the wethers should be kept by themselves. They do not require so good keeping as the ewes and young sheep.

Flocks of sheep thrive the better for being changed frequently, and those will be most benefited which are taken from poor and put into rich pastures. Their pastures should be clear of weeds, briars and burrs. Clover gives the best pasture for sheep. There is a species of laurél which is poisonous to sheep.

Mr. L. Hommedieu says, that an ointment made of gunpowder and brimstone, finely powdered and mixed with currier's oil, and applied to the throats of sheep, will prevent wolves from killing them. This must be renewed as often as the ointment loses its moisture, which may be four or five times in a season.

Merino sheep are the most profitable. Mr. Livingston, in describing the result of his sheep shearing in 1808, says, that from twenty-nine common sheep he had upwards of one hundred and fourteen pounds, which he sold at thirty-seven and an half cents a pound. This, allowing one dollar and fifty cents for the expense of keeping each sheep for a year, fell short three cents on each fleece of paying for their keeping.

Eighty-three half-blooded ewes gave upwards of three hundred and ninety-three pounds; and forty-seven half blooded wethers gave upwards of two hundred and thirty-six pounds. This wool sold for seventy-five cents a pound. Clear profits on the fleece of each ewe two dollars and three cents; on the fleece of each wether, two dollars and fifty-five cents.

Thirty three fourth-blooded ewes gave upwards of one hundred and fifty-six pounds; and three wethers of the same blood gave upwards of sixteen pounds. This wool sold for one dollar and twenty-five cents a pound. Clear profits on the fleece of each ewe four dollars and seventy-five cents; on the fleece of each wether, two dollars and twenty-five cents.

Seven full-bred ewes gave upwards of thirty-six pounds; and one ram, fourteen months old, gave upwards of nine pounds. This wool sold at two dollars a pound. Clear profits on the fleece of each ewe, eight dollars and seventy-five cents; on the ram, seventeen dollars and twenty-five cents. This wool was all sold at the above prices, without being washed.

It has been ascertained that these sheep are as profitable for fatting as for their wool, as they become fat with a less quantity of food, in proportion to their bulk, than any other kind.

Mr. Livingston recommends the engrafting of a Merino stock upon our common ewes, to increase the size of the breed; and in the selection of the ewes to take those which are at least three years old, as large as can be obtained of the sort, with the belly large and well covered with wool; chine and loin broad, breast deep, buttocks full, the eyes lively, the bag large and the teats long, with little hair on the hinder parts.

Mr. Livingston thinks the wool of his Merino sheep, which he has raised in this country, finer than that of those imported. They will endure the rigour of our winters, and are successfully raised in Sweden, where the climate is much colder than ours. He says, that "like all other sheep, they will be found the better for good keeping; yet they will not suffer more than others from neglect; and that they will in every mixed flock be found amongst the most thrifty in the severest weather.

The ram, when put to the ewes, should be better fed than usual. A slice or two of bread, made of Indian corn meal, may be given him by hand two or three times a day. When the ewes are not suckling lambs, they may, at any time, be made to take the ram by feeding them with a little Indian corn.

After Merino wool has been sorted, and is to be manufactured in the family, let it be covered with soft water, mixed with one-third urine, and stand fifteen hours or longer, if the weather be cold. A cauldron is then to be put on the fire with some soft water, and let two-thirds of that which covers the fleeces, be added to it. When so hot as that the hand cannot bear it, take out the wool, put it in a basket, put the basket in the cauldron, and there wash the wool by pressing, without any wringing of it, and then cleanse it in running water. If the water in the cauldron become too dirty, take more from that in which it was first soaked.

Dry the wool in the *shade;* not in the sun; let it then be beaten with a rod, which takes out all seeds, &c. and softens it; then pick it, by opening it lengthwise carefully, and card it with *cotton*, not with wool-cards. Carding machines are not to be used for this wool, unless particularly fitted for it.

The above is the European method of managing this wool before carding, &c. but Mr. Livingston thinks that if the wool be carefully picked and carded, so as to get out most of the dirt, and woven in this way, that it will answer without washing; in which case less oil or grease will be necessary.

We shall here mention some of the most common diseases to which sheep are liable, with the most approved remedies.

Rot.—For a description of this disease, see the article ROT in its alphabetical order. We shall here add somewhat respecting its cure. "Miller recommends *parsley* as being eminently serviceable. Mr. Rice recommends every farmer to remove his sheep, in wet and warm seasons, from such lands as are liable to occasion the rot; but if this be impracticable, he prescribes a spoonful of common salt for each, together with a similar quantity of flour, in a pint of water, once or twice a week, by way of preventive: and if the disorder be in an incipient state, a similar dose administered four or five successive mornings, will, in his opinion, probably effect a cure; as the addition of the flour and water not only abates the pungency of the salt, but also disposes it to mix more gradually, though at the same time more efficaciously with the chyle. Dr. Darwin, however, thinks the salt would be more serviceable, if it were combined into a ball with about sixty grains of iron filings, by means of flour, and introduced into the sheep's throat every morning for one week. See *Massachusetts Agricultural Repository*, Vol. IV. p. 69.

The following recipe for the rot in sheep is from Young's Annals, Vol. XIII. p. 209. "Give to each sheep one spoonful of spirits of turpentine, mixed with two of water, after fasting twelve hours—let them have three doses, staying six days between each dose; this is said to have been used with success, even in cases where the fleece has been nearly gone, and the throat terribly swelled."

Scab.—"This appears by the sheep rubbing the part affected, and pulling out the wool in that part with their teeth, or by loose locks of wool rising on their backs and shoulders.

"The sheep infected is first to be taken from the flock, and put by itself: and then the part affected is to have the wool taken off, as far as the skin feels hard to the finger, washed with soap suds and rubbed hard with a shoe-brush, so as to cleanse and break the scab. Then anoint it with a decoction of tobacco-water, mixed with the third of lie of wood ashes, as much grease as this lie will dissolve, a small quantity of tar, and about an eighth of the whole mass of the spirits of turpentine. This ointment is to be rubbed on the part affected, and for some little distance round it, at three different times, with an interval of three days after each washing. With timely precautions this will always be found sufficient.

Sir Joseph Banks, in a communication to the *Society for the Encouragement of Arts*, directs for this disorder, pure quicksilver, one pound; Venice turpentine, and common oil of turpentine half a pound each; and of hog's lard four pounds, to be triturated in a mortar, till the mercury be completely incorporated with the ingredients.

The method of using this ointment is as follows. The head of the sheep must first be rubbed; after which a furrow is to be drawn with the finger, from the region between the ears, along the back to the point of the tail, so as to divide the wool till the skin be exposed to the touch. Next, the finger, being slightly dipped in the preparation, should be drawn along the skin. Similar lines should farther be opened down the shoulders and thighs, as far as the wool extends; and. if the animal be considerably infected, two other furrows are directed to be traced, parallel to that on the back, and one should likewise be drawn downwards, on each side between the fore and hind legs.

After this application the sheep may be turned among the flock without any danger of the infection's being communicated. Another remedy is an ointment composed of three parts grease, and one of spirits of turpentine.

Tick.—Ticks may be destroyed by oil, a strong decoction of tobacco, or by tobacco smoke. The smoke may be taken into the bellows, the wool opened, the smoke blown in, and the wool closed. This should be repeated at proper distances over every part of the body. A still easier method, though perhaps not so immediately efficacious is to part the wool of the animal on each side of its spine, from its head to its tail, and drop some Scotch snuff in the opening.

Staggers.—"This is a species of apoplexy, arising from too great fulness of blood. It principally attacks young lambs, which fall down; and if not timely relieved speedily perish. The mode of cure generally adopted by shepherds, is to bleed the creatures frequently in the *ege-vein*, and to remove them to a coarse pasture, with a view to prevent a relapse."—*Domestic Encyclopedia.*

Hoven or swollen.—It has been affirmed that a small quantity of ley, made either of wood ashes, or pot or pearl ash, turned down the throat of a sheep that is hoven or swoln, by eating too much green or succulent food, will give immediate relief.

Purging.—A slight purging, when sheep are first turned to grass, will not injure them. But when this is severe, the sheep should be housed, dosed with castor oil, and fed with some crusts of wheat bread.

The fly or maggot is an insect, which breeds in the skin of sheep. If the animal be attacked before shearing, it becomes sickly and indisposed; its wool, not yielding a sufficient quantity of *yolk*, affords a warm nest for the reception of the eggs, which are speedily hatched. The maggots immediately feed on the flesh of the sheep; and if they be not timely destroyed, the vermin will multiply so rapidly as to destroy the animal in a short time. The remedy recommended is corrosive sublimate and turpentine rubbed into the sore. Probably spirits of turpentine, or fine salt, would be equally efficacious.

Mr. Livingston, in the *Transactions of the Agricultural Society*, New-York, observes, that "the legs of sheep are furnished with a duct, terminating in the fissure of the hoof; from which, when the animal is in health, there is secreted a white fluid, but when sickly, ducts are stopped by the hardening of the fluid. He has in some instances found, that the sheep were relieved, merely by pressing out the hardened matter with the finger from the orifice of the duct in in each foot, and thinks that it may, in some cases, be proper to place their feet in warm water, or to use a probe or hard brush, for cleansing this passage."

A writer in the Massachusetts Agricultural Journal, Vol. III. p. 351, observes, that the dysentery or flux in sheep has been cured by "rubbing with a cob between the sheep's hoofs."

Worms in the head of sheep.—The symptoms of this complaint are seen in the animals lopping their ears, shaking their heads, scouring, stupidity, loss of appetite. These generally terminate in consumption and death.

Cure.—Force vinegar by a squirt-gun or syringe into the head of the sheep. This will produce sneezing and convulsions in the sheep, by which the worms will be discharged. See *Massachusetts Agricultural Repository*, Vol. III. page 325., and Vol. IV. pages 30. 33. 140. 145. 190.

In washing sheep, the use of water impregnated with lime, is to be avoided; for this substance decomposes the yolk of the wool; and wool often washed in calcareous water become rough and brittle.

SHELLS, stony coverings, which nature prepares for certain kinds of animals in the sea, and by which they are defended; which are therefore denominated shell fish.

These shells are much of the same nature as limestone, and are one of the best kinds of manure. No length of time deprives those shells of their virtue, which are buried deep in the earth. But shells which lie on the surface of the ground will gradually moulder, and become lime.

This manure is so highly esteemed in some parts of Europe, that the farmers even carry it in bags upon horses to the distance of several miles from the sea.

Shells may be applied to the soil at one season of the year as well as at another; excepting that they should not be carted on at a time when the ground is so wet as to be poachy; because poaching is hurtful to all soils. The farmer may generally do this work at a time when he is most at leisure. Even in winter those may well be removed, which lie lower in the sea than high water mark.

Mr. Weston recommends that shells be ground fine before they are used as manure; and says, the finer they are ground the farther they will go. But it requires so much labour to grind them, that I doubt whether it be worth while to do it, unless it be for gardens. And in the long run, they will benefit the soil as much without grinding. Though the benefit of them, when applied whole, do not appear much in the first and second year, the tillage of every year will help to break and crumble them; and in a course of years, by continual tillage, they will be sufficiently dissolved, and intimately mixed with the soil.

It is chiefly the smaller shells that should be thus used, such as those of clams, muscles, &c. for these will be sooner dissolved than larger ones. As small shells are mostly mixed with sand, or tenacious mud they need not be separated from these substances. Those that are mixed with sand will be a proper dressing for cold, stiff and clayey soils; and those which are mixed with mud should be laid upon soils that are dry and light. For many of the shells will lie with the concave sides upwards in the earth, and will stop the water in its descent, and so assist the soil in retaining moisture.

Mr. Eliot tried a sort of shell sand, which he says he found to be equal to good dung. If it had as much effect as dung at first, it must have been vastly better than dung upon the whole: Because shells are a lasting advantage to the soil.

SITHE, a well known instrument to cut grass. This instrument should consist of tough iron and the best of steel, well wrought together, and nicely tempered. If the temper of a sithe should prove to be too high, it may be lowered by laying it to the hot sun a few days in midsummer.

SLED, *or* SLEDGE, a carriage without wheels, chiefly used to convey loads when the ground is covered with snow. Plank sleds, and framed sleds, are both used. The latter for lightness are rather preferable. But plank sleds are more used for the heaviest loads, as masts and mill logs. The common length of a sled is eight or nine feet; but longer ones are better for carrying boards, and long timber.

SLIPS, twigs taken from a tree, or shrub, to propagate by planting them in a moist soil.

More than half, or even two-thirds of their length, should be buried in the soil. They strike root more easily than cuttings. Early in the spring is the right season to perform it. I have the best success when I do it as soon as the ground is thawed in the spring.

The slips should either be planted immediately after they are taken from the trees; or the lower ends should be enclosed in wet clay till they are set in the ground. This last will be necessary when the slips must be carried to any considerable distance. And in this case, they should lie for a while in water before they are put into the ground.

It is necessary to place them in moist earth, rich, and finely pulverised; and they should be frequently refreshed by a little watering, unless the season be wet.

But it is the surest method to plant slips in pots, especially of those kinds which are least apt to strike root. In this case, it will not be at all difficult to give them continually the right quantity of moisture. Slips from almost any kinds of trees and shrubs may be thus made to grow; but they will never make so large trees as those which come from the seeds. They will be the more fit, however, for the borders of gardens.

SLOUGH, a deep muddy spot of earth.

Soft and hollow places in roads, where puddles of water stand after rain, by means of the frequent passing of loaded wheel carriages, often become deep and troublesome sloughs. The way to prevent their existence, is to make a channel, or a covered drain, where the shape of the ground admits of it, to lead away the superfluous water. For the ground will thus be permitted to dry and harden, so as to prevent the sinking of wheels into it.

To cure a slough in a road, sink pebbles, or any kind of stones into the bottom, and cover them with a thick coat of coarse gravel, or with cinder from a smith's forge, or with rubbish from a brick kiln. But this should be done in a dry season.

SLUICE, a frame of timber, serving to obstruct and raise the water of the sea, or of a river, and to let it pass as there may be occasion for it.

Sluices are required for mills, and for locks to carry on inland navigation. But I shall only consider those sluices which the husbandman may find useful in flooding of low lands, or watering a dry soil with the Persian wheel, or in reclaiming of marshes.

For the first and second of these purposes, sluices with gates to raise and let down are proper. But for the last gates are not needed when the stream is large.

The Persian wheel has floats made hollow, and of such a construction, as to raise the water from a sluice, to the height of two-thirds the diameter of the wheel; where the floats discharge the water into a trough; whence it is conveyed away in such a manner as to water the neighbouring lands. For a particular account of the machine, see *Mills's Duhamel.*

For reclaiming of marshes, boxes with shutters are used, especially when but a small quantity of fresh water will need to pass out through the sluices. A box may be made of four pretty wide and strong planks, either nailed or pegged together. The length of the box must be equal to the thickness of the bottom of the dyke; and rather project a little at each end, that the passages may not be obstructed by dirt or sods falling from the dyke. These boxes should be placed in the lowest hollows of the marsh, or in the creeks, and the ground well hardened beneath them, and on their sides. It is better to place two or three boxes side by side, if needful, than go to the expense of building a more costly kind of sluice. And each hollow or creek, through which a dyke passes, and wherever there is likely ever to be fresh water to convey away, should have one or more of these little sluices.

Each box should have a clapper, or shutter. The shutter is to be fastened to the mouth of the box, at the end towards the sea, with hinges made of iron or wood. The rising tide presses the shutter close to the mouth of the box, so that no water can enter; and at ebb tide the fresh water, when there is any, opens it by its pressure, and passes out.

When it is found necessary to build larger kinds of sluices, *Belidor's Architecture Hydraulique* and *Muller*, should be consulted.

SMUT, a distemper in grain, which dissolves the substance of the kernel, turns it to a black dust, and bu sts the coats of the kernels.

M. Duhamel distinguishes it by its entirely destroying the germ and substance of the grain; by its affecting not only the ear, but the whole plant, and extending itself most commonly to all the ears which arise from the same root. He says he has found it as early as in April, by opening a plant, and taking out a young ear, not more than the sixth of an inch long; that a distempered ear, when it comes out of its hose, looks lank and meagre, and that the black powder may be seen through the thin coat of the grain; that the powder has a fetid smell, and no consistency; that it is easily blown away by wind, or washed away by rain; and that he has never found it to be contagious, like the powder of burnt grain.

M. Tillet observed that the upper part of the stalk of a smutty plant is not commonly straight, from about half an inch below the ear; and that in that part it is stiff and hard, and is almost entirely filled with pith, very different from the stems of healthy plants; whence he concludes, that the ascent of the sap is obstructed in the stems of smutty plants.

The real cause of smut has escaped the researches of many philosophers. M. Duhamel justly observes, that it cannot be a want of fecundation, as it destroys both the male and female organs, long before the time of fecundation.

He confutes the conjectures of its being caused by wet upon the ears, or the violent heat of the sun, by observing that the ears are smutty before they cease to be covered by the blades. And if it were owing to the moisture of the earth, he observes, that there would be more smutty plants in the low and wet, than in the high and dry parts of a field, which is not fact.

He adds, that he never could make it appear that the distemper is caused by insects, though he had been of that opinion; and that Dr. Hales has proved by experiment that it cannot proceed from the seeds being bruised by the flail, by bruising a number of grains with a hammer, which grew well afterwards, and bore sound ears. The same excellent reasoner refutes the opinion of those, who impute smut to dung of sheep or pigeons.

M. Aimen, M.D. has very judiciously observed, "that the smut of corn cannot derive its origin from a defect in the sap, as all the parts of the plant, except the ear, look healthy, and there are plants whose roots are perennial, which appear vigorous, though their seeds are smutty every year. He is of opinion, that whatever weakens the plant, is apt to bring on the smut, and instances, as a proof of this, that it is a frequent custom in his country, to cut rye, as soon as it spindles, for food for their cattle; and that this rye generally produces other ears, which seldom contain any but distempered grain: To which he adds, that seed corn which has been pricked, or run through with a needle; or which is not thoroughly ripe, and that which produces lateral or second ears, is subject to the smut."

He holds "that the distemper proceeds from an ulcer which attacks first the parts which sustain the seeds, and afterwards spreads to the rest of the flower. But some will say, what is the primary cause of that ulcer? In order to discover it, M. Aimen examined several grains of barley with a microscope: Some of them were bigger than others: Some were very hard; and others yielded to the pressure of his nail: Some were of a deeper, and others of a lighter colour; some longer and others rounder, than they ought to have been: Their rind was somewhat wrinkled in several places, whereas in its natural state it is smooth: And lastly, he perceived upon some of them black spots, which, when examined with a magnifying glass, appeared to be covered with mould. These grains were separated carefully, according to their several conditions, and sown apart, though in the same ground. *All the mouldy grains produced smutty ears;* the shrivelled and parched, and those that were attacked by insects, either did not grow at all, or did not produce any smut.

"He then singled out a parcel of sound grains, sowed them, and some time after took them up, in order to examine them again with a magnifying glass. He found some of them mouldy, replanted them all, and observed that the mouldy grains produced smutty ears.

"M. Aimen, without pretending that this is the only cause of the smut of corn, concludes, from these experiments, that mouldiness is a cause of this distemper."

That this philosopher has hit upon the true cause of smut, seems rather probable, when it is considered that mould is a kind of minute moss, and that the things which most effectually kill moss upon land, such as lime, &c. have hitherto proved the best antidotes to this distemper.

The methods of preventing it,

recommended by different writers, are various.

The last mentioned writer thinks, "that the best and ripest corn should be chosen for seed, threshed as soon as possible, and limed immediately after; as well to keep it from growing mouldy, as to destroy the mould already formed, if there be any: Adding, that every method he has tried to make corn so prepared grow mouldy, has been ineffectual, and that he has never known it produce smutty ears."

"As weak plants are found to be most subject to smut, he also recommends good tillage, as a sure means of giving them strength and vigour. And he observes, that the lies made use of, preserve the plants from mouldiness, and of all of them lime seems to him to be the most effectual."

Though liming at the time of sowing, as is the practice in this country, does not always prevent smut, I would recommend it to farmers, to do it in the method that M. Aimen mentions as successful. The lime will probably have a greater effect, when used so early, than when the mouldiness on the kernels is become older and more deeply rooted. The subject I am upon, is of so great consequence to the farmer, and to the public, that I shall make no apology to the reader, for proceeding to lay before him the opinions of other writers; although I shall run out this article to what some readers may call a tedious length.

M. de Lignerolle says, "That the surest means of avoiding smut, and that which he has practised with success ever since the year 1739, on upwards of three hundred acres of land, is, to change the seed every year, to be very careful that the seed corn be well dried, and thoroughly ripe, and that it be not smutty, nor have any smutty powder sticking to it. He then pours boiling water on quick lime, in a large tub; and after the ebullition is over, as much cold water as there was hot, and stirs it all strongly together, in order to dissolve and thoroughly mix the lime. The quantity of wheat intended to be sowed, is sprinkled with this ley, and then well stirred with a shovel, and laid in as high a heap as possible. It is best to keep the grain for a week after this preparation, turning it every day; for otherwise it would heat so as to destroy the germ. By these means he has not had any smut, when the fields around him have been infected with that distemper."

"M. Donat, near Rochelle, thinking the ingredients commonly employed in the steeps too dear for the use of farmers, studied for some years to find out something cheaper, easy to be had every where, and therefore better calculated to be of general use. I have had the good fortune, says he, in a letter to M. Duhamel, to accomplish what I wished; for I now use only pigeons' dung, quick lime, ashes, and sea salt, where this last can be conveniently had. I have sometimes made with these ingredients, steeped in water, so strong a liquor, that it has even destroyed the germ of the grain. But there will be no danger of that, if care is taken to observe the following directions, which are the result of seven years' successful experience, even at times when farmers who have neglected to follow my example have had such wretched crops, as have not paid the charge of reaping.

"Take quicklime and pigeons' dung, of each twenty five pounds, forty pounds of wood ashes, and twenty-five pounds of sea salt, or salt-petre. Put all these into a tub, large enough to hold half a hogshead of common water added to them. Stir them all well with a stick, till the lime is quite dissolved. This ley will keep some time without spoiling. It must be stirred again just before the corn is steeped in it. The grain is then put into a basket, and plunged in the lie, where it remains till it has thoroughly imbibed it; after which it is taken out, and laid in a heap, till it is quite drained of all its moisture: Or, which is a still better way, take a mashing tub, fill it with grain to within four inches of the brim, and then pour in the ley well stirred beforehand. When the tub is full, let the ley run out at the bottom, into some other vessel, in order to use it again for more corn. It is said, however, that a frequent change of the liquor, in which different parcels of wheat are stocked, is advisable. See close of this article. Let the grain be then taken out, and laid in a heap to drain; and continue in this manner to steep all your seed corn. The wheat thus prepared, may be sowed the next day, and must not be kept above five or six days, for fear of its heating. This I say from experience. The quantity of ley above prescribed, will serve to prepare more than twenty bushels of wheat."

Mr. Tull observes, "that brining and changing the seed are the general remedies for smut. The former of these he had heard, was discovered about seventy years before he wrote, by sowing some wheat which had been sunk in the sea, and which produced clean corn, when it was a remarkable year for smut all over England: But he afterwards doubts whether this might not happen by its being foreign seed, and therefore a proper change for our soil. He tells us, that two farmers, whose lands lay intermixed, used seed of the same growth, from a good change of land, and that the one who brined his seed had not any smut, whilst the other, who neglected that precaution, had a very smutty crop. But again he doubts whether this seed might not have been changed the year before, and so might not be greatly infected: Or at least not more than the brine and lime might cure. He adds, that smutty seed wheat, though brined, will produce a smutty crop, unless the year prove very favourable; for that favourable years will cure smut, as unkind ones will cause it: But, above all, he assures us that the drill husbandry is the most effectual cure."

A writer in the *Museum Rusticum*, says, "having observed amongst wheat while green, though shot up into spindle, several black, blighted ears, I examined them, and found these were ears in which, by some accident, the intention of nature was prevented. I suppose, by being detained too long in the hose, and by the natural humidity of the plant, a fermentation was promoted in its ear, destroying the small vessels through which the corns were to receive nourishment; by which means their contents became black, dry, and dusty. These ears growing up with the others, imbibe moisture sufficient to cause the dusty particles in the grains in them to expand, and burst the fine skin which contained them: Being thus set at liberty, the air, if it happen to be a dry season, dries them again; by which means they become light enough to float therein, when separated from the skin which held them. If this happens when the wheat is in the blossom, which it often does, part of the dust enters the stigma of healthy corns, and thereby infects them: The pulp in those becoming black, a fermentation is raised therein, which destroys the life of the grain thus impregnated. Hence the disagreeable smell is acquired peculiar to this disease (the smell in a grain of smut being the same as in a black blighted ear.")

By the black blight, this author seems to mean the same as burnt grain, burnt ear, or ustilago, in which distemper the kernels do not burst, but are converted to a dry black powder. If his hypothesis be just, as it is certainly plausible, it will follow, that there is no more difference between smutty and burnt grain, than between a closed and an open kernel of wheat: And that they are in fact the very same distemper, as indeed many writers have considered them, making no distinction. The antidotes for the one, are certainly proper for the other. For experience has shown in many instances that what prevents the one prevents the other.

The remedy this writer prescribes, appears to be a probable one. "When the corn is shot into spindle, and the ears begin to appear, let some persons go along each furrow in the field, and carefully break off all ears of the black kind; and when broken off, put them into a bag, and carry them away. As it is possible there may be some of these diseased ears which are not bursten, and therefore may escape being gathered, these may be known by the stalk at the neck being crooked backward and forward five or six bends, and the hose nearer to the head of such, than in the ears which are good."

Another writer in the Museum Rusticum, says, "I have for many years past escaped having smutty crops, by a proper care of the seed wheat before it is put into the ground; and the method I pursue, though efficacious, is in itself simple and cheap. I take four bushels of pigeons' dung, which I put into a large tub: On this I pour a sufficient quantity of boiling water, and mixing them well together, let them stand six hours, until a kind of a strong ley is made, which, at the end of that time, the grosser parts being subsided, I cause to be carefully drained off, and put into a large keeve, or tub, for use.

This quantity is sufficient for eighty bushels of seed wheat."

"My next care is to shoot into this steep a manageable quantity of my seed, which is immediately to be violently agitated, with either birchen brooms, or the rudders that are made use of in stirring the malt in the marsh tub, in a brewing office. As the light grains rise, they must be diligently skimmed off; and after the seed has been agitated in this manner, for the space of perhaps half an hour, it may be taken out of the steep, and sown out of hand with great safety: And I can venture to say, that if the land is in good heart, and has been properly tilled, it will not, when sown with these precautions, produce a smutty crop."

Another gentleman, who signs himself A Norfolk Farmer, "declares, he has observed, that if the seed was only well washed, it never failed: That he washed some seed which he knew to be smutty, in a large tub, filled with plain, simple water, stirring it violently with birchen brooms, taking care from time to time to skim off the light. This answered very well, and he has ever since continued the practice." The same practice of washing the seed, is recommended by Mons. de Gonfreville, of Normandy, in the *Foreign Essays on Agriculture.**

It appears very probable, that washing the seed very clean in several waters, may be the best method of preventing both smutty and burnt ears. The bursting of smutty ears in a field at the time of blossoming, may infect the grains in the sound ears; and, may produce a mouldiness, which, if not taken off, may cause the next crop to be diminished and corrupted by one or both of these black distempers.

But a Mr. Powell, in England, writes to the compilers of the *Complete Farmer*, that, in addition to the usual brining and liming of seed wheat, if one pound and a half of red lead were sifted through a cullender upon one bushel, stirring the corn with a shovel, so that every grain may have a spot or two of the lead adhering to it, it will effectually prevent smut: And that fowls will not lie upon it. He is confident, that even smutty seed, so prepared, will produce a sound crop.

A Mr. Marshall, a late British writer on agriculture, says he was informed by a Yorkshire farmer, that he had made use of a solution of arsenic as a preventive of smut, and for twenty years it had proved effectual. The preparation is made by pounding the arsenic very fine, boiling it in water, and drenching the seed with the decoction. The method is to boil one ounce in a gallon of water, from one to two hours. Then add as much water or urine as will increase the liquor to two gallons. This will answer for two bushels of wheat. It may be sowed without drying, or coating with lime. If this will prove an effectual antidote against smut, it may be further said in re-

* There is very little doubt that washing is the most effectual part of all the above recipes. Salt dissolved in water is as good as any complex preparation.

commendation of it, that it will equally secure the seed against birds, and against every kind of insects. Nor need any one be apprehensive that a poisonous taint will be communicated to the crop.

The following observations are principally abridged from the "Code of Agriculture." Steeping and washing seed wheat, as a remedy for smut, is an excellent practice. When the wheat seed is first put into any liquid, to run it very *gently* through a riddle, when not only the smut balls, but the imperfect grains, and the seeds of weeds will float, and may be skimmed off at pleasure, which is not the case when the seed is put hastily into the water. Pure cold water and lime may be effectual, provided the seed be washed in several waters, repeatedly changed, until it be perfectly clean, and then dried by quicklime, slacked either with sea or with boiling water. He recommends *salt water* as being more effectual than even boiling water and lime. The water should be so impregnated with salt that an egg will float in it, or if sea water with such a quantity of salt dissolved in it, as to be equally strong, by which its specific gravity will be so increased that all unsound grains will swim in the pickle. About a bushel of wheat at a time is put into a sufficient quantity of this pickle, in which, when stirred, all the light or diseased grains will rise to the top, and may be skimmed off. The seed wheat is then separated from the pickle, spread upon the floor, and a sufficient quantity of new slacked lime to dry the whole sifted upon it.

The following receipt is likewise from the same author. Dissolve three ounces of blue vitriol in three English gallons of water, (wine measure) for every three bushels of grain to be prepared. Let the liquid be put into a vessel capable of holding from sixty to eighty gallons, in such a quantity that when three or four Winchester bushels of wheat shall be poured into the prepared liquor, it will rise five or six inches above the corn. The grain should be frequently stirred, and all that swims above the surface carefully removed. After the wheat has remained half an hour in the preparation, it should be taken out of the vessel and thrown into a basket, which shall allow the water, but not the grain, to escape. It should then be immediately washed in rain, or pure water, which prevents any risk of injuring the grain. The seed ought afterwards to be dried, either with or without lime, before it is sown. It is proper to observe, that the grain should not be put into the prepared liquor, unless it has been well dressed, *and is thoroughly* dry. It may be kept without injury.

The following miscellaneous particulars respecting smut, and the means of preventing it, merit attention.

1. The same water should never be used but once in washing wheat; even when brine is employed, it is safest to have fresh liquor to each parcel. 2. Lime is not only of service to dry the

seed, but by its caustic and antiseptic qualities, it tends to destroy putridity and animalculæ of every description. 3. If smutty grain is not threshed till the June or July succeeding the year it was reaped, the dust, it is said, will become too volatile to attach itself to the grain when threshed to occasion smut, which by age looses the power of reproduction. 4. Notwithstanding the violence of threshing mills, they do not bruise the smut balls so much as the flail. 5. Great care must be taken, not to thresh wheat on a floor where smutty wheat has been threshed, nor to convey the seed in a sack in which smutty wheat had been formerly put.

On the subject of steeping it may be proper to add, that it would be well to extend that operation to other grains besides wheat. Every sort of seed should be steeped enough to promote a quick vegetation, and to secure a more uniform growth, which would greatly improve both the quantity and quality of the grain; and if the seed of barley and oats, as well as of wheat were clothed with saline and caustic particles, it would either preserve it entirely from the attacks of vermin, or destroy such as may venture to eat of it.—See an "*Essay on Smut in Wheat*," Mass. Agr. Rep., Vol. V. p. 134.

SNEAD, *or* SNATHE, the staff, or handle of a sithe. The right timber for sneads, is white ash that grows on upland, it being light and stiff, which are two very necessary qualities: For if a snead be heavy, it will help to tire the mower; and if it be limber and easy to bend, it will cause the sithe to tremble, which will hinder, in some degree, its cutting; and render the labour of the mower more difficult and fatiguing. It must be naturally of the right crook, and not cut across the grain of the wood.

SNOW, a congealed vapour that falls in little fleeces to the earth.

Snow lies upon the ground commonly in this country, in the winter months, and in March. Snows sometimes fall in November and in April; but they soon melt, and do not remain on the ground unless it be in the thick woods. In some parts of the wilderness, it is not all thawed till July; as on the northern sides of high mountains, where the trees form a deep shade.

Snow is beneficial to the ground in winter, as it prevents its freezing so solid, or to so great a depth as it otherwise would. It guards the winter grain and other vegetables, in a considerable degree, from the violence of sudden frosts, and from piercing and drying winds.

The later snow lies on the ground in spring, the more advantage do grasses and other plants receive from it. Where a bank of snow has lain very late, the grass will sprout, and look green earlier, than in parts of the same field which were sooner bare.

A small snow, that falls level, pretty late in the spring, is better for the soil than rain. As it thaws gradually, it does not run off, but soaks directly into the ground, moistening every part equally, fostering the roots of grass, and other vegetables. And till it is thawed, the growing plants are guarded against the attacks of frosts and winds. If a snow happen to fall after spring grain is sown, it does not injure it at all, but rather assists its vegetating.

In the northern parts of New-England, the ground in some years is covered with snow for four months, even in the cultivated fields. This is not regretted by the inhabitants, as they find it is a great advantage for drawing masts, logs, lumber, and wood, upon sleds, which is much easier than carting them. The roads are also far better, when the ruts and sloughs are filled, and every part paved with ice, or condensed snow. The winters, tedious as they are, seem too short for the teamsters to finish their winter business.

Meat that is killed in December, may be kept tolerably, if buried in snow, until spring. This is an excellent method of preserving fresh and good the carcasses of turkies and other fowls.

Set an open cask in a cold place; put snow and pieces of meat alternately: Let not the pieces touch each other, nor the sides of the cask. The meat will neither freeze, grow dry, nor be discoloured; but be good at the last of March. The surfaces of the pieces should be a little frozen, before they are put into the snow, that the juice of the meat may not dissolve the snow. The cask should be placed in the coldest part of the house; or in an out-house.

SOIL, that part of the earth which lies upon the hard under stratum, over which there is commonly a cover of rich mould, which forms the surface, unless destroyed by severe burning, or washed off by violent rains, or blown away by driving winds.

The original or unmixed soils, in this country, are but few. Clay, loam, sand, gravel, and till, or moor earth, are perhaps all that ought to be reckoned as fit for cultivation. But they are commonly more or less blended together. In places where they are unmixed, it would be a piece of excellent husbandry to mix them, especially where they are contiguous, applying gravel to moor earth, and moor earth to gravel; sand to clay, and clay to sand. And sand upon loam would be an improvement.

A chalky soil is but seldom found in this country. Marle is usually at too great a depth to come under the denomination of soil, and the same may be said of peat. This last cannot easily be reduced to a condition fit for tillage. It is best to destroy it, by digging it wholly out for use, or by draining the land, and burning the peat on the ground. A chalky soil should have sand and hot manures applied to it.

I do not consider a stony soil as distinct from the rest, as removing the stones would bring it under some other denomination And this ought to be done, when land is to be used in tillage, that its operations may be facilitated.

Soils are commonly distinguished into shallow and deep, the latter of which is preferred, as the under-stratum comes not so near to the surface, but that the ground may be stirred to a great depth; and as it is fitted for the growing of long tap-rooted plants, trees, &c.

But the most common distinction of soils is into rich and poor. This difference, which is certainly very great, is not perhaps natural. Richness, I imagine, is rather to be considered as superinduced. All soils have, since the creation, received large quantities of fertilizing substances which were adapted to improve them; and by which, in most places, they have been greatly mended. Not only vegetable substances fallen upon the surface, and changed by putrefaction, have blended their salts and oils in the soil: But the soil has been drinking in vegetable food by the dews and rains, and from the air itself, which is loaded with fertilizing particles. But some spots have retained the added richness better than others.

As to land which has been long tilled, and often plentifully manured, it is not easy always to distinguish what was its original soil; nor how rich or poor it was in its natural state.

It does not follow, that all uncultivated soils ought to be equally rich, by means of the general advantages mentioned above; because some soils are better calculated than others to retain the food of vegetables. Some are destitute of a compact under-stratum; and it is no matter of wonder that such should appear hungry and barren; for whatever richness they receive, is either washed by rains into the bowels of the earth or evaporated again. Some soils are too coarse, or too porous, to be a proper matrix for fertilizing substances. Some are too steep to retain them, so that they are washed into the hollows below. Some are so wet as to sour and corrupt them; and in some, there are are either mineral waters, or steams which are unfavourable to vegetation.

In tillage, the surface mould and the soil beneath are mixed, and the more so the better, as the surface mould is made up as it were of the essences of vegetables.

Soils, by some writers are divided into Sand;—Gravel;—Clay;—Chalk;—Peat;—Alluvial; and Loam.

1. *Sand* consists of small grains of silex, which are not soluble in acids. If it has no mixture it can hardly be cultivated, but it is rarely found without a mixture of other substances. The best mode of improving its texture is by a mixture of clay marl, sea ooze, sea shells, peat, or vegetable earth. See *Sandy Soil*.

2. *Gravel*. Gravelly soils are composed of small soft stones, sometimes of flinty ones; but they often contain granite limestone and other rocky substances. They are improved by deep ploughing; by mixing them with coats of clay, chalk, marl or peat, &c. See *Gravel*.

3. *Clay*. A clay soil is tenacious, smooth and unctuous, and often holds water like a dish. Its texture is improved by a suitable mixture of common sand, sea sand and above all, of limestone gravel, where it can be obtained. Peat moss also, that has for some time been dug up, and exposed to the

action of the atmosphere, is useful, and putrid and calcareous manure is necessary in the course of its cultivation, Dr. Kirwan asserts that "the best manure for clayey soils is marl, or a mixture of marl and dung." See *Clay* and *Clay Soil*.

4. *Peat*. This substance is of vegetable origin, chiefly composed of various sorts of aquatics, immersed in stagnant water. In converting peat into earth it is a rule to plough and dig it in autumn, that it may be exposed to the winter's frost. See *Peat*.

5. *Chalk*. Chalky soils, as before observed, are seldom found in this country. A description of such soils and their mode of cultivation may be found in *The Code of Agriculture*, under the head *Soil*.

6. *Alluvial Soils* are of two sorts; one derived from the sediment of fresh and the other of salt water. See *Marsh* and *Meadow*.

7. *Loam*. Loams are said to be the most valuable of all soils. They are peculiarly well adapted to the convertible husbandry, for they can be altered not only without injury, but generally with benefit from grass to tillage, and from tillage to grass. They should not, however, be kept in tillage too long, nor while they are in cultivation, should two white crops be taken in succession. See *Loam*.

Soils are of various colours; the principal are white, black and red.

White stiff clays are heated with difficulty, and retain their heat only for a short time.

A black soil, containing much vegetable matter, is most easily heated, and soonest cooled.

The red colour in soils, is owing to iron in various combinations. It is favourable or adverse to fertility, according to the nature of the combination.

There are various modes of improving soils.

"Soils with acids, or salts of iron, may be ameliorated by the application of earthy lime or chalk. The sulphate of iron is thus converted into a manure. If there be an excess of calcareous matter in the soil, it may be improved by the application of sand, or clay, or earthy substances. Soils too abundant in sand, are benefited by the use of clay, or marl, or vegetable matter. A deficiency of vegetable or animal matter must be supplied by manure. An excess of vegetable matter is to be removed by burning, or to be remedied by the application of earthy materials. The substances necessary for improving soils are seldom far distant. Coarse sand is often found immediately upon chalk, and perhaps always under it, while beds of sand and gravel are commonly below clay, and clay and marl generally below sand." *Code of Agriculture.*

The constituent parts of soils, chemically considered, are certain compounds of the earths, silica, lime, alumina, magnesia, and of the oxides of iron and magnesium; and animal and vegetable matters in a decomposing state, and saline, acid or alkaline combinations. For a particular description of these, together with the mode of analysing soils by certain chemical substances or reagents, we would refer our readers to Davy's Agricultural Chemistry. Lecture IV.

SOILING, or ASSOILING, feeding animals with new mown grass, or grass not dried, in racks, or otherwise.

This is commonly practised in some countries, where they put it in racks, either under cover or in yards. Thick grass will go much further in this way, than if the cattle were turned in upon it to feed it off; as they would destroy and corrupt more by half with their feet and excrements, than they would eat. But when it is given them in racks, they will eat it up clean, without wasting any of it. An acre of rich land, used, in this way, will summer a number of cows. By the time that it has been once cut over as it is wanted, the first part will be fit to cut again. And the labour of doing it is not to be reckoned as any thing, as the trouble of driving the cows to pasture will be saved. This will be more than a balance for the labour of soiling, if cattle must be otherwise driven to any considerable distance. And it greatly recommends this practice, that a prodigious quantity of manure may be collected by it, which otherwise would be little better than lost, the dung being scattered in pastures, where it evaporates in the air.

In feeding cattle with green food there are many advantages in *soiling*, or supplying them with food, where their manure is preserved out of the field; the plants are less injured when cut, than when torn or jagged with the teeth of the cattle, and no seed is wasted by being trod down. They are likewise obliged to feed without making selection; and in consequence the whole food is consumed; the attachment or dislike to a particular kind of food exhibited by animals, offers no proof of its nutritive powers. Cattle at first refuse linseed cake, one of the most nutritive substances on which they can be fed. *Elements of Agricultural Chemistry*.

A communication by the Hon. Josiah Quincy to the Massachusetts Agricultural Journal, published in Vol. VI. No II. of that work states that "There are six distinct advantages, which those, who advocate soiling, propose to themselves by the practice, and on which they establish the preference of this mode to the common one of pasturing cattle during the summer.

"1st. The saving of land. 2d. The saving of fencing. 3d. The economising of food. 4th. The better condition and greater comfort of the cattle. 5th. The greater product of milk. 6th. The attainment of manure.

"The only offset to all these advantages is the labour of raising and cutting the food, and feeding and taking care of the stock." According to European writers the saving of land differs, according to its management, from *one* to *three*, to *one* to *seven*." That is, *one acre kept for soiling will go as far as three or seven kept for pasture, in the support of stock*." The reason of this diversity of statement, in relation to the degree of saving, among European writers, results from the different ways, in which the land, used for soiling is cultivated.

With regard to *saving of fencing*, Mr. Quincy observes that "the general effect of soiling cattle is *to render all interior fencing absolutely useless*; excepting those, which surround the buildings, and lead from these to the highway.

As respects *economy of food*, Mr. Quincy states that "There are six ways by which beasts destroy the article destined for their food." 1. By eating. 2. By walking. 3. By dunging. 4. By staling. 5. By lying down. 6. By breathing on it. Of these six the first only is useful. All the others are wasteful."

'In pastures, whatever is trodden upon or affected by the dung of cattle, or their urine, is laid upon, or even long breathed upon is lost. And this waste is always in proportion to the richness and the productive power of the pasture; for just in that proportion is the quantities of food injured by all the five modes of destruction above stated. Whereas the same being cut and delivered to them sparingly in point of time, but sufficiently in point of quantity, will every particle of it be consumed. Besides it is found by experience, that, this mode of feeding, beasts will eat many products of the earth, in the stall, which they will absolutely reject in the pasture."

In speaking of "*The better condition and greater comfort of the cattle*," Mr. Quincy observes "The want of sufficient exercise, which is inseparable from this mode of feeding in stalls, is a popular, and when not tested by fact, is deemed an unanswerable objection. Yet all those who have made the experiment, and whose opinions I have seen expressed upon the subject are unanimous in declaring that no ill effect results from this circumstance. One writer asserts that he has kept a large herd several years in this way, and during the whole time "he never had an animal essentially sick, had never one die, and never had one miscarry."

It is to be observed that stall feeding of cattle, does not imply keeping cattle in stalls, or in the house *the whole* time. It only intends always feeding them there, and keeping them there the chief of the time. On the contrary it is an essential part of the system to let them loose in yards. well shaded artificially or by trees, at least two hours in the forenoon, and as many in the afternoon. When cattle are fed in the stall, they are wholly protected from eating noxious vegetables; from drinking bad water; from all injury from being worried by dogs, or one another; they are kept through the heat of the day in the shade protected from flies.

With respect to a *greater product of milk*, it is stated by Mr. Quincy that during the flush of feed, that is, for perhaps the first month after cattle are turned to pasture, there is little difference, so far as respects the product of milk between pasturing and stall feeding. At that time there is generally a great supply of food, the cattle are eager after it. They have great opportunity to select. They feed quietly, and take only the most nutritious and palatable. After this month, if the stall feeder will, this equality will gradually cease, and in his favour. The pasture food almost always grows more or less scarce, according to the particular character of the season. Whereas by taking care to provide a regular succession of succulent crops, he who feeds his beasts in stalls may keep the milk product unaffected by the state of the season to the end of autumn.

The attainment of manures. In pasturing the summer manure is almost wholly lost. It falls upon rocks, among bushes, in water courses, on the sides of hills. It is evaporated by the sun, it is washed away by the rain. Insects destroy a part. The residuum, a dry hard cake, lies sometimes a year upon the ground; often impeding vegetation, and never enriched the earth in any thing like the proportion it would do, if it had been deposited under cover and kept free from the action of the sun, in appropriate and covered receptacles, to be carted out annually in the proper season, and ploughed at once under the surface.

Mr. Quincy gives the following statement of his experience in soiling.

"My stock, consisting at an average of twenty cows, were kept in their stalls through the whole year. The practice was to feed them about six times in a day, and to permit thom to range in a yard about eighty feet square, two hours in the forenoon and two in the afternoon. They were kept well littered and well curried. While they were out of the stable, the attendant took that opportunity to clean the stalls, and to supply fresh litter. During winter, they were fed, as is usual, with salt and fresh hay and vegetables.—From June to November, inclusive may be considered, strictly speaking, as the soiling season; by which is understood, that, in which they are fed with green food in the house. As this is the critical period, I shall be minute in the account of my preparations and proceedings.

"In the autumn preceding I had caused rye to be sown upon an inverted sward, very thick, on about three acres. Early in April I prepared and sowed, in manner as shall be stated afterwards, about three acres and one quarter of land with Indian corn in drills. I also sowed about three acres of oats and buck wheat broad cast, at the rate of three bushels to the acre, about the latter end of the month. The whole quantity of land, I thus prepared to be used in soiling in aid of my grass, did but little exceed nine acres. Of these, that which I sowed with rye turned out so poorly, that I never soiled from it more than five days, so that in fact the land thus prepared did, in efficiency, but little exceed six acres.

"About the first of June, cattle in general were this season turned out to pasture. On the 30th of May, my farmer began to cut the sides of the road leading to my house from the highway and orchard. He continued to soil from this and from grass growing in my orchard until the seventh. On this day he abandoned cutting the grass for soiling and began to cut from the winter rye. This was found too tough, and it was quitted, and my farmer returned to soiling upon grass. Having cut over all the refuse of my grass, by the 24th of June, he then went into the poorest of my mowing land, and afterwards into my clover. From this he continued to soil, until the 6th of July. By this time he had gone over not much short of three acres of mowing land. On the 6th of July, he began to soil from my oats. He continued to soil from these until the 21st of July. On the 21st of July he began to soil on Indian corn, on which he continued until the 30th of July, when he recommenced soiling on corn fodder, and continued upon it until the 30th day of August. On this day he began to cut over the road sides which had been cut early in June. This was continued only to the 2nd of September, when he began to cut the second crop of Indian corn, growing upon the three and fourth acres of Indian corn, which had now shot up in great luxuriance. from the roots of that, which had been cut overbetween the 21st and 26th of July. On this soiling continued until the 8th of September.

On the 9th and 10th he soiled upon about a fourth of an acre of buck wheat. On the 11th soiled on a second crop of clover. From the 12th to the 15th inclusive, on the corn stalks of about an acre of sweet corn, and on the 16th on a patch of millet and oats. This was continued to the 20th, when he began on two acres of Indian corn, sown in drills on the first of August, on land from which a crop of peas had previously been taken. Soiling was continued on this corn until the 3d of October. From this time until the 15th of October, the soiling was wholly upon second crop grass taken from various parts of my mowing land.

From the 15th of October to about the 20th of November, they were kept wholly upon carrot and turnip tops, arising from the topping of about twelve acres of both; being allowed always one foddering of salt hay. This finished the summer feeding. From this time they are kept wholly on salt and English hay. The result then of the experiment, so far as relates to land, is the following:

The twenty head consumed the products of

2½	acres	roadsides and orchards,
3	do.	mowing land,
3¼	do.	Indian corn, cut as fodder,
2	do.	late and light barley,
3	do.	oats,
2	do.	late sown Indian corn after a pea crop,
¼	do.	buck wheat,
1	do.	millet buck wheat and oats,
—		
17	acres.	

This was the whole land which was cut over for soiling; with the exception of the after feed on the mowing land and the tops of carrots and turnips. In comparing this result with the former practice of my farm, I apprehend the following statement to be just.

I offset the keeping from the 11th of September to the 20th of November against the old manner of letting the cattle run at large during the autumn months, on the mowing land, to its great injury by poaching and close feeding.—If this should not be deemed sufficient, I then make no estimate of the difference between keeping fifteen head of cattle, the old stock, and twenty head of cattle my present stock. After these allowances and offsetts, which no man can doubt are sufficiently liberal, then I state, that my experiment has resulted in relation to land, in this, that I have kept the *same amount of stock, by soiling on seventeen acres of land, which had always previously required fifty acres*. [Mr. Quincy must have had bad pastures or badly managed ones; twenty five acres ought to have been enough for his fifteen head of cattle.] The result is, in my opinion, even in this respect greater than what is here stated. This, however, is sufficient to exhibit the greatness of the economy of this mode, so far as relates to land.

With respect to saving of fencing the previous condition of my farm was this. I had at the lowest estimate five miles of interior fence, equal to sixteen hundred rods. which at one dollar the rod was equal to sixteen hundred dollars. And annually for repairs and refitting, cost sixty dollars. *I have now not one rod of interior fence*. Of course the saving is great, distinct and undeniable.

In relation to manures, the effect of soiling is not less apparent and unquestionable. The exact amount of summer product I have not attempted to ascertain, but I am satisfied that every thing considered, it is not less than one buck load per month per head, or on twenty head of cattle, one hundred and twenty load for the six soiling months. In this estimate, I take into consideration the advantage resulting from urine saved, by means of loam, 'sand, or some imbibing recipient prepared to absorb it.

It remains to show that the cost of raising the food, cutting it, and distributing it to the cattle, is compensated by these savings. Upon this point, my own experience has satisfied me that the value of the manure alone, is an ample compensation for all this expence. Leaving the saving of land, of food, and of fencing stuff, as well as the better condition of the cattle, as a clear gain from the system.—As an evidence of this, I state my expences for labour in conducting the soiling process.

During the month of June, I hired a man to do every thing appertaining to the soiling process; that is, cutting the food, delivering it, taking care of the cattle in the day time, for fifteen dollars the month, he finding himself. In this arrangement it was estimated, that I availed myself of half his labour. At the end of the month I had the manure measured, and I found that the manure collected in my receptacle, which was a cellar under the barn, and not including that which had been made, during the four hours each day, in the yard, amounted to fifteen load. A quantity of manure, which I could not have placed on my farm for thirty dollars; and which I could have sold there for twenty dollars, upon the condition it should be carried away. It cost me as above stated fifteen dollars in the labour of the attendant.

During the remaining five months I added another man, because I found that a great economy in vegetable food would result from cutting it into pieces by a cutting knife, and mixing with it about one third of cut salt hay, or straw. This was done, and I kept an accurate account of all the labour of cutting the food in the field, bringing it into the barn, cutting it up there, cutting salt hay or straw to mix with it, mixing this food, and delivering it to the cattle and found that it amounted to one hundred and forty eight days labour. This estimated at a dollar the day is one hundred and forty eight dollars, to which adding fifteen dollars paid for labour, in the month of June, the whole expense was one hundred and sixty three dollars.

The manure at the end of the soiling season, certainly equalled one hundred and twenty loads, and could not have been bought, and brought there for three hundred dollars. Let it be estimated at only two hundred dollars in value. No man can question, I think, the correctness of my assertion, that the value of manure obtained, is a clear compensation for the amount of labour; and this including all the expense of labour, connected with soiling.

In No. 4, vol. vi. of the Agricultural Repository, Mr. Quincy pursues the subject, and observes that, "in soiling, being guided by the nature of the climate, and by the consideration of the vegetables, selected being best known and most successfully cultivated in the neighbourhood, I use but four, 1st. grass; 2d. oats; 3d. Indian corn; 4th. cabbages."

He thinks, however, that it would be wise and practicable to introduce some vegetable, which, sown the preceding year, would enable the farmer to commence cutting earlier. He considers *one quarter of an acre* of his best grass producing land as appropriated to each head of soiling cattle, for its support between the 20th May and 1st of July. "Small farmers, who should top-dress the land every day, cut over, with water leached from the manure heap, would reduce the extent of land required for the process of soiling very considerably." Oats, sowed as early as possible, he uses for soiling nearly, or quite through the month of July. "*One square rod of oats, in full milk, growing on land in proper, (that is high) tilth will support one head of cattle a day.* One quarter of an acre, or forty square rods, for thirty days, is a fair basis of calculation, and making a liberal allowance for accident." The oats, if relied on for the whole month of July should be sowed in succession, one half as early as the seed can be got into the ground, and the other half about a fortnight later. Indian corn should likewise be sowed at about the rate of a quarter of an acre for each head a month. The land to be in good heart and tilth, "to receive at least one ploughing and harrowing about the latter end of April, or beginning of May; after which light furrows should be run three feet asunder, at the depth of three or four inches. In these furrows corn should be sown broadcast, about the thickness, and in the same manner that peas are sown, in the field culture of them. The corn may then be covered by the plough. Although in my experience, a harrow drawn lengthways and crossways, followed by a roller, is sufficient and to be preferred for this operation." Corn should be sown in the middle of May, the beginning and middle of June, and even as late as August, on which soiling may be continued during September. In this month likewise grass of the second crop may be used, from those acres from which soiling was effected in June. The grass of the second crop will generally enable the farmer to soil to the 15th of October, if his grass land is in proper tilth and heart.

"Reduced to a single statement. my experience and system is, for the support of my soiled stock, during the months of July, August and September, to sow in the months of April, May, June and July, equal to three quarters of an acre of land for each head of cattle soiled, in such succession as will give also a regular succession of succulent food in the three first mentioned months."

From the 15th of October to the middle or latter end of November, Mr. Quincy depended upon the tops of carrots and turnips destined either for the market or winter food for his stock. "My practice has been to raise from eight to twelve acres of vegetables. The tops of which, with a single foddering of salt hay, per day, have been, according to my experience, sufficient to support equal to twenty head of cattle from the 15th of October to the middle or latter end of November."

SOOT, condensed smoke, which adheres to the funnels of chimneys. It is replete with the oil and volatile salts which were contained in the fuel. and is therefore an excellent manure, much superior to ashes of any kind.

Both wood soot and coal soot should be carefully saved, and kept from the weather, to be used as top-dressings.

Mr. Worlidge seems to think wood soot the best; but Mr. Mortimer give the preference to that which comes from pit coal, of which forty bushels are allowed to be a sufficient dressing for an acre. But of this kind our farmers can obtain but little; nor indeed plenty of either, unless in the neighbourhood of populous towns, where much of it may be collected for use by those whose business it is to clean chimneys.

Both kinds are to be used only as top-dressings. The coal soot is particularly good for low meadows, or grass lands, which are sour and mossy.

Soot is a good top-dressing for winter grain. But it should be applied early in the spring. Not in autumn, lest it should cause it to grow too fast, by means of which it will be the more liable to be destroyed by the frost of winter. Neither should it be applied late in the spring; because in case of a drought soon after, it will be apt to burn too much.

Mr. Ellis recommends sowing soot over turnips, as soon as they are up. This will tend to prevent flies from attacking them. But that it may have have this effect, it should be finely pulverised; sown early in a morning before the dew is off; and in a moderate quantity, lest its heat should injure the tender plants, to which it will adhere and repel the insects. Sifting is the best way of applying it. See *Manure*.

SOWING, committing seeds to the earth, for the purpose of obtaining a crop.

There are three ways of seeding the ground:—1. In hills as it is called, or in squares: 2. In drills, or continued rows: And 3. In the broadcast method, or at random with a cast of the hand; which last method is always termed sowing. The first requires the least quantity of seed, the last the greatest. But the crops will not be in proportion to the different quantities of seed.

With regard to sowing, several things ought to be attended to; the quality or goodness of the seeds; the time of sowing them; the depth that is best for them; and the quantity, or proportion of seed to the ground.

The quality of seed should be ascertained, in order to determine the quantity that is proper to be sown; for if one tenth part of the seeds, for instance, should be destitute of a vegetative power, a tenth part more of such seeds should be sown than the usual quantity, supposing the seeds to be in perfection.

In order to determine the goodness of the seed to be sown, you should previously take fifty grains at random from the parcel; sow them in good mould, at a proper depth, and carefully observe how great a proportion fail of coming up. They may be sown in a pot, and kept in a warm part of the house, or in a hot-bed, that the farmer may have timely notice of the quality of his seeds, when it is too early in the spring to do it in the open ground. Many have missed of a crop, by not taking this precaution. When seeds are suspected of being too old to vegetate, this previous trial should by no means be neglected.

But if we wish to have seeds in the best condition for sowing, they should be well ripened on their plants before they are gathered in; afterwards they should be kept perfectly dry, that they may not contract the least mouldiness; and never be secluded from the air.

Mr. Miller found that air was absolutely necessary to maintain the principle of vegetation in seeds. Having saved a parcel of fresh seeds of several kinds, he took some of each, and sealed them up in glass phials; the other parts of the same seeds he put into bags, and hung them up in a dry place, in a free air. After a year had passed, he took some of the seeds from each phial, and each bag, and sowed them at the same time, and on different parts of the same bed. The result was, that almost all the seeds he took out of the bags grew well; but, of those which had been kept in the phials, not one came up. This discovery was further confirmed by experiments afterwards. How careful then should both farmers and gardeners be, that no seeds designed for sowing be kept totally secluded from the air!

All kinds of seeds are best kept in their pods, or husks. Especially they should be kept, when they are designed to be transported to distant countries.

Accordingly, some of the best writers recommend the lying of seed wheat in the sheaf to the time of sowing. And, that none but the best of the grain may be sown, instead of threshing, it is advisable to strike a handful at a time gently against a post, and collect what falls out; because the heaviest and best grain is always the most easily detached from the ear.

Being furnished with good seeds, the time for committing them to the earth must in great measure be determined by the judgment of the experienced husbandman; because, from various circumstances, it comes to pass, that the true time admits of some latitude. The time for spring sowing will vary according to the variation of the forwardness of the season; which may be best determined by the respective forwardness of trees and shrubs. See the article *Kalendar*.

But there are other circumstances to be taken into the account, which may further vary the season for spring sowing. A light warm soil may receive the seeds earlier than one that is strong and moist. The former will arrive to the right degree of dryness sooner than the latter, and is earlier fit for the operations of tillage. And this is certain, that seeds that require the earliest sowing, must not be sown before the earth can be well pulverised. Neither should plants that are easily killed by frost, be so early sown as to be up till the spring frosts are past.

I may add, a spot which has a southern exposure may be seeded rather earlier than land which descends to the northward, or than land which is level.

If seeds are sown too early, or when the ground is too wet or cold for them, they will either perish, and fail of coming up; or if they come up, it is slowly, so that the plants become stinted in their growth, and never arrive to a full size.

If the right season for sowing should elapse, the husbandman, being convinced of it, may accelerate vegetation by steeping the seeds in a ley of wood ashes, or any other proper menstruum, so that they may overtake in their growth those which were sown in the right season.

The depth at which different seeds should be buried in the soil is various, according to the difference of seeds and soils. M. Duhamel found by experiment, that but few seeds will come up at all, when buried deeper than nine inches; that some seeds rise very well from the depth of six inches;

and that other seeds do not rise at all when they are more than two inches under the surface. And in general those seeds, the body of which is thrown above the surface in vegetating, should have the less quantity of soil above them, that they may not meet with too much resistance in rising; such as kidney beans and many other sorts. Also the same seeds may, and ought to be buried deeper in a light and dry, than in a heavy and moist soil. When the ground is rolled after sowing, the seeds will vegetate the nearer to the surface; and therefore they do not need to be sown so deep, as when the rolling is omitted.

To determine what is the right depth, in a doubtful case, Mr. Tull has suggested an excellent method. "Take a dozen of sticks for guages; mark the first at half an inch from the end; the next at an inch; and so on, increasing half an inch to each. Then, in the sort of ground you intend to sow, make a row of twenty holes, with the half inch guage; put in twenty good seeds and cover them, and stick up the guage at the end of the row. Then do the like with the rest of the sticks. Observe how the seeds prosper in the different rows, and you will discover at what depth that kind of seed should be buried."

However useful this experiment may be, it can be of little or no use in the old field husbandry; for, in the broadcast way of sowing, the seeds will be differently covered. But sowing fields with the drill, in equi-distant rows, when horse-hoeing is not intended, cannot be too much commended. It is worth while to do it if it were only on account of the seed that may be saved by it. Much seed is wasted in the common way of sowing; for some of the seeds will be so deeply covered, that they will not vegetate: Some will be left on the surface, which is a prey for birds, and perhaps leads them to scratch up some of the rest: Some will lie so near the surface as to be destroyed by variation of weather, being alternately wetted and scorched. And of those seeds that grow, some rise earlier, and some later, so that the crop does not ripen equally. The seeds will fall from the hand of the sower, too thick in some spots, and too thin in others, by means of the unevenness of the surface; and the harrowing perhaps will increase the inequality; so that many will be so crowded as to be unfruitful, while the rest have more room than is necessary.

But when the seeds are put in with the drill, they will all rise nearly together; not so much as one seed will be wasted, or lost, supposing them sown at the right distance; each one may have so much room as is most conducive to its growth; no starved heads will appear, and the whole will ripen together. Half a bushel of wheat, or even a less quantity, in this way, will seed an acre sufficiently. How great must be this advantage at a time of great scarcity of seed!

It is difficult to determine the quantity of seed, that is best to be sown in the broadcast way. Doubt less it should vary according to circumstances.

When seed is very large, and full grown, two bushels may not be more than equal to one that is small and pinched, supposing the seeds equally disposed to vegetate, which is often the case. For the true quantity should be estimated, rather by the number of grains, than by measure or weight. Not that I would recommend the sowing of pinched grain, excepting in case of necessity. For it is to be expected, in general, that the most perfect seeds will produce the best plants.

Rich land will afford nourishment to a greater number of plants than that which is poor. It has been held by many farmers that the poorer the land is, the greater quantity of seed should be sown in it. But Mr. Miller says, "This is one of the greatest fallacies that can be imagined; for to suppose that poor land can nourish more than twice the number of roots in the same space, as rich land, is such an absurdity as one could hardly suppose any person of common understanding guilty of. Where the roots stand close, they will deprive each other of nourishment, which any person may at first sight observe, in any part of the fields where the corn happens to scatter when they are sowing it; or in places where, by harrowing, the seed is drawn in heaps, those patches will starve, and never grow to a third part of the size as the other parts of the same field; and yet, common as this is, it is little noticed by farmers; otherwise, they surely would not continue their old custom of sowing."

The practice of farmers has been various, as to the quantity of seed. In England they sow from two or three to four or five bushels of wheat on an acre; six bushels of oats, and four of barley. But the above quoted author is very positive that a third part of the usual quantities would be better.

The usual quantities in this country are not greater than five or six pecks of wheat or rye, three bushels of oats, and two of barley, for an acre. And from these quantities, in some instances, large crops have been produced. Though, in old countries, the crops are usually larger than ours, I apprehend it is not owing to higher seeding, but to deeper and more perfect tillage, better manuring, and frequent changing of seed, with a judicious rotation of crops.

The sowing of winter grain is perhaps a more difficult matter to manage rightly, than vernal seeding. Farmers certainly mistake their interest, when they persist in sowing winter grain at a certain time of the year, let the weather be ever so hot, and the ground ever so dry. By heat and dryness, the seeds will sometimes be so scorch-in the soil, that not a fourth part of them ever come up. Therefore, if a drought happen at the usual sowing season, it will be needful to defer sowing till some rain has fallen, and the soil has got a due degree of moisture. How long it may be best to wait for such a favourable opportunity, I will leave to the judgment of the experienced farmer.

Also, a spot that has been newly cleared by burning, may be sown later in autumn than other land. It ought to be sown later, if the growth before winter be wished to be only equally forward; for the ashes will so quicken the vegetation, that if it be seeded early, it will attain to too large a growth before winter, and be the more in danger of being killed by frost.

It is doubtless much better to sow winter grain rather early than very late; because that which is sown late, will not be furnished with strong roots before winter, and therefore will not generally so well bear the frost. Though grain sown in December has sometimes prospered well, it ought not to encourage the farmer in sowing so late; because the instances in which it has succeeded have been but few. On the other hand, it should not be sown so early as to give time for the stalks to shoot up before winter. But perhaps it will be found to be a good rule, to sow grain the earlier in proportion as the winters are longer and colder. And yet, confining the true time to certain days or weeks, would be ridiculous.

M. de. Chateauvieux, from many experiments, and long practice, concludes, "that the best time for sowing in such a climate as Geneva, is from the 20th of August, to the end of September, And he thinks the first fortnight in October may answer, if the sowing cannot be done sooner.

But as the experience of persons in other countries may mislead us, it is greatly to be wished that a set of the most accurate experiments were made by some judicious person in this country, in order to ascertain the best time for autumnal sowing.

SPADE, an instrument used in digging. Spades differ in their shape and construction, according to the different operations in which they are to be used.

SPAVIN, a disease of horses, being a tumefaction about the joints, causing lameness.

"There are two kinds of spavin, a blood spavin, and a bog spavin.

"A blood spavin is a swelling and dilatation of the vein that runs along the inside of the hock, forming a little soft swelling in the hollow part, and is often attended with a weakness and lameness of the hock.

"The cure should be first attempted with restringents, and bandage, which will contribute greatly to strengthen all weaknesses of the joints, and frequently will remove this disorder, if early applied. But if, by these means, the vein is not reduced to its usual dimensions, the skin should be opened, and the vein tied with a crooked needle and wax thread passed underneath it, both above and below the swelling, and the turgid part suffered to digest away with the ligatures. For this purpose, the wound may be daily dressed with turpentine, honey, and spirit of wine incorporated together.

"A bog spavin is an encysted tumour on the inside of the hough, or according to Dr. Bracken, a collection of brownish gelatinous matter, contained in a bag, or cyst, which he thinks to be the lubricating matter of the joint altered, the common membrane that encloses it forming the cyst. This case he has taken the pains to illustrate in a young colt of his own, where, he says, when the spavin was pressed hard on the inside of the hough, there was a small tumour on the outside, which convinced him the fluid was within side the joint. He accordingly cut into it, discharged a large quantity of this gelatinous matter, dressed the sore with dossils dipped in oil of turpentine, putting into it, once in three or four days, a powder made of calcined vitriol, alum, and bole. By this method of dressing, the bag sloughed off, and came away, and the cure was successfully completed without any visible scar.

" This disorder, according to the above description, will scarcely submit to any other method, except firing, when the cyst ought to be penetrated to make it effectual. But in all obstinate cases that have resisted the above methods, both the cure of this, and the swelling called wind-galls, should, I think, be attempted after this manner. If, through the pain attending the operation or dressings, the joint should swell and inflame, foment it twice a day, and apply a poultice over the dressings, till it is reduced." *Complete Farmer.*

Mason's Farrier asserts that " a spavined horse may be considered as completely ruined, for a permanent cure can rarely be effected, if attempted even on its first appearance. A blister of Spanish flies applied on the part affected, (after shaving the hair off,) with a bath of strong spirits or vinegar, and a week's rest, will frequently suspend the lameness produced by spavin for a time—but a radical cure may not be expected."

SPAYING, the castration of female animals, to prevent conception, and promote their fattening. It is said, that spayed sows will have a greater quantity of fat upon their inwards than barrows, and that they are, on the whole, more profitable.

SPELT, a species of grain resembling wheat, but smaller, and darker coloured, bearded, with only two rows on an ear. It is used in Germany for bread, and will make malt. Of this grain the ancients are said to have made their frumenty, of which they were very fond. It may be sowed in autumn or spring, and delights in a dry soil.

SPIKY ROLLER, a wooden roller, armed with spikes, of important use in husbandry.

This instrument was formerly just mentioned by Mr. Ellis; but has been of late brought into use by the ingenious Mr. Randall, of York, in England: Who recommends, that the roller be a cylinder of the heart of oak, seven feet long and eighteen inches diameter, with a strong band of iron on each end. Teeth or tines of iron, seven inches long, are driven three inches into the wood, and four inches apart, in the quincunx order, over the whole convex surface. The outer points must be pretty sharp; and the ends which go into the roller should be ragged at the corners, to prevent their coming out. The tines need not be quite so strong as the teeth of a harrow. The whole instrument will weigh near a ton; and a frame is to be annexed to it, for the team to draw by; to which a box may be added for the driver to set on. But beware of putting a wild, restiff or unmanageable team to this fearful instrument. The strength of four oxen or three horses, will be necessary to draw it.

The uses to which the spiky roller is to be applied, are, in the first place, to reduce a stiff, stubborn, and clotty soil, to a fine tilth for sowing. This it will perform with admirable expedition, by only passing forwards, and back again in the same track, reducing it even to a perfect garden mould. And, which greatly recommends it, it is used to advantage when the ground is too dry for ploughing; by means of which, there need not be any delay in preparing land for sowing.

"It is certainly an instrument," say the compilers of the *Complete Farmer*, "that no farm, where the land is stiff, or the least liable to clot, should want. For, besides the constant advantage of saving labour, and bringing land to a better condition for any kind of sowing, than the plough and harrow, with any assistance of the work of hands can make it; in favourable seasons, and under such circumstances as Mr. Randall has mentioned, the loss of the whole crop, by an otherwise unavoidable delay beyond the seed time, may be with certainty prevented."

Another important use of this instrument, is, to renew the fruitfulness of grass land, when it is so bound as to be almost barren, or overrun with moss and bad grasses. Mr. Randall directs, that a good compost be prepared: And in autumn, when the ground is a little moist, that the spikes may enter the soil easily, to pass the roller up and down till the surface is well broken: Then sow hay seeds, and spread the compost over them, to be followed with a smooth roller, with a bush harrow after it. Thus a fine sward will be renewed, and good crops of the best grass will follow.

But it is obvious to remark, that the surface must be pretty level, and the land free from stones, to admit of these operations.

Mr. Randall also recommends passing this roller in the spring, over winter grain, to loosen the surface, and increase vegetation, and smoothing it afterwards with a bush harrow. Though this may seem to be a bold experiment, I think it is probable it might have a better effect than harrowing, which is much approved by many, as the tines would penetrate deeper, and as the plants would be less exposed to extirpation, than by the horizontal motion of the harrow.

SPRING, one of the four seasons of the year, so called from the springing or shooting up of vegetables, which in the winter were in a torpid state.

This season includes, according to common parlance, March, April and May. It is the most busy and hurrying season, for farmers in this country, of any in the year; partly owing to the long continuance of frost, which commonly prevents all kinds of tillage till near the beginning of April; and in the northern parts, till the end of that month. But sometimes it is partly owing also to what we might order otherwise, to sowing a larger proportion than is necessary of our grain in the spring, and neglecting in autumn to cart out so much of our manure as we might, that we complain of being so much hurried with work in the spring.

But besides tillage and seeding, which are enough to employ the whole time, there are other matters to be attended to at this season. The fences are always to be examined, and repaired: For though they were in good order in the fall preceding, high winds, violent storms, and deep snows, may overset, break or settle them, not to mention the gradual decay and rotting of wooden fences. Or the violence of frost may heave and disorder them. Compost dunghills it will often be needful to make at this season, especially if the materials were not all obtainable in the preceding autumn.

But preparing and seeding the ground must not be neglected, nor slightly performed: For as a man soweth, so shall be reap. Sluggishness at the beginning, will be followed with want at the end of the year.

SPRING GRAIN, that which is sown in the spring.

Farmers generally think they cannot sow their wheat and rye too early. But their haste may possibly be too great. It certainly is, if they sow before the ground is sufficiently dry to crumble, and become light and fine by ploughing and harrowing. For working the ground when it is too wet will fail of bringing on the needful fermentation, and tend to make it too close and compact to nourish plants. At least, these operations may leave the soil as unfit to nourish plants as they found it.

It is of no small importance, I confess, that spring wheat and rye be forward, that so they may not be late in ripening: And a good method of quickening these crops, is steeping and liming the seed. The same end may be promoted by top-dressings with warm composts, ashes, &c. But land designed for this use, should be ploughed and laid rough in autumn. It will be the earlier in a condition to receive the seed in the spring. And none but the driest parts of a farm should commonly be employed for crops of spring wheat or spring rye. As to barley and oats they will bear to be sowed rather later: And therefore are more fit to occupy the lower and wetter parts of a farm. Oats especially are often found to produce great crops in such situations, if they be not sown too early; or before the land is sufficiently dry for the operations of tillage.

SPRINGE, a device made of twisted wire to catch birds, or other small animals.

SPUR, a bad distemper in rye.

The grains which are affected, are thicker and longer than the sound ones. commonly projecting beyond their husks, and mostly crooked. They are dark coloured, have a rough surface, and appear furrowed deeply from end to end. They are bitter to the taste; will swim in water at first, and then sink to the bottom. But they are easily distinguishable by their extraordinary bulk and length.

Various have been the conjectures concerning the cause of this distemper. Some think it is occasioned by the bite of an insect: Others ascribe it to fogs, dews, rain, &c. But as I have never found any such distempered grains, but in rye of a rank growth, I rather incline to ascribe it to too great a pressure and flow of sap into the kernels, while they are in their most tender state, by which they are too much distended, and rendered incapable of throwing off the grosser particles of sap; by which means they become fungous and misshapen.

M. Salerne, and others, have given sad accounts of the diseases with which numbers of people have been afflicted, in some years, when they have eaten freely of bread, in which there was much of the spurred rye.

The peasants of Sologne, it is said, sift out these grains, when corn is plenty: But in a time of scarcity, being loth to lose so much grain, they neglect it. And then they are wont to be attacked with a dry gangrene, which mortifies the extreme parts of the body, so that they fall off, almost without any pain.

"The Hotel Dieu, at Orleans, has had many of these miserable objects, who had not any thing more remaining, than the bare trunk of the body, and yet lived in that condition several days.

"As it is not every year that the spur in rye produces these dreadful accidents, Langius is of opinion, that there may be two kinds of this distemper; one which is not hurtful to human constitutions, and the other which occasions the gangrene. It is however probable, that there is but one kind of spur, and that it does not sensibly hurt; first, when sufficient care is taken in sifting the grain; and secondly, when only a small part of the corn is distempered. It is also said, that the spur loses its bad quality after the grain has been kept a certain time: In which case, the reason why some peasants are attacked with the gangrene in years of dearth, may be, that they consume their crop as soon as their harvest is over."—*Duhamel's Culture des Terres.*

STABLE, a house, or lodgment for house.

A stable should have an open airy situation, and be as free as possible from mud and wetness. The floor should be built of pine planks, not on a level, but descending backwards, that the stale may not remain under the horses, so that they may lie dry and clean.

As a horse is a cleanly animal, hen roosts, hog sties, and necessary houses, should not be too near to his apartment. A stable should have windows to open and shut, that fresh air may be let in when the weather is hot: And it should be tight and warm in winter. Otherwise the great vicissitudes of heat and cold will do much hurt to the animals; and the more as, being tied up, they cannot use much motion. Some of the windows should be glass, because horses are fond of light. And it is better for their eyes that they be not confined at all to total darkness in the day time.

A manger is necessary in a stable, to preventing wasting of hay. Some choose their horses should have their hay in racks. Others think it puts a horse into an unnatural posture, as he is used to take his food from the ground. If a rack be used, it should be perpendicular, not leaning towards the horse, nor placed too high: And the manger before it should be two feet wide, or more. The hinder part of the rack should be made shelving, that as the hay settles, it may naturally press towards the horse.

A box for provender may be fixed at one end of the manger, in each stall; or the manger may be made as tight as a box, to prevent loss of grain. But the surest way to prevent wasting, is to give a horse his corn in a pail, with a strap of leather to slip over his head, which will prevent the loss of so much as a single grain. It may be put on or off in an instant. See the article *Horse*.

STACK, a large quantity of hay, grain, or straw, piled up, pointed at the top, and usually covered with long straw, or thatch, to keep out the weather.

Square and oblong stacks are not good. Round ones have a less quantity of superficies in proportion to their contents; and therefore will receive less damage from the weather.

When sheaves of corn are stacked, the heads should be all turned carefully inward. But if designed to stand long, it should be on a floor mounted on blocks, capped with flat stones, to prevent the entrance of vermin.

Farmers should not practise the stacking of good hay, in a country like ours, where timber for building barns is plenty and cheap. For so much of the outside is always spoiled by the weather, that they may soon lose more in this way, than the cost of a barn.

But much of the salt hay in marshes must be stacked, on account of the difficulty of removing it before winter. These stacks must be mounted on what is called a staddle, consisting of piles driven into the ground, of such a height, that the highest tides may not reach the bottoms of the stacks.

STAGGERS, a disorder to which both horses and neat cattle are liable. If the staggering and falling of a horse be owing to hard riding in hot weather, Gibson directs to take without delay a pint of blood from his neck, and then a quart from some vein in his hinder parts, that so an effectual revulsion may be made: And that he should afterwards be kept on a moderate and cleansing diet.

When the disease arises from an apoplectic disorder, he must not only be bled, as in the former case, but be exercised every day with chewing assafœtida and savin, and the most noisome things that can be got, which will put him upon constant action, and forward the circulation in the small vessels. Afterwards recourse must be had to clysters and strong purgatives, rubbing and exercise.

When the disorder arises from vertigo, or swimming of the head, the animal reels, turns round and falls. In this case, take an ounce of senna boiled in five pints of water, with four ounces of common treacle, with the usual quantity of oils, or lard, to throw in as a clyster. And repeat it for two or three days. After which he may have a drench of beer, in which roots of peony, angelica, rue, rosemary, flowers of lavender, and the like, have been steeped. If the disease should continue obstinate, balls of cinnabar and assafœtida with bayberries will be proper here, as well as in apoplectic cases.

Some venture to put ginger, and other stimulating things, into the ear, to give the blood a quicker motion. But this practice, though it may chance to do service, is dangerous.—See *Gibson's Farriery*.

"The staggers, in my opinion, are produced by permitting a horse to feed on grass in the spring and fall; late at night and early in the morning; for early in the morning and late in evening, the fields and pastures are covered with a poisonous webb, which is spun and spread upon the grass by a small spider. So rapidly, so industriously does this little insect work, that in the space of one night, not a blade or spire of grass has been left untouched. This web, catching the dew drops on its bosom, causes the fields in the morning to glisten and sparkle as if covered with a thin sheet of ice. A horse that feeds upon a pasture in this situation, must of course, collect large quantities of this web and dew, and very often the spider itself. They act upon the horse, producing delirium, giddiness, apoplexy, and sometimes death. The lungs appear to be the principal seat of this disease; for in cases of dissection they have been found much enlarged and covered with large brown spots, smell very offensively, and have some appearance of mortification."—*Mason's Farrier*.

This writer is the only one whose works we have seen, who attributes the staggers to the cause above mentioned. We shall hazard no opinion on the correctness of his theory, but think it worth attention and further inquiry. The same writer recommends the following

REMEDY.

Take from the neck vein half a gallon of blood, three times in a week, of sassafras tea three half pints, plantain juice half a pint, assafœtida half an ounce, saltpetre one tea spoonful, mix and give them as a drench three mornings in a week; give an injection, composed of one pint of meal, two quarts of water, one pint of molasses and one spoonful of hog's-lard—let the horse be moderately exercised, and whenever he is standing should be well rubbed—give a mash twice within a week, composed of one gallon of bran, one table spoonful of sulphur, and one tea spoonful of saltpetre, one quart of boiling sassafras tea, and an eighth of an ounce of assafœtida, not permitting the horse to drink cold water for six hours afterwards. Should he be much mended by this treatment, nothing more will be necessary, except feeding him on bran or light food of any kind; but should he appear to receive no benefit from these attentions in four or five days, take of calomel twenty-five grains, of opium two drachms, powdered fennel seed one drachm, of syrup of any kind a sufficient quantity to make the ingredients into a ball, which may be given every morning for four or five days, by which time the horse will get well if his disease will admit of a cure.

"Horses that are confined in a stable never have the staggers; consequently it would be advisable for every person, whose situation will admit of it, to confine their horses, particularly at night, during the spring and fall months."

A gentleman in North Carolina asserts, that the following recipe has been found, after repeated trials, to be effectual.

Take of the expressed juice of garlic six spoonfuls, which pour down the horse's throat by means of a horn, or give it to him in a drench. If the first dose should not relieve him, or he should appear to be maze-headed, repeat it after an intermedium of two or three hours. The juice of the leek or onion, given in rather a greater quantity, will produce nearly the same effect.

STALE, the urine of animals. The urines of different animals are said to possess the same properties as their dungs. They are of more importance as manures than farmers in general are aware of. They may easily convince themselves of this, if they will only consider how much more a piece of ground is enriched by folding than one of the same size would be, by laying the same dung, or an equal quantity, upon it, that is dropped on it by folding.

If the stale of a farmer's stock were all saved, and well applied, it might perhaps be of near as much advantage as their dung. Every possible method should therefore be taken to prevent the loss or wasting of it. Floors where beasts are lodged, should be perfectly tight; and they should be constantly well littered with substances that will absorb and retain the stale; even with common earth, when nothing else is at hand. Pens in which beasts are kept should have a layer of mulch: Or else the surface of the ground should be taken up, and used as a manure. Otherwise the urine is lost.

Mr. Hartlib praises the Dutch for saving the urine of cattle as carefully as the dung, to enrich their lands. The older it is, the better it is for this purpose. When it is deprived of its fiery hot particles by time, it will be a great fertilizer of every kind of soil. Columella recommends old urine as an excellent application to the roots of trees. But care should be taken not to apply too much of it. For trees have been sometimes killed by urine.

STALLION, a stone horse.

One that is kept for covering should be well shaped. See the article *Horse*.

He should be free from every kind of distemper and natural blemish, of a middle size at least, with a good gait, neither addicted to starting and stumbling, and of such a colour as is most coveted. For it may be justly expected, that the colts will inherit the qualities of the horse, let them be good or bad.

STEAM-BOILER. This is an implement which is very simple and so useful that no farmer ought to be without it. The following is a description of its construction.

A steam boiler is commonly made by setting a kettle holding twelve gallons or more in a furnace of brick or stone, and over this a hogshead, with one head taken out and the other bored full of holes, is set so close that the steam of the kettle when boiling, can only rise through the holes, and thence ascend among the articles to be boiled in the hogshead and pass off at the top. In this way a hogs head of potatoes will be nearly as soon boiled as a small part of them could have been, if placed in the kettle underneath.

As the kettle must be so closed as to prevent any steam from passing off, but through the bottom of the hogshead or vat, a pipe or tube must be set one side, through which with the aid of a funnel the water is to be poured into the kettle as often as occasion may require. When poured in, the tube is to be stopped with a plug for that purpose.

Grain of all kinds may be steam boiled to great advantage, for feeding and fatting cattle; but in that case, it is requisite to have the bottom of the hogshead covered with a cloth, to prevent the grain running down through the holes.

By experiments which have been accurately made in Pennsylvania upon Indian corn and potatoes, used for fatting swine, it was found that they increased in weight one third faster on the boiled, than on the unboiled food; or, in other words, they gained three pounds when fed on the former, where they only gained two pounds when fed on the latter.

STERCORARY, a store of dung, or compost, kept under cover: Also the building in which it is kept.

In these places, if they are well constructed, the manure will retain its whole strength. The dung so preserved, if I mistake not, will be of double the value of that which is managed in the common way. For it is secured effectually from washing rains, and drying winds; as well as secluded from the direct influence of the sun, which always causes a copious evaporation from uncovered dunghills.

To make a stercorary, Mr. Evelyn directs, to dig a square or oblong pit, of a size proportioned to the compost wanted, with the side towards the field sloping, so as to receive a cart, to load or unload easily: The bottom to be well paved, and the sides also made so tight as to hold water, the whole being under cover. Then the farmer's care must be to fill the pit with compost suited to the nature of his land.

A good proportion of lime will be proper in these composts, to assist in dissolving the vegetable substances. It is best that the layers of each substance should be thin, as the heaps will need the less turning and mixing afterwards.

It may be very convenient for the farmer to build a stercorary adjoining to the side of his barn, where the cattle are housed, covering it with a continuation of the roof. In this case, the dung may be thrown directly through the windows into the heap; from whence, through doors, it will not be difficult to shovel it into carts.

But, instead of this method, some gentlemen farmers in this country have begun to set the example of making cellars under their barns, into apartments of which, the dung is easily passed through scuttles in the floors. Other substances may be easily mixed with dung from time to time as there may be occasion. Or it may be kept to improve by itself. A cart way must be prepared to go through the cellar, or one entrance at least, not so steep but that a common team may draw out a cart full. The floor should be well paved, and the wall made tight with mortar, to prevent the escaping of the fluid parts of the manure. But the floor overhead needs not to be very tight; because the stale will not be lost, if it should pass through the seams.

This method, in our cold country may be allowed to be preferable to building any other stercorary; especially as part of the cellar may be employed in storing roots, &c. for the cattle. For it is no small advantage, to have the dung deposited, during the winter, in a place where no stop is put to its fermentation by frost. Besides, it is to be remembered, that the freezing causes a plentiful evaporation of the strength of the manure, which in this way is prevented.

It would be well to divide the parts of the cellar under the scuttles into pens, that the heaps may lie the more compact, and be less liable to too much drying. And as the heaps will need mixing with the shovel, it will be convenient to shovel the manure from pen to pen.

But instead of making stercoraries, or dunghills, in the usual modes, the *Society of Improvers*, in Scotland, prefer middens, or middings as they are there called, as it saves labour and expense in the management of manures. "Take, say they, in the field intended to be manured, a head ridge that is conveniently situated. Plough it two or three times, as deep as can be in the cleaving way, if the ridge be high gathered, and harrow it well: Then lay thereon your slimy clay, about a foot thick, a part of the earth uncovered. Next lay a thin layer of dung, another of clay, and after that a layer of unslacked lime, at least a foot thick: Then throw up the earth left uncovered on each side. After this repeat another layer of clay, then lime and finish with clay or sea wreck, covered with earth. The more of the slimy clay the better; for though it may be cold, yet it will not be the worse for a sandy hot ground. If you examine the clay.

we doubt not but you will find it a very fat substance; being, as we conjecture, mostly muscle and other shells mixed with earth, brought by the tide and the river.

"After this bank has stood six weeks or two months, incorporating and fermenting, turn and mix it. Yoke your plough, enter upon your stercorary with a cleaving furrow, and continue repeating the ploughings the same way, until the very bottom be ripped up; then harrow it; it is impossible to overdo it. If it is very cloddy, it should be harrowed between the ploughings. Begin then in the middle, and plough again and again in the gathering way, until it be brought into as narrow bounds, and be raised as high as possible. Let all that the plough has left be thrown up with shovels on the top. Every such turning and heaping occasions a new ferment, and improves the manure. If the first heat should go off before it is reduced to a fine fat mould, it may be turned over again, and will take a new heat. About fifty or sixty cart loads of this compost are used upon an acre of ground."

Any farmer may easily follow this example, and suit his compost to his soil. It will save much carting, especially when the land to be manured with it lies at some distance from the farm yard. At the same time it will reduce those disagreeable ridges that gather in the borders of lots that are long tilled, which are always a richer soil than the rest of the field and more fit for this use.

An operation similar to the above, was experienced by Mr. Eliot. He built a cow yard very long and narrow, at the side of a road, and once in three or four days, he removed the fences from the ends, and gave it a deep ploughing. The consequence was, that all the earth which was stirred with the ploughs became, in his opinion, of equal value as a manure, with good barn dung. The advantage of this method of increasing manure is unspeakably great. The manure of a yard may thus be increased to ten fold.

The 3d volume of Memoirs of the Philadelphia Agricultural Society, pages 222, 3,4,5, contains an account of a Stercorary, erected by Mr. Quincy, of Massachusetts, from which the following is extracted.

"The area of my Stercorary is 90 feet by 40, the cellar is in the shallowest part eight feet deep, in its deepest twelve, and in the well if I mistake not, fifteen.—It is open nearly the whole length of one of its short sides, and one half of the long, viz. at the north and west, besides large openings at the east. There is always four or five feet atmosphere above the top of the manure, and between it and the barn floor, and a constant current setting one way or another. This gives the advantage of a free circulating air, which in general, in such cases is not obtained.

"The great difficulty I have had to encounter, arises from the necessity of *an equal irrigation of the heap*; a difficulty which must attend all permanently covered Stercoraries. For water turned upon the heap through spouts runs in currents, and is not equally sprinkled over a heap like rain, which is nature's process in this business. To obviate this difficulty I have constructed a very simple machine which answers perfectly. The stone piers which support the beams of the barn, divide the cellar length ways into three equal compartments. I have a box six inches deep, four feet wide, and about thirteen feet long, which runs by means of wheels, upon a sort of wooden rail way, made by strips of planks and fixed about a foot from the floor of the barn; this is perforated with suitable holes. A permanent spout extends through the middle of the cellar, and a moveable spout ex tends from this to the perforated box—regular openings are made in the permanent spout, which may be closed at will.--It is also closed at the end. By these means, the box is filled from the reservoir and pump, and each part successively irrigated perfectly and with great ease. A man by two days labour can irrigate my whole cellar, and if effectually done, thrice in a season is sufficient."

The following humerous exhibi tion of the faults of some farmers, as respects the economy of their barn yards and stercoraries, is extracted from "An Address to the Massachusetts Agricultural Society, by Hon. Josiah Quincy.

"As we proceed to the farm, we will stop one moment at the barn yard. We shall say nothing concerning the arrangement of the barn. They must include comfort, convenience, protection, for his stock, his hay, and his fodder, or they are little or nothing. We go thither for the purpose only of looking at what the learned call the stercorary, but which farmers know by the name of the manure heap. What is its state? How is it located? Some times we see the barn yard on the top of a hill, with two or three fine rocks in the centre; so that whatever is carried or left there, is sure of being chiefly exhaled by the sun, or washed away by the rain. Some times it is to be seen in the hollow of some valley, into which all the hills and neighbouring buildings precipitate their waters. Of consequence all its contents are drowned, or water soaked, or what is worse, there having been no care about the bottom of the receptacle, its wealth goes off in the under *strata*, to enrich possibly the antipodes.

"Now all this is to the last degree wasteful, absurd, and impoverishing. Too much cannot be said to expose the loss and injury which the farmer thus sustains. Let the farmer want whatever else he pleases. But let no man call himself a farmer, who suffers himself to want a receptacle for his manure, water-tight at the bottom, and covered over at the top, so that below nothing shall be lost by drainage, and above, nothing shall be carried away by evaporation. Let every farmer, wanting such protection for his manure, be assured that he loses by the sun and rain, ten fold as much as will pay all his taxes, state, town, and national every year. Let not the size of his manure heap be any objection. If it be great he, loses the more, and can afford the expense better. If it be small, this is the best way to make it become greater. Besides, what is the expense? What is wanted? An excavation, two or three feet deep, well clayed, paved, and 'dishing' as it is called, of an area from six to thirty feet square, according to the quantity of the manure; over head a roof made of rough boards, and refuse lumber if he pleases."—*Massachusetts Agricultural Repository, vol.* iv. p. 6.

STOCK. A term used by English farmers, to express the quantity of money or wealth a farmer should have, to enable him to hire and cultivate a farm to advantage.

The larger the farm a man hires, the greater stock he should have in hand. Writers on husbandry reckon the needful stock is equal to the first years rent, and seed; expense of horse keeping, clothes and pocket money; the cost of cattle to be kept, and farming utensils of every kind; besides the labour that must be hired, including fencing and draining.

It would not be amiss, if those who hire farms in our country, would carefully calculate these expenses, before they take farms, and consider their ability. For want of this needful precaution, they often find, when it is too late, that they are unable to carry on the culture to advantage, and are unable to pay their rent. The consequences are unspeakably bad and distressing. Both the landlord and tenant are sufferers.

But the word *stock*, in this country, is commonly used by farmers to express only live stock, or the beasts that are kept upon a farm. These should not be all of one kind, but such an assortment as is best adapted to the convenience and profit of the farmer. The stock should be adapted to the nature and circumstances of the farm.

Young stock, in general, is better than old. The more there are in a growing state, the greater is the profit. And very old cattle, when turned off to fat, do not answer so well as those which are but little past their prime, or full vigour. It costs more to fatten them, and the meat is not so valuable.

It is best to begin with a considerable variety of animals; that the farmer, by observing the profit he gets from each kind, may afterwards vary, as he finds to be best. For this cannot be determined, but by some experience: Because some animals prosper best on one farm, and some on another; some best under one manager, and some under another.

A variety indeed, for other reasons, is always best: One is, because almost every farm produces a variety of food, some of which will answer best for one animal, and some for another. Even in the same pasture, that which one species of animals leave, another will feed upon.

Also, the stock should vary, in some proportion as the lands of a farm do. As some farms contain a large proportion of high and dry pasture grounds, the greater quantity of sheep should be kept. Where low meadow abounds, the kind of stock should be increased, which will do best on coarse water-grasses; which is well known to be neat cattle, that are young and growing. But if a farm yield a plenty of good sweet grass, it is the more suitable for a dairy farm, and the greater proportion of cows ought to be kept.

But on no farm should horses be multiplied beyond the number which are needed, or which can be employed to advantage. For they are great eaters, and require the best of the fodder and pasture. A small farmer can scarcely afford to keep one, unless he puts him to the draught.

Let a farm be what it will, it should never be overstocked. This is an error that too many farmers in this country are guilty of. Doubtless it arises from a covetous disposition; but they sadly miss their aim. Instead of gaining, they lose by it. A half starved stock can never be profitable.

A farm may be said with truth to be overstocked, when a greater number of animals are kept, than can be well fed with its produce, during the whole year. For it is a ruinous practice, to suffer a beast to pine away, and lose, in one part of the year, the flesh he gains in another. And when the farmer is constrained to purchase food for his stock, he too often affords them but a scanty allowance. Sometimes, it is not in his power to obtain it.

The starvation of cattle, or keeping them too short of food, not only prevents their being profitable to the owner, but teaches them to be disorderly, and to break through, or leap over fences; and many times to become absolutely ungovernable; so that they must either be killed, or sold off at a low price; in either of which cases, there is often much inconvenience and loss.

It is far better that some of the stock of food should be left in the spring, than that it should fall short. It is a good reserve against a season of scarcity: And such seasons often happen in this country by drought.

The following general rules, as to the management of stock may deserve attention.

"1. Animals intended for the butcher, should be kept in a state of regular improvement. The finer breeds are highly fed from their birth, and are almost always fat. With other breeds, and on pastures of inferior quality, this is neither necessary nor practicable. But in every case, the same principle of improvement should be adhered to, and such animals ought never to be allowed to lose flesh, in the hopes of afterwards restoring it by better feeding.

"2. The size should never be above that which the pasture can support in a thriving condition. The attempt to raise them to an undue size, by crossing, is censurable. In regard to size, the stock of every kind, and of all the various breeds, should be proportioned to the quantity, and the quality of their intended food.

"3. The best pasture should be allotted to that portion of the stock, which goes first to market; the next in quality to the breeders; and the coarse pasture, to the inferior or growing stock.

"4. Great care should be taken, not to overstock pasture, which is attended with great loss to the farmer, and the community. This ought to be particularly avoided in young and growing animals. If they are kept poor during one part of the year, they will scarcely thrive during the remainder; and whenever ill fed, will never attain to their proper size and proportion.

"Lastly the food, whatever it may be, should not be too suddenly changed. It is seldom profitable to bring lean animals immediately from coarse to rich pastures; and a change from dry, to succulent food, and *vice versa*, should be gradually effected. A change of pasture, however of the same quality, tends to produce a greater accumulation of fat. *Code of Agriculture.*

The following observations relative to the size and form of stock are by Henry Cline, Esq. an English Surgeon. They have met the approbation of the most eminent agriculturists both in America and Great Britain.

"It has been generally understood that the breed of animals is improved by crossing with the largest males. This opinion has done much mischief, and would have done more if it had not been counteracted by the desire of selecting animals of the best forms and proportions, which are rarely to be met with in those of the largest size. Experience has proved that crossing has only succeeded in an eminent degree in those instances in which the females were larger than in the usual proportion of the females to the males; and that it has generally failed when the males were disproportionally large.

The external form of domestic animals has been much studied and the proportions are well ascertained. But the external form is an indication of the internal structure. The principles of improving it must therefore be found on a knowledge of the internal parts.

Of these the lungs are of the first importance. It is on their size and soundness that the strength and health of an animal principally depend. The power of converting food is in proportion to their size. *An animal with large lungs is capable of converting a given quantity of food into more nourishment than one with smaller lungs; and therefore has a greater aptitude to fatten.*

"*Chest.* The size and form of the chest indicate the size of the lungs, of which the form should approach to the figure of a cone having the apex situated between the shoulders, and its base towards the loins; a circular form of chest is preferable to one deep and narrow; for though the latter may have greater girth, the former will have greater internal space in proportion.

"*The Pelvis.* The Pelvis is the cavity formed by the junction of the hip bones with the rump bone. This cavity should be larger in a female that she may bring forth her young with less difficulty; when this cavity is small, the life of the mother and her offspring is endangered.

"The size of the pelvis is indicated by the width of the hips, and the space between the thighs; the breadth of the loins is always in proportion to that of the chest and pelvis.

" *Head.* The head should be small, by which the birth is facilitated to the offspring, it also indicates the animal to be of a good breed, and occasions less weight of unprofitable substance to the consumer.

" Horns are useless to domestic animals, and occasion a great weight of bone in the head. The skull of a ram with horns weighed five times as much as that of one without horns, each being four years old. A mode of breeding, which would prevent the production of horns, would therefore afford a considerable saving.

" The length of the neck should be proportioned to the height of the animal, that it may collect its food with ease.

" *Muscles.* The muscles and tendons, which are their appendages, should be large, by which an animal is enabled to travel with greater facility.

" *Bones.* The strength of an animal does not depend on the size of the bones, but on that of the muscles; many animals with large bones are weak, their muscles being small. Animals imperfectly nourished during growth have their bones disproportionally large. If this originates from a constitutional defect, they remain weak during life; large bones may therefore indicate an imperfection in the organs of nutrition."

Of the improvment of form. The chief point to be attended to for the improvement of form, from Mr. Cline's principles, is the selection of males for breed of a proportionally smaller size than the females, both being of approved forms; the size of the fœtus depends on the size of the female, and therefore when the female is disportionally small, her offspring has all the disproportion of a starvling from want of due nourishment.

The larger female has also a greater supply of milk, and her offspring is therefore more abundantly provided with nourishment after birth.

When the female is large in proportion to the male, the lungs of the offspring will also be greater. By crossing in this manner, there are produced animals with remarkably large chests, as has been often noticed; the advantage of large lungs has been already pointed out.

In animals where activity is required, this practice should not be extended so far as in those which are intended for the food of man.

The size of animals is commonly adapted to the soil which they inhabit; when the produce is scanty the breed is small; the large sheep of Lincolnshire would starve. where the small sheep of Wales find abundant food.

Crossing may be attended with bad effects, even when begun on good principles, if the above rule be attended to throughout; for instance, if large ewes were brought to Wales, and sent to the rams of the country, the offspring would be of improved form; and if sufficiently fed of a larger size than the native animals, but the males of the breed would be disproportionately large to the native ewes, and therefore would produce a starveling ill formed race with them.

The general mistake in crossing has arisen from an attempt to in crease the size of the native race of animals; being a fruitless effort to counteract the laws of nature; which from theory, from practice, and extensive observation, Mr. Cline, concluded to be decidedly wrong; for in proportion to this unnatural increase of size, they become worse in form, less hardy, and more liable to disease.

The Massachusetts Agricultural Repository, vol. vi. p. 78., contains some valuable remarks on the subject of " Dairy Stock," by S. W. Pomeroy, Esq. We shall give the following extract, which presents an important fact, not sufficiently known or attended to by writers who have treated on the same or similar subjects.

" In the selection of bulls, most farmers confine their attention to form and colour only, instead of tracing their descent to a valuable dairy stock. It has been observed by Linnæus that those properties of animals which relate to the vessels, or in scientific terms, the *cortical substance, or vascular system*, are der ved from the *males*," and among other examples tending to confirm this opinion, he states " that a cross from the male Angora goat, with the common female goat produces that fine wool or substance, called Camel's hair; but that the progeny from the male common goat with the female Angora, is productive of nothing but the same worthless hair of the sire." See likewise Agricultural Repository, vol. iv. p, 257.

STONES, well known hard and brittle bodies, which abound in some lands. Those of the slaty kind, or which are flat or square cornered, are fit for building wall fences, and should be applied to that use. And many of the pebble kind may go into walls among others of a better shape; especially if the wall is built double, as it always should be where stones are plenty. Where there are more stones than are needed, the walls may be made thicker and higher than is needful on other accounts; and lots should be made the smaller; for there are certain conveniences in having small sized lots, though they may not be thought necessary, in any other view than for disposing of the stones.

Pebbles are a greater annoyance on a farm, as they need removing, but are not very good for any kind of building. Picking them off very minutely, for common field tillage, is not needful. But the largest pebbles should be taken away.

Stones that are very large, and which cannot with ease be removed whole, may be blown to pieces with gunpowder. They will be not only more handy for removing, but far better to put into walls. For the blowing of round stones will make some square and regular faces. They will often come cheaper in this way than if they were dug out of quarries. As the soil that is occupied by a large stone is better than the rest of the field, it is purchased at an easy rate by removing the stone.

But another method of breaking rocks, which ought to be generally known, and which sometimes turns out cheaper is this : Drill two holes in a stone, ranging with the grain, when that can be discovered by the eye. Then filling each hole with two semi cylindrical pieces of iron, drive a long steel wedge between them. The stone will thus be split open. And, commonly, very regular shaped pieces for building may be thus obtained.

Another method is, to burn an inflammable piece of dry wood, laid on the part where you wish a flat rock to open. Thus the rock is heated in a straight line, and may be made to open in that part, by a smart blow of a maul. This method often answers well when the stones are flat shaped, and not too thick.

That stones which are so large as to obstruct the operations of husbandry, ought to be removed from land in tillage, all will agree. But it has been long a disputed point, whether the smaller stones should be taken away. Some have contended that they add fertility to the soil.

That the moisture of the soil is as much greater, as the proportion of room the stones take up in the soil is undeniable ; unless the stones occasion some evaporation. But many fields need not any increase of moisture, but would rather be improved, by being made as much drier as they can be, by removing the stones from the surface.

M. Duhamel is of opinion, that no stones increase fruitfulness, unless they be lime stones, marle, or those that are of a calcareous nature. These, by rubbing against each other, &c. in the operations of tillage, do probably yield a dust that increases the richness of land.

But all stones in tillage land are so troublesome, and so much increase the labour of tillage, that, when they are not calcareous, they should be taken away, or at least so much thinned, that ploughing and hoeing may be comfortably performed, and without much injury to the tools used by the farmer. Fixed stones under the surface should be removed, or so sunk by diging under them as to put them out of the plough's way, that ploughing may be performed without danger of destroying the plough.

To know whether stones are calcareous or not, they should be tried with aqua fortis, or spirit of sea salt. For stones on which the spirit does not effervesce, can be of no advantage to the soil. By the way, I do not expect that calcareous stones will be found in many fields in this country.

Ground that is laid down for mowing must have even the small stones taken out of the way of the scithe. But, instead of picking them up, some recommend driving them down into the soil, when the ground is so soft in the spring that it can be easily done. In this case a field will not be disfigured with the heaps, nor any of the surface lost.

STOOKING, *or* SHOCKING, setting sheaves into shocks to guard corn from wet.

Farmers have various methods of doing this. But I would propose for their consideration, a method inserted in the *Museum Rusticum*, Vol. II. page 250.

"Ten sheaves are disposed in two rows, each row leaning against the other; then two sheaves are laid on the top, so as to meet at the centre with their tails, and to slope downwards."

The writer thinks, and with good reason, that this method is very favourable to drying the corn, if it needs it, as well as to defending it from rains. Thicker built shocks, if they chance to get wetted, will need opening to dry the sheaves.

In general, it is better for the corn to stand in shocks a few days in the field, than to carry it sooner to the stack or mow. There will be less danger of its taking damage by heating.

STOVER, fodder for cattle. See *Fodder*.

STRAIN, vulgarly called *Sprain*, a violent extension, or stretching of the sinews, or tendons, by which the fibres are sometimes broken.

All sorts of animals, and particularly horses, are liable to lameness by strains.

My designed brevity will not permit me to treat fully on this subject. But let it be noted, that when a horse is lamed by straining, he should be permitted to rest, and be secured from wet and cold. Rest alone will sometimes recover the tone of the fibres, and complete the cure.

But bad strains should have some suitable applications to the parts affected. Oily medicines are in general to be avoided, on account of their relaxing quality: But oil of turpentine may be admitted.

A part that is lamed by straining should be bathed thrice a day. with hot verjuice or vinegar, in which a small piece of soap may be dissolved.

Early in the disease, if the part be swelled, a poultice should be applied after bathing. It should be made of oatmeal, rye meal or bran, boiled in vinegar, strong beer, or red wine lees, with lard enough to prevent its growing stiff.

After the swelling is down. bathe with camphorated spirits of wine, mixed with half as much oil of turpentine. Or, instead of the oil, take sharp vinegar, and spirit of vitriol, in equal quantities.

Keep on a linen bandage, drawn pretty tight, if the part affected will admit of it. But long resting from labour, will in some cases be needful. For further direction, the reader may see *Bartlet's Farriery*.

Mason's Farrier prescribes the following remedies for strains.

NUMBER I.

"Take of sharp vinegar, one pint, spirit of any kind half a pint, camphor one ounce; mix them well together, and bathe the part injured twice a day; a piece of flannel wet with the mixture, and wrapped around the part, will be very beneficial; take from the neck vein half a gallon of blood.

NUMBER II.

"Take of opodeldoc a piece the size of a marble, and rub it on the strained part with the naked hand, until the hand becomes dry, twice a day; should the injured part resist both of those remedies, you may conclude the injury is a very serious one, which nothing but time can relieve, and the horse must be turned out upon grass a sufficient length of time for nature herself to perform the great operation.

STRANGLES, "a swelling under the throat of a horse, between the two jaw bones, which seems not to differ very much from that which in a human body is called the *quinsy*. Its seat is not so much upon the glands as on the muscles; and therefore it comes the more readily to an imposthumation.

"If the swelling has a tendency forwards between the jaws, so that the passages of the throat are not in danger of being choked up by it, the safest way is to ripen, and bring it to a suppuration; and for that end anoint the part with ointment of marsh mallows, covering them up warm. Or take oil of bays and fresh butter, of each a like quantity, ointment of marsh mallows the weight of both: Or the poultice recommended for the glanders may be applied warm twice a day.

After the swellings are ripe, and that you perceive matter in them, but that they do not break, which perhaps may be hindered by the thickness of the skin, you may open them with a lancet; but if they do not ripen as you could wish, you had better make use of a hot iron, and sear the outside pretty deep. But whether you open them by incision, or by the iron, you must be sure to make your operation in the lowermost dependent part, that the matter may more easily run off. While you observe this method your incision need be but small.

"As soon as the matter is nearly discharged, you may press out what remains with your thumb. Then make a dossil of fine flax, dip it in warm basilicum, and introduce it into the orifice, but not too far, nor must it be continued above three or four days in any common case: For keeping the orifice too long open will derive too great a quantity of matter upon the parts, causing them to ulcerate, or to turn fistulous. Therefore when the running abates, only apply smooth flat pledgets of lint, armed with the same ointment, and above them thick compresses of soft canvass, in several doubles, to fill up the vacant place between the jaws, that the divided parts may again be united. If hard lumps remain after the sores are healed up, they may be removed by the following plaster. Take diachylon and red lead plaister, of each four ounces, pitch two ounces; dissolve them with a sufficient quantity of oil or lard. Then take bole in fine powder an ounce and an half, and stir into this mixture, and make it to the consistency of a plaister. This must be spread on leather, or thin dowlas, and after the hair has been clipped off very close, it may be notched and applied all under his chops, where it is to lie as long as it will stick on: And by the help of this all the little hardnesses will be dissolved.

"Sometimes this distemper is cast off chiefly by the nose: and sometimes the discharge is inward, about the roots of the tongue. In this case most of the matter issues also from the nose. In either case, the horse should be moderately ridden. But it will be very proper to wash his mouth sometimes with honey of roses; for that will keep it clean, and prevent ulcers. But if sores are like to continue, dissolve a quarter of an ounce of crude sal ammoniac in a pint of water, and wash his mouth with it once or twice a day.

"If the cure seems imperfect, and the horse does not thrive upon it, recourse may be had to purging; for which purpose I chiefly recommend the preparations of aloes; because these are the most effectual to work upon the blood, &c." *Gibson's Farriery.*

STRAWBERRY, *Fragaria*, a well known fruit which is much esteemed.

Mr. Miller reckons four sorts; the wood strawberry, the Virginia, or scarlet strawberry, the hautboy strawberry, and the strawberry of Chili.

It is the scarlet strawberry that is most common, and perhaps most worthy of cultivation. Our grass fields often produce these strawberries in plenty. But it is better to have a spot of ground devoted to the culture of them; as they will be much larger and better flavoured; and as the trampling of the grass in the mowing grounds may be thus in some measure prevented.

A light loamy soil is best for them: And but little dung should be applied to the soil, as a large quantity will cause them to run much, and to be less fruitful.

The time to remove these plants, is said by the above mentioned author to be September, or the beginning of October. But they are known to do well in this country when removed early in the spring, and watered a few times after it. But they will bear little fruit that year

He directs. "That the ground should be cleaned from the roots of all bad weeds. For as the plants are to stand three years, before they are taken up, those weeds would overbear the plants.

"The usual method is to lay the ground out into beds of four feet broad, with paths two feet, or two and a half broad between them. These paths being for the convenience of gathering the strawberries and for weeding and dressing the beds. The plants should be in the quincunx order, and fifteen inches apart, so that there will be but three rows in each bed.

"The plants should never be taken from old neglected beds, where the plants have been suffered to run into a multitude of suckers, or from any plants which are not fruitful; and those offsets which stand nearest to the old plants should always be preferred to those which are produced from the trailing stalks at a greater distance.

"During the summer, the plants should be constantly kept clean from weeds, and all the runners should be pulled off as fast as they are produced. If this is constantly practised, the plants will become very strong. Where proper care is taken of the plants the first summer, there is generally a plentiful crop the second spring; whereas, when this is neglected, the crop will be thin and the fruit small.

"The old plants are those which produce the fruit; for the suckers never produce any till they have grown a full year; therefore it appears how necessary it is to divest the old plants of them; for wherever they are suffered to remain, they rob the fruitful plants of their nourishment, in proportion to their number; for each of these suckers sends out a quantity of roots, which interfere, and are so closely matted together, as to draw away the greatest part of the nourishment from the old roots, whereby they are weakened. And the suckers also render each other very weak, hence the cause of barrenness. For I have known, where the old plants have been constantly kept clear from suckers, they have continued very fruitful three years, without being transplanted.

"In autumn divest the plants of any strings or runners which may have been produced, and of all the decayed leaves, and clear the beds of weeds. Then the paths should be dug up, and the weeds buried which were taken from the beds, and some earth laid over the surface of the beds, between the plants. This will strengthen and prepare them for the following spring. And if after this, there is some old tanners' bark laid over the surface, between the plants, it will be of great service to them. In the spring, after the danger of hard frost is over, the beds should be forked, to loosen the ground and break the clods. And in this operation the tan which was laid over the surface will be buried, which will be a good dressing to the strawberries. Then if the surface is covered with moss, it will keep the ground moist, and secure a good crop of fruit; and the moss will preserve the fruit clean. When heavy rains fall, after the fruit is full grown, there will be no dirt washed over them, which frequently happens where this is notpractised." *Miller.*

Gypsum is said to produce powerful effects as a manure for strawberry-plants. Straw may be laid under the vines to keep the fruit from the ground instead of moss.

STRAW CUTTER. A machine for cutting straw and corn stalk was invented by Mr. Elihu Hotchkiss, of Brattleborough, Vermont, which was recommended by the Trustees of the Massachusetts Agricultural Society as being the best and cheapest for the purpose known in this country. The Society's premium of seventy five dollars was paid to the inventor, and the patent right for Massachusetts was purchased by the Society. For a description of this machine with drawings, see Massachusetts Agricultural Repository, vol. iv. p. 93.

STRING HALT, a kind of lameness peculiar to the hind quarters of a horse, which occasions a sudden jerking of the legs upwards in his going. When it seizes the outside muscles the horse straddles and throws his legs outwards. But when the inside muscles are affected, his legs are twitched up to his belly. Sometimes it is only in one leg, sometimes in both. The cure is difficult, and rarely accomplished. Rubbing and fomentations are recommended, with daily moderate exercise: By which the blood and spirits may be equally derived into the disordered muscle and its corresponding one. See *Gibson and Clark.*

STUBBLE, the stumps of reaped grain, or the parts of the stems left standing in the field.

When the land is light and sandy, the stubble of wheat and rye should be ploughed into the soil to enrich it. For this will have as good effect as a moderate manuring with dung; especially if it be ploughed soon after the crop is taken off; for then the stubble is in its best state. But after it has been exposed for some time to the sun and wind, it has much less virtue in it to enrich the soil. If this operation be performed in season. the stubble, together with what weeds are growing among it, will be equal to the best green dressing.

But with ploughs of the common construction, the ploughing of stubble ground is disagreeable work; neither can it be well performed. The plough is so apt to choak up, that it is more than one man can well do to keep it clear. Ploughs for this work should be much deeper built than the common ones. And this work might be greatly facilitated, if a heavy roller were passed over the stubble, to lay it flat to the ground before ploughing. When this is doing, great care should be taken to pass the roller the same way that the plough is to go. By means of this, the coulter will but seldom be clogged with the stubble. If this rolling be neglected, a small roller annexed to the fore end of the plough beam, in the place of a foot, or even a foot itself, will greatly help to clear the way for the coulter.

When stubble is on a soil that is stiff, it is not accounted so advisable to plough it in. For such land is not apt to cover the stubble so closely as to cause it speedily to putrefy. It will often lie in a sound unaltered state for a long time, and be very troublesome at the next ploughing.

But if the ground should be seeded after one ploughing, it might be expected the stuble would render the ground so hollow and cavernous as to starve many of the plants that grow upon the surface. At the same time these hollows would be receptacles for noxious insects and vermin. But in a light sandy soil, the stubble is soon reduced to a condition to nourish vegetables.

The better way, therefore, to dispose of the stubble on stiff lands, is, to mow it, collect it, and carry it into the farm yard; where, by the trampling of the cattle, and mixing it with their excrements, it will be converted, before the following spring, into a rich and valuable manure. And it is almost incredible how much a farm may be, in this way, improved and fertilized.

While in the old countries they are under the necessity of making use of part of their stubble, in thatching the roofs of their buildings, and of part of it for fuel, the farmers in this country have the privilege of converting the whole of theirs to manure. And perhaps it may be as well to do so, on whatever soil it is, as to turn it in with the plough; for it is not easy to bury it completely by ploughing; and the part which is not covered is of little or no advantage to the ground.

An English writer states, that "Mr. W. Curtis, of Lynn, Norfolk, found very beneficial effects from burning the stubble of oats, which was left eighteen inches high for that purpose. On a field broken up from old pasture the same year, he afterwards sowed wheat and oats in succession on the same ground the stubble of both which was burned in the same manner. The ashes were in every case ploughed into a small depth, and the verges of the field mowed to prevent accidents. After the third crop of corn, all of which were abundant and remarkably free from weeds, the field was laid down with clover and grass seeds, and the ensuing crops both of hay and grass proved infinitely finer than those before the ground was broken up.

Another piece of land was cropped for three successive years in the same manner as the first, to which it was similar in every respect of soil, aspect and previous management, but in which the stubble was ploughed in, instead of being burned; the produce of each crop on it was much inferior to that of the first experiment, and the weeds increased so greatly, that in laying it down to grass, they quite overpowered the grass seeds, so much so, that it was necessary to re sow it; and ever after, while Mr. Curtis held it, the grass and hay produced were coarse and full of weeds; and consequently inferior both in value and quantity to those of the other field, on which the stubble had been burned."

In burning stubble, the danger which is to be apprehended from the spreading of the flames may, perhaps, be obviated by tracing a furrow round the field, and setting fire to the stubble on the inner edge of the furrow.

STUMP, the part of a tree which remains fast in the ground after felling.

Stumps are very troublesome for some years after trees are removed, unless they be taken out. But doing this is commonly thought to be too much labour, especially when they are of any considerable bigness.

Mr. Evelyn's engine for pulling up large roots, may be thought useful by those who are in haste to have their land thoroughly cleared, and do not grudge the expense of doing it. See *Complete Farmer*, under the article *Stump*.

M. de Turbilly advises to the blowing up stumps with gun powder. But I think my countrymen will hardly go into this method, unless it be in particular cases.

Most of the stumps of hard wood trees will be so decayed in their roots in six or seven years, that they may be easily taken away. So will some of the soft woods. But the stumps of white pine trees, that are large, will last more than half a century. However, when they are well dried, and have somewhat decayed, they may be conquered by fire, where there is plenty of wood to pile on them. But before this is attempted, the earth should be removed from their sides, and if practicable from underneath them, allowing them a few days to dry. A good method of destroying the stumps of white pine trees is, to dig up the smaller ones, and pile them round the larger; and when they have become dry, set fire to them.

The stumps of trees are such hindrances to the operations of tillage, as greatly diminish the profit of farming. Where land abounds with them, they are equal to a heavy tax. This should be considered by those who have the power of taxing new settlements. It is horrid oppression, to tax lands that are full of stumps of trees, equally with the oldest and most improved.

STY, a small house, or hut, in which hogs are kept, or lodged.

Hogs that are not confined should always have a sty, or cot, to lodge in, placed in a convenient situation, and easy of access. It should be very tight, warm and dry, of whatever materials it may be built, and kept well littered: For if swine have not a warm lodging, or are much pinched with the cold, they will be injured in their growth.

That a sty may be the warmer, the door of it should be no larger than is needful for the swine to pass through. And a door should be hung in this passage by the top, that it may be pushed up by their noses, either inwards or outwards, as there shall be occasion. The swine will not fail to open it, and it will shut of itself. See *Hogsty.*

SUBSOIL. The value of a soil depends much upon the nature of the subsoil, or under stratum. A retentive clayey subsoil is injurious to land, which is naturally moist. But a clayey subsoil is serviceable to a sandy soil, by retaining moisture, and supplying what is carried off by evaporation, or consumed by plants. When the surface soil is thin, a stony subsoil is generally prejudicial, unless it be composed of limestone, and then the soil, though thin, is remarkably fertile.

SUCKER, a young twig, or shoot, from the stock, roots, or limbs, of a plant or tree.

SUNFLOWER, *Helianthus*, one of the largest of annual plants, so well known as to need no description.

The sunflower is a native of America; but has been carried into England, where it flourishes. It bears very large discous flowers, and produces a large quantity of black seeds, which are of use for feeding poultry.

The seeds should be sowed early. They will grow in almost any soil. The young plants may be transplanted at any time, before they are six or seven inches high, only observing to take up a ball of earth about their roots. They should be placed in the northern borders of gardens, if in gardens at all, where they will do the least harm to other plants by their shadow; and they should stand not less than two feet asunder. They will rise to the height of a dozen or fourteen feet.

It is said that the seeds of this plant will afford an oil equal to olive oil, and that seventy bushels may be raised from an acre. A bushel of seed, we are told, will yield a gallon of oil as fine as that which we import from Florence; and the mass which remains after pressing out the oil is useful as food for hogs, poultry, &c. This plant gives out great quantities of oxygen gas, or vital air, and, on that account, it has been recommended to cultivate it in unwholesome places. In England it has been applied to other valuable uses. The fibres of the bark have been substituted for hemp; and the stalks, when dried, are used for fuel, the pith affording tinder. The leaves, when green, are a good food for rabbits. It may be raised in rows like Indian corn.—See *Massachusetts Agricultural Repository*, Vol. III. pages 70, 71.

A correspondent of the "American Farmer" states (Vol. II. p. 46), that he planted sunflower seed on the ninth of May, in a rich black soil, drill fashion, rows four feet apart, plants from two to three feet asunder when thinned, amounting to eight square rods. They grew well under a common corn cultivation, ripened, and the seed heads were cut off the stalks and thresher about the 11th of October. The crop was at the rate of ninety bushels to the acre.

SURFEIT, a disease to which cattle,. and horses especially, are liable.

It is produced by various causes; it comes from intense labour, from overheating, and from diseases not well cured.

"A horse is said to be surfeited, when his coat stares, and looks rusty and durty, though proper means have not been wanting to keep him clean. The skin is full of scales and dander, that lie thick and mealy among the hair, and is constantly supplied with a fresh succession of the same for want of due transpiration. Some horses have hurdles of various sizes like peas or tares: Some have dry fixed scabs all over their limbs or bodies; others a moisture attended with heat and inflammation; the humours being so sharp and violently itching, that the horses rub so incessantly, as to make themselves raw. Some have no eruptions at all, but an unwholesome look, and are dull, sluggish, and lazy; some appear only lean and hidebound; others have flying pains and lameness, resembling a rheumatism; so that in the surfeits of horses, we have almost all the different species of the scurvy, and other chronical distempers.

"The following method is usually attended with success in the dry species. First, take away about three or four pounds of blood; and then give the following mild purge, which will work as an alterative, and should be repeated once a week or ten days, for some time.

Take succotrine aloes six drams, or one ounce; gum guaicum half an ouuce; diaphoretic antimony, and powder of myrrh, of each two drams; make into a ball with syrup of buckthorn.

"In the intermediate days, an ounce of the following powder should be given morning and evening in his feeds.

Take native cinnabar, or cinnabar of antimony finely powdered, half a pound; crude antimony in fine powder four ounces; gum guaicum also in powder four ounces; make into sixteen doses for eight days.

"This medicine must be repeated till the horse coats well, and all the symptoms of the surfeit disappear. If the horse is of small value, two or three common purges should be given, and half an ounce of antimony, with the same quantity of sulphur, twice a day, or the alterative balls with camphire and nitre.

"If the little scabs on the skin do not peel off, anoint them with the mercurial ointment; during the time of using which, it will be pro per to keep the horse dry, and to give him warm water. This ointment properly rubbed into the blood, with the assistance of purging physic, has frequently cured these kind of surfeits, without any other assistance.

"The wet surfeit, which is no more than a moist running scurvy, appears on different parts of the body of a horse, attended sometimes with great heat and inflammation: The neck oftentimes swells so in one night's time, that great quantities of a hot briny humour issue forth, which, if not allayed, will be apt to collect on the poll or withers, and produce the poll-evil or fistula. This disease also frequently attacks the limbs, where it proves obstinate, and hard to cure; and in some horses shews itself spring and fall.

"In this case bleed plentifully, avoid externally all repellers, and give cooling physic twice a week; as, four ounces of lenitive electuary, with the same quantity of cream of tartar, with four ounces of glauber salts, quickened, if thought proper, with two or three drams of powder of jalap, dissolved in water gruel, and given in a morning fasting.

"After three or four of these purges, two ounces of nitre made into a ball with honey, may be given every morning for a fortnight; and if attended with success, repeated for a fortnight longer.

"The powders above mentioned may also be given with the horse's corn; or a strong decoction of guaicum shavings, or logwood, may be given alone, to the quantity of two quarts a day. These, and indeed all alterative medicines, must be continued for a long time, where the disorder proves obstinate.

"The diet should be cool and opening; and if the horse is hidebound an ounce of fenugreek seeds should be given in his feeds for a month or longer. And as this disorder often proceeds from worms, give the mercurial physic too, and afterwards the cinnabar powder, as above directed; but as in general it is not an original disease, but a symptom only of many, in the cure, regard must be had to the first cause; thus as it is an attendant on fevers, worms, &c. the removal of this complaint must be variously effected." *Complete Farmer.*

SWAMP, wet sunken land. See the article *Bog*.

SWARD, the surface of grass ground.

The sward is formed of a web of the roots of grass, mixed with the best mould. The common English grass, or poa, forms a very firm and tough sward, which may be cut up and used as turf. Herdsgrass has a weaker sward; and clover, being tap rooted, forms a very weak and and crumbly sward. The first kind bears the tread of cattle, and the pressure of wheels far better than the others.

The sward is wisely contrived by the Author of Nature, for the recruiting and enriching of land. At the same time that it prevents the descending of the food of vegetables too deep into the soil, it is continually collecting it from the at mosphere. So that for a long course of years, some grass lands will afford good crops of hay with little or no manuring.

The sward always contains the richest part of the soil. Accordingly it is always found, that land newly ploughed out of sward is more fruitful than that which has been longer in tillage. Sometimes it will bear as good a crop without dung, as it will in succeeding years, well dunged.

Other things being equal, a sward that is always fed very close does not gather richness so fast as one that has more fogge. In the one case, the vegetable food in the atmosphere is blown away; in het other, it is entangled in the fogge, retained by it, and carried into the soil by rains, and melting snows.

A sward on which cattle lie much, or where swine frequently run. as in the borders of many of our roads, becomes very rich. So that if the surface be pared off, to the thickness of two or three inches, and laid in heaps to ferment, with the grassy sides inward, it will soon become an excellent manure. A dressing of this will surprisingly renew the fertility of an old worn out field. But a mixture of dung with this manure will be no small advantage.

By the sides of fences, a sward gathers richness faster than in the other parts of a field. The vegetable food which floats in the air, the earthy parts especially, may be supposed to fall, for the same reason that snow does, and remain, under the lee of fences. And the banking up of snow in these situations is another cause of enriching the surface of the ground. By lying for some time after the ground is bare, being peculiarly adapted to catch and retain the food of plants, it conveys the more of it into the sward. High and close fences produce these effects more observably than low and open ones But this is to be understood with limitation; for it is observed that a close fence seven or eight feet high has often a much smaller bank of snow under its lee than a fence of but four or five feet.

When the sward of mowing ground binds too much, it should be broken up and tilled. But to prevent binding, it should not be fed very close after mowing; and especially a sprinkling of well rotted compost, applied in autumn, will not only prevent binding, but increase the next crop, to such a degree, that manure cannot be better applied.

SWARTH, *or* SWATH, a line or row or grass, or corn, &c. cut down by the mower.

The double swarth is where the grass of two swarths is thrown together in mowing. It saves a little labour in raking to make the double swarth the centre of a windrow. But when the crop is thin, this is an object unworthy of attention.

SWINE, the name of a species of tame quadruped animals, well known in all countries.

Sowen is the old English plural of *sow*, whence the name swine, which custom has applied to the whole species of hogs.

The keeping of swine is of essential advantage to the husbandman; because they feed much upon things, which would otherwise be of no considerable service to him. The roads, and commons, in the farming towns of this country, afford swine, excepting in winter, the greatest part of their scanty living. For they feed heartily, not only on grass, but many sorts of weeds, the tops and roots of fern, the roots of several kinds of aquatic plants, &c. They pick up grain and seeds that are necessarily scattered about the barn and out houses, besides eating worms and many kinds of insects.

Besides, the farmer's house affords many things which contribute to their support, which would otherwise be lost, such as whey, sour skimmed milk, and butter milk, the washing of tubs and dishes; animal and vegetable food that has accidentally got corrupted, de cayed and rotten fruit, the offal of beasts, fowls and fish, and the grounds of cider, beer and other liquors.

As this animal is much disposed and adapted to do mischief, those that are permitted to go at large, should be well yoked, that they may not break through fences. And whether they go at large or in pastures, they should always be ringed in the nose, to prevent their tearing up the ground too much, in search of roots. When kept in pens, they need ringing also, that they may live the more quietly together, and not tear and wound each other.

Their running at large is, perhaps, not advisable, unless it be in wide roads, or in places where there is a large outlet for them; or where grass, brakes, acorns, or nuts of beech, chesnut, or hickory, are not to be found in plenty. For they are exposed to the more accidents; and in some of our settlements, which border on the wilderness, the bears are apt to catch them. And the most of our public roads are so much frequented by other hungry beasts, that the swine can have but little advantage from the grass.

But in case of necessity, when the farmer happens to have no suitable pasture for his swine, let the wash from the house be regularly given them, morning and evening, to prevent their wandering away too far from home. This will induce them always to lodge at home, especially if they have a good sty to receive them in inclement weather; by means of which they will be the less in danger of being lost, or of straying too far from home.

I am sensible that the method of management I here recommend, is liable to one considerable objection; which is, that if they be fed at all at home, they will be apt to haunt the house all the time, crying for more, and never go way far in quest of food.

Instead of attempting to answer this objection, I wish I could make it appear a sufficient argument to convince farmers of the great advantage, or the necessity of having good enclosed pastures for their swine. They will be safer, and fare better; and the cost of it is not great.

The hog pasture should be so near to the dwelling house, that it may not be troublesome to carry the wash to the swine: And yet so far off that the people in the house may not be stunned with their noise. And a warm cot must be made in some convenient part of their pasture, for them to lodge in.

To prepare a pasture for them, let the ground be broken up, tilled and manured, and then laid down with clover. For swine are more fond of this grass than of any other which our country produces. Let the quantity of land be so proportioned to the number of swine, that they may keep the grass from running up to seed. For this will prevent waste; and the shorter the grass is, the sweeter it will be, and the more tender and agreeable to their palates.

I suppose that one acre of rich land in clover, will support twenty or more swine, large and small together, through the summer; and bring them well forward in their growth, but they should have rings in their noses to prevent their rooting out the clover.

It has been proved, by many trials that hogs, in such a pasture, may be kept in good plight, without any other food. Some say they may be half fattened.

Arthur Young, Esq. of Great Britain, in the summer of the year 1766, pastured sixty four swine of various sizes, on two acres of clover ground. And allowing two pence half penny per week, one with another, their feedings amounted to seventeen pounds, sixteen shillings sterling. Their keeping was set at a low rate, six months feeding for one swine being *5s. 5d.* and the profit of the clover put to this use is astonishing. He assures the public that all these swine grew very fast. And in his opinion, this use of clover is greatly preferable to making it into hay. I think this is not to be doubted, though the crop of hay were supposed to be the greatest that is ever obtained.

It should be remembered, that the pasturing with swine will enrich land more than pasturing with other beasts, and hereby the profit of the farmer will be increased. And if a common clover lay will produce a good crop of wheat, much more may be expected of the same kind of ground, after pasturing swine upon it; as their dung adds much to the fertility of the soil.

Hogs may be turned into their pasture about the first of May, and kept in it till the last of October. And if, in May and October, the grass should not be quite sufficient for their support, some potatoes or other roots may be thrown to them.

The fence about the pasture should be so tight and strong that the swine will not need to be yoked: Because yokes do much towards preventing their growth, as I have found by letting yoked and unyoked ones of the same litter run together in a pasture.

It will be a great advantage to a hog pasture, to have plenty of water in it through the summer. Running water is best, as it will afford them the most wholesome drink, and at the same time, serve as well as any other for them to wallow in; and it will keep them clean, which is no small advantage.

But the most dirty puddle water is better than none, as they can cool themselves in it in hot weather, which is greatly refreshing to them, and conducive to keep them in health.

The best way of managing swine is, to keep them always in middling plight: Not too fat, lest their health should be in danger, especially when the weather is hot: Not too lean, because this is apt to give them a ravenous appetite, and tempt them to eat things that are not wholesome for them. Those that have been long starved, can not be made fat without great expense: Sometimes more than they will repay with their flesh.

When it can with convenience be so ordered, it is an excellent piece of husbandry, to make a hog pasture of an orchard. Their dung is allowed to be the very best of manure for the trees. They will keep the ground light and loose; destroy insects that infest the trees, and feed heartily on the premature apples that fall, which the farmer is too often tempted to grind up for cider. And the shadow of the trees will be very grateful and comfortable to them in summer. An orchard may be prepared with clover as well as any other spot of ground. But it should be remembered that, when the trees in an orchard are young and small, swine should not be permitted to go among them: For there will be danger of their wounding them, and stripping off some of the bark.

I would not wish to have the keeping of swine wholly engrossed by the farmer. In populous towns, where there is so much as room for gardens, every family should keep at least one, to take off the refuse of the kitchen, dish water, &c. which would be otherwise thrown away.

Let him be kept in a pen, or sty, perpetually. For the weeds even of a small garden, thrown into his pen, will be ten times better for him, than all he can pick up in the streets, though there may happen to be a little grass in the bye lanes.

He should be fatted and killed in the fall; and a half year old shoot, bought in, to supply his place. They who keep but one swine, or even two, had better purchase their pigs than attempt to breed them.

It is perhaps of more consequence than some are aware of, to be furnished with the best breed of swine. There has been very little care hitherto, in our country, taken about this matter.

It would far exceed our limits to give a description of the different breeds of this animal. Among the best are the Byfield breed, the Bedford breed, and the recent importations from England, which seem to be a cross of the Chinese breed, with the larger and coarser breeds of Great Britain.

In the breeding of swine care should be taken to have them pigged in the right season, about March or April. These will bear the rigour of the following winter, much better than those which come later: Though a sow will farrow three times in a year, it is better that they should do it only twice.

The above mentioned writers say, "It is common for sows to have thirteen or fourteen pigs at a litter: But the sow can rear no more than she has teats to suckle them with: the rest must therefore be destroyed, or put to other sows. If a sow miss the time of going to boar, that she might in course have done, give her some oats parched in a pan, in her wash, or the small end of a runnet bag, and it will cause her quickly to go to boar. The pigs which you rear, after you have chosen out the best for boars and sows, the males must be gelt, and the sows spayed."

The most eligible food for store hogs in winter ought to be known, regard being had to the cost of the food, and the advantage gained in the growth of the swine. Raw potatoes seem to be the most fashionable food in the northern parts of New England. But if swine are wholly confined to this food, they will but just live, without increasing much in their growth. But boiled potatoes will make them grow and even fatten them. Roasted potatoes are still better for them than boiled, as they are less watery.

From some late experience I have had of feeding them with raw carrots, I judge them to be a more fattening food than even boiled potatoes. The swine are exceedingly fond of them, sometimes preferring them to Indian corn. And I think the cost of raising carrots will not be found greater on the whole, where the soil is suitable for them, than that of raising potatoes.

The feeding of store swine constantly with any kind of corn, seems to be too expensive. Possibly it will be found, upon further trial, that carrots are the best substitute. Red beets are also a good food for them, and parsneps excellent: But turnips and cabbage are improper.

Mr. Young made many accurate experiments in order to find out the cheapest and best method of feeding swine in sties. The result of the whole was, that boiled carrots had the preference.

Sows and pigs on a farm, as he justly observes, should have the benefit that arises to swine from the dairy. The rest of the swine may be fed chiefly on clover in the summer, and on roots in the winter, if this rule be observed, the dairy farmer may keep a very great number of swine to advantage.

But it should not be forgotten, that swine are, in one respect at least, like human creatures, apt to be cloyed when confined wholly to one kind of food. They should therefore be, in some measure gratified with variety. It has been often remarked, that potatoes suit hogs better in summer than in winter. The secret is, that in summer no hog is wholly confined to potatoes; for he gets at least some grass and weeds.

I know not whether it has ever yet been determined, at what age it is best, or most for the owner's interest, that swine should be fatted and killed. That it is in no part of the first year, excepting the China breed, I suppose will be generally granted. For as they have not near attained to their full growth, it cannot be expected that they should be so quickly, or so easily fatted by feeding. But as the young pork is more palatable, some will prefer it on the whole, though it should be more dearly obtained.

The more common practice is, to kill them at about the age of a year and a half. But I suspect the profit would be greater, if they were kept one year longer. For it is well known, that they bear the cold of the second winter much better than that of the first. As their growth is nearly or quite completed, they are the more easily fattened; and I never could perceive but tha the pork was equally good and palatable, as those killed at eighteen months old.

According to the opinion of the Rev. Mr. Eliot, the best time in the year to shut up hogs to fatten them, is the month of August. I rather prefer the month of September, when it may be depended on, that they will not suffer at all by the heat in their confinement: And there will be time enough to make them fat, before the weather comes to be extremely cold.

He that attempts to fatten his hogs in winter will be a loser: For it has been found by long experience, that they do not gain in their flesh near so fast in a frosty, as in a temperate season. I therefore take care to get them fit for the knife by the beginning or middle of December, and I should choose to kill them still earlier. were it not for the advantage of keeping the lean part of the pork for some time without salting; as it most commonly may be done by exposing it to frost, in the coolest part of a house.

But a very important question is, what food and management is best in fattening swine? Pease answer well, when the price of them is low. But I am constrained to give the preference to Indian corn. Let them be fed in September with green ears from the field. There is nothing they will devour more greedily than this corn, and even the cobs with it.

In Indian harvest, the unripe ears should be picked out, and given to the hogs that are fatting, without delay: Or as fast as they can eat them: For it will do them four times as much good in this state, as it will after it is dried, it being difficult to dry it without its turning mouldy, or rotten; so that they will scarcely eat any of it in this state, unless they be kept shorter of food than fatting hogs should be.

After the unripe corn is used, that which is ripened must be given them.

If it be thought most convenient to feed them with corn of the preceding year, it should not be given them without soaking, or boiling, or grinding it into meal. For they will not perfectly digest much of the hard kernels; it being often too hard for their teeth. It has been thought by good judges, that the corn will be at least a sixth part more advantage to the swine, for soaking it in water. But there is, if I mistake not, still more advantage in grinding it. What new corn is given them, may be in ears, as it is not hardened enough for grinding. I know of nothing that will fatten hogs faster than a dough of meal and water. But as this is expensive food, the dough may be mixed with boiled potatoes, or boiled carrots. They eat these mixtures as well as dough by itself; and it appears to make no material difference in their fattening. In this mixture, barley meal will answer instead of Indian: Which should be attended to in our more northern parts, where two bushels of Barley may be as easily raised, as one of Indian corn. Both kinds of meal I have found to be a good mixture with boiled potatoes; but it should by all means be a little salted to give it a good relish.

While hogs are fatting, little or none of the wash from the kitchen should be given them. Their drink should be fair water, which they relish better than any other drink, and of which they will drink a good deal, when they are fed only on corn, or stiff dough.

To prevent measles, and other disorders in hogs, while they are fatting, and to increase their health and appetite, a dose or two of brimstone, or antimony, given them in their dough, is useful, and should not be neglected.

Some change of food may be advisable, in every stage of their existence, as it always seems to increase their appetite. But while they are fatting, laxative food in general should be avoided, as these animals are seldom known to suffer by costiveness, especially when they are full fed, but often from the contrary disorder. If they chance to be costive, a little rye will help them.

In feeding, steady care should be taken that not one meal should be missed, nor mistimed, and their water should never be forgotten. They should always have as much food as they will eat up clean; but never more than that quantity, lest they defile it, and it be wasted. A little at a time, and often, is a good rule.

If their skins be scurfy, or inclining to manginess, a little oil poured upon their backs, will cause it to come off. And some say, a small mess of rye now and then, as a change in their food, is good against these and other disorders.

If the issues in their fore legs should chance to get stopped, every attempt to fatten them will be in vain. These therefore should be watched; and if found to be stopped, they should be rubbed open with a corn cob.

Rubbing and currying their hides very frequently, is of advantage to keep up perspiration. It is grateful to the animals, as well as conducive to their health and growth. A proper scrubbing post in the middle of their pen will not be amiss. And during the whole time of their fatting, they should have plenty of litter. They will lie the more dry and warm, and it will be more than paid for, by the increase of good manure.

When hogs are killed, a single one should not be left to live alone in a pen. He will be apt to pine too much after his former companions: And in cold weather he will suffer for want of lodging so warm as he has been accustomed to do.

The fat part of pork should be plentifully salted with the best and strongest clean salt. It will take three pecks for a barrel. The pork should be kept continually under pickle; for if it be exposed ever so little to the air, it will become rusty and unpalatable. See *Hogsty*.

Boiled or steamed clover hay will serve to keep hogs during winter, but the addition of potatoes or carrots, boiled or steamed with the hay will be an improvement. For the best mode of boiling or steaming, see *Steamboiler*.

Mr. Young directs soiling swine in a yard in preference to feeding them on clover in the field during summer. But Judge Peters, of Pennsylvania, says, "In summer my hogs chiefly run on clover. Swine feeding on clover in the fields will thrive wonderfully; when those, (confined or not) fed on cut clover will fall away." (*Memoirs of the Philadelphia Agricultural Society*, Vol. II. p. 33.) The same gentleman asserts that hogs, while fatting, should constantly have some dry rotten wood, kept in the pen, which they will eat occasionally, and it proves very beneficial to them. It is also declared, as well by that gentleman as other writers, that food when soured by a proper degree of fermentation is much the best for fatting swine, and that one gallon of sour wash will go as far as two of sweet for that purpose. Mr Young says, that the best method of feeding all kinds of grain to hogs is to grind it to meal, and mix it with water in cisterns for that purpose, at the rate of five bushels of meal to a hundred gallons of water.

"Mr. Timothy Kirk, of Yorktown, Pennsylvania, fed one pig with boiled potatoes and Indian corn, and another with the same articles unboiled. The two animals were weighed every week, and the difference between them was as 6 to 9. The experiment was continued several weeks, and the animals alternately fed upon boiled and unboiled food, with an uniformity of result, which sufficiently proved the very great profit arising from boiled food."—*Domestic Encyclopedia.*

We shall here take notice of some of the diseases to which swine are liable, and point out their remedies.

The *Mange* is an eruption in the skin, caused by the want of cleanliness in the hogstye. It causes so great an irritation, that the animals rub themselves till a scab is produced and in a short time ulceration ensues.

Dr. Norford (*Annals of Agriculture*, Vol. XV.) recommends the following ointment, which seldom fails to effect a perfect cure, provided it be properly applied, and the animals kept clean after the disease is removed. Take three ounces of hog's lard, one ounce of fine flour of sulphur, two drachms of white hellebore, newly pulverized, and half an ounce of the water of kali, prepared in the shops. These ingredients are to be thoroughly incorporated, so as to form an unguent; the whole of which is directed to be rubbed on to the animal at one time, and is said to be sufficient for a hog of six or seven stone: if the ointment be properly applied, there will be no need of a repetition. If the hog has a slight cough, doses of antimony from half an ounce to an ounce and an half, according to the size of the animal, should be finely pulverized, and mixed with his food, for ten days or a fortnight. When from long neglect, the neck, ears, and other parts become ulcerated, they should be anointed every third or fourth day with an ointment made of equal parts of tar and mutton suit, melted together, till the cure is completed.

The *Murrain* is known by the animal's hanging down his head which is swollen; short and hot breathing; palpitation of the heart, staggerings, and an abundant secretion of viscid matter from the eyes.

The 36th volume of *Annals of Agriculture*, gives the following remedy: A handful of nettles is to be previously boiled in a gallon of small beer, when half a pound of flour of sulphur, a quarter of a pound of elecampane, three ounces of liquorice, and a quarter of a pound of aniseeds are to be added in a pulverized state. This preparation should be administered in milk, and the quantity here stated is said to be sufficient for six doses.

Staggers.—In this disorder the animal turns round rapidly, and if not assisted will die in half an hour.

Remedy.—You will see a bare knob in the roof of the mouth. Cut it and let it bleed. Take the powder of loam and salt, rub it with it, and then give the hog a little urine. See a letter from J. P. De Gruchy to Hon. Richard Peters, published in "*Memoirs of the Philadelphia Agricultural Society*, Vol. II, p. 28.

T.

TAIL SICKNESS, a distemper attended with weakness and sluggishness, to which horned cattle are liable in the spring. The end of the tail becomes hollow and relaxed, but not, as some have asserted, destitute of feeling. A cure is easily effected, by the amputation of a small piece of the tail, which will be attended with a discharge of some blood. But when the tail is but little affected, and near to the end, a slit of one inch, or an inch and a half, in the end of the tail, is preferable to amputation.

TEAM, the beasts that are used together in draught.

The right ordering and management of a team is a matter of no little consequence to the farmer, not only in ploughing and harrowing, but in carting and sledding.

In some countries, teams consist only of horses, but in a new country, where roads are bad, and the ground full of obstacles, this will not answer. A horse team travels so much quicker than oxen, that where there are many obstacles, no harness will hold them; and violent jerks would injure and discourage if not ruin, the horses.

But in fields that are perfectly cleared, ploughing with horses may be performed with much greater despatch and advantage. An acre and a half is but a moderate day's work for a team of horses, whereas an ox team seldom does more than an acre. And there is a proportionable advantage in teaming on a good road with horses.

But when it is considered how much more expensive the support of horses is, than that of oxen, and that an ox, when past labouring, is valuable for beef, the general use of horses in the draught is not to be accounted eligible, in any country. Mr. Marshall computes that a million sterling annually is lost in Britain, by the using of horses for draught instead of oxen: And that a hundred thousand persons might be supplied with a pound of animal food per day, without consuming one additional blade of grass, if oxen were used in general.

The slowness of oxen is partly natural, and partly acquired. That excessive slowness may be prevented, care should be taken never to overload them. When their work is easy, they may be quickened without danger of hurting them; and their contracting a habit of moving slowly may be thus prevented.

It is also found, that old oxen are always apt to be slower than young ones. It is not advisable, therefore, to continue to work them till they are old; but to turn them off at six or seven years old, at which age they will be better beef than older cattle.

Perhaps a team consisting partly of oxen and partly of horses, may be found to be, upon the whole, of the greatest advantage, especially among small farmers. For they find it needful to keep one horse or more, which will be idle most of the time, unless they work him in the team with the oxen.

The methods of harnessing a team, are so well known to farmers by their experience, that they would perhaps scorn to be instructed in it. But I would recommend it to them to take it into consideration, whether improvements might not be made in this article of rural economy. Particularly, whether the hard and heavy wooden yokes with which oxen are worked, should not give place to a mode of harnessing similar to that of horses. Collars, to open and shut at the top, with hames and traces, for oxen, are used in England, and have been recommended by some of its ingenious writers. But if this advise should be generally followed in this country, the use of two-wheeled carts must be laid aside, the wooden yokes being necessarily connected with them. See the article *Waggon*.

TEASEL, CARDUUS FULLONUM, or *Fuller's Thistle*, *Dipsacus*, a species of thistle, the heads of which are of use to raise the nap on woollen cloth. They are fitter to work on fine than on coarse cloths.

This is a plant which ought to be cultivated in this country, in order to facilitate and improve the manufacture of woollen. And from some trials that have been made it appears that it may be done without difficulty.

Mr. Miller says, "This plant is propagated by sowing the seeds in March, upon a soil that has been well prepared." Any time in April will answer in this country. "About one peck of seed will sow an acre; for the plants should have room to grow, otherwise the heads will not be so large, nor in so great quantity. When the plants are come up, they must be hoed in the same manner as is practised for turnips, cutting out all the weeds, and singling out the plants to about eight inches distance. And as the plants advance, and the weeds begin to grow again, they must be hoed a second time, cutting out the plants to a wider distance; for they should be left, at last, a foot asunder, and should be kept clear from weeds, especially the first summer: For when the plants have spread so as to cover the ground, the weeds will not so readily grow between them. The second year after sowing, the plants will shoot up heads, which will be fit to cut about the beginning of August; at which time they should be cut, and tied up in bunches, setting them in the sun, if the weather be fair; but if not, they must be set in rooms to dry them. The common produce is about a hundred and sixty bundles, or staves, upon an acre, which they sell for one shilling a stave."

Those who would see a more particular account of this useful plant, may consult the *Complete Farmer*.

TETHER, *or* TEDDER, a rope with which a horse is tied in the field. He may be fastened by the neck, or the foot, or head, shifting his situation as often as needful. Where grass grows in part of a lot, with a tillage crop growing on part of it, it is often convenient to feed off the grass in this way.

THATCH, straw laid on the top of a building, or stacks, to keep out the weather. In this country it is used only for stacks, sheds and hovels.

THILL HORSE, the hindermost horse in a team, which goes between the thills or shafts.

THISTLE, *Carduus*, a prickly weed found in tillage and pasture lands. It indicates a rich soil; but it is a very troublesome weed, as it exhausts much of the strength of the soil, and easily propagates itself far and wide, by its downy seeds, which are wafted by the wind to considerable distances.

An effectual way to subdue thistles is, to pull them up by hand while they are small, or before the usual time of cutting them. It may be easily done after a considerable rain, while the ground is soft by being moistened. But the operator must have his hands well defended by gloves of stout leather. They may be taken out at any time with a strong spade. Or thistles may be cut up in autumn, about the beginning of September, and burnt before they have scattered their seed. A thorough tillage of the land, or a good summer fallowing, will subdue them.

But it signifies little for one farmer to destroy the thistles in his ground, so long as they are suffered to flourish in contiguous fields; because the seeds are wafted in the air from field to field, to a considerable distance. For this reason it is, that in France, a man may sue his neighbour for neglecting to thistle his ground in the proper season; or may employ people to do it at his expense.

Not only the fields, but the borders and hedges around them, should be cleared of thistles, or a complete conquest over them will never be obtained. And this is the most difficult part of the work.

Thistles, cut an inch above the ground, will not be so difficult to subdue as those cut at the same time with the hoe, and below the surface. In the former case, the remaining stub of the thistle gets filled with water, which rots and destroys the plant.

THRASHING, beating out corn, seeds, &c.

This is commonly done with a flail. But it is greatly suspected that many seeds are so bruised by this instrument, that they will not vegetate. That which is to be sowed, should therefore be extricated from the heads in some better method. Striking the grain by handfulls, against a beam or post, is recommended. Some sorts of seeds should be beaten out with a staff, or a slender rod.

Turning a heavy wooden wheel upon grain, by a horse, might be the means of saving a good deal of labour. This is the common method of getting out the seeds of clover: And it might answer equally well for any other kind of smooth and slippery seeds. The wheel holds the chaff in its place and drives the seeds before it.

An engine was once made by a gentleman in the state of New-Hampshire, consisting of a broad framed wheel, filled with heavy fliers, which, as the wheel turned, fell out forwards upon the grain, with considerable force. There was thus two kinds of force at once applied to the grain, the pressure of the wheel, and the strokes of the fliers.

The ancient practice of treading out corn with oxen and horses must have been exceedingly inconvenient. It is no wonder that it has been long disused in most places.

When the work is performed, with a flail or otherwise, the workman should either be bare footed, or have soft shoes, or Indian mockasins on his feet, that he may not crush the corns by treading on them. Wheat, when it is new, is particularly liable to be crushed.

Mr. Mortimer thinks it a day's work for a man with a flail to thrash four bushels of wheat, or rye; six of barley, or five of oats, beans, or pease. But Mr. Lisle says, a good thrasher assured him, that twelve bushels of oats, or barley, are reckoned a good day's thrashing and five or six bushels of wheat. But no certain rule can be given, by reason of the difference in grain. Every one knows that large plump grain is more easily thrashed, than that which is poor and blighted.

This work should be performed when the weather is dry, both on account of the ease of the labourer, and the grain itself, which will keep the better afterwards.

The beards of barley come off the more easily in thrashing, when the swaths of this corn have taken the dew before it is housed. It often requires much thrashing after it is extricated from the straw. It will keep well in a mow unthrash ed, for a year, or longer.

Beans and peas always thrash best after sweating in the mow, which they are apt to do. After kiln drying, or drying in the sun, they will keep a long time in the granary, though laid ever so thick.

Threshing Machines, of various kinds have been invented and put in use since Mr. Deane wrote. Sir John Sinclair says that the threshing machine is considered to be the most valuable implement that modern times have pro duced, and states the following advantages as resulting from them.

"1. From the superiority of the mode one twentieth part more is gained from the same quantity of straw than by the old fashioned method. 2. The work is done much more expeditiously. 3. Pilfering is avoided. 4. The grain is less subject to injury. 5. Seed corn can be procured without difficulty from the new crops. 6. The market may be supplied with grain more quickly in times of scarcity. 7. The straw, softened by the mill is more useful in feeding cattle. 8. If a stack of corn be heated, it may be threshed in a day and the grain preserved from injury. 9. The threshing mill lessons the injury from smutty grain; and 10. By the same machine, the grain may be separated from the chaff and small seeds, as well as from the straw. Before the invention of threshing mills, farm servants and labourers endured much drugery; the large corn farmer sustained much damage from bad threshing; and had much trouble, vexation and loss from careless and wicked servants; but now, since the introduction of this valuable machine, all his difficulties, in these respects are obviated."

The Edinburgh Encyclopedia, New York edition, gives a description of a threshing machine, invented by Mr. Mickle, which, on a large scale, and driven by water, threshes, fans and cleans wheat, at the rate of sixty bushels an hour. On a smaller scale, worked by two horses and three hands to attend to it, this machine will thresh and clean sixty bushels of wheat, or double that quantity of oats in eight hours. Rollers, or small mill stones are added to many of these machines for crushing or grinding grain. Knives for cutting hay might be added.

Mr. Elihu Hotchkiss, of Brattleborough, Vermont, has invented, and taken a patent for a threshing machine, which obtained a premium from the Massachusetts Agricultural Society, in 1818. The Society likewise purchased the patent right for Massachusetts, and it may be found among the implements belonging to that Society.

TIKE, a most disagreeable insect, bred in the woods, and found on sheep, dogs, cattle, &c. By drawing nourishment from animals, they swell themselves up to a large size. See *Insect*.

TILLAGE, the work or business of tilling, or working the ground, including ploughing, hoeing and harrowing. See those articles.

The same field should not be kept in tillage perpetually. It answers a better purpose to lay a field sometimes to grass. No rotation of tillage crops can be so good, as to have some years of grass taken into the course. It makes the varity greater, and it checks the increase of certain insects, as well as destroys some kinds of weeds.

High lands are usually selected for tillage. But low and moist lands, well drained, ridged, and water furrowed, are often more productive; and there is less danger of exhausting them, and reducing them to a barren state.

A correspondent of the editors of the *Museum Rusticum*, says, "The inhabitants of Market Weighton have five fields, two of a sandy soil, and three of a strong clayey soil: The two former destined to rye, and the others to wheat. Their sandy lands are disposed in ridges, lands, or beds to four swaths breadth: And finding by experience, that considerable parts of each land, towards each furrow, are starved by the coldness of the water dripping from the higher parts of the lands, they have for many years altered their former method: And only ploughed the half of each land, viz. the two middle swaths: So that they have now excellent rye growing on the higher and drier half of every land, and excellent meadow on the lower and wetter half, which being just two swaths, is mowed with great ease and exactness.

"It will perhaps be thought by some, that by making narrower lands, they might have more dry land, and consequently more corn. But I apprehend that these industrious husbandmen find by experience, that when they make their lands narrower, and consequently with less descent, the water stagnates in the higher parts, and consequently spoils their whole crop. Nor could they, I suppose, sow more corn on their lands, in their present disposition, with convenience, as their present method allows them just one swath on each side.

"They have rye and meadow in one of their two sandy fields every other year, and a fallow the next year. The saving half of the field in grass affords good grass for their sheep, &c. in that year; and allows them to keep a good stock thereon; and this stock, in return, manures the ground considerably, both the fallow and the swath. One of their fields affords them plenty of spring corn; and thus they are supplied with wheat, rye, spring corn, meadow and summer grass, from their five fields, which in any other management they could scarcely be." *Museum Rusticum*, Vol. VI. page 83.

To TILLER, to spread, or mulply shoots.

Grain that is sown thin, in a good soil, will produce a considerable number of stalks and ears from one root. Winter grain will tiller more than that which is sown in the spring; for which reason spring grain should be sown thicker.

TILTH, the state in which ground is after tilling. When it is well pulverised, and made light to a sufficient depth, it is said to be in good tilth.

TIMBER, wood for building, mechanism, &c.

It has been and still continues to be a very general opinion, that timber should be felled in the winter, while the sap is down.

This has been for many years past disputed, and an abundant mass of evidence has been produced to shew that timber should be cut or felled in mid summer, with a view to its durability. It is so very much opposed to vulgar opinion that it requires much inquiry, and frequent experiments to settle it. But it is proper, in such a work as this, to announce the modern opinion, in order to induce farmers to make the experiment.

When the bark of timber trees is wanted for use, the bark may be stripped off from the trunks in May or June, while the trees are standing, and the trees felled in autumn following, or rather in winter.

M. Buffon, of the Royal Academy of Paris, has proved by a variety of experiments, that the timber of trees thus barked is more solid, hard, weighty and strong, than the timber of trees felled in their bark, and thence concluded, with probability, that it is more durable. I suspect that the increased strength and solidity of the wood were in the blea, or what is vulgarly called the sap of the tree. The surface so exposed will soon be too hard to permit the entrance of any worm. The saving of the bark for tanning is sometimes so important an object, where wood has become scarce, that it may well be worth while to go into this practice. It is doubtless better than the practice in England, where felling begins about the end of April; a statute requiring it to be done then, that the bark may be saved for the advantage of tanning leather.

The ancients chiefly regarded the age of the moon in felling their timber. Their rule was to fell it in the wane, or four days after the new moon, or sometimes in the last quarter. But this is of little consequence. For the sap will be down when the weather continues day and night to be frosty, be the moon's place as it may.

Timber should be cut at the right age. For if the trees be too young, or too old, the timber will be less durable. It is said that oak trees should not be cut under sixty years old, nor above two hundred. Trees should however be cut in their prime, when almost fully grown, and before they begin to decay; and this will be sooner or later according to the quality of the soil, or the nature of the cli mate.

While timber is seasoning, it should not be much exposed to the weather, or the heat of the sun, that it may not dry too rapidly, and spring, warp, or crack. Neither should pieces be piled too close on each other, but kept apart by short sticks across, to prevent mouldiness and decay, by permitting the air to pass through the pile. The same observations will apply to all kinds of valuable boards.

"In order to preserve timber from cracking while seasoning, let it be blocked out for the purposes wanted, and laid in a hay-mow, when the hay is carted in. When the hay is dealt out the next winter, the timber may be taken out, well seasoned, and free from cracks. This is an excellent plan for seasoning all kinds of timber for carriages, &c. When this is to be done, if the trees be felled in the winter, let them lie in logs, until hay time arrives."—*Farmer's Assistant.*

It is published in the 'Memoirs of the Philadelphia Society, for the promotion of Agriculture,' that satisfactory experiments have proved that timber used for posts will last considerably longer, by setting the end in the ground, which was uppermost as it grew.

TIMOTHY GRASS. *See grasses.*

TOBACCO. *Nicotiana*, a well known narcotic plant, which has become very important, in Europe and America, since Sir Walter Raleigh imported it into England, and is very generally used, in one way or other, by persons of both sexes. It need not to be told how much many abuse themselves with it. It has its name from Tobago, one of the Caribbee islands.

To cultivate this vegetable successfully, burn the surface of a piece of ground, as early as possible in the spring; rake it well, and sow the seeds pretty thin. Or if the goodness of the seeds be suspected, they may be sown a little thicker. When the leaves are as large as the nails of one's fingers, the plants may be removed.

A sandy or gravelly soil should be chosen, as most suitable to the nature of this plant, which requires much heat, and is not apt to suffer by drought; and a southern exposure is best.

The ground should be prepared for the plants, by several ploughings and harrowings; and be made rich by folding. Or, if the folding be omitted, old dung of the hottest kinds should be put in the holes.

But the common way of raising tobacco in cow pens, and barn yards, without pulverising the soil, is detestable. The taste of such tobacco is intolerable.

The ground being well prepared, let the young plants be transplanted into it in a wet day about three feet asunder, or three feet and a half.

After this it will be needful to keep the ground clear of weeds, and the plants should be daily viewed and examined, to clear them of the worms that eat them; for there is a sort which have a voracious appetite for this kind of food, though it is a sovereign antidote to all other insects.

The tops of the plants should be broken or cut off, at the height of three feet, or a little less or more, according to the greater or less vigour of the different plants ; those excepted which are to bear seed, which should be some of the stoutest and most thrifty. It should be done so early in the summer, as to allow time for the upper leaves to grow to the same size as the lower ones ; of which the cultivator must be his own judge, attending to circumstances. All the plants should be topped about the same time, let their height be greater or less ; for if this be done too late, though there will be a greater number of leaves, the tobacco will be of a worse quality, nor will the quantity in weight be much, if at all, increased, because the leaves will be thinner and lighter.

The suckers which shoot out at the foot stalks of the leaves, should be broken off as often as they appear ; that so the leaves may have all possible advantage of the sap of the plants to perfect them.

The maturity of tobacco is known by certain small dusky spots appearing on the leaves. When it is in this state, it should be cut down carefully with a strong knife, below the lower leaf, on the morning of a sunny day, and the plants laid singly in the sun to wither, which if they do not sufficiently in one day, must be in the same manner exposed the next day.

Being entirely withered, the plants should be laid in close heaps, in the barn, or some other building, to sweat, for the time of forty-eight hours at least. After which let holes be made with a gimblet in the lower ends of the stems, and the plants connected by two and two, with sticks about eight inches long thrust into those holes ; then hang them upon smooth poles, placed about sixteen inches apart, in an apartment which is pretty tight.

As the tobacco turns dry and brown, the plants should be slipped nearer together on the poles. But this should be done only when the air is damp, and when the leaves do not crumble. It is often found convenient to reduce them to close order to make room for the remainder of a crop, which will be ripe later.

From the roots of plants which are cut early, suckers will arise, and give a second crop ; but it will be of an inferior quality. It may stand out till late in autumn, as a small degree of frost will not injure it.

When the tobacco has hung till all the greenness is gone out of the leaves, and at a time when the air is damp, the leaves should be stripped from the stalks, tied up in hands, packed in casks or chests, well pressed down, and kept in a dry place. But by no means in a cellar, which would soon spoil it. It will not be so fit for use the first year as afterwards.

That is the best tobacco which is raised with the least assistance from manure. And, as high manuring is required when it is cultivated in cold climates, I cannot wish to see many attempts to do it, in any place that is north of the forty second degree of latitude. In cold countries, the leaves are apt to be very thin, and so weak that they will scarcely hold together in the curing; and it is far from being so well flavoured as that which comes from the southern states; from whence I wish it may be always imported. For I suppose we ought in general to cultivate only those vegetables, to which our climate is most suitable.

TOP DRESSING. Dung or other manures, spread over the surface of the ground, for the nourishing of plants that are growing in it. These manures should be well pulverised, that they may be spread evenly.

Top dressings are used with advantage, for grain, grass, flax, &c. The timing them judiciously is a matter of no small importance. They should not be too freely given to winter grain in autumn, lest they unseasonably produce a luxuriant growth, at a time when it exposes the tender plants to be the more injured by frost. The right time to give this culture to grain, is just at the season when it is earing; for then is the time when it seems to require the greatest supply of nourishment.

If the application of top dressings to mowing grounds were generally practised in this country, and yearly repeated as it ought to be, instead of the present general, or rather, universal neglect of it, it would put a new face upon things. A vast plenty of hay, double crops, two cuttings in a year, and much increase of wealth to farmers, and the country in general, would soon appear to be the happy consequences.

The materials used for top dressings are numerous, and various. See the article *Manure*.

There is scarcely any question, on which farmers are more divided, than as to the policy of applying manure as a top dressing to grass lands, in the spring or fall. The reasoning seems to be in favour of spring dressing, and it is supported by many excellent names. But, it ought to be known, that intelligent farmers, near the metropolis, most generally dress their lands in autumn. Besides, the reason stated above, that grass lands are less injured by carting over them in the fall; it may be added, that it is a season of greater leisure, and although it is confidently asserted, that the manure is wasted by rains and snows, yet much ought to be allowed on the other side, for the protection afforded by the top dressing, to the tender roots of the plants during winter, and ought we not to add something for the low temperature of the atmosphere in winter, which prevents evaporation? Whatever principles of fertility exist in manure, are in winter carried down into the soil. We are fully convinced, that a scorching sun, and drying air, are more pernicious to manures, spread thinly over the surface than any drenching rains can be, unless on declivities, where top dressings are unquestionably of less value than on level grounds. The fact, that farmers who grow rich by supplying the great towns with hay, generally adopt the practice of fall dressing their grass lands, deserves weight.

Sir John Sinclair recommends top dressing the growing crop, when it is suspected that the land is not rich enough to bring a full crop to perfection, and says, 'this should be done early in the spring, when the land is sufficiently dry to bear the treading of a horse without poaching; and after the manure has been applied, the land should generally be harrowed or rolled. Soot, ashes, and other light manures, are thus most advantageously made use of.'

TRANSPLANTING. Removing plants from their seed bed, and setting them in other places, where they will have better room to perfect their growth.

Some vegetables indeed, need to be transplanted twice, especially some kinds of trees, first from the seed bed into the nursery, afterwards from the nursery into orchards, groves, forests, &c.

The first thing in the latter transplanting of trees is, to have the ground prepared before the trees are taken up, that so they may remain out of the earth as short a time as possible; the next is, to take up the trees. In doing this, carefully dig away the earth round the roots, so as to come at their several parts to cut them off: For if they are torn out of the ground without care, the roots will be broken and bruised, to the great injury of the trees. When you have taken them up, prepare them for planting, by pruning the roots and heads. All the small fibres are to be cut off, as near to the place from whence they are produced as may be, excepting perhaps when they are to be replanted immediately after they are taken up. But it will require great care to plant them in such a manner as not to distort, or entangle the fibrous roots, which, if done, will be worse for the plant than if they were cut off. Then prune off all the bruised or broken roots, all such as are irregular, and cross each other, and all downright roots, especially in fruit trees. Shorten the lateral roots in proportion to the age, the strength, and nature of the trees; observing that the walnut, mulberry, and some other tender rooted kinds, should not be pruned so close as the more hardy sorts of fruit and forest trees. In young fruit trees, such as pears, apples, plums, peaches, &c. that are one year old from the time of their budding or grafting, the roots may be left only about eight or nine inches long: But in older trees, they must be left of a much greater length. But this is only to be understood of the larger roots.

"The next thing is the pruning of their heads, which must be differently performed in different trees; and the design of the trees must be considered. Thus, if they are designed for walls or espaliers, it is best to plant them with the greatest part of the heads, which should remain on till they begin to shoot in the spring, when they must be cut down to five or six eyes, at the same time taking care not to disturb the roots. But if the trees are designed for standards, you should prune off all the small branches close to the place where they are produced, as also irregular ones, which cross each other; and after having displaced these branches, you should also cut off all such parts of branches as have by any means been broken or wounded; but by no means cut off the main leading shoots, which are necessary to attract the sap from the root, and thereby promote the growth of the tree.

"Having thus prepared the trees for planting, you must now proceed to place them in the earth: But first, if the trees have been long out of the ground, so that the fibres of the roots are dried, place them eight or ten hours in water, before they are planted, with their heads erect; and the roots only immersed therein, which will swell the dried vessels of the roots, and prepare them to imbibe nourishment from the earth. In planting them, great regard should be had to the nature of the soil; for if that be cold and moist, the trees should be planted very shallow; and if it be a hard rock or gravel, it will be better to raise a hill of earth where each tree is to be planted, than to dig into the rock or gravel, and fill it up with earth, as is too often practised; by which means the trees are planted, as it were in a tub, and have but little room to extend their roots.

"The next thing to be observed is, to place the trees in the hole, in such a manner that the roots may be about the same depth in the ground, as before they were taken up: Then break the earth fine with a spade, and scatter it into the hole, so that it may fall in between every root, that there may be no hollowness in the earth. Then having filled up the hole, gently tread down the earth with your feet, but do not make it too hard; which is a great fault, especially if the ground be strong and wet.

"Having thus planted the trees, they should be fastened to stakes driven into the ground, to prevent their being displaced by the wind, and some mulch laid on the surface of the ground about their roots. As to such as are planted against walls, their roots should be placed about five or six inches from the wall, to which their heads should be nailed, to prevent their being blown up by the wind." *Dict. of Arts.*

The smaller species of vegetables, and particularly annuals, are removed but once, if at all. A rainy or damp season, if such a one happens, should be chosen for this operation, as the plants will need the less watering by hand, or sheltering from the heat of the sun.

But when it is found necessary to do it in dry weather, the evening should always be preferred to the morning, as the coolness and dampness of the night will do much to prevent the withering of the plants.

Plants which are only to be carried a few steps should be removed with a gardener's trowel, for when a good ball of earth is taken up, and put into a hole with a plant, the roots are but little disturbed, or altered, and the plant not at all affected by its removal.

Mr. Forsyth says "in transplanting trees, especially large ones, I consider it to be of great consequence, that they be placed in the same position, (that is, having the same parts facing the same points of compass) as formerly. If you take notice when a tree is cut down, you will find that three parts in four of the growth are on the north side."

For the mode of transplanting turnips, which will apply to cabbages, &c. See article *Turnip*.

TREE, a large vegetable, with one woody stem, arising to a considerable height.

Trees are distinguished into evergreens and deciduous; the former hold their leaves during the winter, the latter shed them in autumn. They are otherwise distinguished into standards and dwarfs; and again into timber and fruit trees.

TROWEL, a tool which is of great use and advantage, in gardening; especially in transplanting small and tender plants, as by taking up a ball of earth about their roots, it prevents injury to the plants. It is made like the trowel used by bricklayers, excepting that it is hollowed into the shape of a large gouge.

TUMOUR, "a preternatural swelling in any part of a hòrse, arising from external injuries, or internal causes.

"Swellings caused by external accidents, as blows and bruises, should at first be treated with restringents. Let the part be bathed frequently with hot vinegar or verjuice, and, where it will admit of a bandage, let a flannel wetted with the same be rolled on. If by this method the swelling do not abate, apply, especially to the legs, a poultice of red wine lees, or beer grounds, and oatmeal; or vinegar, oil and oatmeal; either of these may be continued twice a day, after bathing, till the swelling abates; when, in order to disperse it entirely, the vinegar should be changed for camphorated spirit of wine, to four ounces of which may be added one of spirit of sal ammoniac; or it may be bathed with a mixture of two ounces of crude sal ammoniac, boiled in a quart of chamber ley, twice a day, and rags dipped in the same may be rolled on.

"Fomentations made by boiling wormwood, bay leaves, and rosemary, and adding a proper quantity of spirits, are often of great service to thin the juices, and fit them for transpiration; especially if the injury has affected the joints.

"But in bruises, where the extravasated blood will not by these means be dispersed, the shortest way is to open the skin, and let out the grumes.

"If the swelling fixes under the jaws, behind the ears, on the poll, withers, or in the groins and sheath, &c. it should be encouraged and forwarded by ripening poultices, wherever they can be applied. Oatmeal boiled soft in milk, to which a proper quantity of oil and lard is added, may answer this purpose, applied twice a day, till the matter is perceived to fluctuate under the fingers, when it ought to be let out. For which purpose, let the tumour be opened with a knife or strong lancet, the whole length of the swelling, if it can be done safely, for nothing contributes so much to a kind healing, as the matter's having a free discharge, and the opening being big enough to dress to the bottom.

"Pledgets of tow spread with black or yellow basilicon (or the wound ointment) and dipped in the same, melted down with a fifth part of oil of turpentine, should be applied to the bottom of the sore, and filled up lightly with the same without warming. It may be thus dressed once or twice a day, if the discharge is great, till a proper digestion is procured, when it should be changed for pledgets spread with the red precipitate ointment, applied in the same manner.

"Should the sore not digest kindly, but run a thin water and look pale, foment as often as you dress with the above fomentation; and apply over your dressing the strong beer poultice, and continue this method till the matter grows thick, and the sore florid.

"The following ointments will generally answer your expectations in all common cases, and may be prepared without, as well as with the verdigrise.

Take Venice turpentine and bees wax; oil of olives one pound and a half: yellow rosin twelve ounces; when melted together, two or three ounces of verdigris finely powdered may be stirred in, and kept so till cold, to prevent its subsiding.

Take of yellow basilicon, or the above ointment without verdigris, four ounces; and red precipitate finely powdered half an ounce; mix them together cold, with a knife or spatula.

"This last, applied early, will prevent a fungus, or proud flesh, from shooting out; for if you dress too long with the above digestive, the fungus will rise fast and give some trouble to suppress it; when it will be necessary to wash the sore as often as you dress, with a solution of blue vitriol in water, or to sprinkle it with burnt alum and precipitate. If these should not be powerful enough, touch with a caustic, or wash with the sublimate water, made by dissolving half an ounce of corrosive sublime in a pint of water.

"But this trouble may in great measure be prevented, if the sore is on a part where bandage can be applied with compresses of linen cloth; for even when these excrescences regerminate, as it were, under the knife, and spring up in spite of the caustics above mentioned, they are to be subdued by moderate compression made on the sprouting fibres by these means." See more on this subject in *Bartlet's Farriery*, page 236.

TURF, a clod filled with grass roots, taken from the surface of the ground.

That which is used as fuel in some countries, is properly the sward of a wet and boggy soil, and consists of a sulphureous earth, and the roots of aquatic vegetables.

In Flanders, they pare their turf from the surface of the earth, and cut it in the form of bricks.

The Dutch take their turf from the bottom of the canals which divide their lands; by means of which they keep their dikes clear and navigable.

In the north of England and Scotland, turf is dug out of soft, moist, rotten earth, which they call *peat moss*. It is decayed moss mixed with moory earth, and aquatic grass roots.

Some writers confound turf with peat, as if they were the same substance. Peat consists of decayed wood, large trees in a sound state being often found in a peat soil, and those that are changed into peat retain their shape. Nut shells and leaves are observed in it, which indicate that peat was originally wood. Turf is therefore quite a different substance, of much less value as fuel; and yields a weaker kind of ashes.

Turf also differs from peat, as in places where turf is cut out it will in some years be renewed; but this is not the case with peat, which being once dug out is never renewed.

TURKEY, a large domestic fowl, brought from Turkey, and is called by the name of its country.

As many of them are reared in the farming towns in this country, I shall here give directions from a good writer, how it may be done with success.

" Most of our housewives, says a Swedish author on husbandry, have long despaired of success in rearing turkies; and complained that the profit rarely indemnifies them for their trouble, and loss of time: Whereas, continues he, little more is to be done than to plunge the chick into a vessel of cold water, the very hour, or if that cannot be, the day it is hatched, forcing it to swallow one whole pepper corn, and then restoring it to its mother. From that time it will become hardy, and fear the cold no more than a hen's chick. After which it must be remembered, that these useful creatures are subject to one particular malady whilst they are young, which carries them off in a few days. When they begin to droop, examine carefully the feathers on their rumps, and you will find two or three, whose quill part is filled with blood. Upon drawing these the chick recovers. and after that requires no other care than what is commonly bestowed on poultry that range in the court yard.

" These articles are too true to be denied; and in proof of the success, three parishes in Sweden have, for many years, gained several hundred pounds by rearing and selling turkeys."—*Rural Economy*, page 739.

Buck wheat is accounted a good food for turkeys; but insects contribute much to their living in summer. When grasshoppers are plenty, they will fatten upon them.

R. Weston, Esq. recommends fattening turkeys with walnuts, given them whole. See his *Tracts*, page 190.

TURNIP, a white esculent root.

The sorts, according to Mr. Miller, are three; the flat, or round shaped turnip, the long-rooted, and the French turnip.

Of the first sort some are green topped, others red purple topped; the yellow; and the early Dutch turnip, which are not of so much value. The last sort is sown early in the spring, to supply the markets in the beginning of summer. The green topped turnip is preferred to the rest, as it grows to a larger size.

Turnips love a light sandy or gravelly soil, or a sandy loam. It should be made soft and fine, but not too rich, lest the turnips be rank and ill tasted.

Ground that has been newly cleared, yields the largest and sweetest turnips; and on such a spot there is the least danger from insects.

Next to new land, swarded ground is to be chosen for a crop of turnips; and the way to prepare it is, to plough it pretty deep in the spring, and fold it by turning in the stock for a good number of nights. For there is scarcely any of our fields sufficiently rich to produce turnips without manuring: And folding hitherto appears to be the best method of enriching the ground for this purpose. It should be well harrowed as often as once a week, while the folding is continued, to mix the excrements of the cattle with the soil.

The ground should be cross ploughed as soon as the soil is sufficiently rotten, and reduced by harrowing to a fine tilth, before it is sowed. Where a good stock is kept, as much as an acre may be sufficiently folded.—See *Folding*.

The time for sowing the seed is about the middle of July. Doing it on a set day is ridiculous: For a time should be chosen when the ground has the right degree of moisture to make the seed vegetate; and if this should happen a week earlier, or a fortnight later than the usual time, it need not be regretted; but the opportunity ought to be embraced.

I have sown them in drills the first week in August, and had a good crop. One great advantage of sowing so late is, that the turnips will escape insects. And if the crop should not happen to be quite so large as if the sowing had been earlier, the roots will not fail of being better for the table.

One pound of seed is the common allowance for an acre of land. But to guard against the fly, the quantity may be a little increased. And it is recommended by judicious writers, that it be a mixture of equal parts of new and old seed, that the plants coming up at different times, the one sort or the other may chance to escape the insects. With this view, Mr. Tull constructed his turnip drill to lodge the seeds at different depths, which it seems had the desired effect.

The seed sown broadcast must be harrowed in with a short tined harrow, and then rolled with a wooden roller, to break the clods, and level the surface.

In a week, or thereabouts, the young plants will be up: And if it be a dry season, the fly will be apt to destroy them: To prevent which, some powdered soot, or lime, may be sowed very thinkly over them, by sifting, in a dewy morning. This will quicken the growth of the plants, as well as otherwise defend them. And the faster the plants grow, the sooner they will unfold their rough leaves, and be out of danger of the fly. Or it may answer well to sprinkle the ground with an infusion of elder, wormwood, or tobacco. But it must be done seasonably, as soon as the plants are up.

But if the young plants cannot be saved, as it may sometimes so happen, the ground may be harrowed, and sowed again, the cost of seed being but little, to compare with the loss of a crop.

When the plants have got five or six leaves, they should be hoed, and the plants cut out to six or eight inches asunder. In the second hoeing, which should be three or four weeks after the first, they should be further thinned, to the distance of fourteen or fifteen inches; especially if they are designed for the feeding of cattle. The roots growing at such a distance will be large, so that what is wanting in number, will be more than made up by their bulk. But if they are designed for the table, they need not be more than from six to ten inches apart, as overgrown ones are not so fit for this purpose.

But few have been hitherto raised in this country, for the feeding of cattle. But if our farmers would follow the directions given above, they would find it easy to raise hundreds of bushels for their stocks. Forty-seven tons have been the crop of an Irish acre, as Mr Winn Baker testifies, under his culture. Those who have made no spirited trials, will hardly conceive how much the hoeings will increase their crops. Even without hoeing, where the weeds are suffered almost to stifle them, a crop of turnips sometimes turns out to be profitable: How much more profitable may it be expected, if they had sufficient room, and were not robbed of their nourishment by standing too near together?

In England, the drill husbandry has been applied to turnips, and the produce has exceeded those sown broadcast, which have been hand-hoed. The late Lord Viscount Townsend made a fair trial, and found that the crop of an acre of drilled turnips weighed a ton and a half more than that of an acre in the old husbandry, though the latter were well hand-hoed. For two years past, I have sown turnips in the drill way, in the poorest part of my garden, where a crop of pease had grown the same summer, and never had better turnips. They were sufficiently large for the table, though they grew so near together in the rows that the roots crowded each other, and were not sown earlier than about the tenth of August. The earth was hoed into ridges three feet apart, and a single channel seeded on each of the ridges. This is the more observable, as I have often sown turnips in the broadcast way, on the same spot, and at the usual time of sowing fall turnips, and never before raised any that were fit to eat. I have also for several years raised turnips in the field in the drill way. The ridges were raised in May with the cultivator, about three feet apart. They were kept clear from weeds till about the last of July, by the cultivator and the hand-hoe, and then sown in single drills. Nothing more was necessary afterwards, except thinning and once hoeing. The crops were so much better than I have obtained from broad cast sowing, that I am induced to persist in, and recommend this method.

A crop of turnips in the old husbandry prepares the ground excellently for a crop of wheat, or flax, the following year. But it would be in much better order, if the turnips were horse-hoed.

In other countries, they feed the turnips off of the ground with sheep; or draw them up for neat cattle, through the winter, as fast as they are wanted; and even let them stand until spring, when it is convenient.

But in this country, they must be harvested in autumn, about the end of October, or even earlier in some places; and then stored in cellars, out of the way of the frost; which must needs be a drawback on the profit of this crop. Those that are designed for the table in winter, should be buried in, or covered with, dry pit sand, to prevent their becoming corky.

The most excellent mutton is fatted on turnips, and they are a good food for horned cattle. But milch cows should not be fed very plentifully on turnips, as there is some danger of their giving the milk an ill taste.

To produce good turnip seeds, some of the best roots, of the middling size, which have begun to sprout, should be planted early in the spring, in a good spot, free from shade. They should be in rows, eighteen inches asunder, and the ground must be kept clear of weeds till the seed is ripe. Stakes and laths may be needful round the outside, to keep the branches from falling to the ground before the seed is fully ripe.

The right culture of the French turnip is much the same as the above, excepting that they should be allowed more room, and that the ground should be tilled to a greater depth than is necessary for the other sort, and sowed at the end of June.

They should not be sown in the spring; for this will make them hard and sticky; nor more than about a month earlier than other fall turnips. I have known it answer well to sow both kinds mixed together, where the soil has been mellow and deep. But I rather prefer sowing each kind by itself; because the one requires to be earlier sowed than the other.

As there are some other insects which prey upon turnips, besides the fly, while they are in seed leaf or afterwards, it has led some to set plants of tobacco, perhaps six or eight feet apart, among their turnips, which is thought to have an excellent effect.

Mr. M. Mahon gives the following directions for the field culture of turnips.

"The ground intended for the production of turnips in the ensuing season ought to be deeply ploughed in October or November, and to be left in that rough state, to receive the benefits of the winter frosts, &c. In the April following, when it is perfectly dry, harrow it, and let it lie so till the middle of May, when it should have another deep ploughing across, or contrawise to the former; the first week in June harrow it, and towards the end of that month, give the field a light coat of well rotted manure, and immediately plough it in lightly; after which the soil and manure are to be incorporated, by harrowing the ground effectually with a weighty harrow.

"The time of sowing depends much on the application; when turnips are intended for early consumption, they may be sown soon after the middle of July. The general mode in the middle states, is to begin to sow about the twentieth of July, and to continue sowing as convenient, from this time to the middle of August, or a few days after.

"If the farmer could insure his first crop, I would not advise him to sow till the last four days in July, or even to the third or fourth day of August, but liable as the turnip crop is to numerous accidents and miscarriages, it is prudent to have a week or two in reserve for a second sowing, in case the first should fail.

The quantity of seed sown on an acre, by the great turnip farmers, is never less than one pound; more frequently a pound and an half, and by some two. If every grain was to come to perfection, a quarter of a pound would be more than sufficient; but having to encounter so many accidents, a pound is the least quantity that ought to be sown on an acre.

"The period in which you intend to sow being arrived, plough your prepared turnip ground once more, lightly; give it one or two strokes with the harrow all over, and sow the seeds immediately on the fresh surface. The method of sowing is generally by broadcast, with a high and even hand; but some sow it in rows by means of a machine called a *turnip drill*, which method is greatly approved of, particularly, as by it much labour is saved in hoeing and thinning the plants.

"In the former method the seed is covered by drawing a *light* harrow backward, that is, wrong end foremost, to prevent the tines, which are generally set somewhat pointed forward from tearing up the sods, and burying the seed too deep.

"One of the most important parts of the treatment due to the cultivation of the turnip yet remains to be done, that is, to roll the field with a heavy roller immediately after harrowing in the seed, provided the ground is sufficiently dry, or as soon after as it is in a fit condition. By this means the clods are broken, and much of the seed that would otherwise be exposed to birds, &c. will be covered and the surface rendered smooth and compact thereby, and consequently more retentive of moisture, which will greatly promote the vegetation of the seed and growth of the plants.

"But the all important point is, that the rolling of the ground, is experimentally found to be the most effectual method, hitherto discovered for the preservation of the rising crop from the destructive depredations of the fly. The turnip fly is always found most numerous in rough-worked ground, as there they can retreat and take shelter under the clods or lumps of earth, from such changes of weather as are disagreeable to them, or from the attacks of small birds and other animals.

"Experiments have been tried, on coating the seed with sulphur, soot, &c. and of steeping it in train oil and in solutions of various kinds, as a security against the fly, but the result has not been such as to establish any practice of this nature.

"Hoeing the plants, and setting them out, as it is called, comes next under consideration. The method of doing this dexterously is difficult to describe; nothing but practice can teach it. A boy in the turnip counties, by the time he is the height of the hoe, begins to make use of it; consequently every man who has been bred there to country business, is a turnip hoer, yet not always, even with this advantage, an expert one.

"The operation to be performed dexterously and well, requires a quickness of eye and a dexterity of hand that every man is not favoured with; while some men catch the proper plants to be singled, and set them out with a rapidity and neatness of execution very pleasing to the observer.

"The critical time of the first hoeing is, when the plants as they lie spread upon the ground, are nearly the size of the palm of the hand; if, however, se d weeds be numerous and luxuriant, they ought to be checked before the turnip plants arrive at that size; lest by being drawn up tall and slender, they should acquire a week, sickly habit.

"A second hoeing should be given when the leaves are grown to the height of eight or nine inches, in order to destroy weeds, loosen the earth, and finally to regulate the plants; a third, if found necessary, may be given at any subsequent period."

The common white turnips should be used before spring, as they are apt to become spongy; but the ruta baga will keep longer, and may be used after the first mentioned are exhausted.

For further information relative to the culture of this plant, see Massachusetts Agricultural Repository, Vol. V. p. 21. Vol. VI. p. 36. 39. 265.

A treatise upon the Ruta Baga, or Swedish turnip, has been written by William Cobbet, from which the following notices are extracted.

Description of the Plant.—"The leaf of every other sort of turnip is of a *yellowish* green; but the leaf of the Ruta Bage is of a *blueish* green, like the green of peas when of their full size, or like the green of a young and thrifty Yorkshire cabbage; hence it is called the *Cabbage Turnip.* The outside of the bulb of the Ruta Baga is of a greenish hue, mixed towards the top with a colour bordering upon the red; and the inside, if they are true and pure, is of peep *yellow,* nearly the colour of gold."

Mode of saving and preserving the seed.—"The Ruta Baga is apt to degenerate, if the seed is not saved with care. In England, we select the fairest roots, and of the best form, for seed, rejecting all such as are of a whitish colour, or greenish towards the neck, preferring such as are of a *reddish cast.* These, when selected, should be carefully preserved over the winter, and set in the month of March or April in a rich soil, remote from any roots of the turnip or cabbage kind, to preserve the seed pure and unmixed. Two or three roots, if they do well, will yield seed sufficient for an acre of land. Let the seed remain in the pods until the time of sowing."

Time of sowing.—" The time of sowing may be from the 25th of June to the 16th of July, as circumstances may be."

Quality and preparation of the Land.—" As a fine, rich, loose garden mould, of great depth, and having a porous substratum, is best for every thing that vegetates, except plants that live best in water, so it is best with Ruta Baga. I know of no soil in the United States, upon which this root may not be cultivated with the greatest facility, excepting a *pure sand*, and a *stiff clay*, which are very rare in this country.

Manner of sowing.—" My ploughman puts the ground up in little ridges, having two furrows on each side of the ridge, so that each ridge consisted of four furrows, and the tops of the ridges were about four feet from each other; and as the ploughing was performed to a great depth, there was of course a very deep gutter between every two ridges.

" I took care to have the manure placed so as to be under the middle of each ridge, that is to say, just beneath where my seed was to come, which was sown principally in this manner:—A man went along by the side of each ridge, and put down two or three seeds in places ten or twelve inches distance from each other, just drawing a little earth over, and pressing it lightly upon the seed, in order to make it vegetate quickly, before the earth became too dry. In this method four pounds of seed sowed seven acres. Two men sowed the whole seven acres in two days.

" Broadcast sowing will, however, generally be preferred; but when I have spoken of the after-culture, I shall compare the two methods, that the reader may decide for himself.

After-Culture.—" When the plants were fairly up, we went with a small hoe, and took out all but one in each ten or twelve inches, and thus left them to stand single. We next went with a hoe, and hoed the tops of the ridges, about six inches wide on each side of the rows of plants, and then horse-hoed between the rows, with a common horse-plough, after the manner of tilling Indian corn, or potatoes, by first turning the earth from the plants, and next towards the plants at the second hoeing. There is no ground lost in these wide intervals, for the lateral roots of the large turnip, as well as the Ruta Baga, will extend six feet from the ball of the plant; and my crop of thirty-three tons, or thirteen hundred and twenty bushels to the acre, taking the whole field together, had the same intervals; and less than this, as was practised by my neighbours, always diminished the crop. Wide as the invals were, the leaves of some of the plants would nearly meet across the rows, and I have had them frequently meet in England.

"In the broad cast method, the after culture must of course be confined to hoeing, or as Tull calls it, *scratching*.

"In England, the hoer goes in when the plants are about four inches high, and hoes all the ground over, setting out the plants at the distance of about eighteen inches; and if the ground becomes foul, he is obliged, in about a month afterwards to hoe the ground over again. This is all that is done, and a very poor all it is, as the crops on the very best lands invariably show, when compared with the ridge-crops."

Transplanting.—"This is a third mode of cultivating the Ruta Baga, and in certain cases far preferable to either of the others. My large crops at Botley, (England,) were from roots transplanted.

"I prepared one field of five acres, and another of twelve, with ridges, in the manner described for sowing, and on the 7th of June in the first field, and on the 20th of July in the 2nd, I set my plants, as in sowing, twelve inches asunder. I ascertained to an exactness, that there were thirty three tons to the acre, throughout the whole seventeen acres, and after this, I have never used any other method.

"In my usual order, the rows four feet asunder on the ridges, there are ten thousand eight hundred and thirty turnips, on each acre of ground and therefore, for an acre of ground to produce thirty-three tons, each turnip must weigh nearly seven pounds.

"From a large field I afterwards set on the 13th of July, I weighed one waggon load, which averaged eleven pounds each, and several weighed fourteen pounds, which would probably give fifty tons to the acre.

"The plants will succeed best when set in fresh earth, or *earth recently proved by the plough*.

"When we have our plants, and hands all ready, the ploughman begins, and turns in the ridges, (which have been prepared as before stated;) that is, he turns the ground back again, so that the top of the new ploughed ridge stands over the place where the deep furrow was before he began. As soon as he has finished the first ridge the planters begin to set, while he is ploughing the second, and so on through the field. This process is not very tedious, for in 1816, I had fifty-two acres of Ruta Baga planted in this way, and a crop of more than fifty thousand bushels. A smart lad will set half an acre per day, with a girl or boy, to drop the plants, and I had a man who would set often an acre a day. Care must be taken not to bury the heart of the plant. I observed how necessary it was to fix the plant firmly in the ground; and as the planter is strictly charged to do this, he is apt to pay little attention to the *means* by which the object is accomplished. The thing is done easily enough, if you cram the butts of the leaves down below the surface. But this brings the earth with the first rain at least over the heart of the plant; and then it will never grow at all."

"Observe well what has been said about fresh earth, and never forgetting this, let us talk about the art of planting. We have a setting stick, which should be the top of a spade handle cut off about ten inches below the eye, and pointed smoothly, the planting is then done in the manner of setting cabbages. Choose a dry time for your transplanting for this reason; if your plants are put into wet ground, the setting stick squeeses the earth up against the plant in *a mortar like state;* the sun comes and bakes this mortar into a hard glassed clod; the hole also, made with the stick is smooth upon its sides, and presents an impenetrable substance to the roots and fibres of the plant, and thus the vegetation is greatly cherished; but when plants are set in dry earth, the reverse of all this is true, and the fresh earth will supply proper moisture under any degree of drought. The hole thus being made in dry weather, set your plant *without bending the point;* support it with one hand firm in the hole, and with the other hand, apply the setting stick to the earth on one side of the hole, so as to form a sharp triangle with the plant, then thrust the stick down a little below the bottom or the point of the plant, and with a little twist, press the earth up to the plant, at the point or bottom of the root, and it is done. But if a vacancy remains below the bottom of the plant, it will not thrive well. This is true in all transplantings, both of roots and trees."

The reasons which Mr. Cobbett gives, in favour of transplanting are as follow.

1. "Time may be gained for one or two extra ploughings, between the 25th of June, and the 25th of July, or even the 25th of August.

2. "This transplanted crop, may follow some other crop, such as early June cabbages, early peas, or potatoes, kidney beans, peas, white onions, &c.

Time and manner of harvesting.—Mr. Cobbett, here recommends, that the earth be turned off from the roots by an ox plough, in dry weather, before the hard frosts set in, and then gathered by hand, which will save the trouble and expense of loosening them by the spade, and greatly expedite the work.

"The crop when pulled, may be secured over winter in the usual manner of securing potatoes, either in the cellar or in the pits dug in some dry part of the field and covered close and secure."

Uses and mode of applying the crop.—1. "As food for cattle, cows, sheep and hogs both raw and boiled or steamed, (which is preferable,) or even poultry when boiled or steamed.

2. "Cows that give milk, breeding sows, ewes with lamb, their lambs and even pigs, at weaning are greatly benefitted by the use of Ruta Baga, especially when boiled or steamed.

3. "This root by far surpasses the turnip when fed off by sheep with hurdles, or otherwise, or by hogs upon the field.

4. "The tops when cut before the roots are gathered, are valuable as green feed for all the stock mentioned above. An acre will yield about four waggon loads."

Mr. Cobbett says, that the Ruta Baga is not so good till it arrives at a mature state, which will be in February.

Common salt is said to be an excellent manure for turnips, but it would probably do best to be mixed with dung or compost.

The ruta baga, may be considered as an important acquisition to our Agricultural productions, affording an excellent fresh forage for horses, cattle, sheep, and swine, throughout the season. The requisites for its cultivation are; 1st, deep ploughing, and plentiful manuring; 2nd, pulverising the earth in the most effectual manner; 3rd, planting the seed. and transplanting the plants as speedily as possible, after manuring and ploughing, that the germs and roots may receive all the advantage of a fresh fermentation, and rolling down and pressing down the earth after sowing; 4th, deep ploughing between the roots, by which the mould is exposed to the influence of the air, and those chemical changes facilitated, on which the growth of vegetables essentially depends. Ridging on land naturally dry is injurious, and intervals between the rows of two and an half, or three feet, instead of four feet as recommended by Mr. Cobbett, are preferred by some. See remarks. by Mr. James Thacher, Massachusetts Agricultural Rep. vol. vi. p. 39, 40.

TURNIP CABBAGE. "A species of cabbage, so called, because the stalk, at some distance from the ground, after rising of the usual thickness, and in the manner of those of other cabbages, enlarges suddenly to such a degree, that it forms a knob of a very large turnip, of which likewise it has sometimes the figure, though it is in general more oblong.

"By this peculiar formation of the stalk, or production of the turnip like knop, together with its being perennial, this species of cabbage is distinguished from all others. From the top of this turnip rise a number of leaves, of a greenish red, or sometimes greenish purple colour; which answer to the radical leaves in other plants. They do not, though this plant is truly of the cabbage kind, ever close together, and form a compact globular, or oblong mass, as in the common species; but keep their erect growth, or turn outwards.

"From among these leaves spring a number of other stalks, of which those that are nearer the extremity, branch, and send out flower stalks, spreading horizontally; and those that are more in the centre grow erect, and without branches. On these stocks are leaves, springing out alternately, and of the same colour with the others. The flowers are small and yellow, and succeeded by long cods, full of seed, of the size of that of mustard, and a lighter brown colour.—*Complete Farmer.*

It grows wild near Dover, in England; but it is doubted whether it be indigenous. When it is cultivated in gardens, it is rather as a curious than as an esculent plant: Yet it is eatable, and is recommended by Mr. W. Baker, for the use of seamen: And he thinks it of importance as winter food for cattle. I have not yet known trials enough of it in this country, to be able to ascertain its value. But its bidding defiance to all inclemencies of weather, after it is once firmly rooted, is a circumstance that ought to incline us to make trial of it.

"The turnip cabbage," says a Mr. North, "is one of the hardiest roots that grow; and I dare affirm, might be propagated to great advantage, for feeding sheep, &c. For in the most severe winter that I can remember, when cabbages, turnips, &c. have all been demolished by the extremity of the weather, the turnip cabbages have not been hurt. They are a very solid and juicy root, and do not grow spongy when they are old, as turnips do. The tops may be cut off, and given to sheep in the spring, and the root laid by in an out house, to feed them in April and May, when no other roots can be had. Sheep are so fond of these roots, that they will leave the best turnips for them. They will eat them tops and bottoms as they are growing in the fields." Mr. North is not very accurate, in calling the turnip part of this cabbage a root, as it is only an enlargement of the upper part of the stem, and several inches above the ground.

In the Bath Society papers, is the following account of Sir Thomas Reevor's method of cultivating this root. "In the first or second week of June, I sow the same quantity of seed, hoe the plants at the same size, leave them at the same distances from each other, and treat them in all respects like the common turnip. In this method I have always obtained a plentiful crop of them. On the 23rd of April last, having two acres left of my crop, I divided them by hurdles into three equal parts. Into the first part I put twenty four small bullocks and thirty middle sized wethers, which, at the end of the first week, I shifted into the second division, and then put seventy lean sheep into what was left of the first: These fed off the remainder of the turnips left by the fat stock: And so they were shifted through the three divisions, the lean flock following the fat, till the whole was consumed. The twenty four bullocks, and the thirty fat wethers, were fed four weeks, and the seventy lean sheep as long. So that the two acres kept twenty four bullocks and one hundred sheep four weeks. The value, at the rate of keeping at that season, cannot be less than 4d. a week for each sheep, and 1s. 6d. a week for each bullock, which amount together to £ 14 10 8, for the two acres. Thus you see that in providing a most incomparable food for cattle, in that season of the year in which the farmer is most distressed, and his cattle almost starved, a considerable profit may be likewise obtained.

"The land on which I sow turnip rooted cabbage is a dry mixed soil, worth only fifteen shillings per acre."

The Bath Society have subjoined, "that this account is as interesting as any they have been ever favoured with, and recommend it to farmers in general, that they adopt a mode of practice so decisively ascertained to be highly judicious and profitable."—*Encyclopædia.*

Whether this plant, which has but newly found its way into our country, is hardy enough to bear the frost of our winters, I suppose is yet to be proved.

V.

VETCH. A small species of pulse or pea, much used in England, as well for a green crop to be ploughed in, to enrich the ground as for a crop sometimes made into hay, and sometimes suffered to ripen, and the seeds given to animals. In the northern parts of this country, though often tried they have not been found to answer. It is doubted whether they can be successfully cultivated in any part of our country.

VINE, *vitis*, an important plant of the creeping kind, famous for its fruit, the grape, and the generous liquor it affords for the use of mankind.

Those who would cultivate vines in gardens, should procure those sorts which have been found to prosper well in cold countries: Particularly those which ripen their fruit well in Great Britain, Germany, or the northerly parts of France. The white sweet water, the Chasselas blanc, the white muscadine, and other white grapes, may be best to cultivate in gardens, as they are in general more palatable for eating.

My account of the culture of vines, shall be an abridgment of what Mr. Miller has written on this subject, in his *Gardener's Dictionary.*

"All sorts of grapes are propagated either from layers or cuttings: The latter is preferred; because the roots, being slender, are apt to get dried, and die by transplanting.

"Make choice of such shoots as are strong, and of the last year's growth. Cut them from the old vine, just below where they were produced, taking a knob, or piece of the two years wood, to each, which should be pruned smooth. Then cut off the upper part of the shoots, so as to leave the cutting about sixteen inches long. When the piece of old wood is cut at both ends, near the young shoot, the cuttings will resemble a little mallet. In making the cuttings after this manner, there can be but one taken from each shoot; whereas most persons cut them into lengths of about a foot, and plant them all, which is very wrong; for they will not be so fruitful.

"Cuttings, thus prepared, if not then planted, should be placed with their lower part in the ground, in a dry soil, with litter over their upper parts to prevent their drying. They may thus remain till the beginning of April, (May in this country) then take them out, and wash them from the filth they have contracted, and if you find them very dry, let them stand with the lower parts in water six or eight hours, which will distend their vessels, and dispose them to take root.

"Then, the ground being prepared, the cuttings should be planted. Open the holes about six feet distance from each other, putting one good strong cutting into each hole, which should be laid a little sloping, that their tops may incline to the wall: It must be put in so deep, that the uppermost eye may be level with the surface of the ground.

"Having placed the cutting in the ground, fill up the hole gently, pressing down the earth with your foot close about it, and raise a little hill just upon the top of the cutting, to cover the upper eye quite over, which will prevent its drying. Nothing more is necessary, but to keep the ground clear from weeds till the cuttings begin to shoot; at which time look over them carefully, to rub off any small shoots, if such are produced. You must continue to look over them once in three weeks during the summer season, and rub off all lateral shoots that are produced, and keep the ground clear from weeds.

"The Michaelmas following, if your cuttings have produced strong shoots, prune them down to two eyes. Being cut thus early in autumn, the wounds will heal before the bad weather comes on, and the roots will be strengthened.

"In the following spring, gently dig the borders, to loosen the earth, but be careful not to injure the roots of your vines. Also raise the earth up to the stems of the plants, so as to cover the old wood, but not so deep as to cover either of the eyes of the last year's wood. After this they will require no further care till they begin to shoot, when you should look them over carefully, to rub off all weak dangling shoots leaving no more than the two shoots which are produced from the two eyes of the last year's wood, which should be fastened to the wall. And till the vines have done shooting look them over in three or four weeks, to rub off all the lateral shoots as they are produced, and to fasten the main shoots to the wall as they are extended in length, which must be shortened before the middle or end of July, when it will be proper to nip off their tops, which will strengthen the lower eyes. And during the summer keep the ground clear from weeds; and permit no plant to grow near the vines, which would not only rob them of their nourishment, but shade the lower parts of their shoots, and prevent their ripening: Which will not only cause their wood to be spongy and luxuriant, but render it less fruitful.

"As soon as the leaves begin to drop in autumn, prune these young vines again, leaving three buds to each of the shoots, provided they are strong: Otherwise shorten them down to two eyes: For it is a very wrong practice to leave much wood upon young vines, or to leave their shoots too long, as it weakens their roots. Then you should fasten them to the wall, spreading them out horizontally each way, that there may be room to train the new shoots the following summer. And in the spring the borders must be digged as before.

" The third season, you must go over the vines again, as soon as they begin to shoot, to rub off all danglers as before, and train the strong shoots in their proper places, which may be supposed to be two from each shoot of last year's wood. But if they attempt to produce two shoots from each eye, the weakest of them must be rubbed off. If any of them produce fruit, as many times they will the third year, you should not stop them so soon as is generally practised upon the bearing shoots of old vines; but permit them to shoot forward till a month after mid-summer, at which time you may pinch off the tops of the shoots: For if this were done too soon, it would spoil the buds for the next year's wood. which in young vines must be more carefully preserved than on older plants.

" During the summer, you must constantly go over your vines, and displace all weak lateral shoots as they are produced, and carefully keep the ground clear from weeds, that the shoots may ripen well."

After three years, the vines are considered as grown up; and concerning the management of grown up vines, the same writer says; " Vines rarely produce any bearing shoots from wood that is more than one year old; therefore great care should be taken to have such wood in every part of the trees; for the fruit is always produced upon the shoots of the same year, which come out of the buds of the last year's wood.

" The best method is to shorten the bearing shoots to about four eyes in length, because the lowermost seldom is good, and three buds are sufficient; for each of these will produce a shoot, which generally has two or three bunches of grapes: So that from each of those shoots may be expected six or eight bunches, which are a sufficient quantity. These shoots must be laid about eighteen inches asunder, for if they are closer, when the side shoots are produced, there will not be room enough to train them against the wall.

" At the winter pruning of your vines, you should always observe to make the cut just above the eye, sloping it backward from it, that if it should bleed the sap might not flow upon the bud. And where there is opportunity of cutting down some young shoots to two eyes, in order to produce vigorous shoots for the next year's bearing, it should always be done; because in stopping those shoots which have fruit on them, as soon as the grapes are formed, which is frequently practised, it often spoils the eyes for producing bearing branches the following year, and this reserving of new wood is what the Vignerons abroad always practice in their vineyards. The best season for pruning of vines is about the middle or end of October.

" The latter end of April, or the beginning of May, when the vines begin to shoot, carefully look them over, rubbing off all small buds, which may come from the old wood, which only produces weak dangling branches, as also when two shoots are produced from the same bud, the weakest of them should be displaced, which will cause the others to be stronger; and the sooner this is done the better for the vines.

"After one month, go over them again, rubbing off all the dangling shoots as before; at the same time fasten up all the strong branches, that they may not hang from the wall. Towards the middle of June, stop the bearing branches, which will strengthen the fruit, provided you leave three eyes above the bunches: For if you stop them too soon it will injure the fruit, by taking away that part of the branch which is necessary to attract the nourishment to the fruit, as also to perspire off the crudities of the sap.

"But though I recommend the stopping those shoots which have fruit at this season, it is not to be practised upon those shoots which are intended for bearing the next year; for these must not be stopped before the middle of July, lest you cause the eyes to shoot out strong lateral branches.

"During summer rub off dangling branches, and train the shoots to the wall as before, which will accelerate the growth of the fruit, and admit the air to them, which is needful to ripen, and give them a rich flavour. But you must never divest the branches of their leaves."

The same directions should be followed, when the vines are trained to espaliers, or to wooden fences; and the pruning, stopping, &c. in vineyards, are the same as in gardens, or green houses.

The vines in vineyards must be supported by stakes; by two short ones the first year, by longer ones the next, and so on as there shall be occasion. But as to those things which are most peculiar to vineyards, the above author directs that the soil which is to be chosen for the purpose, is *that* the surface of which is a light sandy loam, not more than a foot and a half or two feet deep, with a gravelly or chalky bottom; but if the soil have a bottom of strong clay or loam, it is not fit for this purpose. As to the situation of the place, it should incline to the south, with a gradual descent, that the water may drain off; but a steep slope is not good as it will not so well hold the manure.

To prepare the soil for planting, he says, "In the spring, if the ground is green sward, it should be ploughed as deep as the surface will permit; then well harrowed to break the clods, and cleanse it from the roots of noxious weeds. After this, it must be frequently ploughed and harrowed for one year, to render the surface light. The following spring the ground should be ploughed again, and after making the surface even, the rows should be marked out from south-east to north-west, at the distance of ten feet from each other; and these rows should be crossed again at five or six feet distance, which will mark out the places where the plants should be placed." But as we are favoured with a drier atmosphere, I should think the rows may as well be planted nearer together, as it will be a saving of ground and labour.

He adds, 'the proper kinds of grapes should be chosen. The Vignerons abroad always observe that the grapes which are good for eating never make good wine; and therefore make choice of those, whose juice, after fermenting, affords a noble rich liquor. These grapes are always austere, and not so palatable.' He supposes the sort most proper for vineyards in England, is the *Auvernat*, or true Burgundy grape; and thinks it best that a vineyard should produce only one kind of grapes; because the mixing of the juice of several together, will cause the wine to ferment at different times."

After the cuttings are planted, he recommends keeping them clean from weeds; and as the rows are at a great distance from each other, he allows that the spaces between may be sown, or planted with any kinds of esculent roots, which do not grow tall, provided there is proper distance left from the vines, and care taken that the vines be not injured by the crops, or in gathering, or carrying them off the ground. And this husbandry may be continued till the vines come to bearing; after which there should be no sort of crop put between them, because the cleaner the ground is kept, the more heat will be reflected to the grapes.

"The ground should be yearly well dug or ploughed, and always kept free from weeds; because the roots of the vines will otherwise be robbed of their nourishment.

"When a vineyard is arrived to a bearing state, it should be so pruned, that there should be never too many branches left on a root, nor those too long; for though there may be a greater quantity of fruit produced, yet the juice will not be so good as when there is a moderate quantity; and the roots will be weakened, which is found to be of bad consequence. The number of branches which the Italians leave upon a strong vine are four; two of the strongest have four eyes; and the two weaker are shortened down to two eyes each. Shoots that have borne fruit, are either cut quite away, or reduced to two eyes.

"The ground of a vineyard should not only be constantly well tilled, but a dressing of some good manure applied to it every third year, in the spring, and well mixed with the soil.

"When a vineyard is carefully dressed, it will be as pleasing a sight, as any plantation of trees or shrubs whatever. If the rows be regular, and the stakes exactly placed, and the upright shoots stopped to an equal height, there is nothing in nature which will make a more beautiful appearance. And during the season that the vines are in flower they emit a most grateful scent, especially in the morning and evening. And when the grapes begin to ripen there will be fresh pleasure in viewing them."

A vineyard that is well managed will bear fruit the third year from the planting; the crops will be growing larger till the seventh or eighth year; and remain fruitful till about the fiftieth year; after which it will begin to decline, and die away.

But instead of the cutting culture of the vine some have recommended that a plantation of trees of low growth be made, placed at proper distances, with a vine planted at the root of each, which will climb up the trees, and bear fruit without cutting. The ground between the rows should be continually well tilled and manured.

Much labour might be saved by this mode of culture; and that it might answer, the fruitfulness we often observe in wild grapes may lead us to conclude. It seems to be following nature; but the fruit must be waited for, a considerable number of years after planting. And after all, it never will yield so good a wine, as vines that are kept low by cutting; for the low vineyards in France afford a much richer wine than the high ones, though less in quantity. The former are but three or four feet high, the latter seven or eight, and both cultivated by cutting.

For further directions respecting the culture of vine, see *Forsyth* on Fruit Trees, chap. VIII. Massachusetts Agricultural Repository, vol. VI. p. 66.

VINEGAR, sour wine. But the name is also applied to any acid penetrating liquor, prepared from wine, cyder, beer, &c.

The process of turning vegetable matters to vinegar, is thus delivered by Dr. Shaw: "Take the skins of raisins, after they have been used in making wine; and pour three or four times their own quantity of boiling water upon them, so as to make a thin aqueous mixture. Then set the containing cask, loosely covered, in a warmer place than is used for vinous fermentation; and the liquor, in a few weeks time, will become a clear and sound vinegar; which being drawn off from its sediment, and preserved in another cask, well stopped down, will continue perfect, and fit for use."

All such things as håve undergone, or are fit for, a vinous fermentation, will afford vinegar. Our common summer fruits, sap of maple trees, and other vegetables, and even the pomace from which cyder has been pressed, will make vinegar, by means of only the addition of water, and exposing it to the open air, and warmth. But some spirit should be added to weak vinegar, to give it a body, and fit it for keeping.

"Liquor to be changed into vinegar, being kept warmer than in vinous fermentation, it begins in a few days to grow thick and turbid; and without throwing up bubbles, or hissing, as happens in vinous fermentation, deposits a copious sediment. The effect of this separation begins first to appear on the surface of the liquor, which gathers a white skin that daily increases in thickness, till at length it becomes like leather; and now if it continues longer in this state, the skin turns blue, or green, and would at last grow fetid and putrefy. Therefore, in keeping down this skin, as it grows, and thrusting it gently to the bottom of the vessel, consists much of the art of vinegar-making, especially from malt."

Mr. Genet, of New-York, gives the following directions for making vinegar of whey.

"After having clarified the whey, it is poured into casks with some aromatic plants or elder blossoms, as suits the fancy, and exposed in open air to the sun, where it soon acquires an uncommon degree of acidity."

VIVES, a disease in horses, which differs from the strangles only in this, that the swellings of the kernels under the ears of the horse (which are the parts at first chiefly affected) seldom gather, or come to matter, but by degrees perspire off, and disperse, by warm clothing, anointing with the marsh mallow ointment, and a moderate bleeding or two. But should the inflammation continue notwith standing these means, a suppuration must be promoted.

"When these swellings appear in an old or full aged horse, they are signs of great malignity, and often of an inward decay, as well as forerunners of the glanders.

Take of crude mercury, or quicksilver, one ounce; Venice turpentine, half an ounce; rub together in a mortar till the globules of the quicksilver are no longer visible; then add two ounces of hog's lard.

"Some authors recommend this ointment to be used at first, in order to disperse the swellings, and prevent their coming to matter; bleeding and purging at the same time for that purpose; but as in young horses they seem to be critical, the practice by suppuration is certainly more eligible and safe: For want of properly affecting which, the humours frequently settle, or are translated to the lungs, and other bowels, or falling on the fleshy part of the hind quarters, form deep imposthumes between the muscles, which discharge such large quantities of matter as sometimes kill the horse, and very often endanger his life."—*Bartlet's Farriery*, page 99.

ULCER, "a solution of the soft part of an animal body, together with the skin.

"The first intention in the cure of ulcers, is bringing them to digest, or discharge a thick matter; which will in general be effected by the green ointment, or that with precipitate. But should the sore not digest kindly by these means, but discharge a gleety thin matter, and look pale, you must then have recourse to warmer dressings, such as balsam, or oil of turpentine, melted down with your common digestive, and the strong beer poultice over them. It is proper also in these sores, where the circulation is languid, and the natural heat abated, to warm the part, and quicken the motion of the blood, by fomenting it well at the time of dressing; which method will thicken the matter, and rouse the native heat of the part, and then the former dressings may be reapplied.

"If the lips of the ulcer grow

hard or callous, they must be pared down with a knife, and afterwards rubbed with the caustic.

"Where soft fungous flesh begins to rise, it should carefully be suppressed in time. otherwise the cure will go on but slowly. If it has already sprouted above the surface, pare it down with a knife, and rub the remainder with a bit of caustic; and to prevent its rising again, sprinkle the sore with equal parts of burnt alum, and red precipitate; or wash with the sublimate water, and dress with dry lint even to the surface, and then roll over a compress of linen as tight as can be borne; for a proper degree of pressure, with mild applications, will always oblige these spongy excrescences to subside, but without bandage the strongest will not so well succeed.

"All sinuses, or cavities, should be laid open as soon as discovered, after bandages have been ineffectually tried; but where the cavity penetrates deep into the muscles, and a counter opening is impracticable or hazardous; where, by a continuance, the integuments of the muscles are constantly dripping and melting down; in these cases injections may be used, and will frequently be attended with success. A decoction of colcothar boiled in forge water, or solution of lapis medicamentosus in lime water, with a fifth part of honey and tincture of myrrh, may be first tried, injecting three or four ounces twice a day, or some rosin melted down with oil of turpentine may be used for this purpose. If these should not succeed, the following, which is of a sharp and caustic nature, is recommended on Mr. Gibson's experience.

Take of Roman vitriol half an ounce, dissolve a pint of water, then decant and pour off gently into a large quart bottle; add half a pint of camphorated spirit of wine, the same quantity of the best vinegar, and two ounces of Ægyptiacum.

"This mixture is also very successfully applied to ulcerated greasy heels, which it will both cleanse and dry up.

"These sinuses, or cavities, frequently degenerate into fistulæ, that is, grow pipey, having the inside thickened, and lined as it were with a horny callous substance. In order to their cure, they must be laid open, and the hard substance all cut away. Where this is impracticable, scarify them well, and trust to the precipitate medicine made strong, rubbing now and then with caustic, butter of antimony, of equal parts of quicksilver and aqua fortis.

"When a rotten or foul bone is an attendant on an ulcer, the flesh is generally loose and flabby, the discharge oily, thin and stinking, and the bone discovered to be carious, by its feeling rough to the probe passed through the flesh. In order to a cure, the bone must be laid bare, that the rotten part of it be removed; for which purpose, destroy the loose flesh, and dress with dry lint; or the dossils may be pressed out of tincture of myrrh or euphorbium. The throwing off the scale is generally a work of nature, which is effected in more or less time, in proportion to the depth the bone is affected, though burning the foul bone is thought by some to hasten its separation.

"Where the cure does not properly succeed, mercurial physic should be given, and repeated at proper intervals: And to correct and mend the blood and juices, the antimonial and alterative powders, with a decoction of guaiacum and lime water, are proper for that purpose."—*Bartlet's Farriery*, p. 253.

URINE, a serous and saline fluid, separated from the blood, and emitted by the canal of the urethra.

Considered as a manure, this is an important liquor; it is supposed to be richer than the stale of beasts. The difference may be as that of human ordure to barn dung.

Mr. Bradley relates, as of his own knowledge, that human urine was thrown into a little pit constantly every day, for three or four years. Two years after some earth was taken out of this pit, and mixed with twice as much other earth, to fill up a hollow place in a grass walk. The turf which was laid upon this spot grew so largely and vigorously, besides being much greener than the rest, that by the best computation he could make, its grass in a month's time, was above four times as much in quantity as that of any other spot of the same size, though the whole walk was laid on very rich ground.

And Mr. Hartlib instances a widow woman near Canterbury in England, who saved in a pail all the urine she could, and when the pail was full, sprinkled it on her meadow, the grass of which looked yellow at first, but afterwards grew surprisingly.

Human urine therefore should be considered as of great value to the farmer. A good method of using it is, to throw it upon compost dunghills which are under cover.

Also, old urine is preferable to dung for manuring of trees, as it penetrates better to their roots; and it is said to remove divers infirmities of plants.

W

WAGGON, a carriage, or cart, mounted on four wheels.

Waggons answer much the same ends as common carts; but are greatly to be preferred, especially in journeying; and they are quite necessary for horse-teams, especially when the horses are harnessed by two and two, as they ought to be in large teams. And if oxen are used they should be harnessed in the same manner as horses. The advantages of waggons are especially these two.

1. They are far less fatiguing to the oxen than carts; because they have to bear only the weight of the tongue upon their necks; whereas, in carts, and on descending ground, the oxen are crushed to death, as it were, with bearing a great part of the load; or in ascending a hill, the load pulls their necks upwards, so that they are almost choked and rendered incapable of drawing the load. Bulky loads, such as hay

and the like, produce these bad effects in the greatest degree.

2. Waggons are better for the roads, as the wheels do not make so deep impressions in the ground as those of carts. The load bearing equally on four wheels, each wheel bears but half so much weight as one wheel of a cart. Consequently, a waggon wheel presses the ground with but half the force of a cart wheel; and therefore penetrates to but half the depth; supposing the tire in both to be of equal breadth. If teamsters used only waggons upon our roads, the roads would soon be found to be greatly mended, as they would be freed from those deep ruts, which are so disagreeable and dangerous to travellers.

WALL, the principal part of a building, serving both to enclose and support it.

The walls with which farmers are most concerned are fences of stone upon their farms. See the article *Fence*. The walls of cellars, even in the poorest farm houses, ought to be so constructed as to keep out all frost. Therefore the upper half of the wall should be laid in good lime mortar. This will render banking to keep out frost unnecessary. And banking should be avoided, as it causes the sills of houses to decay. It requires yearly attention and labour, and gives to a house an ill appearance.

When brick walls are to be built for houses, &c. particular care should be taken in laying the bricks. In summer they should be laid as wet, and in winter as dry as possible, to make them bind the better with the mortar. In summer, they should be covered up as fast as they are laid, to prevent the wall from drying too fast. In winter, they should be well covered to protect them from snow, rain, and frost, which are all hurtful to the cement. But in the coldest part of winter mason work can hardly be performed at all in this country. Walls laid in this season are not expected to be durable.

WARBLES, small hard tumours on the saddle part of a horse's back, occasioned by the heat of the saddle in travelling. They are usually called *Saddle Boils*.

A hot greasy dish clout frequently applied will sometimes cure them. When this fails, camphorated spirit of wine will be found more effectual; especially if a little of the spirit of sal ammoniac be added.

WATER, a simple, transparent fluid, which becomes solid with a certain degree of cold.

A general division of this fluid is into salt and fresh. But according to Dr. Shaw, it seems divisible into as many different species, as the earth is into strata or beds. Thus there are mineral waters, of various kinds, according to the mineral substances through which they pass, and by which they are by any means impregnated. Water may therefore be as compounded a body as earth; and perhaps neither of them can any where be found perfectly pure or unmixed.

Water is of infinite use in all the works both of nature and art; as without it there could be no generation, nutrition, or growth in any animal, vegetable, or mineral bodies. The blood could not flow in the veins, the sap in the vessels of vegetables, nor the particles of minerals concrete or grow together, without water. Neither could there be any corruption, fermentation, or dissolution performed without it. But I must not attempt to speak of all its uses, being limited by my general subject.

Those however have grossly mistaken the matter, who have supposed water to be the sole food by which vegetables are nourished; though it must be allowed to be absolutely necessary to their nourishment and growth. It is an important agent in mixing and dissolving the other ingredients of the food of plants; gives them the fermentation that is necessary, and is an essential part of the nutritive steam that enters the roots and other parts of vegetables. And the purest water contains in itself many earthy particles, as appears by the crusts that adhere to the insides of vessels in which water is often boiled.

Nor is water wholly destitute of salts and oils. It is no wonder, therefore, that a plant will live, and grow for some time, with its roots thrust into a vessel of water. For even in this situation, it partakes of every necessary ingredient of its natural food, though not in the right proportions.

Water not only serves thus as a vehicle to the food of plants, but carries with it where it runs in streams many particles of vegetable mould, &c. which enrich the soil. After heavy rains, it is found that water deposits a fertilizing sediment, which changes the surface of the soil to a dark colour.

Watering the ground also promotes the putrefaction of every substance, whether animal or vegetable, contained in the soil, or lying on the surface.

Where a good head of water can be made without too great an expense; or where a brook or part of a river, can be so diverted from its course, as to be made to spread its waters over the neighbouring grass grounds, it should be attended to, as a matter of no small importance. But the water should be perfectly under command; otherwise it may prove rather hurtful than beneficial. Too much of it would be as great an evil as too little.

Particular care should be taken that the water which is led over grass grounds, be only such as is fit to nourish plants, which chalybeate waters, or springs impregnated with a mineral acid, certainly are not. Such water is rather poisonous to plants, and prevents their growth.

The quantity of water thrown over the land should be rightly proportioned; a light sandy soil will bear more, a stiff soil a less quantity, without overcharging the vessels of the plants, or cooling them too much.

The channels should be so made that the water may be communicated to every part, excepting where there are little hollows which are naturally wet. The main channel should be broad, and its descent but just enough to cause the water to run; and from the main channel, a great number of very narrow ones should be cut, that the water may be well spread through every part of the surface.

Watering the ground in this manner should be avoided when the air is extremely hot; because heat draws the water too hastily into the plants, which renders them weak. The night time should be preferred to the day for letting out the water, or a day that is cool and cloudy may be chosen, rather than one that is hot and dry.

No water should be admitted in the spring, till after the ground which has been hoven by the frost is well settled. But afterwards, and in a dry spring, watering may be plentifully used, till the grass begins to spring up. After the shooting of the grass, the water should be administered more sparingly, or not at all if the weather proves rainy. And when the grass comes to be tall, no water should be applied, but in case of necessity, as when a drought prevails, which would otherwise shorten the crop; for if it were applied at this time in a large quantity, the grass would be lodged; or if the water be not clean, it will foul the grass, and give the hay an ill taste.

After the second crop of hay is taken off, water may be thrown over the ground in plenty; for the ground is then very dry, and the weather so cool, that vegetables will not be injured by plentiful watering.

Ground that is thus watered will produce plenty of grass, unless it should happen to be too much chilled by watering; to prevent which it should have a sprinkling of some warm compost, soot, or other hot dressing, each year, in autumn.

Care should be always taken to leave off this watering before the month of December, when strong frosts are expected; because frost destroys all vegetables much more when they are full of sap, and where the soil is very wet.

When you wish to water land which is above the level of an adjacent stream, an engine may be used to raise it to the proper height. They who are willing to be at this expense may find descriptions of *Archimedes' watering screw, the Persian wheel, M. Belidor's wheel, and an engine invented by M. de la Bay, in the Complete Farmer*, article *Water*.

WATER FURROWING, drawing furrows in the lowest parts of a field of wheat, or other winter grain, as soon as it is harrowed in, in order to draw off the superfluous water, that none may stand on the surface during the winter.

This piece of good husbandry is considered, in the old countries, as indispensably necessary; unless when a field is on such a declivity that the water cannot stand on it. For if water stand long upon corn, in a frosty season, or almost any season, it will either be killed, or so stinted in its growth, that it will produce nothing.

If this practice were adopted by my countrymen, the labour of doing which is but a trifle, I am persuaded they would find so great an advantage in it, that they would not afterwards incline to lay it aside. It is probable it might prevent the winter killing of much grain, an evil that is greatly complained of.

WATERING, applying water to plants to nourish them.

Vegetables that are newly transplanted, as they have their roots more or less diminished, or otherwise injured, often need watering till they have taken new root. But this should be done with caution. If a dry season follow the transplanting, let them be watered if they appear to droop, only on evenings, and in cloudy weather, and with water that has been exposed, one day at least, to the shining of the sun; not with water directly from a well, or a cold spring, as it will give a chill to the plants. Only a small quantity should be applied at once, that it may have an effect similar to that of a refreshing rain. For water, applied too plentifully, sometimes washes away the finest of the mould from the roots; or makes little cavities about them, which admit too much air.

In a dry season, whole gardens sometimes need watering; and in doing it the above precautions are to be regarded. They are happy who have a piece of standing water in their garden or a rivulet near at hand, from whence the garden may be watered without much labour.

WEATHER. The state or condition of the atmosphere with regard to heat, cold, wind, rain, frost, snow, &c.

Innumerable advantages would arise to the husbandman from a foreknowledge of the changes of the weather; and even from a foreknowledge of the general characters of the approaching seasons. In the former case, he would be able to order his business from day to day in the best manner, and so prevent much hurry, perplexity and loss; especially in the seasons of hay making and harvesting; in the latter, he would be happily directed in his choice of crops, and the best methods of cultivating them. And as this knowledge is not to be obtained, the ability to make very probable conjectures is nextly to be coveted, as it will be found to answer very valuable purposes.

Virgil, and other early writers on husbandry, pointed out many rules of prognosticating the changes of weather in their own countries. But these will not universally, and perhaps not generally, apply in other regions. Persons ought therefore to be careful that they do not lay too much stress upon them.

Journals of the weather, winds, and state of the atmosphere, should be made and kept in every climate in this country; from the comparing of which for a course of years, it is profitable that some valuable prognostics might be formed, which have not yet been thought of.

The observations made in one climate, will not perfectly answer for another, even in the same country. In some parts of the state of Massachusetts, for instance, a south wind in a summer morning forebodes a hot day, and in other parts a cool one; and the changes of weather are not just the same in the maritime as in the inland parts. The latter are hotter in summer and colder in winter than the maritimes.

That our farmers may enable themselves to form judicious prognostications, some parts of knowledge, falsely so called, ought to be unlearned, or exploded; as weeds must be extirpated that useful plants may be nourished.

Those who draw their prognostics from the motions and aspects of the celestial bodies are certainly less apt to attend to more sure objects in the earth and atmosphere.

If any celestial influence is to be much regarded in this case, one would think it ought to be only that of the earth's satellite, the moon, because of its nearness. That she influences the tides is well known; and these may have a small influence on the winds and weather. And probably vapours may be more copiously raised when she is nearest to the earth, which affords a greater probability of approaching rain or snow.

But what reason can there be assigned, for supposing that her *place* in the heavens at the moment of her full, change, or quartering, can have any influence on our atmosphere? Or that the weather in the latter quarters, should be governed by that in her former quarters? These things, though believed by many, have never been confirmed by any set of accurate observations. The few that I have made have rather tended to refute them.

Barometrical observations may be of some real use. "In general, it may be expected, that when the mercury rises high, a few days of fair weather will follow; if it falls, rain may be expected."

Birds of passage have always been supposed to indicate the nearness of approaching seasons, by their removing to different climates.

It is well known that the weather is much governed by winds. Rain is very often preceded, and accompanied by a south-east wind, and snow by a wind from the north-east, or north north-east: And an east wind, continued twelve hours, seldom fails to bring rain; and yet some rain often times comes from every other quarter. When the winds blow from any point betwixt north and west the weather is expected to continue fair and dry.

The weather is oftener unsettled about the times of the equinoxes, than at other seasons: And high winds and storms are more to be expected.

Falling weather oftener happens a little before or after the full and change of the moon than at other times, especially if she happen to be near to her perigree, at the time of these changes. For the attraction of vapours from the surface of the earth is then greatest.

The redness of the sky at the rising of the sun, has ever been considered as a sign of foul weather approaching; but it is far from being infallible.

The falling of heavy dews is a sign of the continuance of fair weather.

The setting of the sun behind a black, watery cloud, betokens approaching rain.

The wading of the sun, as it is called is a pretty sure sign that foul weather is very near.

When the disc of the sun, either at rising or setting, appears very broad and dim, the atmosphere is charged with plenty of vapour, which will soon condense and fall upon the earth. The contrary happens when the sun appears brilliant, small and dazzling, at rising and setting, as these appearances indicate a dry state of the air. The copious evaporation of boiling water is a pretty sure sign of falling weather. For I suppose this arises from the same cause, (whatever it be) as the increased ascent of vapours in general.

The wind commonly blows from the point from whence meteors are seen to shoot in the night preceding.

When swine are busy in collecting sticks and straws, foul weather is approaching.

Many more signs of changes in the weather might have been mentioned: But I study brevity.

WEDGE. One of the mechanical powers, the force of which is extremely great.

The wedge is a triangular prism, whose basis are acute angled triangles.

The wedge is a needful implement among farmers, for the cleaving of their wood and timber. It should be made of a soft kind of iron, that it may not be broken near the point, by driving it in frosty weather, or by forcing it into hard knots. The head should be about two inches thick, and the length of the wedge from eight to nine inches.

WEEDING. The operation of destroying weeds among useful plants, or of freeing from weeds.

There are several methods of doing it. Where plants in beds, whether sown broad cast or in rows, are young and small, and are infested with weeds, the weeds must be pulled up with the thumb and finger; or else cut out with the weeding hook. When weeding is done by hand, care should be taken that the roots be not left in the ground. Therefore the fingers of the weeder should go into the soil more or less, according to the strength of the roots. The weeds may be saved in baskets, and thrown to the swine. Where the plants are larger, the weeds may be killed by a hoe with a sharp edge, which should be wider or narrower, according to the distance at which the plants ought to stand from each other after hoeing. But it is necessary that the hoeing be performed in dry weather; otherwise many of the weeds will revive, and grow: And the forenoon is better than the afternoon for this work, because a dewy night coming on before they are withered, some of them will get rooted before the next morning. Weeders should be careful that they tread upon weeds that they hoe or pull up, as little as possible; because doing this will close the earth about them and prevent their dying.

When small plants stand in rows

according to the drill husbandry, the plants in the rows must be once weeded, and thinned as in gardens; but the weeds in the intervals are to be more expeditiously destroyed with the plough, horse hoe, or cultivator.

WEEDS, useless or noxious plants, or vegetables not to be cultivated.

Weeds infest all kinds of land more or less, and occasion abundance of labour for the farmer, and the gardener, so much that we seldom find it perfectly accomplished.

Weeds are several ways hurtful to the crops that are cultivated.

1. They rob other plants of the food that should nourish them. For they require as much nourishment from the earth as better plants do; and in general they are fed with the same kind of juices: For it has been proved, that the food of all plants is nearly the same. So that wherever weeds are suffered to stand and grow among plants, the crop will receive proportionably the less quantity of nourishment from the earth. It will also receive the less nourishment from the air, unless the weeds are of a much lower stature than the plants that were meant to be cultivated, and nearly covered by the plants.

2. Weeds also lessen the quantity of pasture for plants in the soil. For their roots occupy and fill many of those interstices in the soil, which would have been occupied by the roots of the cultivated plants. And many kinds of weeds have such a multitude of strong and binding roots, that they render the soil stiff and hard; and so compact that the roots of tender plants cannot extend themselves in it. This bad quality is possessed, in a remarkable degree, by the quitch grass, and by some other weeds.

3. Weeds prevent plants from tillering, or branching out from their roots, as many kinds, and particularly the farinaceous, which are the most valuable, are inclined to do, when they have room. But the growing of a multitude of weeds among them reduces them to a crowded situation; the consequence is, that they shoot up only single stalks, by which means the crop is greatly diminished.

4. Weeds deprive plants of that free circulation of air about them, which is necessary to their being in a healthy and vigorous state. For want of this they run up weak, remain of a loose and spongy contexture, and bend down and lodge by their own weight, unless the weeds happen to be so strong as to hold them up.

5. Weeds, besides the general evil of taking away the food of plants, rob the soil, particularly of its moisture, and speedily reduce it to such a dry state that neither weeds nor other plants can receive from it any vegetable food, for want of that proportion of moisture which is necessary to give it fluidity. Accordingly it is observable that the abounding of weeds brings on an early appearance of drought.

Lastly, some weeds of the creeping kind twine about the plants in such a manner as to prevent their growth, and the circulation of their sap. Others overtop them, and shut out the direct influence of the sun, and further oppress them with their unwholesome dripping on them. And there are still others, the dodder in particular, which draw their nourishment directly out of the bodies of plants, by striking their fibres into them, and cause them to languish.

So that, on the whole, the mischiefs done by weeds are so great and many, that when they are suffered to grow unmolested among useful plants, whatever culture has been given to the soil to prepare it for a crop is thrown away; and the seeds that are sown are entirely wasted and lost.

Weeds may be divided as plants in general are, into annuals and perennials; or, as Mr. Dixson has done it, into those which are propagated by the seed, and those that are propogated by the root. But I shall pursue a method which is different from both, and which better suits my present inclination: And only consider those weeds in a general way, which chiefly infest our lands that are in tillage, and those that prevail in our grass land; and how to prevent the existence, or prevalence of these kinds; and the best methods of preventing, subduing and destroying them.

1. As to those which infest land that is employed in tillage, and which are very numerous, there are several ways to prevent their existence in fields; at least, to prevent their arising in such plenty as to be very troublesome or detrimental to the crops.

The first thing that I would insist upon is, that no seeds of weeds should, by any means, be carried into the fields. And that this evil may be avoided, no dung nor compost, should be laid upon the soil, until it has undergone such fermentations in heaps, as to allow opportunity to destroy the vegetative quality of all the seeds that are contained in it. Therefore all dunghills intended for the manúring of sowed fields, should be shoveled over two or three times in a summer; by means of which, most or all of the seeds contained in the heaps will vegetate, and be destroyed. But when land is laid down for grass, this precaution is not necessary: Because tillage weeds will not flourish in grass lands; at least not so much as to do any great damage.

But when there is a necessity of using new dung, which abounds with the seeds of weeds, on land to be continued in tillage, let it, if possible, be applied only to a hoed crop, in the tilling of which, the weeds will be destroyed as fast as they rise, during the summer.

Or, if low ground hay has been used by itself, in feeding the young part of the stock (as indeed it ought always to be) let the dung that is made of that be laid on the driest parts of the farm. There will be no danger in laying it on while it is new; for if the seeds sprout and come up, they will not prosper, as the soil does not suit their nature, being mostly aquatics.

As to other manures, such as marle, mud, ashes, lime, soot, sea weeds, &c. these having no seeds in them that will grow upon land, there is no danger of their causing the ground to become weedy. If they do this, in any measure, it must be only by vivifying latent seeds in the soil.

Another thing which is indispensably necessary to prevent the abounding of weeds, is, to suffer no weeds, neither in gardens nor in tillage lands, to ripen their seeds in autumn, and scatter them out upon the ground. The practice of sluggards must be laid aside: For *as a a man soweth, so shall he also reap.* If weeds are thus sown on every part of a field, it is no wonder if the ground be filled with them in the ensuing year: Nothing short of a miracle, unless it be want of strength in the soil to nourish them, can prevent it. The prudent husbandman, therefore, before the seeds of the weeds are ripened, will go over his grounds, and destroy all the weeds that appear. If there should happen to be a plenty of them, let them be piled in heaps in the borders of the fields, and a little earth thrown on them; in which situation they will ferment and putrefy, and become good heaps of manure, in the succeeding year.

If any should object to the labour of doing this, let them consider whether it will not save them ten times as much labour in future, in subduing the weeds, after the ground is filled with them, besides procuring them the advantage of having much better crops.

Another good preventive of the increase of weeds, is burning the stubble as it stands after reaping. On land that is designed to be sowed the next year, this is a good piece of husbandry; for it will destroy so many of the seeds of weeds, as to prevent the ensuing crop from being so very weedy as it otherwise would be; at the same time, it will destroy many insects, clean the ground, and render it fitter for the operations of tillage; not to mention how much the ground will be fertilized by the ashes of the stubble.

But, to prevent the increase of weeds, as well as to answer other good purposes, it is best that two broad cast, sowed crops, when it can be otherwise ordered, should not succeed each other. Where a hoed crop is well tilled every other year, the weeds are not wont to increase, in such a degree as to be very troublesome.

Another precaution, and which has not been sufficiently attended to, is taking care to avoid sowing the seeds of weeds with grain and other crops. A thorough cleaning of the seed is of great importance. This may be accomplished by winnowing, sifting, swimming and repeated washings. But the seeds of weeds are often sowed, after they have been separated from grain by winnowing. This will be the case when the chaff and rubbish at the tail of the sheet is thrown upon a dunghill, which is to be removed and applied to the soil, before the seeds contained therein have had opportunity to vegetate and get destroyed.

If a farmer begins on a farm that is not much infested with weeds, and will carefully observe the directions given above, I think he will be always favoured with clean crops, and that weeds will never prevail much in his tillage land.

It is true, that some seeds may be wafted by the winds from other fields into his enclosures. The best preventive of this is having close fences. But those seeds to which a light down adheres, are often mounted high in the air; so that there is no possibility of fencing against them. Of this kind are the seeds of thistles, &c. However, they seldom tarry in the places where they first fall, but are driven to the borders of fields.

But when, by means of a negligent or slovenly culture, lands in tillage are become weedy, the farmer should perfectly know, and spiritedly practise, the most effectual methods of subduing the weeds, which are his most formidable enemies. For they diminish the husbandman's crops, more than most people are aware of.

The most effectual way of destroying them is, by a good summer fallowing, alternately ploughing and harrowing the ground in dry weather, as often as the weeds appear in plenty on the surface. At each of these stirrings a large quantity of weeds is destroyed, and converted into manure; and the seeds of another crop so exposed to the air, that they will spring up, and at the next stirring of the soil be destroyed. And if these operations are continued until no more weeds arise, it may be concluded that the ground is in excellent order for a crop. It is not only cleaned from weeds, but greatly enriched: And it will not be liable to be infested with weeds, to any great degree, for several years after.

Next to fallowing, which may be thought rather expensive, the best way to get the better of weeds, is to employ the land in hoed crops, for two or three years in succession, not omitting autumnal hand-weeding. Or a crop of pease, sowed thick, or any crop that forms a close cover to the surface, may do much towards subduing the weeds. The few weeds that vegetate under such a cover are so cramped and kept under, as to be prevented from going to seed; or at least from bringing their seeds to maturity.

Extraordinary deep ploughing, or trench ploughing, has been recommended as effectual. But in this practice, care should be taken to avoid turning in too many of the seeds of weeds. The autumn, therefore, is not the right time for this, but rather the spring or summer. For if many seeds are thus buried, the next deep ploughing that there may be occasion for, will bring them again up to the air, and cause not a few of them to vegetate, though it should happen to be not till many years after.

As deep ploughing destroys weeds, or rather puts them out of the way, so, on the other hand, nothing can more encourage their growth than the shallow ploughing, which is commonly given for sowed crops in this country.

But if all other methods were to prove ineffectual for the destruction of annual weeds, laying tillage land down to grass, and sowing it with cleaned grass seeds, will not fail to subdue these weeds; the most of which depend upon tillage to promote their growth; and the rest will be overtopped and stifled by a good crop. of grass. And a strong sward will soon be formed, through which the tender kinds of annual weeds will not penetrate, of which sort the most are, that are found in land that is tilled. Even the quitch grass, one of the most vexatious of all weeds, is thus either destroyed or rendered harmless: For when it flourishes in grass ground, it makes an excellent hay.

2. The weeds that appear in *grass lands* in this country, may be divided into upland weeds and aquatics, some few of which are annuals, but a greater number, especially of those that are most noticed, are perennials.

Of the upland weeds those which have proved to be the most troublesome are the upright crow foot, Ranunculus, commonly known by the name yellow weed; ragweed, ragwort, or Roman wormwood, Senecio; the greater daisy, ox eye, or piss abed, Chrysanthemum; yarrow, dandelion, dock, thistles, sorrel and John's wort.

Some of these, particularly the two last, and the daisy and ragweed, are conquered by a plentiful manuring of the ground; for where the sand is rich they are not found to flourish. Pasturing the land with sheep is said to be fatal to the daisy, and the crow foot.

But the most effectual way to destroy these weeds, is to break up the land, and employ it in tillage.

When it is not found convenient to take this method, the weeds may either be dug out or else pulled up by hand, which, when the ground is moistened by rain, may be easily done. It is to be remembered, that this should be done at or before midsummer, that none of their ripened seeds, or any that may possibly vegetate, may be scattered on the ground.

The aquatic weeds, such as flags, rushes, and the like, are not easily subdued by any of the above methods, some of which have often been tried in vain. Draining the land, which deprives them of that degree of wetness which is suitable to their nature, is an infallible method, and, perhaps, the only effectual one, of destroying them. But liming the ground at the same time, renders the operation more sudden and effectual. Or if lime cannot be had, ashes and soot are good substitutes.

But the most speedy method of getting rid of these weeds is, digging out the roots, or using the land in tillage.

WEEVIL, an insect of the beetle kind, resembling the Maybug, with a long sharp-pointed head, to the hinder part of which are fixed two antennæ. It is black, and therefore easily distinguished in any corn; but its principal and favourite food is wheat, of which, either old or new, it devours great quantities, without however communicating any bad smell to it.

Upon thrusting one's hand into a heap of corn, one may easily perceive by its heat, whether it contains many of these insects, which generally lie pretty much collected; and the particular places where they are most numerous, feel much warmer than the rest.

This observation led M. Duhamel to think, that a considerable heat is probably necessary for the hatching of their eggs; and that in this case, even if they should live, they will not be able to breed in his ventilating granaries.

To prove this he made repeated experiments, the results of which made it evident that this insect cannot multiply in corn that retains a proper degree of coolness, which it may be made to do by frequent ventilating. When corn is sifted in a sieve fine enough to retain the grain, the weevils then agitated shrink up their legs, and are in that posture so much smaller than the grain, that most of them drop through the sieve.

But of all the methods M. Duhamel tried, that which he preferred was, to dry the corn on a stove, or oven, heated to eighty or ninety degrees of M. de Reaumur's thermometer. See *Insect*.

WEIGHT OF CATTLE. In the Picture of London, for the present year, it is stated, that about the year 1700, the average weight of oxen killed for the London market, was 370 pounds; of calves, 50 pounds; of sheep, 28 pounds; and of lambs, 18 pounds. The average weight at present is—of oxen, 800 pounds; calves, 140 pounds; sheep, 80 pounds; and lambs, 50 pounds. The whole value of butcher's meat, as sold in Smithfield, is about *eight millions sterling*.

WHEAT, *Triticum*, the most useful of all farinaceous plants, the seeds of which are the best grain for bread.

The different species of wheat, according to Mr. Miller, are, the winter wheat, without awns, or beards; the summer, or spring wheat; grey pollard, or duck bill wheat, with long beards; the cone wheat, so named for the shape of the ears, which resemble a cone; and the polonian wheat. Other kinds he supposes to be only varieties occasioned by culture and soil.

The Smyrna wheat is very different from all other, producing one large central ear, and several smaller lateral ones from the bottom of the large one. This suits a rich and strong soil, as there is no danger of its being over fed. It seems to be better adapted than any other species of wheat to the horse-hoeing husbandry.

The only species or varieties that are cultivated in this country are, the winter and spring wheat, the bald and bearded, the red and white, and the Siberian wheat, which is bald, or without awns.

About a bushel and a half of seed is a sufficient quantity for an acre. Oftentimes it produces very large crops from a less quantity. The larger and fuller the seed is, the greater quantity by measure will be required; the smaller, the less quantity: For the number of grains is to be regarded more than the measure, or weight.

Changing the seed yearly, or at least every other year, is proper, or even necessary: For it has always been experienced, that the same seed, sown several years in succession, degenerates; so that the crops continually grow worse. But seed should not only be shifted from one place to another, but also sown on a soil that is different from that on which it grew. It should be taken from a stiff soil for a light one, and from a light for a stiff one.

As seed wheat cannot always be procured from distant countries in season farmers at a few miles distance should at least change with each other. But in the northerly parts of this country, bringing seed from the southward will not do well, as it ripens the later.

A soil of good loam, well warmed, and stirred, is proper for wheat. But other soils, well prepared, sometimes answer well enough. Wheat succeeds upon the strongest lands well tilled, when they have been drained, and laid dry.

The way of preparing the soil for a crop of wheat (I mean that which has been most practised in the old countries and has succeeded well in this) is, by a good summer fallowing. Three or four ploughings, and as many harrowings, at proper intervals, will bring almost any soil, that is not very stubborn, into a good tilth. And if it be thought needful to apply any manure to the soil before sowing, let it be old dung or compost; and let it not be laid on till just before the last ploughing, or seed furrow, as it is called; so that the corn may have the greatest possible benefit of the steam produced by the manure in the soil.

Where summer wheat is intended to be sown, the ground being previously clear of weeds, by means of a hoed crop preceding, one good ploughing in autumn, and two in the spring, may be sufficient.

But another good method of preparing the land for a wheat crop, is sowing it with clover. After it has borne two crops of clover, it will be in excellent order for wheat. Give only one ploughing, sow the seed and harrow it in.

But whether you sow winter or spring wheat, let the seed be always well washed before sowing. I need not here repeat what has been said on the affair of steeping the seed. See the article *Smut*.

The time of sowing winter wheat admits of some latitude. Some time about the latter part of August, or beginning of September, is to be chosen. It may be a little later or earlier, provided the ground have the right degree of moisture for tillage, and for promoting vegetation. The latter should be carefully attended to, and the seeds should never be sowed in ground when it is very dry. Rather than do this, rain should be waited for as much as two or three weeks.

Wheat that is sowed in autumn, a clover lay excepted, should, instead of harrowing, be covered with a shallow furrow, and the surface left rough. It will be less in danger of being killed by the frost in winter, and less injured by drying winds in the following spring. The furrows should be left without harrowing; for, the more uneven the ground is, the more the soil will be pulverized and mellowed by the frost. But it will often be of advantage to pass a roller over the ground in the following spring, to close the earth about the roots, and prevent the earth from being dried by the spring winds, to too great a depth.

Wheat sown in the spring should be only covered with the harrow, as it has no time to lose, and ought to be up early. For the same reason it should be sowed as early as the ground can be got into the right order to receive it.

If weeds arise in plenty among the wheat, they should be destroyed in the spring; otherwise the crop will be much robbed of its nourishment. This matter has not yet been sufficiently attended to in this country; though in other countries, particularly in Scotland, the people make as much account of weeding their fields of grain, as their gardens. This should by all means be performed before the time when the plants begin to send out their ears; because, after this, they will be the more in danger of being hurt by people's passing among them. Especially the wheat ought not to be touched while it is in blossom. As it is in the critical season of fœcundation, it may be very detrimental to the crop. And if the weeds be not taken out before the grain is become large and tall, many of them will be covered and hidden under the plants; so that only the largest weeds will be pulled out, while the rest are suffered to stand and perfect their seeds. For these reasons, Mr. Miller recommends to farmers a practice among kitchen gardeners, which he says has been found to be of great benefit to crops, and has also been a great saving to them in the operation of destroying weeds, which is, making use of a small kind of hoes for cleaning the wheat early in the spring, before the ground is covered with the blades of the corn. By this instrument, all the weeds, small and great, will be cut up; and if it be done in dry weather, will die. He recommends a second hoeing about three weeks after the first, which will subdue those weeds which happen to remain alive. And at the time of hoeing, he recommends thinning the plants, where they happen to stand too thick, which he thinks will do much to prevent the lodging of the corn, and render it less liable to be smutty. Whether these hoeings will greatly increase the crop I think there can be no reason to doubt, unless they should cause it to ripen so much later as to expose it to blight. But I should not expect it from such a superficial hoeing.

Some farmers are apt to think there is no inconvenience in suffering sheep to feed on the young plants of winter wheat, in autumn or winter, or even early in the spring. But who can rationally suppose, that the blades with which nature has furnished these plants, are not of use, to draw in nourishment from the air and dews, for the increase of the stalk and the ear? In order to be satisfied of this, the above mentioned writer cut off some plants of wheat alternately, early in the spring, and always found the stalks of these plants much smaller and shorter, and the grain poorer, than those of the intermediate plants whose blades were not cut.

There certainly is but one case in which it can be right that winter grain, whether wheat or rye, should be fed in autumn; and that is, when, by being sowed very early, it is in danger of earing before winter; for this should by all means be prevented. But in no case should a beast be suffered to feed on it in the spring.

" Wheat is ripe for cutting when the straw is turned yellow, its ears hang, no greenness appears in the middle of them, and the grain is hard when bitten."

The farmers in England cut their wheat greener than they did formerly, because they find it makes a rather whiter flour, which sells at a higher price. One of their skilful farmers, who cuts his wheat early, says, that he found upon trial, his wheat early reaped was heavier by the bushel, than the same wheat when it stood till it was thoroughly ripe. This might be owing to its greater smoothness, by means of which it lay closer in the bushel; for it is hardly credible that the whole crop will be heavier for cutting it green. But as loss by shattering out the corn is thus prevented, it may be a balance to it shrinking the more, as I think it certainly does. The prudent farmer must use his own judgment in this matter.

Wheat and other grain that is lodged, may, and ought to be, cut the earlier: For after the straw is broken or corrupted, it conveys no nourishment to the grain, or as bad as none.

The ancients reaped their corn, as Pliny says, before it was fully ripe. And it is certain that great inconveniences arise from letting some sorts of grain stand till they come to their utmost maturity. The chaff and straw are the worse fodder. And if such corn chance to take wet in harvesting, it suffers the more for being very ripe. But corn cut in a greener state will bear a good deal of wet without damage.

When corn is blighted, it should be cut the sooner, or even before it is half ripe. For though it ceases to receive nourishment through the straw, it is said it will improve by lying in the ear, and that it threshes out the more easily.

One acre of grain is a large day's work for the reaper. The usual price of reaping an acre in this country, has been a bushel of the grain.

From three to four quarters, that is, from twenty-four to thirty-two bushels of wheat on an acre, is reckoned in England to be a good crop.

To prevent loss by shedding, the corn which is laid in grips, is to be bound up in sheaves, either in the cool of the evening, or before the dew is quite gone off in the morning. And it should never be removed either into the shock, the barn, or the stack, in the middle of a sunny day, but rather at a time when the slight dew is just sufficient to prevent its shedding. See the article *Harvest*.

As to sowing wheat in drills, see New Husbandry.

But setting of wheat is reckoned by some of the English as a great improvement in husbandry. A Norfolk farmer one year set fifty-seven acres. The superiority of his crop, both in quantity and quality, was so great that it induced him the following year to set three hundred acres, and he has continued in the practice of setting ever since. This noble experiment was the means of introducing the practice in the vicinity, and to a considerable extent. Though the set crops appear very thin during autumn and winter, the plants tiller and spread prodigiously in the spring. The ears and the grain are larger, and specifically heavier per bushel than other wheat.

The lands on which this method is most properous are, either after clover stubble, or on which trefoil and grass seed were sown the spring before last. These grounds, after the usual manuring, are once turned over by the plough in an extended flag or turf, ten inches wide; along which a man, who is called a dibbler, with two setting irons, somewhat bigger than ramrods, but considerably bigger at the lower end, and pointed at the extremity, steps backward along the turf, and makes the holes, about four inches asunder each way, and an inch deep. Into these holes the droppers drop two grains, which is quite sufficient. Thus three pecks of grain is enough for an acre. The regularity of its rising gives opportunity for weeding or hand-hoeing.

This method is advantageous when seed corn is dear. Sir Thomas Bevor found the produce to be two bushels per acre more than from sown wheat, having much less small corn mixed with it; and it fetches a higher price three pence per bushel. The expense of dibbling an acre is 10s.

Another new method of cultivation is, propagating wheat by dividing and transplanting its roots. "On the second of June 1766, Mr. C. Miller sowed some grains of the common red wheat; and on the eighth of August a single plant was taken up and separated into eighteen parts, and each part planted separately. These plants having pushed out several side shoots, by about the middle of September, some of them were taken up and divided, and the rest of them between that time and the middle of October. This second division produced sixty-seven plants. These plants remained through the winter, and another division of them, made between the middle of March and the middle of April, produced five hundred plants. They were divided no further, but permitted to remain. The plants were in general stronger than any wheat in the fields. Some of them produced upwards of one hundred ears from a single root. Many of the ears measured seven inches in length, and contained between sixty and seventy grains.

The whole number of ears, which, by this process, were produced from one grain of wheat, was 21,109; which yielded three pecks and three quarters of clear corn, the weight of which was 47 lb. 7 oz.: And from a calculation made by counting the number of grains in an ounce, the whole number of grains was about 576,840." *Encyclopedia.*

Though this method is curious, it is attended with so much work, that few or none will be disposed to follow Mr. Miller's example.

For about thirty years immediately preceding the year 1813, few attempts were made to raise wheat in parts adjacent to the sea coast in Massachusetts; and a belief generally prevailed, that it could not be made to thrive there, owing to peculiarity of climate or some other inexplicable cause. Since that time, however, it has been considerably cultivated. We shall add to this article some abridged statements of some experiments, which led to the more general introduction of that kind of culture.

Mr. Dudley Hardy sowed, on three quarters and an half quarter of an acre of land, in Brighton, near Boston, twenty-eight quarts of spring wheat originally from Londonderry. The land the preceding year had been planted with Indian corn. It was ploughed in the fall; and in the month of March, before the frost was all out of the ground, was ploughed over again two or three times, and then harrowed with an iron tooth harrow. The grain was prepared by steeping it in ley made of ashes twenty-four hours, and on the 7th of April sowed, and harrowed in with the same harrow. "After this," says Mr. Hardy, "I bruised the ground smooth with a brush harrow." The twenty-eight quarts produced eighteen bushels, weighing sixty pounds a bushel. One bushel, ground and boulted, gave forty-six pounds and an half of flour. Mr. Hardy thinks, that spring wheat should be sowed in the month of March, if the frost will permit.—*Massachusetts Agricultural Repository*, Vol. III. p. 31.

Mr. J. Lowell gives the following account of a trial of the same kind of wheat.

"I had but one small piece of ground in a proper state to receive wheat. It measured one third of an acre. The soil was very thin over a bed of gravel, extremely subject to drought, and incapable, as I thought, of bearing a large crop of any sort. Potatoes had been cultivated on it for two years preceding. It had been twice ploughed the fall before, after the potatoes were dug. In the spring, four horse cart loads of horse dung were spread upon it and ploughed in. On the seventh day of April, I sowed upon it three quarters of a bushel of Mr. Hardy's wheat. This wheat was of small size, and rather shrivelled. It is said to be the same known and cultivated as Londonderry wheat.

"The crop looked extremely well; none of it was blighted; and on the second of August it was reaped. It weighed from fifty-six to fifty-eight pounds the bushel."—*Massachusetts Agricultural Repository*, Vol. III. p. 216.

The same publication, pages 217, 218, contains the experiments of Hon. J. Quincy, and Hon. P. C. Brooks, which, though somewhat less successful than the preceding, (Mr. Quincy having raised fifteen, and Mr. Brooks fourteen bushels to the acre) yet as the grain was of good quality, and free from blast or smut, establish the fact that the climate of Massachusetts is not un favourable to the cultivation of wheat.

Bezaleel Taft, Jun. Esq. of Uxbridge, likewise states in substance, that his father, about fifteen years since, procured a bushel of spring wheat from Barry, on the Onion river, in the state of Vermont. The produce of this was fifteen bushels. He continued to procure his seed from that quarter for several years, but at length sowed the seed of his own growth prepared by washing it clean, stirring it well in two or three changes of water. After washing it was soaked about twelve hours in a weak ley; and after turning off the ley, about two quarts of slacked lime was stirred into a bushel of wheat.

The ground selected for the cultivation of wheat was such as would be most sure to produce a good crop of Indian corn, and the wheat was sowed as early in the spring as the soil could be stirred and remain light.

Five pecks of seed were sowed to an acre, and the crops have been from twelve to twenty-two bushels; and about sixteen upon an average to an acre.

This communication was dated the 19th November, 1814, and the writer says, "For the last three years, I believe this town has produced annually about a thousand bushels, and the last season we had at least four times as many bushels of wheat as of rye from the same numbers of acres, in the same state."

The writer considers wheat as more favourable to a future crop of grass than rye or oats, as it shades the grass less; and when the crop is removed, it is not so apt to be scorched, having been more accustomed to the rays of the sun. In that quarter of the country, wheat is invariably washed before it is sent to the mill, not to free it from smut or mildew, but from dust which adheres to it in consequence of the sandy nature of the soil. He says, that "an active man will wash ten bushels in two or three hours. Care ought to be taken not to have it remain longer than necessary in the water. We usually dry it on blankets or sheets in the sun. Care should likewise be taken not to have it get too dry, as the flour in that case is not so nice. One day's sun is sufficient to dry it in the summer, and two in the fall. If suffered to become too dry, the hull or bran is brittle, and cuts to pieces in grinding, so as to mix with the flour. When only dried sufficient to prevent its clogging in the mill, the flour separates much better from the bran, and is far preferable for use."—*Massachusetts Agricultural Repository*, Vol. III. pp. 218, 219, 220.

Mr. John Jenks gives a statement of his experiment relating to the same subject. His ground was two acres, and seed four bushels. A part of this seed was soaked in weak ley, and part in sea water, and the parcels kept separate. Both were soaked eight hours, the water drained off, the wheat spread on a tight floor, lime sprinkled on it and raked over, until it was all covered with the lime and dry, and appeared all over as white as rice. There was no discernible difference in the growth of the grain soaked in sea water. and that soaked in ley. One peck of his seed was soaked in weak ley only an hour, and then limed and sowed by itself upon a quarter of an acre adjoining the land already sown, part of the two acres. This was much blighted. and produced little in comparison with the other part of the two acres. The produce of an acre and three quarters, which was sowed with the wheat longest soaked was forty-four bushels, weighing sixty-two pounds the bushel when first sowed. The land on which this grew was a good dark rich mould, and herds'-grass and clover was sowed with the wheat. The grass was well grown, and covered the ground when the wheat was reaped, and the writer supposes that if no grass had been sown, the crop of wheat might have been better.'—*Agricultural Repository*, Vol. III. p. 221.

In the spring of 1814, Gorham Parsons, Esq. sowed four acres and thirty-two rods with four bushels; ploughed in eight loads of manure to the acre, and laid it down with herds'-grass, red-top, and red clover. It produced eighty-four bushels and an half of wheat, besides a good crop of grass, which was mowed in September. Mr. Parsons thinks, that his crop would have been better had double the quantity of wheat been sown, or two bushels to the acre.—*Agricultural Repository*, Vol. III. p. 271.

In 1816 J. Lowell, Esq. observes, that "I selected a piece of land, measuring nearly three quarters of an acre; its quality is but indifferent, a light thin soil on a gravelly base. It had been leased to different tenants for six years, and was very much exhausted, having been constantly in tillage, and never surcharged with manure.

"I sowed only one bushel of wheat, at least one half less than I should have sown; it was steeped in brine for three days, limed, and sown with one bushel of plaster of Paris. The crop was sixteen bushels of very fair, full and heavy wheat; that is, about twenty-two bushels to the acre."—*Agricultural Repository*, Vol. IV. 272.

Mr. Justin Ely writes to the President of the Massachusetts Agricultural Society as follows:—

"The largest crop of winter wheat was raised in Springfield last summer (1816) that is known ever to have been raised in this vicinity, and perhaps larger than was ever before raised in New-England.

"Four acres of land, one of the house-lots in Springfield-Street, belonging to the distillery company, three years ago last spring were English mowing. It was manured and ploughed up and planted with Indian corn, and dung put in the hills. The crop was abundant. The next spring it was covered with very rich manure from the distillery, and hemp seed sowed thereon. The crop was large and heavy. The land was then ploughed twice and sowed with the bald wheat, one bushel and three quarts to the acre. The produce was two hundred bushels of good, clean, heavy wheat, from the four acres.

" About half an acre of the hemp was not pulled with the other hemp, but was suffered to stand till the hemp seed was ripe, whereby the sowing of the wheat, on that part, was delayed too late in the season, which diminished the crop of the wheat on that part, six or eight bushels, as supposed.

" It is the opinion of many people, that the unprecedented cold and draught of the last summer checked and retarded vegetation so far as to prevent the destruction of the crop by blasts and other causes, and that if the last summer had been as warm and wet as usual, the whole crop would have been blasted and lodged, so as to have been wholly ruined."

For further experiments relating to the culture of wheat, see *Massachusetts Agricultural Repository*, Vol. IV. p. 195. 267. 278. 345. Vol. V. p. 65. 192. 265. Vol. VI. p. 239.

Mr. Payson Williams, of Fitchburgh, Massachusetts, gives the following account of his method of raising a crop of spring wheat, being twenty-eight bushels and thirty quarts on one acre and an eighth part of an acre, for which he received the Massachusetts Agricultural Society's premium of forty dollars, October 1819.

The land on which the wheat was sown, was in 1818 planted with potatoes, (for one acre of which I obtained your premium) which, after harvesting, was ploughed a short time before the setting in of winter. In the spring of 1819, as soon as practicable. (after spreading on six load of fermented manure) it was again cross ploughed—26th April sowed on the furrows two bushels of what is known by the name of the *Gilman* wheat, (which I procured of the Hon. P. C. Brooks of Boston) on one acre and twenty square rods, and cross-harrowed the same, following the harrow at the same time with the clover seed, which in turn was cross-harrowed in. The wheat before sowing was washed in water until perfectly clean, then immersed in a liquor, or ley, made in the proportion of four pints of water to every pound of wood ashes, then add one pound of unslacked lime to every bushel of seed. as recommended by M. Du Hamel. When the wheat plant was out of ground two inches, I sowed on a part of the field plaster of Paris, at the rate of ten bushels to the acre, which I never have been able to discover, has had the least effect, (I had the like ill success in the use of a ton, on various parts of the farm.) The amount of the wheat by actual measure, was twenty-eight bushels and thirty quarts. It may not be improper here to state, that on the most close examination, I could not discover one kernel of smutty grain in the whole crop; and had it not been for the ravages of the grasshopper in this field (in many parts of which they cut off one fourth part of the heads, which were of course lost,) there would probably have been thirty four bushels. I esteem this kind of wheat a valuable acquisition to this part of the country. The grain weighing sixty-two pounds to the bushel, and yielding at the mills in this quarter, forty-five pounds of flour, in quality equal, I think, to the best Baltimore."—*Massachusetts Agricultural Repository*, Vol. VI. pp. 32, 3.

WHEEL, a simple machine, consisting of a round piece of matter, revolving on an axis. It is one of the principal mechanic powers.

As the farmer has much occasion for using wheel carriages, something ought to be said of wheels in a work of this kind. And perhaps the following observations may be advantageous to those who have been most destitute of information.

1. The wheels of carriages must be exactly round. For if the nave were out of the centre, the wheel in turning would be affected in the same manner upon plain ground, as other wheels are where the ground is uneven; and the difficulty of the draught is in the same manner increased. For this reason, wooden wheels, which are designed to go on the roads, should always be shod with iron; because those which are not, by wearing faster in some parts of the rim than in others, soon lose their circular form, and become harder to draw, jolting up and down on the most level way.

2. The felloes or the rim, must be at right angles with the naves, or hubs, notwithstanding the inclination of the spokes; for otherwise the wheels will not move regularly in the ruts, but form a crooked track upon the ground, just as it does when the axis is too small to fill the hole, or boxes, in the nave. This is as really detrimental to the going of wheels as their being out of round. The spokes, in this case, will not be all equally inclined to the nave, and the wheel will have the less strength, and will be more liable to be broken in moving upon an uneven place.

3. The spokes must be so set in the nave as to incline outwards. This would be needless and wrong, if wheels were always to be moved on a perfect plane; for they would bear the load perpendicularly. But as wheels must often go upon uneven places, and fall into holes and ruts, and as the wheel that is lowest bears a greater part of the load than the other, and sometimes almost the whole of it; it is necessary that the wheel should become perpendicular at the moment of its sinking in the part under the nave. Dishing wheels are also less liable to be overset than perpendicular ones. But this might be prevented by using a longer axle.

4. The axletree must be at right angles with the shaft, or tongue, of a cart or waggon. Otherwise it will not move straight forwards in the track, or directly after the team, but sideways. But the axle should be so set in the naves, that the wheels may be gathering forward, as it is called, that is, that the fore parts of the rims may be a little nearer together than the hinder parts. Otherwise the wheels will not so easily keep in the ruts, as will appear evidently to any one that takes notice of their going. And the axle is usually so inserted

into the naves, that the wheels gather at the bottom. This evidently gives an advantage to the axle; as the wood, in fitting it to the holes of the naves, is not cut across the grain on the under side, it is not so liable to be broken by the weight of the load. Whether there be any other advantage in it, I do not determine. But it is certain that, in this case, the spokes in the lower part of the wheel, which have the whole pressure of the load, will be the nearer to a perpendicular situation, while on level ground, which seems to be some advantage. But the lower spokes ought in no part of a sidling road to lean outwards. For this will put both the axle and the wheel to a great strain.

5. High wheels are more easily drawn than low ones. They have less resistance from friction, supposing the axle to be of the same size as in low wheels, which ought to be supposed, the load being the same. For a wheel that has double the circumference of another, will make but one turn while the other makes two; consequently the first will have but half so much friction to overcome as the second. High wheels have also the advantage of low ones in surmounting obstacles, as every spoke acts as a lever, in moving the wheel forward, and as a longer spoke removes the power to a greater distance from the weight. And high wheels easily pass over holes into which small ones would sink; and the impression they make upon soft ground is not so deep as that which is made by small ones, as a greater portion of their rim is always in contact with the soil.

It is objected, that a cart with high wheels is more apt to overturn. This inconvenience might be easily removed, by fastening the lower timbers of the cart to the under side of the axle, as in some parts of this country has been practised for a long time, particularly in coal carts.

Another objection to high wheels is, that they make the cattle draw too high. This may receive the same answer as the former. Or they may be made to draw low, by fixing the traces to arms made for the purpose, reaching as far below the axle as any one pleases.

6. The fore and hinder wheels of a waggon should be of equal height, in order to render the draught as easy as possible, as Dr. Desaguliers proved by experiments.

7. All persons who use the same roads should place their wheels at the same distances from each other. For he who does not observe this caution, but makes them go two or three inches wider or narrower, will find that he must carry a less load than others do, in proportion to the strength of his team.

Lastly, broader rims than those which are in common use, would be more conducive to the goodness of the roads, if all who drive teams in the same roads would agree to be confined to the same breadth. For the broader the rim, the less the wheels will sink into the soil; but the narrower the rim, the deeper ruts will be made. See the article *Cart*.

WHEEZING, *or* PURSIVENESS, a distemper in horses, otherwise called *Broken Wind*, in which the horse makes a hissing or whistling sound in his throat in breathing, and has a greater heaving in the flanks than in common colds.

This disorder is commonly caused by surfeiting, violent exercise when the belly is full, or by being ridden or driven into the water when he is hot and sweaty, or from an obstinate cold not well cured.

For the cure of this disorder, Dr. Braken advises, 'that the horse should have good nourishment, much corn and little hay, and that every day the water given him be impregnated with half an ounce of salt petre, and two drams of sal ammoniac.' Some say the hay made of white weed will cure this distemper.

WHEY, the serous part of milk, or that which remains fluid when the curd is taken for cheese.

Whey is an important liquor for swine. They are extremely fond of it, and it conduces much to their growth and fattening. It is particularly proper for sows that suckle pigs, as it greatly increases their milk. But care should be taken not to overfeed swine with this liquor; for it has often happened, that after drinking plentifully of it, especially in very hot weather, they swell up and die.

For a method of making vinegar from whey, *see vinegar*.

WHITE SCOUR, a disease with which sheep are often affected, and is fatal to them in other countries, though I have not known it to be so in this.

"The following medicine has been often given with success, provided the sheep are at the same time removed into a dry pasture.

Take a pint of old verjuice, half a pound of common or bay salt, dried well before the fire, pounded, and sifted through a sieve. Then mix the verjuice with the salt by degrees; and add half a pint of common gin, and bottle it up for use. When any of your sheep are seized with this disorder, separate them from the flock, and give each of them three large table spoonfuls of the mixture for a dose, repeating it two days after, if they are not better."—*Complete Farmer*.

WHITE WEED, has a five cornered stalk. The leaves are jagged and embrace the stalk. The flowers are discous, large and radiated. The ray is white, and the disk yellow; the seeds have no down. It flowers in June, and is perennial in the roots.

When this weed has got possession of the ground, no good grasses grow with it; because, perhaps, the roots bind the soil in such a manner as to cramp other roots. Or, being a strong feeder, it deprives other roots of their food.

When it is in its green state, neither neat cattle nor horses will eat it. But if it be cut while in blossom, and well dried for hay, the cattle will eat it freely in winter, and live well on it. The crop however is always thin and light. If it is mowed late, or not well cured and preserved, the hay will be of very little value.

Dunging the ground is an enemy

to this weed; and it is said that pasturing with sheep kills it. But to conquer it effectually, there can be no better way than to use the land in tillage, for hoed crops, several years in succession.

WILLOW, *Salix*, called also *sallow* and *osier*, a well known tree, of which there are many species.

Some willows may be propagated to great advantage for fire wood, as they are very rapid in their growth.

Other kinds have twigs that are long and tough, and are useful for hoops, basket work, &c. These are commonly kept low by cutting; when this is neglected they grow into large trees.

A wild shrubby willow, natural to this country, is common in our wet lands. This is called white willow, *Salix alba*, and in medicine is a good substitute for the *Cortex peruvianus*.

The *salix viminalis*, the toughest kind of willow, has not, that I know, been yet propagated in this country. The weeping willow, so called, has been imported from Europe: But whether it prospers I am unable to say.

No more than two sorts of foreign willow have been much propagated in this country, a yellow and a green sort. The former grows well, even in a soil that is pretty dry. The latter flourishes finely in a wet situation.

Mr. Miller says, "all sorts of willows may be easily propagated by cuttings or sets, either in the spring or autumn, which readily take root. Those sorts which grow to be large trees, are cultivated for their timber, so are generally planted from sets which are about seven or eight feet long; these are sharpened at their larger end, and thrust into the ground by the sides of ditches and banks, where the ground is moist; in which places they make a considerable progress, and are a great improvement to such estates, because their tops will be fit to lop every fifth or sixth year. This is the usual method now practised in most parts of England, where the trees are cultivated, as they are generally intended for present profit. But if they are designed for large trees, or are cultivated for their wood, they should be planted in a different manner; for those which are planted from sets of seven or eight feet long, always send out a number of branches toward the top, which spread, and form large heads fit for lopping: But their principal stem never advances in height; therefore, where regard is paid to that they should be propagated by short young branches, which should be put almost their whole length in the ground, leaving only two, or at most but three buds out of the ground. And when these have made one year's shoot they should be all cut off, except one of the strongest and best situated, which must be trained up to a stem, and treated in the same way as timber trees. If these are planted with such a design, the rows should be six feet asunder, and sets four feet distance in the rows: By planting them so close they will naturally draw each other upward; and when they are grown so large as to cover the ground and meet, they should be gradually thinned, so as at the last to leave every other row, and the plants in the rows, about eight feet asunder. If they are so treated, the trees will grow to a large size, and rise with upright stems to the height of forty feet or more.

"When these cuttings are planted, it is usual to sharpen those ends to a point, which are put into the ground, for the better thrusting of them in; but the best way is to cut them horizontally, just below the bud or eye, and to make holes with an iron in the ground where each cutting is to be planted, and when they are put in, the ground should be pressed close about the cuttings with the heel, to settle it. and prevent the air from penetrating to the cuttings.

"The after care must be to keep them clear from weeds the two first seasons, by which time they will have acquired so much strength, as to overpower and keep down the weeds: They will also require some trimming in winter, to take off any lateral shoots, which, if suffered to grow, would retard their upright progress."

WIND GALL, "a flatulent swelling, which yields to the pressure of the finger, and recovers its shape on the removal thereof. The tumour is visible to the eye, and often seated on both sides of the back sinew, above the fetlocks, on the fore legs; but most frequently on the hind legs; though they are met with in various parts of the body, wherever membranes can be so separated, that a quantity of air and serosities may be included within their duplicatures.

"When they appear near the joints and tendons, they are generally caused by strains, or bruises on the sinews, or the sheath that covers them; which, by being overstretched, have some of their fibres ruptured; whence probably may ooze out that fluid which is commonly found with the included air; though where these swellings shew themselves in the interstices of large muscles, which appear blown up like bladders, air alone is the chief fluid; and these may safely be opened, and treated as a common wound.

"On the first appearance of wind galls, their cure should be attempted by restringents and bandage, for which purpose let the swelling be bathed twice a day with vinegar, or verjuice alone, or let the part be fomented with a decoction of oak bark, pomegranate, and alum boiled in verjuice, binding over it, with a roller, a woollen cloth soaked in the same. Some for this purpose use red wine lees, others curriers' shavings wetted with the same, bracing the part up with a firm bandage.

"If this method, after a proper trial, should not be found to succeed, authors have advised that the swelling be pierced with an awl, or opened with a knife; but mild blistering is in general preferred to these methods; the included fluids being thereby drawn off, the impacted air dispersed, and the tumour gradually diminished. A little of the blistering ointment should be laid on every other day for a week, which brings on a plentiful discharge, but generally in a few days is dried up, when the horse may be put to his usual work, and the blistering ointment renewed in that manner once a month or oftener, as the horse can be spared from business, till the cure is completed. This is the only method to prevent scars, which firing of course leaves behind, and unless skilfully executed, too often likewise a fulness of the joint, with stiffness. The mild blistering ointment, where the sublimate is left out, is the properest for this purpose."—*Bartlet's Farriery*, page 276.

WINE, a general name given to any brisk and cordial liquor drawn from vegetable bodies, and fermented; but it is the more appropriate name of the juice of the grape.

My general design, will permit me only to give an abstract of the making and management of wines, as practised in France.

The wine presses that are used in that country are similar to the screw presses that we use in making cider, but of more nice and exquisite workmanship.

"In order to make good wines, the grapes of the same vine must be gathered at different times. The first gathering should be the ripest clusters, cut close to the fruit, to avoid the sharpness and austerity of the stalk, and all rotten and green grapes should be taken away from the branches.

"The second gathering is some time after the first, when all that are ripened are taken. The third and last gathering will consist of the refuse of the two former; but no rotten grapes should be admitted in either.

"A method of making wine in the greatest perfection is, strip the grapes from the stems, before they go into the vat. Wine thus made is the mellowest, best coloured, soundest, and fittest for keeping.

"The wine of black grapes may be made of almost any colour; and the French make all their wines, both white and red, from black grapes.

"They who make white wine go into the vineyard in a damp misty or dewy morning, with a sufficient number of hands to gather a whole pressing of grapes in a few hours, beginning very early that they may not be belated. They cut off the best bunches, lay them gently in hand baskets, in which they carry them to the panniers, and in the panniers on gentle beasts to the press, taking care not to tumble nor bruise them. The dew increases the quantity of the wine, but renders it weaker. When the heat is not great, the vintagers continue their work till eleven o'clock, otherwise they leave off at nine because of a hot sun.

"As soon as the grapes arrive they are thrown into the press, and the first pressing is given without delay. The wine that runs from this pressing is the most delicate of any, but not the strongest.

"After this first pressing, which is gentle for fear of discolouring the liquor, the press is raised, the scattering grapes are laid upon the cake, and the second pressing is given. The press is screwed down with greater force than before;

and this second running is but little inferior to the first, in flavour or colour. It has the advantage in this, that it has a stronger body, and will keep a longer time. Sometimes they mix the wine of the first and second pressings.

" After the two first pressings, the sides of the cake of grapes are cut down perpendicularly, with a steel spade, so far as they exceed the breadth of the upper part of the press, which is let down upon the cake. The grapes that are cut off are laid on the top of the cake, and the third pressing, commonly called the first cutting, is given. The juices of this first cutting are excellent.

" A fourth pressing, a fifth, &c. which are called a second and third cutting, &c. are afterwards given, the sides of the cake being cut down and laid up each time, till the grapes cease to yield any more juice.

" The liquor of the cuttings becomes gradually more red, because the press becomes more forcible on the skin of the grapes, particles from which render the wine red.

" The wines of these different cuttings (as the latter pressings are called,) are collected separately, and afterwards mixed according as they contain more or less of the quality that is wanted.

" The pressings for white wine should be performed as quick as possible one after another, that the grapes may not have time to heat, nor the liquor remain long upon the murk. Particular attention is paid to this for the two first runnings, because they are the finest wine.

" Of the same black grapes, the black morillons, the pineaus, and the auvernats, of which white wine is made in Champaign, red wine is made in Burgundy.

" As much as the heat of the sun is avoided by the vintagers who make white wine, so much it is sought after and chosen by those who make red wine. These gather their grapes when the sun shines hottest; because its action on the outside of the berries has more effect than several days steeping in the vat, as the grapes then ferment very speedily. The other cautions in gathering grapes for white wine are observed here.

" Some express the juice of these grapes in the open air, in the vineyard, or near it, by throwing the bunches into large tubs, and there mashing and bruising them to pieces with sticks, or putting children into the tubs, to tread out the juice. Others carry them home, observing not to bruise them by the way, and put them in a vat, in which they are trodden and mashed. This is repeated, in either case, till the vessel is full; after which the broken grapes lie in the liquor more or less time, according to the heat of the weather, the flavour of the must, and the degree of colour intended to be given to the wine. During this time, the whole is frequently stirred together, the better to raise a fermentation, and tinge the liquor with a due degree of red.

" Some say the grapes should lie in the liquor forty-eight hours, while others talk of letting them infuse seven or eight days, or longer.

But the authors of the *Maison Rustique* say that the duration of the infusion of the husks should be proportioned to the heat of the weather, the quality of the grapes, and the intended colour of the wine. For the Coulange wine four hours only are allowed. The wine is apt to contract a roughness from the stalks, if it remains too long on them; and too much fermenting of the must renders the wine harsh and coarse, depriving it of its most volatile parts. Others make it a rule not to draw off their must till its head begins to fall; but this is evidently wrong, because a great part of the most active spirits of the liquor is thus evaporated.

"Indeed, for thicker, heavier and coarser wines than those of Burgundy and Orleans, which are chiefly intended above; such, for example, as the Bourdeaux claret, a whole day is frequently allowed for steeping of the husks, and sometimes more, before the press is recurred to.

"When the must has fermented upon the husks in the tubs or vats, as long as is thought proper, it is poured or drawn off, strained, and put into casks. Afterwards the murk, that is, the remainder of the grapes at the bottom of the tub or vat, is collected together, and put into the press, and pressed three or four times, till it is perfectly dry, the sides being cut down each time, as in making white wine. The liquor thus obtained, especially if the press be screwed so tight as to crack the seeds of the grapes, has a stronger body than the former running, but not any part of its fine, high and delicate flavour. Some, however, mix a little of it with their other wines, to strengthen them, and make them keep the better.

"Others, from a spirit of economy, pour as much water as they think proper upon the murk in the vat, immediately after the must has been drawn off. They do it without delay, lest the murk should turn sour, as it is apt to do. Then they stir it about, and leave it upon the murk, a longer or shorter time, according to the heat of the weather, till they find it pretty high-coloured, and judge that it has incorporated all the remaining particles and spirit of the wine. They then draw this water off into another vessel, and carry the remaining murk to the press, where they squeeze it till no moisture remains in it. The liquor thus expressed is mixed with that which was drawn off before, and is then barrelled. This is chiefly intended for common use, or rather for servants; but it will keep no longer than during the following winter, though it is brisk and pleasant enough while it keeps good.

"Another use the French make of the murk, is, to mend such of their wines, whether old or new, as are deficient in colour or strength. They turn them out of the casks into the vat immediately after the must has been drawn off, stir the murk up so as to mix it thoroughly with the wine, and let it stand twenty-four hours if it be new wine, and twelve hours if it be old. When they find it has taken a sufficient degree of colour, and that it is not sweet to the taste, but agreeable to drink; they draw it off; barrel it up separately, that they may know it again, and put the remaining murk to the press.

"The white unripened grapes, that were rejected at the former gatherings, are to hang on the vines till the latter end of October, or beginning of November, that they may be a little bitten by the frost. They are then gathered, and a wine is made of them, which is sold quite warm from the vat, and does well enough to mix with a coarse red wine.

"When the murk has undergone the utmost dint of pressing, it will be as hard as a stone; but even then it will yield, when diluted with water, fermented and distilled, a spirit for medical and domestic uses.

"In several parts of Germany, where the grapes seldom come to full maturity, the makers of wine have iron stoves in their wine cellars, and keep a constant fire in them during the time of their fermentation. This, by heightening the fermentation, ripens and meliorates the wines, and renders them more palatable. Exposing the casks to the sun will have the same effect on wines that are too acid to ferment sufficiently.

"As the wines of the last gathering are backward in fermenting, the people of Champaign and Burgundy, after their wines have been drawn off from their first lees, three weeks after being first put up, practise rolling them backwards and forwards. They do this five or six times a day, for four or five days successively; then two or three times a day for three or four days; afterwards twice a day for four days; then once a day for about a week; and afterwards once in four or five days. If the grapes were pressed very green, rolling in this manner is continued, in all, for about six weeks. But if they were tolerably ripe, rolling once in four or five days, for about a month or six weeks, is found to be sufficient. This rolling supplies the defect of the first fermentation.

"New wines will generally ferment of themselves, in a few days after they have been put into casks. Those that do not may be helped, by putting into them a little of the froth, or yeast, which works from the others. The finest wines will work the soonest, and the fermentation will continue for about ten or twelve days, according to the sort of wine, and the season of the year.

"While the wine ferments the bung hole of the casks must be left open, or only covered with a thin linen cloth, to prevent any dirt from falling in; and this cloth should be laid hollow, so that the froth arising from the fermentation may work off.

"When the fermentation is abated, which is known by the froth's ceasing to arise so fast as before, the bung may be closed down, after filling up the cask with liquor to within two inches of the top, and a vent hole should be opened and left, to carry off whatever may be thrown up by any subsequent fermentation. This filling up of the cask should be regularly performed, every two days, for about twelve days; for the fermentation will continue a considerable time in some degree: And if the casks be not kept so full as that the foulness thrown up by the fermentation may be carried off at the vent hole, it will fall back again into the wine and prevent its becoming clear. Afterwards it must be filled to within an inch of the bung, every fifth or sixth day for a month: After this, once a fortnight for three months.

"Though the fermentation will be over in a shorter time, yet the casks must be filled up once a month, so long as they remain in the cellar. For as the wine will insensibly waste in them, it will grow flat and heavy, if it be not constantly kept filled up. They should be filled up with a wine of the same kind, kept in some small vessel, or in bottles, for this purpose; and the vent holes must be stopped when the fermentation is over.

"The first drawing off from the lees is done about the middle of December. The casks should stand without the least shaking or other disturbance, till the middle of February, when it will be right to draw the liquor off again into other casks. If the quantity of lees is then found to be so considerable as to endanger their contracting a putrid taint, by remaining too long mixed with the wine, it may be advisable to draw it off again, after a proper interval of time: Or sometimes it may be necessary to repeat the racking many times. But in racking, though care should be taken to keep the casks full, wines of dissimilar qualities should not be mixed.

"The lees of several casks that have been racked off may be collected together, and when the thicker part has subsided, a spirit may be drawn from the thinner.

"When wine is to be transported, and thereby necessarily exposed to a warmer air, brandy is often added, to check the propensity to a new fermentation. It is also sometimes done to give strength to the wine; but the practice is not to be recommended, unless applied before the fermentation is completed.

"It is necessary for the preservation of some wines, which are apt to be on the fret, to fumigate the casks with burning brimstone. This resists fermentation. But the colour of red wines is said to be destroyed by it. The colour of wine is frequently artificial. A deep yellow is made by burnt sugar: A deep red, which is not natural to any wine, is almost always made by red woods, elderberries, bilberries, &c.

"Turbid wines are fined by a mixture of the whites and shells of eggs, powdered alabaster, and isinglass. The shells and alabaster may correct a small degree of acidity. Isinglass alone will fine it in a few days.

"If wine is grown very sour, the best way of correcting it, so as to preserve the spirit and flavour, is, adding a quantity of salt of tartar, sufficient to neutralize the acid, just before the wine is used.

"If it be intended that wine should not froth, the best time for drawing it off, whether into bottles or casks, is when the weather is extremely fair and clear, the barometer high, and the wind northerly; because the air is then coolest, and the wine least apt to be upon the fret.

Raisin Wine is a wine made from raisins steeped in water.

"Take thirty gallons of soft water, either rain or river water, in as clear a state as possible. Put it into a vessel at least one third bigger than will contain that quantity; and then add to it one hundred weight of Malaga raisins picked from the stalks. Mix the whole well together, that the raisins may not remain in clotted lumps: And then cover it partly, but not entirely, with a linen cloth; and let it stand in a warm place, if the season be not hot. It will soon ferment, and must be well stirred about, twice in four and twenty hours, for twelve or fourteen days. It must then be examined by the taste, to try if the sweetness be nearly gone off; and if that be found so, and the fermentation be greatly abated, which will be perceived by the raisins lying quietly at the bottom, the fluid must be strained off, and pressed out of the raisins, first by hand, and afterwards by a press, if it can be easily procured. But, instead of a press, two boards may be used, with the assistance of a large weight, or other strong force, which must be continued as long as any fluid can be made to drop from the mass. The fluid being thus separated from the skins of the raisins, must be put into a good sound wine cask, well dried and warmed, together with eight pounds of Lisbon sugar, and a little yeast. But some part of the liquor must be kept out, to be added from time to time, as the abatement of the fermentation, that will come on again, may admit, without the wine rising out of the cask. In this state it must continue for a month, with the bung-hole open: And then, the whole of the liquor kept out having been now put into the vessel, it must be closely stopped up, so that no air may enter: And in this state it must be kept a year, or longer; then bottled off.

"The wine may be drunk, and will be very good, at the end of a year and a half: But it will be much better, if kept longer, and will improve for four or five years. When it has a proper age, it will equal any of the strong cordial foreign wines; and may, by the addition of proper substances to flavour and colour it, be made to resemble them."

"This is the most perfect kind of what may be called artificial wine; but others may be made cheaper. Expense may be saved two ways: One is, substituting sugar for raisins, leaving out four pounds of raisins for one pound of sugar added. Or the proportion of sugar and raisins, and a proportion of clean malt spirit added, when the bung of the cask is closed up.

"Any other kind of large raisins may be used, instead of the Malaga. The thinner the skins are, and the sweeter the pulp, the stronger the wine will be."

"*To make Birch Wine.* After collecting the sap of the birch, it is to be made into wine before any fermentation takes place; a pint of honey, or a pound of sugar, is to be added to every gallon of the sap, the whole to be well stirred up, and then boiled for about an hour, with a few cloves and a little lemon peel; during which the scum is to be carefully taken off. When cool, a few spoonfuls of new ale or yeast is to be added, to induce a due degree of fermentation; and after this has ceased, or nearly so, the liquor is to be bottled and put away in a cool place in the cellar, for use; though no doubt it becomes improved by age.

"When properly made, the liquor becomes so strong that it frequently bursts the bottles unless they are placed in spring-water. Stone bottles are said to be the best for containing the liquor, as they are stronger than glass."—*Farmer's Assistant.*

Mr. Cooper gives the following directions for making wine of cider, and other ingredients.

"Take cider of the best running of the cheese, and of the best quality, and add to it as much honey as will make the liquor bear an egg; strain the liquor through a cloth as you pour it into the cask; fill the cask full, with the addition of two gallons of French brandy to a barrel; set it away in a cool place, with the bung-hole open to ferment; as the fermentation proceeds, it will throw out considerable froth and filth; keep filling it frequently with more of the same kind of liquor, kept for the purpose, until the fermentation has nearly subsided; then put in the bung; but not tightly, in order that the liquor may have some further vent, and, as soon as the fermentation ceases, close up the vessel. The next spring rack off the liquor into a new clean cask; and in order to clarify it, Mr. Cooper directs a mixture of sweet milk, the whites of eggs and clean sand to be beat up, and well stirred into the cask."

The *Farmer's Assistant* observes, that "it is believed that about a quart of sweet milk to a barrel well stirred and mixed with the liquor as it is poured in, will answer equally well, and perhaps better. This operation alone will not only clarify liquors, but, by repeating it several times, the highest coloured wines may be nearly or quite divested of all their colour. After the liquor has been thus clarified, let it be drawn off again into bottles, or into fresh clean casks, and kept in a cool cellar for use. Mr. *Cooper* says that his liquor, thus prepared, has often been taken, by good judges of wine, for the real juice of the grape; and has been pronounced by them superior to most of the wines in use. Age, however, is essential in perfecting this kind of wine, as well as all others."

WINNOWING, clearing corn from its chaff by wind.

WINNOWING MACHINES. The invention of winnowing machines was first brought from China to Holland, and was introduced into Scotland more than a century since, by Andrew Fletcher, of Salton. Threshing mills have, gene rally, a set of fanners attached to them. From the inequality of motion, however, attendant on those machines, it is thought best in general to have them separate. A simple winnowing machine, invented by John James, 3d. is mentioned as belonging to the Massachusetts Agricultural Society, and deposited in their hall at Brighton.

WINTER, one of the four quarters of the year.

According to some, winter begins at the time when the sun's distance from the zenith of the place is greatest, and ends at the vernal equinox. But it is more usually considered as including December, January, and February. Notwithstanding the cold of winter, it is proved by astronomers, that the sun is nearer to the earth in winter than in summer. The reason of the cold is the increased obliquity of the rays of the sun, together with the increased length of the nights.

Winter is the season when the days are shortest: But the shortness of the days is little regretted by our farmers, as they have then the most leisure, or are least hurried in their business. For, in this country, the ground is so continually frozen in winter, that none of the operations of tillage can be performed. The good husbandman, however, is not idle; his year's stock of wood is to be procured, and his stock must be daily and carefully tended: Stones which have been piled may in the easiest manner be removed on sleds to the places where they are wanted, for fencing or other uses. Such of the produce of his farm as he can spare may be carried to market; which may be more easily done than at any other season. See the article *Employment*.

WINTER GRAIN, see the articles *Wheat* and *Rye*.

WITHE, a twig or shoot of tough wood, used instead of a rope, to tie things together, &c.

Young shoots of walnut wood are proper for withes; those of birch, and of some kinds of willow, answer well enough. Withes in fences will last only two years at the longest; and some of them are apt to fail sooner. Those which are cut in autumn, after the leaves are fallen, will last considerably longer than those which are cut in the spring or summer. If they lie for a day or two exposed to the sun, after they are cut, they grow tougher, and are more easily twisted. Steeping them in salt water will render them more durable.

WOAD, called by botanists *Isatis sativa, vel latifolia*. The leaves of this plant are cut off in their full sap, sweated in heaps, and ground to a paste, made up into balls and dried, to be used in dying blue.

The following description of woad, and the manner of cultivating it, are from Miller's Gardener's Dictionary.

The common woad, (*Isatis Tinctoria*,) is a biennial plant, with a fusiform, fibrous root. Stem upright, round, smooth, woody at bottom, branched at top. Root-leaves ovate lanceolate, on long foot-stalks, down which they run a little. Stem-leaves alternate, quite entire, embracing, smooth. from two to three inches long, and scarcely half an inch in breadth. These are sometimes very slightly tooth letted; and a few hairs are sometimes found both on the stem and leaves. Flowers small, terminating the stem and branches in a close raceme. Both corolla and calyx yellow; petals notched at the end. Seed vessels on slender peduncles, hanging down. Chestnut-coloured or dark brown and shining when ripe, of an oblong elliptic form, near half an inch long and two lines wide, compressed at top and on the sides into a sharp edge, swelling like a convex lens in the middle, with a straight longitudinal suture on each side, one celled, two valved, but hardly opening spontaneously; valves of spongy substance like cork, and boat-shaped.

Seed smooth, striated a little, two lines long, and three quarters of a line wide, yellow, or brownish yellow when ripe; it has only a single membranaceous coat. Embryo curved, yellowish. Cotyledons ovate, fleshy, plano-convex.

Mr. Miller thus describes the cultivated plant, which however differs little from the wild one except in luxuriance. The lower leaves are of an oblong oval figure, and pretty thick consistence, when growing in a proper soil; they are narrow at their base, but broad above, and end in obtuse roundish points; are entire on their edges, and of a lucid green. The stalks rise near four feet high, dividing into several branches, with arrow-shaped leaves sitting close; the ends of the branches are terminated by small yellow flowers, in very close clusters. The pods are shaped like a bird's tongue, half an inch long, and one eighth of an inch wide, turning back when ripe. It flowers in July, and the seeds ripen the beginning of September.

A fine blue colour is obtained from woad. It is also the basis of black and many other colours.

As the goodness of woad consists in the size and fulness of the leaves, the only method to obtain this is to sow the seed upon ground at a proper season and allow the plants proper room to grow as also to keep them clear from weeds. The method practised by some of the most skilful gardeners in the culture of spinach, would be a great improvement to this plant, for some of them have improved the round-leaved spinach so much by culture, as to have the leaves more than six times the size they were formerly; and their fatness has been in the same proportion, upon the same land, which has been effected by thinning the plants when young, and keeping the ground constantly free from weeds.

A hazel loam, whose parts will easily separate, is the best soil for woad. The ground should be ploughed and laid in narrow high ridges just before winter;—it should be ploughed again in the spring; a third time in June; and lastly towards the end of July or early in August. In the intervals between each ploughing, it will be necessary to harrow the soil, so that all weeds may be destroyed.

Woad is sown in England early in August, and generally broadcast, though the drill-husbandry is the most advantageous. At the end of two or three weeks, the plants must be hoed at the distance of at least six inches; after which they will require no further attention, except a careful weeding in October, and particularly in the month of March.

The proper time for gathering the leaves is determined by their full growth, and the first change of colour at their points; they are cut with an edged tool, and collected into baskets by women and children. If the land be good three or four successive crops may be taken; but the two first are the finest. After the leaves are gathered they are submitted to the action of mills, similar to those employed in grinding oak bark; and in which they are reduced to a kind of pulp. The woad is then laid in small heaps, which are closely and smoothly pressed down. As often as the crust formed on the outside cracks or breaks it is again closed, in order to preserve the colouring matter. In this state it remains for a fortnight; at the expiration of which the heaps are broken up; the external part is worked into a mass, and the whole is formed into oval balls, either by the hand or by the means of moulds. The balls are now exposed to the sun under shelter; when perfectly dry they are ready for sale. Such is the process which woad undergoes before it becomes fit for dying *blue* colours. But Mr. Astrue is of opinion, that if this vegetable were cured in the same manner as indigo, it would produce a colour of equal lustre to that obtained from such an expensive drug. Dambourney directs to boil the fresh leaves of woad with diluted bullock's blood, or more effectually with caustic soap-boiler's ley; in this simple manner, a dark green decoction of a bluish shade will be obtained; and after clarifying the liquor, it will form a blue precipitate; which dissolved in oil of vitriol, and properly diluted, imparts a beautiful colour to woollen cloth. Farther, even the leaves, in a state of fermentation, with pure water on adding a small portion of caustic alkaline ley, afford a fine blue sediment, resembling the true indigo.

[This useful article abounds in the Western States, and also in Pennsylvania, according to the report of an experienced native of Great Britain, who is settled near the head of Ohio, in Washington county, Pennsylvania. It appears that the farina is much richer than that in England; and that instead of two crops, which are obtained in England, five are yielded in the United States.]—*Massachusetts Agricultural Repository*, Vol. III. p. 136—139. See likewise the same work, Vol. IV. p. 289—298.

WOLF, a wild beast of the dog kind.

This animal is very fierce, equal in size to a large mastiff, and has much the same appearance.

Wolves are gregarious, go in droves, and surprize the nightly traveller with their hideous yelling. No beast of prey in this country is more formidable; they sometimes attack men.

New-England, even from its first settlement, has been much infested with wolves. And, notwithstanding the bounties that have been given by government for destroying them, the settlements bordering on the wilderness are still subject to their mischievous incursions; so that there is little safety for sheep in these situations. Almost whole flocks in a night are sometimes destroyed by them. This exposure to wolves is equal to a heavy tax upon our frontier plantations.

To secure the sheep from this enemy, it is necessary that they be pastured in the open fields by day, and housed in strong places every night: And even these precautions do not always prove effectual.

Some say, that smearing the heads of sheep with a composition of tar and gunpowder will prevent their being attacked by the wolves; but I cannot certify this from my own experience.

WOOD, a solid substance, of which the trunks and limbs of trees, as well as their roots, consist.

" The wood is all that part of a tree included between the bark and the pith."

Dr. Grew has discovered, by the assistance of the microscope, that what we call wood, notwithstanding its hardness and solidity, is only an assemblage of minute pipes, or hollow fibres, some of which rise from the root upwards, and are disposed in form of a circle; and the others, which he calls insertions, tend horizontally from the surface to the centre; so that they cross each other, and are interwoven like the threads of a weaver's web.

Each year's growth in wood is marked by circles visible to the naked eye, by counting of which circles the age of a tree is known.

WOOD LAND, ground covered with wood, or trees. They are mostly designed for fuel and timber. In felling them, care should be taken to injure the young growth as little as possible. Fire wood, as well as timber, should be felled, when the sap is down. Otherwise it will hiss and fry upon the fire, and not burn freely, although it should be ever so long dried. To thicken a forest, or to increase the number of trees in a wood lot, it should be well fenced, and no cattle be permitted to be in it. And something may be done, if needful, by layers and cuttings.

The practice of the populous nations of Europe, whose forests have been cut off centuries ago, and who are compelled to resort to measures of the strictest economy, to supply themselves with fuel, ought to have great weight with us. France, in an especial manner, ought to be looked up to for wise lessons on this subject. Her vast and thickly settled population, her numerous manufactures, her poverty in mineral coal, the eminence, which she has attained in all economical arts, entitle her to great respect. It is the practice of the French people not to cut off their woods oftener than once in twenty or twenty-five years, and by *law*, when they are cut over, the owner is obliged to cut the *whole smooth*, with the exception of a very few trees, which the officers of the government had marked to be spared for larger growth—without giving any opinion, as to the propriety of the direct interference of the government on such a topic, we should say that the example proves that in the opinion of the French scientific and practical men, it is expedient, when wood lands are cut, that they should be cut smooth, in order that the new growth might start together, not overshaded by other trees of larger growth. We have no favourable opinion of the utility of cutting down trees in a scattered manner, as they appear to fail, and still less of planting acorns in thinner spots of the forest. The growth thus produced must remain forever feeble.

WOOL, the covering of sheep.

Each fleece consists of wool of several different qualities. The English and French usually divide wool into three principal sorts; 1. that of the neck and back, which they call mother wool; 2. that of the tails and legs; 3. that of the breast and belly. These different kinds should not be mixed in spinning.

The two best qualities in wool are length and fineness.

To improve the wool of our flocks, no lambs should be kept for breeders, but such as bear the best wool; and fine woolled rams should be procured from distant places, or from foreign countries.

It will also conduce to the improvement of the wool, if the sheep are kept in dry pastures, upon short and sweet grass in the summer, and upon the best hay in the winter. Great care should also be taken to prevent their being overheated, either by obliging them to feed in the hottest part of hot days in summer, or by housing them in too hot, small, and close places, at any season. They should also be kept, as much as possible, from fouling their fleeces, by lying on their excrements, without litter.

The Spanish wool is preferred to any other that is produced in Europe. A writer in that country says, "There are two kinds of sheep in Spain, namely, the coarse woolled sheep, which remain all their lives in their native country, and which are housed every night in winter; and the fine woolled sheep, which are all their lives in the open air, which travel at the end of every summer from the cool mountains of the northern parts of Spain, to feed all the winter on the southern warm plains of Andalusia, Manca, and Estremadura. It has appeared from very accurate calculations, that there are not fewer than five millions of the fine woolled sheep in Spain. Special ordinances, privileges, and immunities, are issued for the better preservation and government of the sheep, which are under the care of twenty-five thousand men. These sheep pass the summer in the cool mountains of Leo, Old Castile, Cuença, and Arragon. The first thing the shepherd does when the flock returns from the south to its summer downs, is, to give the sheep as much salt as they will eat. Every owner allows his flock of a thousand sheep twenty-five quintals of salt, which the flock eat in about five months. They eat none in their journey, nor in winter. From the territory called the Montana, at the extremity of Old Castile, from whence they set out, to Estremadura, is one hundred fifty leagues, which they march in less than forty days. As soon as April comes, the sheep express, by various uneasy motions, a strong desire to return to their summer habitations. The sheep of Estremadura, which never travel, have coarse, long, hairy wool. The itinerant sheep have short, silky, white wool; the fineness of which is owing to the animal's passing its life in the open air, of equal temperature; for it is not colder in Andalusia or Estremadura in winter, than it is in Montana or Molina in summer." The sheep, by these judicious removals, are grazing both in summer and winter, and are never exposed to extreme heats or colds. Next to this management insular situations in temperate climates are the best places for the constant feeding of sheep, where the extremes of heat and cold are not so great as on a continent in the same latitude.

Although we are accustomed to consider wool as the exclusive production of the sheep, and as characteristic of that species of animal, yet there are sheep, which are as destitute of wool as a deer, or a grey hound. Sheep of this kind are found in Tartary Africa, the East Indies, and the Chinese Empire. There are likewise animals, which yield a fur in many respects resembling wool, and which is used for the same purposes, such as the goat, the camel, the dromedary, and the vicuna of South America.

The covering with which Providence, always attentive to the wants of his creatures, has furnished most quadrupeds is adapted to the climate which they inhabit. In those where a heavy coat would be an incumbrance we find even the sheep destitute of its usual fleece. On the other hand, from latitudes where the rigours of winter are very severe we import those furs, which contribute to the comfort and elegance of our own attire.

The amelioration of flocks of sheep has always been closely connected with the progress of the arts, and of civilization, for we uniformly find in countries where these have flourished, a race of sheep, which yield wool much superior to that we find all around them.

The particular shape of the filament is most probably determined by that of the pore in the skin, through which it is protruded. In some families of sheep we observe the pile perfectly round and even, like a very nicely drawn wire, and in others it is uniformly flat and smooth, like a small bar of finely polished steel. This difference is frequently discovered by inspection alone, but becomes more obvious if one end of the filament be held fast while the other is rolled round its own axis between the fingers. It then reflects the rays of light, if it be flat in the same way that the metallic shreds, which we obtain from gold and silver lace, do when they are treated in a similar manner, and concerning whose shape we can have no reasonable doubt.

Another topic which deserves attention is a substance intimately blended with the pile of the fleece, which on account of its yellowness and consistency is denominated the yolk. The best breeds of sheep produce this in the greatest abundance. Those parts of the sheep which produce yolk most copiously furnish the best wool. M. Vauquelin, who has submitted this yolk to chemical analysis, declares that the greater part of it is a soap with a base of potash, and that the remaining portion consists of salt in a state of combination with other substances.

The manner in which this yolk acts upon wool is unknown. The most plausible conjecture appears to be that it is a secretion, exuding from the skin, which by mingling with the pile renders it soft, pliable and healthy, in the same way as oil does a thong of leather.

It would be vain to undertake to point out any particular kind of wool, which, in all circumstances might be considered as best, as this would depend on the kind of manufacture contemplated, but whether the wool be long or short of a coarser or finer pile, it will be good or bad in proportion to the quantity of sound and healthy yolk in which it is produced.

The breed of the sheep is the basis upon which all improvements of the flock must be founded. Though English and Spanish sheep have degenerated, when exported to other climates it is said that the deterioration has been the consequence of the intermixture of blood with sheep of an inferior quality. Mr. Lackner says, "the sheep of England, when transported to Jamaica, yield the same kind of 'burly fleece,' as they did at home, and if prevented from mingling their breed with the native stock their offspring afford a wool exactly similar to what they would have done in this climate."

The mode of washing sheep sometimes practiced where water is scarce and the shepherd careless instead of separating from the fleece all the sand, clay and other dirt with which it is incumbered, supplies it with a still larger proportion. They are too often washed in muddy pools, or small rivulets of water, which are rendered turbid by the process, or driven along dusty roads, or lodged in sandy or dusty places. Sheep too are often allowed to carry about them loads of their own excrement, and thus impeaching the humanity as well as the cleanliness of their owners. What are called the tag locks should be carefully cut away, and the sheep should be kept in clean dry pastures after having been washed before shearing.

Graziers may increase the length of their staple by various means. The management of the breed is not only the most natural and easy method, but that also which is most usually adopted. Its effects are more permanent than others which are sometimes resorted to, but less pure from deleterious influences; for it is not unfrequently observed, that the ram communicating to his offspring an increased length of staple, gives to it also a coarser pile. Feeding the sheep upon the richer grasses, upon turnips and oil cake, thus forcing both the carcase and the fleece, seems to be a method of increasing the wool free from contaminating influence, but requires the animal to be constantly supported, even to the point of luxurious feeding; and the effects of the system remain no longer than it is continued. Another method is to keep the wool upon the back of the sheep two whole years; which requires care that the animal be not injured by cold, or by hunger, while the sheep is growing.

The softness of the pile is an essential quality of the fleece; and this depends very much on the breed of sheep, and the quantity of yolk, which they constantly afford. Mr. Luccock says, that "The Spaniard is so thoroughly aware of the value of this property and the means to promote it, that he not only attends with peculiar care to the breed which travels to the mountains, but before shearing encloses the sheep in sudatories, in order to saturate and soften the pile with the yolk." The softest pile is collected, if the breed be similar from flocks, which have been kept in good condition, upon loamy soils, and into whose fleeces no particles of absorbent earth have been admitted. Lime or chalk, it is said, will injure fleeces by absorbing or altering the quality of the yolk.

"Formerly," says an English writer, "the manufacturer in showing his cloth was more anxious to exhibit the fineness of the thread than prove the softness of the pile. For this purpose he used to scrape off the pile from a small part of the cloth to display the smallness and regularity of the spinning: since he has been enabled by machinery to spin a small even thread with great facility, the practice is discontinued. Cloth is now finished without that hard, shining surface, which was given to it a few years since by hot-pressing, which prevented the softness of the pile from being felt. By the present mode of cold pressing, the softness of the pile becomes immediately perceptible to the touch, and is considered as one of the most distinguishing and essential qualities of a good cloth. It may be affirmed, that taking two packs of sorted wool of the same apparent fineness, one possessing in an eminent degree the soft quality, the other of the hard kind, the former will, with the same expense to the manufacturer, make a cloth, the value of which shall exceed the latter full twenty-five per cent."

Housing the sheep at night, and providing for them a shelter during the day from the rain and sun preserves and improves the wool, and conduces to the health of the animal; and in proportion to the regularity of the temperature in which sheep are kept, and to the regular supply of nourishment which they receive, will the hair or fibre of the wool preserve a regular degree of fineness.

M. Monge gives the following account of that extraordinary property in wool, called the felting quality.

" The felting of wool or hair, is an effect resulting from the external conformation of their fibres, which appear to be formed either of small lamina placed over each other, in a slanting direction from the root towards the end or point of each fibre, like the scales of fish lying one over the other in succession from the head to the tail; or of zones placed one upon another, as in the horns of animals; from which structure each fibre, if drawn from its root towards its point, will pass smoothly through the fingers, but if it be drawn in a contrary direction from the point towards the root, a sensible resistance and tremulous motion will be felt by the fingers. This peculiar conformation disposes the fibres to catch hold of each other; and as they cannot recede when acted upon by other bodies, they naturally advance by a progressive motion from the root towards the end."—See *Ann. de Chymie*, tom. VI. p. 300, &c.

The hairs of wool, when carded and spun, are laid in every direction, and when they are compressed and agitated, this disposition to catch each other, and move from the point to the root, must inevitably bring the whole mass closer together. This is the case when cloth is fulled or milled, by which it is shortened both in length and breadth. Not being perfectly satisfied with the account given by M. Monge, I adopted a very simple experiment to ascertain its truth. I took a staple of coarse wool, of considerable length, with the hairs lying regularly in one direction. At the distance of an inch from each end, I made a tight ligature with a thread. I measured the middle of the staple between each end, and then proceeded to mill it, by compressing it in my hand in a solution of soap and warm water. I continued the operation until each end of the staple beyond the thread was felted into a hard knob, or button, which could not be separated by the fingers. The middle of the staple remained unfelted, the hairs quite distinct from each other; and it was not in the least shortened in the process, either in the wet state or when dried. In the middle part of the staple, between the two threads, the hairs were kept in the same direction by being tied, and could not acquire the retrograde motion, or adhere by the surfaces catching hold of each other. The hairs at the end of the staple being at liberty to double and move in different directions, were soon felted together into a smooth and round knob, in which the ends or points of the wool were entirely buried.

See Hints to Wool Growers, Agricultural Repository, Vol. IV. page 57. Likewise a paper on " British Wool," &c. in the same work, vol. IV. page 145, and Luccock's *Essay on Wool*.

WORMS, a well known species of insects.

Fields and gardens are often infested by worms. The best antidotes are, dressings of sea manures, wetting the ground with sea-water. Soot, lime, and ashes sprinkled on the ground, oppose their ravages. The refuse brine of salted meat, or water in which walnut leaves have beeen steeped for two or three weeks, sprinkled on gardens, will subdue the worms. The water of salt springs may be used with great advantage by those who are near them. Those who are remote from them, and from the sea, may use salt and water, and such other antidotes as they can most easily procure.

When worms breed in horses, it arises from weak digestion, or a vitiated appetite. To cure a horse of bots:

"Take quicksilver two drams; Venice turpentine half an ounce; rub the quicksilver till no glistening appears; then take an ounce of aloes, a dram of grated ginger, thirty drops of oil of savin, and syrup of buckthorn enough to make the whole into a ball.

"One of these balls may be given every six days, with the usual precautions with regard to mercurial physic: And the following powders intermediately.

"Take powdered tin and Æthiops mineral, of each an ounce; and give it every night in a mash, or in his corn.

"These medicines, or any of the various preparations of antimony and mercury, should be continued several weeks together, in order to free the animal entirely from these vermin."—*Bartlet's Farriery.*

WOUND, a hurt given by violence.

"In all fresh wounds made by cutting instruments, there is nothing more required than bringing the lips of the wound into contact by suture or bandage, provided the part will allow of it; for on wounds of the hips, or other prominent parts, and across some of the large muscles, the stitches are apt to burst on the horse's lying down and rising up in the stall. In such cases, the lips should not be brought close together. One stitch is sufficient for a wound two inches long; but in large wounds may be at an inch or more distance; and if the wound is deep in the muscles, care should be taken to pass the needles proportionably deep, otherwise the wound will not unite properly at the bottom.

"Should the wound bleed much from an artery divided, the first step should be to secure it, by passing a crooked needle underneath, and tying it up with a waxed thread. If the artery cannot be got at this way, apply a button of lint or tow to the mouth of the bleeding vessel, dipped in a strong solution of blue vitriol, styptic water, oil of vitriol, or hot oil of turpentine, powdered vitriol, or colcothar, &c. and remember always to apply it close to the mouth of the bleeding vessels, and take care that it is kept there by proper compress and bandage, till an eschar is formed; otherwise it will elude your expectations, and frequently alarm you with fresh bleedings.

"In a healthy and sound constitution, nature furnishes the best balsam, and performs herself the cure, which is so often attributed to medicine. When it is otherwise, and the blood is deprived of its balsamic state, as will appear from the aspect of the wound, and its manner of healing, it must be rectified by proper internal medicines, before a good foundation for healing can be laid by any external applications.

"The lips of the wound being brought together by the needle or bandage, it needs only to be covered with rags dipped in brandy, or a pledget of tow spread with the wound ointment, and the wounded part kept as much as possible from motion.

"Remember to dress all the wounds of the joints, tendons, and membranous parts, with terebin thinate medicines: To which may be added honey and tincture of myrrh; and avoid all greasy applications whatever. Fomentations and poultices are also generally here of great use; the former thin and attenuate the impacted fluids, greatly promote a free perspiration of the limb, and facilitate the unloading the surcharge of the vessels, by quickening the motion of the fluids; while the latter, by relaxing the vessels, abate their tension, and relieve the obstruction by promoting digestion."—*Bartlet's Farriery.*

Y.

YARD, a measure of three feet.

YARD, a small enclosure for cattle. See the article *Barn Yard.*

YEAR, the time the sun takes to go through the twelve signs of the zodiac.

YEAST. Dr. Mease recommends yeast made after the following recipe, as preferable to any other kind.

Boil twelve clean washed, middle sized potatoes; and at the same time, boil, in another vessel, a handful of hops in a quart of water; peel and mash the potatoes in a marble mortar, pour part of the hop water while hot upon the potatoes, mix them well, and pass them through a sieve; then add the remainder of the hop-water, and half a tea cup full of honey, beat all well, and add a small portion of leaven to bring on the fermentation. Put the whole in a stone jug, and set it by the fire, (in the winter.) All the utensils must be scalded every time they are used, and washed perfectly clean. One tea cup full of the above potatoe yeast will answer for two quarts of flour. In summer the yeast ought to be made every second day.—*Domestic Encyclopedia.*

YELLOW WEED, meadow crow foot, *Ranunculus.*

This weed is known in England by the names, king cob, king cup, gold cups, gold knobs, butter cups, and butter flowers. The flowers, which appear in June, and are of a bright yellow colour, give our mowing lands and pastures a very gay appearance. It is of a very hot and acrid nature, and is eaten in its green state by neither neat cattle nor horses. Therefore, the opinion of its increasing the yellowness of butter in summer is groundless. See the article *Weeds.* This weed, when dried, is so good fodder, that hungry cattle seldom leave any part of it.

YELLOWS, a disease incident to horses and neat cattle, more properly called the jaundice. The vulgar name of this disease in neat cattle is, *the Overflowing of the Gall.* See that article.

Horses are frequently subject to this distemper, which is known by a dusky yellowness of the eyes: The inside of the mouth and lips, the tongue and bars of the roof of the mouth looking also yellow. The creature is dull, and refuses all manner of food; the fever is slow, yet both that and the yellowness increase together. The dung is often hard and dry, of a pale yellow, or light pale green. His urine is commonly of a dark dirty brown colour; and when it has settled some time on the pavement, it looks red like blood. He stales with some pain and difficulty, and if the distemper is not checked, soon grows delirious and frantic. The off side of the belly is some times hard and distended; and in old horses, when the liver has been long diseased, the cure is scarce practicable, and ends fatally with a wasting diarrhœa: But when the distemper is recent, and in young horses, there is no fear of a recovery, if the following directions are observed.

First of all bleed plentifully, and give the laxative clyster, as horses are apt to be very costive in this distemper; and the next day give him this purge:

Take of Indian rhubarb powdered, one ounce and a half; saffron two drams, succotrine aloes six drams, syrup of buckthorn a sufficient quantity.

"If the rhubarb should be found too expensive, omit it, and add the same quantity of cream of tartar, and half an ounce of Castile soap, with four drams more of aloes. This may be repeated two or three times, giving intermediately the following balls and drink:

Take of Æthiops mineral half an ounce; millepedes the same quantity, Castile soap one ounce; make into a ball, and give one every day, and wash it down with a pint of this decoction:

Take madder root and turmeric, of each four ounces; burdock root sliced, half a pound; Monk's rhubarb four ounces; liquorice sliced two ounces: Boil in a gallon of forge water to three quarts; strain off and sweeten with honey.

"Balls of Castile soap and turmeric may be given also for this purpose, to the quantity of three or four ounces a day, and will in most recent cases succeed.

"By these means the distemper generally abates in a week, which may be discovered by an alteration in the horse's eyes and mouth; but the medicines must be continued till the yellowness is entirely removed. Should the distemper prove obstinate, and not submit to this treatment, you must try more potent remedies, viz. mercurial physic, repeated two or three times at proper intervals; and then the following balls:

Take salt of tartar two ounces, cinnabar of antimony four ounces, live millepedes and filings of steel, of each four ounces, Castile or Venice soap half a pound: Make into balls of the size of pullets' eggs, and give one night and morning, with a pint of the above drink.

"It will be proper, on his recovery, to give him two or three mild purges, and, if a full fat horse, to put in a rowel."—*Bartlet's Farriery.*

YEOMAN, the addition of the first or highest degree among the plebeians of England. The yeomen are properly freeholders, who cultivate their own lands.

YEST, *or* YEAST, *or* BARM, a head or scum rising upon beer, or ale, while it is working or fermenting in the vat.

It is used as a leaven or ferment by the bakers, serving to swell or puff up their dough, which renders the bread lighter, softer, and more wholesome, as well as more palatable. But when there is too much yest in bread, it gives it a bad taste.

YEW TREE, the name of a tree, well known and common in England. Mr. Miller says, it grows naturally also in North America.

Yew trees may be easily propagated by sowing the seeds in beds, and removing the young plants into nurseries, as many other trees are managed. It is an evergreen, and a tonsile tree, that may be easily cut and pruned into any form that is desired. The trees grow slowly, but sometimes come to be large trees; and their timber is highly valued for various uses.

YOAK, *or* YOKE, a wooden instrument, with which oxen are coupled together by their necks, and by which they draw.

The parts of a yoke are the cross tree, which should be made of white maple, the bows, and the staple and ring.

The cross tree should be of a crooked form, that the oxen may draw partly by the bows; for if it were straight the bows would not press their shoulders at all.

If one of the oxen be stronger than the other, as it commonly happens, the staple should not be inserted in the centre, but a little nearer to the bow of the stronger ox.

Short yokes are best for the ease of drawing: But if the oxen learn to crowd, or to draw from each other, they should be put to work in a long yoke till they are cured of such ill habits.

The bows should be made of the toughest of wood, and that which is very strong and stiff when seasoned, such as hickory or white oak, and be well fitted to their necks. And a bow should be large, about an inch and a half in diameter for large oxen, not only for strength, but because it will be easier to the ox.

The principal use of the ring is, to receive the end of the tongue of a cart or sled. This gives the oxen command of the carriage.

THE END.

INDEX.

ERRATA.

Page 47, 1st column, 24 from the top, for "farms," read *forms*.

Page 53, 2d column, line 6 from the bottom, for "latter," read *former*.

Page 62, 1st column, top line for "these," read *there*.

Page 171, 1st column, line 15 from the bottom for "but," read *best*.

Page 214, 2d column, line 14 from the bottom, for "saving," read *sowing*.

Page 222, 2d column, line 12 from the top, for "changeable," read *chargeable*.

Page 301, 2d column, line 15 from the top, for "stopping," read *sloping*.

Page 304, 1st column, line 4 from the bottom, for "tar," read *tan*.

Page 327, 1st column, line 22 from the top, after "gradually," add *increases*.

Page 332, 1st column, line 5 from the bottom, for "care," read *ease*.

Page 388, 2d column, line 12 from the bottom for "ege," read *eye*.

Page 439, 1st column, line 10 from the bottom, for "thresher," read *threshed*.

Page 462, 2d column, line 15 from the top, for "sublime," read *sublimate*.

Page 470, 2d column, line 17 from the top, for "proved," read *moved*.

Page 471, 1st column, line 28 from the top for "cherished," read *checked*.

TECHNOLOGY AND SOCIETY

An Arno Press Collection

Ardrey, R[obert] L. **American Agricultural Implements.** In two parts. 1894

Arnold, Horace Lucien and Fay Leone Faurote. **Ford Methods and the Ford Shops.** 1915

Baron, Stanley [Wade]. **Brewed in America:** A History of Beer and Ale in the United States. 1962

Bathe, Greville and Dorothy. **Oliver Evans:** A Chronicle of Early American Engineering. 1935

Bendure, Zelma and Gladys Pfeiffer. **America's Fabrics:** Origin and History, Manufacture, Characteristics and Uses. 1946

Bichowsky, F. Russell. **Industrial Research.** 1942

Bigelow, Jacob. **The Useful Arts:** Considered in Connexion with the Applications of Science. 1840. Two volumes in one

Birkmire, William H. **Skeleton Construction in Buildings.** 1894

Boyd, T[homas] A[lvin]. **Professional Amateur:** The Biography of Charles Franklin Kettering. 1957

Bright, Arthur A[aron], Jr. **The Electric-Lamp Industry:** Technological Change and Economic Development from 1800 to 1947. 1949

Bruce, Alfred and Harold Sandbank. **The History of Prefabrication.** 1943

Carr, Charles C[arl]. **Alcoa, An American Enterprise.** 1952

Cooley, Mortimer E. **Scientific Blacksmith.** 1947

Davis, Charles Thomas. **The Manufacture of Paper.** 1886

Deane, Samuel. **The New-England Farmer,** or Georgical Dictionary. 1822

Dyer, Henry. **The Evolution of Industry.** 1895

Epstein, Ralph C. **The Automobile Industry:** Its Economic and Commercial Development. 1928

Ericsson, Henry. **Sixty Years a Builder:** The Autobiography of Henry Ericsson. 1942

Evans, Oliver. **The Young Mill-Wright and Miller's Guide.** 1850

Ewbank, Thomas. **A Descriptive and Historical Account of Hydraulic and Other Machines for Raising Water,** Ancient and Modern. 1842

Field, Henry M. **The Story of the Atlantic Telegraph.** 1893

Fleming, A. P. M. **Industrial Research in the United States of America.** 1917

Van Gelder, Arthur Pine and Hugo Schlatter. **History of the Explosives Industry in America.** 1927

Hall, Courtney Robert. **History of American Industrial Science.** 1954

Hungerford, Edward. **The Story of Public Utilities.** 1928

Hungerford, Edward. **The Story of the Baltimore and Ohio Railroad, 1827-1927.** 1928

Husband, Joseph. **The Story of the Pullman Car.** 1917

Ingels, Margaret. **Willis Haviland Carrier, Father of Air Conditioning.** 1952

Kingsbury, J[ohn] E. **The Telephone and Telephone Exchanges:** Their Invention and Development. 1915

Labatut, Jean and Wheaton J. Lane, eds. **Highways in Our National Life:** A Symposium. 1950

Lathrop, William G[ilbert]. **The Brass Industry in the United States.** 1926

Lesley, Robert W., John B. Lober and George S. Bartlett. **History of the Portland Cement Industry in the United States.** 1924

Marcosson, Isaac F. **Wherever Men Trade:** The Romance of the Cash Register. 1945

Miles, Henry A[dolphus]. **Lowell, As It Was, and As It Is**. 1845

Morison, George S. **The New Epoch:** As Developed by the Manufacture of Power. 1903

Olmsted, Denison. **Memoir of Eli Whitney, Esq.** 1846

Passer, Harold C. **The Electrical Manufacturers, 1875-1900.** 1953

Prescott, George B[artlett]. **Bell's Electric Speaking Telephone.** 1884

Prout, Henry G. **A Life of George Westinghouse.** 1921

Randall, Frank A. **History of the Development of Building Construction in Chicago.** 1949

Riley, John J. **A History of the American Soft Drink Industry:** Bottled Carbonated Beverages, 1807-1957. 1958

Salem, F[rederick] W[illiam]. **Beer, Its History and Its Economic Value as a National Beverage.** 1880

Smith, Edgar F. **Chemistry in America.** 1914

Steinman, D[avid] B[arnard]. **The Builders of the Bridge:** The Story of John Roebling and His Son. 1950

Taylor, F[rank] Sherwood. **A History of Industrial Chemistry.** 1957

Technological Trends and National Policy, Including the Social Implications of New Inventions. Report of the Subcommittee on Technology to the National Resources Committee. 1937

Thompson, John S. **History of Composing Machines.** 1904

Thompson, Robert Luther. **Wiring a Continent:** The History of the Telegraph Industry in the United States, 1832-1866. 1947

Tilley, Nannie May. **The Bright-Tobacco Industry, 1860-1929.** 1948

Tooker, Elva. **Nathan Trotter:** Philadelphia Merchant, 1787-1853. 1955

Turck, J. A. V. **Origin of Modern Calculating Machines.** 1921

Tyler, David Budlong. **Steam Conquers the Atlantic.** 1939

Wheeler, Gervase. **Homes for the People,** In Suburb and Country. 1855